AF412633

ROBOTICS
From Manipulator to Mobilebot

ROBOTICS
From Manipulator to Mobilebot

ZIXING CAI

Central South University, China
& ZIXING Academy of AI, China

NEW JERSEY · LONDON · SINGAPORE · BEIJING · SHANGHAI · HONG KONG · TAIPEI · CHENNAI · TOKYO

Published by

World Scientific Publishing Co. Pte. Ltd.

5 Toh Tuck Link, Singapore 596224

USA office: 27 Warren Street, Suite 401-402, Hackensack, NJ 07601

UK office: 57 Shelton Street, Covent Garden, London WC2H 9HE

Library of Congress Cataloging-in-Publication Data
Names: Cai, Zixing, 1938–   author.
Title: Robotics : from manipulator to mobilebot / Zixing Cai,
    Central South University, China & Hunan ZIXING AI Academy, China.
Description: New Jersey : World Scientific, [2023] | Includes bibliographical references and index.
Identifiers: LCCN 2022007584 | ISBN 9789811253461 (hardcover) |
    ISBN 9789811253478 (ebook for institutions) | ISBN 9789811253485 (ebook for individuals)
Subjects: LCSH: Robotics.
Classification: LCC TJ211 .C26824 2023 | DDC 629.8/92--dc23/eng/20220622
LC record available at https://lccn.loc.gov/2022007584

British Library Cataloguing-in-Publication Data
A catalogue record for this book is available from the British Library.

For any available supplementary material, please visit
https://www.worldscientific.com/worldscibooks/10.1142/12756#t=suppl

Desk Editors: Balamurugan Rajendran/Amanda Yun

Typeset by Stallion Press
Email: enquiries@stallionpress.com

Foreword by Song Jian

Robotics — A Glorious Chapter in Automation[*]

The advancement of modern automatic control technology has opened up new possibilities for scientific research and exploration work, and opened up new scientific undertakings that cannot be accomplished by manpower. In the 1990s, deep sea exploration from 6,000 meters to 10,000 meters was achieved. It has realized the detection of Venus, Mars, Jupiter and some moons and comets in the solar system. The orbiting of the Hubble Space Telescope provides astronomers with unprecedented tools and opportunities to study the universe. In 1997, the Pathfinder trolley developed by American scientists successfully completed the field survey of the surface of Mars, which is one of the highest achievements of automation technology in this century.

The advancement and application of robotics is the most convincing achievement of automatic control in this century, and it is the highest level of automation in the contemporary era. It took only 20 years for the robot to learn to walk on two legs from crawling and become an upright robot, while it took millions of years for humans

---

[*]This is an excerpt of the academic report "Intelligent Control — Goal Beyond the Century" delivered by Academician Song Jian, Chairman of the 14th World Congress of the International Federation of Automatic Control (IFAC), and President of the Chinese Academy of Engineering at the opening ceremony of the conference (July 5, 1999, in Beijing).

to go from crawling to standing upright. Robots can already use tools by hand, can see, listen, and speak in multiple languages. The robot does the dirtiest and tiring work safely and reliably. It is estimated that there are now nearly one million robots working on production lines in the world. Nearly 10,000 factories are producing robots, with sales increasing by more than 20% every year. Robots are ambitious and are preparing to enter the service industry in the 21st century. They will be taxi drivers, nurses in hospitals, family care for the elderly, and cashiers in banks.

If microelectronics goes further, you can squeeze IBM/6000SP into its head and run Deep Blue software, just like defeating world champion Gary Kasparov in May 1997, letting the world chess masters intimidated. Isaac Asimov once envisioned, "Robots have a mathematical genius, can mentally calculate triple integrals, and do tensor analysis problems like snacking," and it is not difficult to do.

The fear and social mentality against automation and robotics that emerged in the 1960s has proven to be unfounded. Today, in some countries that use robots the most, unemployment rates have not risen significantly. Even where they have, no one accuses scientists and engineers of controlling scientists and engineers. That is the fault of financiers and politicians. On the contrary, the widespread entry of smart technology into society is conducive to improving human quality of life, increasing labor productivity, improving the cultural quality of the entire society, and creating more employment opportunities.

Standing on the threshold of entering the 21st century, looking back at the modern history of human civilization and progress, if the 19th century realized the mechanization of social manual labor and extended human physical strength, then the main feature of the 20th century was the realization of labor production automation, which greatly improved social labor productivity, has created much more social wealth than at any time in the past, completely changed human production and lifestyle, improved quality of life, and prolonged the average life span of human beings. This is entirely the credit of modern science and technology.

We can be proud that cybernetic scientists and engineers have made important contributions to this. It is predicted that in the 21st century, automation technology will remain the frontier of high technology and continue to be the core force for advancing the new technological revolution. Manufacturing and service industries are still the main areas where it has made brilliant achievements.

Driven by life sciences and artificial intelligence, a strong trend has emerged in the field of control theory and automation to improve the intelligence of control systems. A new academic group, the International Federation on Intelligent Automation (IFIA) was established in 1992, marking that the research on intelligent control has entered the scientific frontier. Some consensus has been reached on the future development direction and path of this new discipline. The following points can be listed:

First, studying and imitating human intelligence is the highest goal of intelligent control. Therefore, people call an automatic control system that can automatically recognize and remember signals (images, language, text), learning, reasoning, and have automatic decision-making capabilities as intelligent control systems.

Second, intelligent control must rely on multidisciplinary cooperation to achieve new breakthroughs. Life sciences and brain sciences are indispensable for more in-depth knowledge of the human body and brain function mechanisms. Uncovering the evolutionary mechanism of the biological world, the precise structure of self-organization ability, immune ability and genetic ability in the living system is extremely important for the construction of intelligent control system. This is mainly the task of biochemists and geneticists, but cybernetic scientists and engineers can contribute to this.

Third, the improvement of intelligence cannot rely solely on the accumulation of subsystems. To achieve "the whole is greater than the sum of the components," nonlinear effects alone are not enough. The more intelligent the system will be, the more complex it will be. The behavior and structure of complex giant systems must be hierarchical. The harmonious unity of the interests of the subsystem and the whole is the basic principle for the survival and development of organisms. Each level has new characteristics and status descriptions. It is necessary to establish a structure compatible with each level up and down and a friendly interface with the surroundings. In statistical mechanics, the extraction of parameters from the thermal motion of molecules to the macroscopic state of gases is an example of hierarchical division. This is the best description of what physicists call "coarse-graining extraction."

Fourth, the evolution of all creatures in the world is gradual. Human beings have gone through 10,000 years from the Neolithic Age to robots. It only took 100 years from mechanical automation to electronic automation. It will take decades to achieve intelligent automation and raise the intelligence quotient of robots to the level of *Homo sapiens*. This is an insurmountable process of scientific and technological progress. In the second half of the 20th century, microelectronics, life sciences, and automation technology advanced by leaps and bounds, creating good starting conditions for the realization of intelligent control and intelligent automation in the next century. To achieve this goal, not only technological progress, but also breakthroughs in scientific ideas and theories are needed. Many scientists insist that this requires the discovery of new principles or the modification of known basic theorems of physics in order to thoroughly understand and imitate human intelligence, and to design and manufacture automatic control systems with advanced intelligence. In any case, the process has already begun. It can be imagined that in the 31st IFAC meeting after 50 years, human production efficiency will be ten times higher than it is now, and no one will go hungry anymore. Elderly people all over the world can have a robot waiter by their side to help with their lives. Everyone attending the meeting may bring a robot secretary in the file box, just like the current electronic notebook.

The 21st century is a particularly important historical period for mankind. The world's population will stabilize at a relatively high level, such as twelve billion, which is double the current level. The scientific community must make necessary contributions to guarantee the survival and sustainable development of human beings and our homeland, the earth, and cybernetic scientists and engineers should undertake the main tasks. To further develop and vigorously promote the application of cybernetics and automation technology, and to ensure that future generations live happily in a world free from shortage, hunger and pollution, it is our gift. As the physicist Murroy Gellmann said, in the foreseeable future, natural evolution, including humans, will give way to the advancement of human science, technology and culture. The term Cybernetics comes from Greek and originally meant "helmsman." We are at least qualified to be the scientific advisers and assistants of the helmsman and play a greater role in promoting social progress. This is our honor.

Foreword by Toshio Fukuda

In 1962, the first robot of Unimation of the United States was put into operation at General Motors Company (GM), marking the birth of the first generation of modern robots. Since then, robots have become a reality of human production and life.

In the past 60 years, significant progress has been made in robotics science and technology. As Academician Jian Song pointed out: The advancement and application of robotics is the most convincing achievement of automatic control in the 20th century, and the highest level of automation in the contemporary era. Since mankind entered the 21st century, robotics has shown many notable features such as systematic and modularization, miniaturization and heavy-duty, mobility, agility, digitization, networking, and intelligence, and have become the most typical frontier integrated cross-technology field. Robot digital and modular design technology has been widely used; Various micro-robots have played a key role in many micro-assembly operations; Micro-nano robots and bionic robots can realize the analysis of single-cell biological characteristics, spatial detection and micro-nano operation of biology cells; Giant and heavy robot Hercules are showing their talents in shipyards and docks; The number of mobile robots coming from behind has far exceeded that of traditional robot operators; Many manipulators are also equipped with mobile platforms to become mobile manipulators with greater mobility and high precision; The highly flexible agile processing robot unit has doubled the capacity of intelligent manufacturing; Robot application system engineering has been promoted; Networking robots have already become a reality for teleoperation, including telemedicine operations; Various types of intelligent robots have been used in industry, agriculture, construction, service industries such as space and deep-sea exploration, medical rehabilitation, and entertainment have been increasingly widely used, serving all walks of life and benefiting millions of households.

As far as I know, international robotics research is very extensive and in-depth, and many societies affiliated to IEEE are waiting to conduct high-level robotics-related research and achieve high-level research results. With the rapid development of robotics, its research and application results have been fully reflected in numerous robotics monographs and scientific papers. These robotics monographs are completely rich and colorful. Most of these monographs are dedicated to the study of manipulators. In recent years, there are some monographs dedicated to the study of mobile robots. Some of them discuss the kinematics, dynamics, sensors, control, planning and programming of manipulators; It is also discussions on the architecture, dynamics model, state estimation, environment modeling and positioning, navigation and target tracking, perception and obstacle avoidance of mobile robots and intelligent vehicles. There are many wonderful chapters in these robotics monographs, and they have made outstanding contributions to the continuous development and technological upgrading of international robotics. However, it is still relatively rare to systematically and comprehensively study and summarize traditional robots (manipulators) and mobile robots at the same time. Professor Zixing Cai's monograph *Robotics: From Manipulator to Mobilebot* comprehensively discusses the main scientific and technological content of manipulators and mobile robots with a completely new look.

The monograph *Robotics: From Manipulator to Mobilebot* covers a wide range of topics in the field of robotics and is a high-level work with distinctive features. First of all, the book is comprehensive and involves the main core technologies of robotics, including the basic theories and technologies of robotic manipulators, and the basic principles and methods of mobile robots. Secondly, this book focuses on recent innovations, including those of the author, and intelligent methods such as computational intelligence and deep learning technology that are widely used in robotics. Third, based on national scientific research projects, theory and practice are highly integrated, with theory, technology and methods, and application examples, which will help readers to understand the development and application of robotics theory. Due to the above characteristics, this book will definitely be welcomed by readers in the field of robotics.

Robotics: From Manipulator to Mobilebot is a brand-new robotics monograph with a high technological and academic level and important application value. I am convinced that the publication of this book will have a positive impact on the international robotics community, and make an important contribution to the promotion and application of robotics knowledge, and to the promotion of the development of international robotics.

Toshio Fukuda
Nagoya University, Japan
Beijing Institute of Technology, China
October 2021

Foreword by Volker Graefe

The birth of robots made the dream of using machines to serve humans and society a reality, and it was a major technological event in the 20th century. Since the beginning of the 21st century, robotics and robot technology have accelerated their development from industrial robots for welding and assembly to a variety of robots in various industries and fields. Robots exist in many forms, from giant transport vehicles in shipyards to micro-nano robots for processing cells in nano-space. Since their beginnings over 60 years ago, robotics and robot technology have made great progress. Robots have been integrated into all walks of life and have entered thousands of households. They have played an important role in the economic development of countries around the world, improved people's well-being and promoted social development. It may be said that robots have been integrated into human society and have become mankind's right-hand man and friend.

Zixing Cai has been known as an outstanding teacher and researcher of robotics and artificial intelligence for many years. He is a professor of the Central South University and the founder and the chief scientist of the Hunan ZIXING Artificial Intelligence Academy, both located in Changsha, the capital city of Hunan Province in China. As we are both engaged in robotics research and teaching and have many common interests, especially in the fields of intelligent mobile robots and artificial intelligence embodied in robots, I consider him a close colleague and, moreover, a personal friend. Being an Honorary Professor of the National University of Defense Technology and a senior Advisor of the Hunan ZIXING Artificial Intelligence

Academy I have often visited Changsha to give lectures at these institutions and at the Central South University. On such occasions I had frequent and fruitful exchanges with Professor Cai and his associates about our many common research fields and research interests, which has given me a good understanding of his excellent research and teaching.

One result of Zixing Cai's lifelong work in robotics and artificial intelligence is his latest book, *Robotics: From Manipulator to Mobile Robot*. It will be of great interest and benefit for students and experts in a wide variety of areas within robotics. Readers who are mainly interested in manipulators may first learn the mathematical foundation, dynamics and kinematics of manipulators, and then study and discuss manipulator planning and control in combination with the actual situation. Readers studying mobile robots and intelligent driving may study the architecture, dynamics model, stability and tracking of mobile robots first, and then study and discuss the positioning, modeling, navigation and control of mobile robots in combination with actual topics. Of course, if readers are only interested in traditional manipulators or mobile robots, they can choose to read just the relevant chapters of the book. Readers who conduct in-depth research may have an opportunity to gain from this book some professional knowledge in visual perception, intelligent control, machine learning and intelligent programming.

Many robotics monographs have been published already, but they hardly present a comprehensive and systematic research and analysis of traditional manipulators and intelligent mobile robots at the same time. Zixing Cai's present book fills this gap. It introduces and discusses traditional manipulators and mobile robots simultaneously and at a high academic level. It would be difficult to find anybody who would be more qualified to write such a book that is truly useful for the reader than Zixing Cai who has been awarded the title "State Eminent Educator" in recognition of his excellent teaching and writing skills.

This new monograph by Professor Cai is a high-level and comprehensive book on robotics. It possesses systematic, advanced, novel, readable, and theory-to-practical, and it will surely provide a valuable source of knowledge and teaching reference for a vast number of robotics researchers, university teachers and students. Most importantly, it will make positive contributions to the development and application of robotics and robot technology in the world.

Volker Graefe
Bundeswehr University Munich
Germany
October 2, 2021

Preface

Mankind's fantasy and pursuit of robots has a history of more than 3,000 years. It is expected that mankind can create a kind of human-like machines or artificial humans in order to support or even replace people in various labor and work.

As a highly intersectional frontier discipline, robotics has aroused widespread interest from more and more people with different backgrounds who conducted in-depth research, achieved rapid development and remarkable achievements. Since the beginning of the 21st century, the development of the industrial robot industry has accelerated, the prospects of the industrial robot market are promising, and the rapid development of service robots has entered thousands of households and benefited the people. In recent years, artificial intelligence has entered a new era of development, promoting international robotics research and application into a new period, and pushing robotics technology to a higher level.

I have been engaged in robotics research and teaching for 40 years. In the early 1980s, I started to teach myself the basics of robotics. From 1983 to 1985, I studied in the United States, successively at the University of Nevada (UNR) and Purdue University, under the guidance of professors such as K.S. Fu, R.P. Paul, Eugene Kosso and Y.S. Lu, to study robotics and artificial intelligence. After studying and returning to China, I carried out basic research and teaching of artificial intelligence and robotics, and contribute to the knowledge dissemination and talent training of related disciplines.

The long period research and teaching in the robotics field have received many important achievements and have provided abundant materials and laid a solid foundation for the writing of the English monograph *Robotics: From Manipulator to Mobilebot.*

This book is a systematic and comprehensive introduction to the basic principles of robotics and its applications. The book consists of three parts and 12 chapters. Chapter 1 briefly describes the origin and development of robotics, discusses the definition of robotics, analyzes the characteristics, structure and classification of robots, and discusses the research field of robotics. Part I — the "Robot Manipulator" consists of five chapters, involving the mathematical foundation, kinematics, dynamics, control and planning of the manipulator. Chapter 2 discusses the mathematical basis of the manipulator, including the position and posture transformation of the space particles, coordinate transformation, homogeneous coordinate transformation, object transformation and inverse transformation, and general rotation transformation. Chapter 3 describes the representation and solution of manipulator motion equations, that is, manipulator forward kinematics and inverse kinematics problems, including the motion equations of manipulator motion posture, direction angle, motion position and coordinates, as well as the representation of link transformation matrix, and explain the solution methods of the analytical solution, numerical solution and other solution methods, as well as the differential motion of the manipulator and its Jacobian matrix. Chapter 4 deals with the dynamics equations of the manipulator; focuses on the analysis of the two methods of solving the dynamics equations of the manipulator, namely the Lagrangian functional balance method and the Newton-Euler dynamic balance method, and then summarizes the steps to establish the Lagrangian equation, and take the two-link and three-link manipulators as examples to derive the dynamic equations of the manipulator. Chapter 5 studies the position and force control of the manipulator, including robot control principles and control methods, transmission system structure, position servo control, position and force hybrid control, decomposition motion control, and adaptive control and intelligent control of the manipulator. Chapter 6 discusses robotic planning issues. After explaining the role and tasks of robot planning, start with the robot planning in the block world, and gradually carry out the discussion of robot planning, including rule deduction, logic calculation, general search method, and expert system-based planning. In the aspect of robot path planning, it focuses on robot path planning based on approximate Voronoi diagram, robot path planning based on immune evolution and example learning, etc. Part II — the "Mobile Robot" consists of four chapters. Chapter 7 summarizes the architecture of the mobile robot and discusses the dynamic model, stabilization and tracking of the wheeled mobile robot and its design examples. Chapter 8 discusses the positioning and modeling of mobile robots, focusing on issues such as mobile robot map construction, dead reckoning positioning, synchronous locating and mapping (SLAM), mobile robot SLAM data association methods, and mobile robot SLAM in a dynamic environment. Chapter 9 is about mobile robot navigation, introduces the main methods and development trends of mobile robot navigation, explores local navigation strategies and composite navigation strategies of mobile robots, and analyzes mobile robot path navigation based on ant colony algorithm, navigation strategies based on feature points, and examples of robot navigation based on machine learning. Chapter 10 summarizes the intelligent control of mobile robots. After an overview of intelligent control and intelligent control systems, it focuses on the control of mobile robots based on neural networks and the control of mobile robots based on deep learning. Part III — the "Applications and Prospects of Robotics" consists of two chapters, and Chapter 11 introduces the application and market of robotics. First of all, the robot application areas introduced involve industrial robots, explore robots, service robots and military robots. Then, it discusses the latest development status of the domestic and foreign robotics market and the current status and forecast of the robotics market. Chapter 12 is the outlook for robotics, analyzing the development trend of robotics, including the ambitious development plans of robots in various countries, various social problems brought about by robot applications, and the challenges of cloning technology to intelligent robot technology.

I hope that this book will be particularly suitable as a learning reference for scientific and technical personnel engaged in robotics research, or in development and application of robots, and that it will also be used as a textbook for postgraduate and undergraduate students in colleges and universities.

Acknowledgements

In the process of writing and publishing this book and the related books, I have received enthusiastic encouragement and help from many leaders, experts, professors, friends and students. Academicians of the Chinese Academy of Sciences and the Chinese Academy of Engineering, such as Song Jian, Zhang Zhongjun, Xiong Youlun, Zhang Qixian, and Cai Hegao, *et al.* wrote foreword for the abovementioned works, reflecting the great support and love for the author and readers. Central South University, Hunan ZIXING Academy of Artificial Intelligence and Hunan Association for Artificial Intelligence, as well as many robotics experts, university teachers and students, editors and readers have concerned about my work and the compilation and publication of this book, and they have also put forward some very pertinent suggestions. These are all great help to the author, and I am deeply encouraged. I would like to express my heartfelt thanks to relevant leaders, academicians, experts, editors, cooperating teachers and students, and readers.

Dr. Song Jian, an academician of the Chinese Academy of Sciences and the Chinese Academy of Engineering, a foreign academician of the Russian Academy of Sciences and the US National Academy of Engineering, Academician of the Royal Swedish Academy of Sciences, Academician of the Eurasian Academy of Sciences, has long given me strong support, unremitting guidance and enthusiastic encouragement for my robotics and artificial intelligence research. As the chairman of the conference and the president of the Chinese Academy of Engineering, his report at the opening ceremony of the 14th Conference of the International Federation of Automatic Control (IFAC) was far-sighted, philosophical, forward-looking and enlightening, and has been used as the foreword of my *Robotics* book (4 editions in 2000, 2009, 2015 and 2020), and adds luster to the book. Now, this foreword has become one of the forewords of this new book on robotics. I would like to pay high tribute to Academician Song Jian.

I would like to especially thank Toshio Fukuda and Volker Graefe for writing the forewords for this book in their busy schedules. Toshio Fukuda is a professor of the Meijo University in Nagoya (Japan), President of the IEEE (2019–2021), an academician of the Japanese Academy of Sciences and a foreign academician of the Chinese Academy of Sciences, while Volker Graefe is an IROS Fellow, a professor of the Bundeswehr University Munich (Germany) and the director of the Intelligent Robots Laboratory there. Both of them are well-known experts and leaders in international robotics research. They have in-depth research and profound knowledge in various important fields of robotics, especially intelligent robots. Professor Toshio Fukuda is known as the "Father of International Micro-nano Robots," leading the research of international micro-nano robots and bionic robots, Professor Volker Graefe has a solid research foundation in the fields of vision, navigation, communication and learning of intelligent robots, and is one of the pioneers in the research of international autonomous vehicle driving. The foreword written by the two masters is important support and love to the author.

I sincerely thank many of my collaborators in robotics books and papers for their great cooperation and outstanding contributions in the past 25 years. They are Chen Aibin, Chen Baifan, Chen Hong, Dai Bo, Duan Zhuohua, Gong Tao, Gu Mingqin, Guo Fan, He Hangen, Hu Dewen, Jiang Zhiming, Lai Xuzhi, Li Meiyi, Li Yi, Liu Juan, Liu Lijue, Peng Zhihong, Tang Shaoxian, Timofeev A V,Wang Lu, Wang Yong, Wen Zhiqiang, Xiao Xiaoming, Xie Bin, Xie Guanghan, Xu Xin, Yu Jinxia, Yu Lingli, Zhao Hui, Zheng Minjie, Zhou Xiang, Zou Xiaobing, etc. I would also like to extend my sincere thanks to Chen Qijun, Chen Rujun, Cui Yi'an, Fu Yili, Gao Pingan, Gao Yan, Huang Xinhan, Kuang Linai, Liao Yongzhong, Liu Hong, Liu Limei, Liu Xianru, Liu Xingbao, Liu Zhiyong, Liu Chaoyang, Lu Weiwei, Pan Wei, Pan Yi Xiao, Peng Meng, Peng Weixiong, Ren Xiaoping, Shi Yuexiang, Song Wu, Sun Guorong, Tan Min, Tan Ping, Tang Suqin, Wang Lei, Wang Wei, Wen Sha, Xie Jin, Yang Qiang, You Zuo, Zhang Quan, Zheng Jinhua, Zou Lei, Zou Yiqun, Zuo Fujun etc. the professor and postgraduates, for their support and help in my robotics work.

In addition, I would like to thank the authors of relevant referenced robotics monographs, textbooks and related papers at home and abroad. Their writing results have provided valuable reference materials for this book.

Experts and readers are welcome to criticize and correct this book.

Cai Zixing
July 19, 2022
In Deyi Garden, Changsha, China

About the Author

Zixing Cai is an IEEE Fellow and IEEE Life Fellow, academician of the International Academy of Navigation and Motion Control, CAAI Fellow, AAIA Follow, academician of the New York Academy of Sciences, United Nations Expert (UNIDO), Professor and Doctoral Supervisor of the School of Information Science and Engineering at Central South University, China, Chief Scientist of Hunan ZIXING Academy of Artificial Intelligence. He has served as the Vice President of the Chinese Association for Artificial Intelligence (CAAI) and Founding Director of the Intelligent Robotics Professional Committee of CAAI; Honorary Chairman of Hunan Society of Artificial Intelligence, China; Director of the Chinese Association of Automation (CAA) and member of Intelligent Automation Professional Committee of CAA; Member of the Artificial Intelligence and Pattern Recognition Professional Committee of the Chinese Computer Federation (CCF); Member of the Evaluation Committee of the IEEE Computational Intelligence Society (CIS); and Member of the IEEE CIS Evolutionary Computing Technology Committee. Prof Cai has also served as an Adjunct Professor at Rensselaer Polytechnic Institute (RPI), USA; St. Petersburg Technical University, Russia; Technical University of Denmark; Peking University, China; National University of Defense Technology, China; Beijing University of Aeronautics and Astronautics, China; Beijing University of Posts and Telecommunications, China; etc., and a Visiting Researcher at the Institute of Automation, Chinese Academy of Sciences; St. Petersburg Automation and Informatics Institute, Russian Academy of Sciences; and so on. Prof Cai is an academic leader in the fields of artificial intelligence, intelligent control and robotics in China.

In the robotics field, he has presided over or participated in national-level scientific and technological research projects, mainly related to key projects and major projects of the National Natural Science Foundation of China, such as "Theories and Methods of Navigation and Control of Mobile Robots in unknown Environments," "Key Technologies and Integrated Verification Platform for Autonomous Vehicles," "Cognition Computing of Audiovisual Information," "Key Scientific Issues in Intelligent Driving of Highway Vehicles," "Technical Basis for Cooperative Identification and Reconstruction of Heterogeneous Multi-moving Objects," "Fuzzy Control of High-speed Train Operation," "Evolutionary Control of Autonomous Robots" and "Robot Planning Expert System" and other researches that have achieved internationally advanced research results.

He has either independently or taken the lead in writing and publishing robotics-related monographs and textbooks, including *Robotics: Principles and Applications* (1988), *Robotics* (2000, 2009, 2015, 2022), *Fundamentals of Robotics* (2009, 2015, 2021), *Theories and Methods of Navigation and Control of Mobile Robots in Unknown Environments* (2009), *Principles and Techniques of Multi-Mobile Robot Cooperation* (2011), *Self-positioning Technology of Mobile Robots in Unknown Environments* (2011), *Key Techniques of Navigation Control for Mobile Robots under Unknown Environment* (2016), *Exploring the Kingdom of Robots* (2018), *Autonomous Vehicle: Perception, Mapping and Target Tracking Technology* (2021), etc. Among them, *Robotics* won the first prize of the National Excellent Textbook Award of the China Ministry of Education, and *Multi-Mobile Robot Collaboration Principles and Technology*, and *Mobile Robot Self-Positioning Technology in Unknown Environments* were funded and published by the National Science and Technology Academic Publication Fund. *Theory and Method of Navigation and Control of Mobile Robots in Unknown Environments* was published by the Science Publishing Fund of the Chinese Academy of Sciences.

Professor Cai has persisted in the teaching of and reformation of robotics for a long time. He has successively taught tens of thousands of undergraduate students, thousands of graduate students, and hundreds of overseas master's students, as well as guided of theses and course theses. People have known these as Professor Cai Zixing's "Hundreds-Thousands-Ten Thousands Education Project," which aims to make positive contributions to the training of robotics and artificial intelligence talents. He has presided over eight national quality engineering projects of the Ministry of Education, China, including national quality courses, quality resource sharing courses, and national teaching teams etc.

He has successively won the Outstanding Robotics Education Award of the China Robotics Congress, and the Outstanding Robotics Education Award of the Chinese Association for Artificial Intelligence. Prof Cai is also the winner of the first National College Teachers Award, the Wu Wenjun Artificial Intelligence Technology Achievement Award, Xu Teli Education Award, and BaoSteel Grand Prize of National Excellent Teacher Award. He was also awarded the First Prize of the Science and Technology Progress Award of the National Ministry of Education, and the First Prize of the National Excellent Teaching Material Award of Higher Education and other important awards.

As a result, his rich achievements have made him being known as "the first person in artificial intelligence education in China," "founder of the Chinese intelligent robotics discipline" and "founder of Chinese intelligent control."

Contents

Part II　Mobile Robot　287

7. Architecture and Dynamics Model of Mobile Robot　289

12. Robotics Outlook　　　　　547

List of Tables

List of Figures

Chapter 1

Basic Concepts

Since mankind entered the 21st century, in addition to devoting itself to its own development, it has had to pay attention to issues such as robots, aliens, and human clones. This book will not discuss the issue of aliens and human cloning, but will focus on the issue of robots [4].

People of today are no strangers to the name "robot." From ancient myths and legends to modern science fiction, drama, movies and TV, there are many wonderful depictions of robots. Although many important results have been achieved in robotics, the vast majority of robots in the real world is neither as wise and brave as described in myths and literary works, nor as versatile as some entrepreneurs and propagandists preach. At present, the capabilities of robots are still relatively limited. However, robotics is developing rapidly, and it can be said to be changing daily, and is beginning to have an increasing impact on the entire economy, industry, agriculture and service industry, space and ocean exploration, and all aspects of human life [14, 51].

1.1 The Origin and Development of Robotics

1.1.1 *The origin of robotics*

The origin of robots can be traced back to more than 3,000 years ago. "Robot" was a newly coined word that appeared in many languages and scripts in the 1920s. It embodies a long-held dream of mankind, that is, to create a human-like machine or artificial human to replace all kinds of jobs of people [24].

It was not until more than 60 years ago that "robot" was cited as a professional term. However, the concept of robots has existed in the human imagination for more than 3,000 years. As early as the Western Zhou Dynasty (1066 BC–771 BC) in China, there was a story about the artisan Yanshi who dedicated a singing and dancing robot to King Zhou Mu. In the 3rd century BC, the ancient Greek inventor Daedalus used bronze to create Talos, a guardian for King Minos of Crete. In a book of the 2nd century BC, there was a description of a mechanized theater with robot-like performers that could perform dances and line-ups in court ceremonies [24]. During the China's Eastern Han Dynasty (25–220 AD), the guide car invented by Zhang Heng was the world's earliest prototype of a robot [10].

In the modern era, mankind expects to invent all kinds of mechanical tools and power machines to assist or even replace various kinds of manual labor. The steam engine invented in the 18th century opened up a new age of machine power instead of manpower. With the invention of powered machines, the first industrial and scientific revolution appeared in human society. The advent of various automatic machines, powered machines and power systems has transformed robots from fantasy to reality. Many mechanically controlled robots, mainly various exquisite robot toys and handicrafts, emerged at this period. For example, from 1768–1774 AD, the famous Swiss watchmaker Jaquet-Droz and his sons designed and manufactured three life-sized robots-writing dolls, drawing dolls, and organ dolls; these were automatic machines controlled by cams and springs. In 1893, the walking robot Android designed by Canadian Moore was powered by steam. These treasures of robotic craftsmanship marked a big step forward for mankind on the long road of robots from dream to reality.

At the beginning of the 20th century, the mother fetus of human society and economy drew closer to the verge of giving birth to robots, and humanity viewed this with unease, unsure if the robot to come would be a darling or a monster. In 1920, the Czech playwright Karel Čapek first proposed the term "robot" in his science fiction play "Rosam's Universal Robot" (R.U.R) [22]. The drama anxiously foretells the tragic impact that the development of robots will have on human society and aroused widespread concern. What Čapek highlighted was the safety, intelligence and self-reproduction of robots. The advancement of robotics technology was likely to cause unwanted problems and undesirable results.

In response to social anxiety about the robots to come, the famous American science fiction novelist Asimov proposed the famous "Three Rules for Robots" in his novel "I am a Robot" in 1950 [5]:

(1) The robot must not endanger humans, and it is not allowed to stand by while watching people suffer;
(2) Robots must absolutely obey humans, unless such obedience is harmful to humans;
(3) The robot must protect itself from harm, unless it is to protect humans or humans order it to make sacrifices.

These three rules gave new ethics to the robot society and make the concept of robots more acceptable to human society.

The American George Dvor designed the first electronically programmable industrial robot in 1954 and published the robot patent in 1961. In 1962, the first robot Unimate made by the US Universal Automation (Unimation) was put into use at General Motors (GM), which marked the birth of the first generation of robots [23]. Since then, robots have become a reality in daily life and humans have continued to write a new chapter in the history of robots with their own wisdom and labor.

1.1.2 *Development of robotics*

In the first decade after the advent of industrial robots, from the early 1960s to the early 1970s, the development of robotics technology was relatively slow, and the efforts of many research units and companies were unsuccessful. The main achievements of this stage were the mobile intelligent robot Shakey developed by the Stanford International Research Institute (SRI) in 1968 and the first robot made by Cincinnati Milacron in 1973, the robot T3 etc. which were suitable for putting on the market [6].

In the 1970s, the artificial intelligence (AI) academic community began to have a keen interest in robots. They found that the emergence and development of robots brought new vitality to the development of artificial intelligence, provided a good test platform and application place, and was a potential field in which artificial intelligence might make significant progress. This understanding was quickly endorsed by the scientific and technological circles, industry circles and relevant government departments in many countries. With the rapid development of automatic control theory, electronic computers and aerospace technology, by the mid-1970s, robotics technology entered a new stage of development. By the end of the 1970s, industrial robots had a greater development [8, 41]. After entering the 1980s, robot production continued to maintain the momentum of development in the late 1970s. By the mid-1980s, the robotic manufacturing industry became one of the fastest growing and best economic sectors [44, 50].

By the late 1980s, due to the saturation of industrial robot applications by traditional robot users, resulting in a backlog of industrial robot products, many robot manufacturers closed down or were merged, causing a downturn in international robotics research and the robotics industry. By the early 1990s, the robotics industry showed signs of recovery and continued development. However, the good times did not last long, and there was another slump in 1993–1994. The number of industrial robots in the world increased every year, but the market moved forward in waves. From 1980 to the end of the 20th century, there were three saddle-shaped curves. After 1995, the number of robots in the world increased year by year, and the growth rate was also high. Robotics has entered the 21st century with a good momentum of development.

Entering the 21st century, the development of the industrial robot industry has accelerated, with an annual growth rate of about 30%. Among them, the growth rate of industrial robots in Asia is as high as 43%, which is the most prominent.

According to statistics from the United Nations Economic Commission for Europe (UNECE) and the International Federation of Robotics (IFR), more than 1.75 million industrial robots were installed globally from 1960 to the end of 2006; and more than 2.3 million were installed from 1960 to 2011. The industrial robot market is promising. In recent years, the global robotics industry has developed rapidly. In 2007, the total sales volume of the global robotics industry increased by 10% over 2006. Humanization, heavy, and intellectualization have become the main development trends of the robot industry in the future. In addition, there are tens of millions of service robots in operation [33, 34].

According to statistics from the International Federation of Robotics (IFR), the global robot market is expected to reach more than 30 billion USD in 2019, of which industrial robots are 13.8 billion USD and service robots are 16.9 billion USD. The total number of industrial robots in operation in the world now exceeds 2.7 million [35, 47].

Since the 1970s, robotics in China has experienced a development process from 0 to 1, from small to large, and from weak to strong. Today, China has become the world's largest robotics market, and an unprecedented boom in robotics development and application is being carried out across the country, making new contributions to China's rapid and sustainable economic development and continuous improvement of people's well-being [8, 13, 15, 16, 23, 26].

In 2019, China's newly installed industrial robots amounted to 140,500 units, with a cumulative installed capacity of 783,000 units, ranking first in Asia in total, with an annual growth rate of 12%. However, the density of robots in China is still low [3, 11].

In the past 50 years, robotics has achieved remarkable developments, which are embodied in: (1) The rapid development of the robot industry in the world; (2) The scope of application of robots in various fields of industry, science and technology and national defense; (3) A new discipline of robotics has been formed; (4) Robots are developing in the direction of intellectualization; (5) Service robots have become a rookie of robots and have developed rapidly [25, 38, 48].

Most robots in industry today are not intelligent. With the rapid growth of the number of industrial robots and the development of industrial production, higher requirements are put forward on the working capabilities of robots, especially robots and special robots with different degrees of intelligence are required. Some of these intelligent robots can simulate humans walking on two legs and can walk and move on uneven ground; some have visual and tactile functions, capable of independent operation, automatic assembly and product inspection; some have autonomous control and decision-making capacity. These intelligent robots not only use various feedback sensors, but also use various learning, reasoning and decision-making techniques of artificial intelligence. Intelligent robots also apply many of the latest intelligent technologies, such as telepresence technology, virtual reality technology, multi-agent technology, artificial neural network technology, genetic algorithm, bionic technology, multi-sensor integration and fusion technology, and nanotechnology [36, 39].

Robotics has a very close relationship with AI. The development of intelligent robots is based on AI and complements AI. On the one hand, the further development of robotics requires the guidance of the basic principles of AI and adopts various AI technologies; on the other hand, the emergence and development of robotics has brought new vitality to the development of AI, generated new impetus, and provided a good test bed for AI experimentation and application. In other words, AI wants to find practical applications in robotics, and to further develop basic theories such as knowledge representation, problem solving, search planning, machine learning, environmental perception, and intelligent systems [17]. Roughly speaking, AI, or machine intelligence, is imitated by machines to imitate human intelligence. The new type of robot that applies various AI technologies is an intelligent robot.

Mobile robots are a class of robots with high intelligence, and they are also a frontier and key field of intelligent robot research. Intelligent mobile robot is a kind of robot system that can perceive the environment and its own state through sensors, realize target-oriented autonomous movement in an environment with obstacles, and complete certain tasks. The difference between mobile robots and other robots is that the mobile robots emphasize the characteristics of "movement." Mobile robots can not only play an increasingly important role in production and life, but also are effective tools and experimental platforms for studying the generation of complex intelligent behaviors and exploring human thinking patterns. The intelligence level of robots in the 21st century has been developed to a new height and will be raised to an amazingly higher level.

1.2   Definition and Characteristics of Robots

1.2.1   *Definition of robot*

So far there has been no uniform definition of robots. In order to define technology, develop new capabilities for robots, and compare the results of different countries and companies, it is necessary to have some common understanding of the term robot. Now, the definition of robots in the world is quite different [19, 42].

Regarding the definition of robots, there are mainly the following [1, 10, 12]:

(1) The definition of the *British Concise Oxford Dictionary*: A robot is "a human-like automaton, an intelligent and submissive but non-personal machine."

(2) The definition of the American Robotics Industry Association (RIA): A robot is "a multifunctional manipulator that is used to move various materials, parts, tools, or special devices, perform various tasks through programmable actions, and has programming capabilities."

(3) The definition of Japan Industrial Robot Association (JIRA): An industrial robot is "a general-purpose machine equipped with a memory device and an end effector that can rotate and automatically complete various movements to replace human labor."

(4) The definition of the National Bureau of Standards (NBS): A robot is "a mechanical device that can be programmed and perform certain operations and mobile tasks under automatic control."

(5) The definition of the International Organization for Standardization (ISO). "The robot is an automatic, position-controllable, multifunctional manipulator with programming capabilities. This manipulator has several axes and can handle various materials, parts, tools and special devices with the help of programmable operations to perform various tasks."

(6) *Chinese Encyclopedia* defines a robot as a multifunctional manipulator that can flexibly complete specific operations and motion tasks and can be reprogrammed. The definition of a manipulator is: an automatic machine that simulates manual operation, which can grasp and carry objects or operate tools to complete certain specific operations according to a fixed procedure [45].

The definition in China's *AI Dictionary* is: a robot is an automated machine with some intelligent abilities similar to humans or organisms, such as perception, planning, movement, and coordination. It is an automated machine with a high degree of flexibility [20].

With the evolution of robots and the development of robot intelligence, these definitions need to be revised, and even the robot needs to be redefined.

The category of robots should include not only "human-like machines made by humans," but also "creatures made by humans" and even "artificial humans," although we do not approve of making such people. It seems that there is no uniformly defined robot, and it will be more difficult to give it an exact and accepted definition in the future!

1.2.2   *Main features of the robot*

Robots have many characteristics, and versatility and adaptability are the two main characteristics of robots.

1. Versatility

The versatility of the robot depends on its geometric characteristics and mechanical capabilities. Versatility refers to a certain actual ability to perform different functions and complete a variety of simple tasks. Versatility also means that the robot has a variable geometric structure, that is, a geometric structure that can be changed according to the needs of the production work; in other words, the mechanical structure allows the robot to perform different tasks or complete the same work in different ways [23]. Most existing robots have varying degrees of versatility, including manipulator maneuverability and control system flexibility.

It must be pointed out that versatility is not determined by degrees of freedom alone. Increasing degrees of freedom generally improve the degree of versatility. However, other factors must also be considered, especially the structure and capabilities of the end devices, such as whether they can be used with different tools.

2. Adaptability

The adaptability of a robot refers to its ability to adapt to the environment, that is, the designed robots can perform tasks that are not fully specified by themselves, regardless of unexpected environmental changes that occur during task execution [43]. This ability requires the robot to recognize its environment, that is, to have artificial perception. In this regard, the robot uses its following capabilities:

(1) The ability to use sensors to sense the environment;
(2) The ability to analyze task space and perform operational planning;
(3) The ability of automatic command mode.

Compared with human's ability to explain the environment, the robot perception developed so far is still relatively limited. Some important research work in this field has made major breakthroughs.

For industrial robots, adaptability means that their programmed program mode and movement speed can adapt to changes in the size and position of the work piece and the work site. Among them, two kinds of adaptability are mainly considered:

(1) Point adaptability, which involves how the robot finds the position of the point. For example, find the position of the operating point to start the program.

Point adaptability has four kinds of searches (allowing automatic feedback adjustment of the program), namely approximate search, delayed approximate search, precise search and free search. Approximate search allows the sensor to interrupt robot motion along the program direction under program control. The delayed approximate search can interrupt the robot's motion after the programmed sensor is activated for a certain period of time. The precise search can make the robot stop at the precise position where the sensor signal is changed. Free search enables the robot to find a position that satisfies the display of all programmed sensor signals.

(2) Curve adaptability, which involves how the robot uses the information obtained by the sensor to work along the curve. Curve adaptability includes speed adaptability and shape adaptability.

Speed adaptability involves the selection of the best movement speed. Even with a completely definite movement curve, it is still difficult to choose the best movement speed. With speed adaptability, the speed of the robot can be adjusted according to the information provided by the sensor.

Shape adaptability involves the problem of requiring tools to track a curve of unknown shape.

Comprehensive use of point adaptability and curve adaptability can automatically adjust the program. What is initially compiled is only a rough program, and then the system adapts itself to the actual position and shape.

1.3   Composition and Classification of Robots

This section discusses the system composition, degrees of freedom and classification of robots one by one [12, 21, 36].

1.3.1   *Composition of the robot system*

A current robot system generally consists of the following four interacting parts: manipulator, environment, task and controller, as shown in Figure 1.1(a), and Figure 1.1(b) [10, 23, 24].

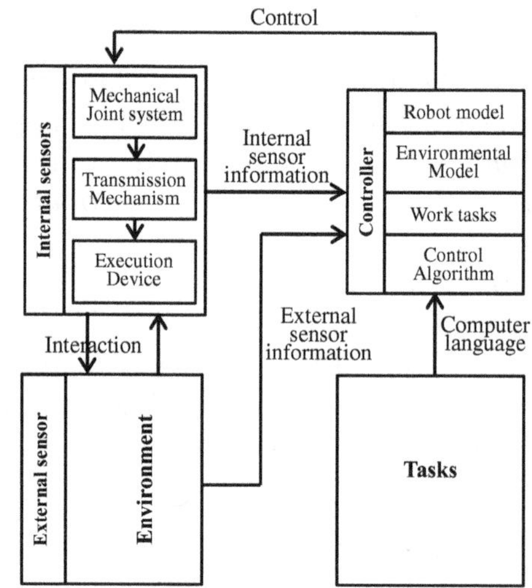

(a) Composition of the robot system

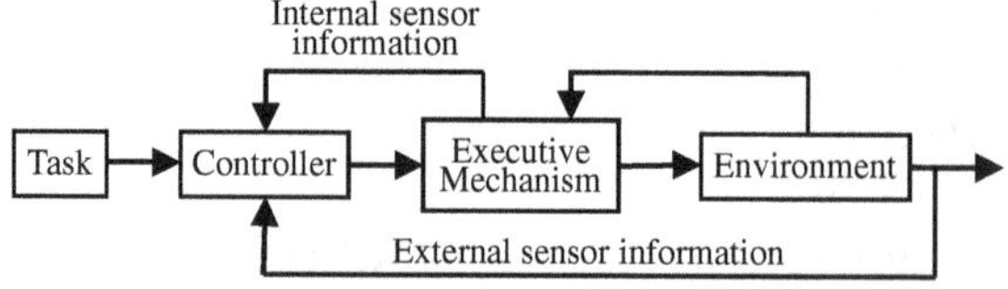

(b) The simplified form of robot system

Figure 1.1.   Basic structure of the robot system.

The manipulator is a machine with a transmission actuator. It is composed of arms, joints and end effectors (tools, etc.), combined into an interconnected and interdependent movement mechanism. The manipulator is used to perform specified tasks. Different manipulators have different structure types. Figure 1.2 shows a simplified geometric structure of the manipulator.

Most manipulators are articulated mechanical structures with several degrees of freedom, and generally have six degrees of freedom. Among them, the first three degrees of freedom guide the gripping device to the desired position, and the last three degrees of freedom are used to determine the direction of the end effector, as shown in Figure 1.2. The robot structure will be discussed further later.

The environment is the surrounding environment where the robot is located. The environment is not only determined by geometric conditions (accessible space), but also by all the natural characteristics of the environment and everything it contains. The inherent characteristics of robots are determined by the interaction between these natural characteristics and their environment.

In the environment, the robot will encounter some obstacles and other objects. It must avoid collisions with these obstacles and act on these objects.

Some sensors in the robot system are set somewhere in the environment and not on the manipulator. These sensors are part of the environment and are called external sensors.

Environmental information is generally deterministic and known, but in many cases the environment has an unknown and uncertain nature.

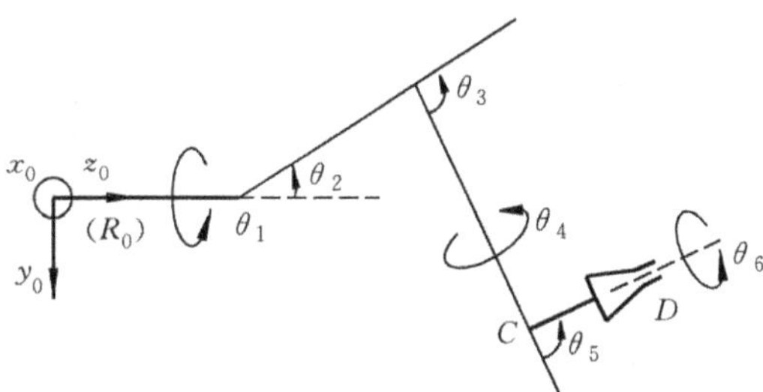

Figure 1.2. Sketch of the geometric structure of the manipulator.

Define the task as the difference between the two states of the environment (initial state and target state). These tasks must be described in an appropriate programming language and stored in the control computer of the robot system. This description must be understandable by the computer. Depending on the system used, the language description can be graphical, spoken (speech) or written text.

The computer is the controller or brain of the robot. The robot receives the signal from the sensor, processes the data, and generates control signals to drive each joint of the robot according to the pre-stored information, the state of the robot and its environment, etc.

For robots with relatively simple technology, the computer only contains fixed programs; for robots with relatively advanced technology, small computers, microcomputers or microprocessors with fully programmable programs can be used as their computers. Specifically, the following information is stored in the computer:

(1) Robot action model, which represents the relationship between the excitation signal of the actuator and the subsequent robot movement.
(2) The environment model, which describes everything in the reachable space of the robot. For example, explain which areas are unable to work due to obstacles.
(3) Task program, which enables the computer to understand the task to be executed.
(4) The control algorithm, which is a sequence of computer instructions, provides control of the robot in order to perform the work that needs to be done.

1.3.2   *Degrees of freedom of the robot*

The degree of freedom is an important technical indicator of the robot. It is determined by the structure of the robot and directly affects the maneuverability of the robot.

1. Degree of freedom of rigid body

Any point on the object is related to the orthogonal set of coordinate axes. The number of independent motions of an object in the coordinate system is called the degree of freedom (DOF). The movements that an object can perform (see Figure 1.3) are:

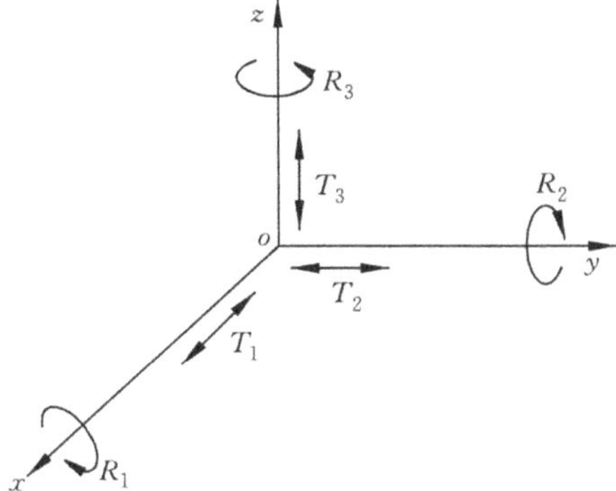

Figure 1.3.   The six degrees of freedom of a rigid body.

Three translational motions T_1, T_2 and T_3 along the coordinate axes O_x, O_y and O_z;

Three rotational movements R_1, R_2 and R_3 around the coordinate axes O_x, O_y and O_z.

This means that the object can use three translations and three rotations to orient and move relative to the coordinate system.

A simple object has six degrees of freedom. When a certain relationship is established between two objects, each object loses some degree of freedom to the other object. This relationship can also be expressed by the movement or rotation between two objects that cannot be carried out due to the establishment of a connection relationship.

2. Robot's degree of freedom

People expect a robot to move its end effector or a tool connected to it to a given point in an accurate position. If the purpose of the robot is unknown in advance, then it should have six degrees of freedom; however, if the tool itself has a special structure, then six degrees of freedom may not be needed. For example, to place a ball at a given position in space, three degrees of freedom are sufficient (see Figure 1.4(a)). For another example, to locate and orient a rotary drill bit requires five degrees of freedom; the drill bit can be expressed as a cylinder rotating around its main axis (see Figure 1.4(b)).

Generally, the arm of the robot has three degrees of freedom, and the other degrees of freedom are possessed by the end effector. When a robot is required to drill, its drill bit must be rotated. However,

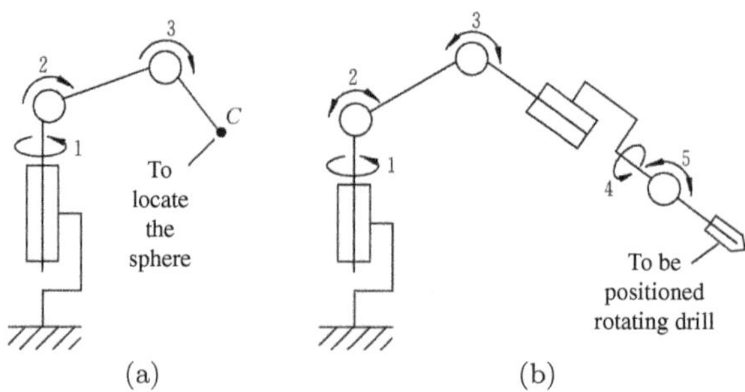

Figure 1.4.   Examples of robot degrees of freedom.

this rotation is always driven by an external motor. Therefore, it is not regarded as a degree of freedom of the robot. The same applies to manipulators. The gripper of the manipulator should be able to open and close. However, the degree of freedom used for the opening and closing of the gripping hand cannot be regarded as one of the degrees of freedom of the robot, because this degree of freedom only affects the operation of the gripping hand. This is very important and must be remembered.

3. Degree of freedom and mobility

The degree of freedom cannot be described as an attribute of one thing to another. Figure 1.5(a) is an example. In the figure, for a fixed base, point A has no degrees of freedom, point B has two degrees of freedom, and point C has three degrees of freedom. If the position of point D is determined, then the joint C used to move D will be redundant in theory, although there is no such need in practice. At this time, it can be considered that joint C has no degree of freedom anymore, but has a degree of mobility. However, if the CD is oriented by the anchor point C, then the joint C becomes a degree of freedom, which can make the CD oriented within a certain range. If you want to make the CD point in any direction, you need two other degrees of freedom.

There are two points worth remembering:

(1) Not all mobility constitutes a degree of freedom. Considering the function performed, a joint may become a degree of freedom, but it is not static. For example, in Figure 1.5(b), although there are many joints (five joints), the independent degrees of freedom of this robot are not more than two in any case.

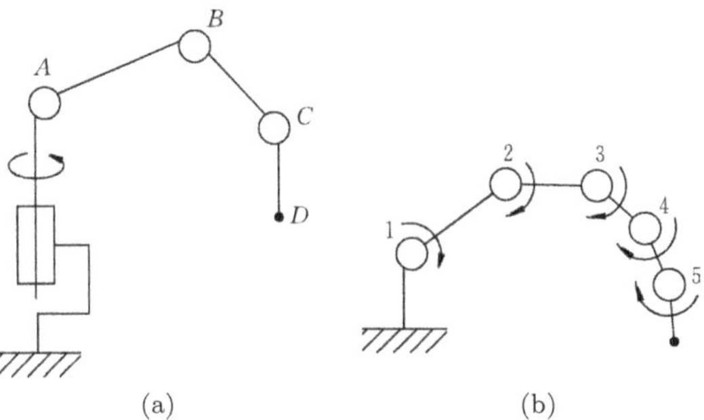

(a)   (b)

Figure 1.5.   Degrees of freedom and mobility.

(2) Generally, robots are not required to have more than 6 independent degrees of freedom, but much more mobility can be used. Understanding this is very important for establishing the control of the robot. Too many degrees of freedom may produce redundant degrees of freedom. Nevertheless, there are still people who are studying robots with 9 degrees of freedom in order to obtain greater mobility.

1.3.3   *Classification of robots*

There are many ways to classify robots. Here we first introduce three classification methods, namely, according to the geometric structure of the manipulator, the control method of the robot, and the information input method of the robot.

1. Classified by the geometric structure of the manipulator [26]

The mechanical configuration of the manipulator is varied. The most common structural form is described by its coordinate characteristics. These coordinate structures include Cartesian coordinate structure, cylindrical coordinate structure, polar coordinate structure, spherical coordinate structure and articulated spherical coordinate structure. Here is a brief introduction to the three most common robots: cylindrical, spherical and articulated spherical coordinate structures.

(1) **Cylindrical coordinate robot** is mainly composed of vertical column, horizontal arm (or manipulator) and base. The horizontal manipulator is installed on the vertical column, which can

expand and contract freely, and can move up and down along the vertical column. The vertical column is installed on the base and can move on the base together with the horizontal manipulator (as a part). In this way, the working envelope of this robot forms a cylindrical surface. Therefore, this kind of robot is called a cylindrical coordinate robot.

(2) **Spherical coordinate robot** is like the turret of a tank. The manipulator can telescopically move inside and outside, swing on a vertical plane, and rotate on a horizontal plane around the base. Therefore, the working envelope of this robot forms part of the spherical surface and is called a spherical coordinate robot.

(3) **The articulated spherical coordinate robot** is mainly composed of a base, upper arm and forearm. The upper arm and forearm can move on a vertical plane passing through the base. Between the forearm and the upper arm, the manipulator has an elbow joint; between the upper arm and the base, there is a shoulder joint. The rotation movement on the horizontal plane can be realized either by the shoulder joint or by rotating around the base. The work envelope of this robot forms most of the spherical surface, which is called an articulated spherical robot.

2. Classified by robot control method [38]

According to the control mode, robots can be divided into two types: non-servo robots and servo-controlled robots.

(1) **Non-servo robots**. This kind of robot works in a preprogrammed sequence, using terminal limit switches, brakes, latch plates and sequencers to control the movement of the robot manipulator. Among them, the latch plate is used to pre-determine the working sequence of the robot, and it is often adjustable. The sequencer is a sequence switch or stepping device, which can turn on the energy of the drive device in a predetermined correct order. After the driving device is connected to energy, it drives the robot's arms, wrists, and grippers to move. When they move to the position specified by the terminal limit switch, the limit switch switches the working state, sends a "work task or prescribed movement has been completed" signal to the sequencer, and causes the terminal brake to act and cut off drive energy to stop the manipulator.

(2) **Servo-controlled robots**. Servo-controlled robots have stronger working capabilities than non-servo robots. The controlled quantity of the servo system can be the position, speed, acceleration, force, etc. of the actuator (or tool) at the end of the robot. The feedback signal obtained by the feedback sensor and the integrated signal from the given device are compared with a comparator to obtain an error signal. After being amplified, it is used to excite the robot's driving device, and then drive the end effector to move in a certain regularity to reach the specified position or speed, etc. Obviously, this is a feedback control system. Servo control robots can be divided into point servo control and continuous path (track) servo control.

3. Classified according to the information input method of the robot controller [24]

When using this classification method to classify, it is slightly different for different countries, but they can have a unified standard. Here we mainly introduce the classification methods adopted by the Japan Industrial Robot Association (JIRA), the American Robot Association (RIA) and the French Industrial Robot Association (AFRI).

(1) JIRA classification

The Japan Industrial Robot Association divides robots into six categories.

Type 1: Manual operating hand. It is a processing device with several degrees of freedom that is directly operated by the operator.

Type 2: Sequencing robot. It is a manipulator that gradually repeats a given task in accordance with a predetermined sequence, conditions, and position. The predetermined information (such as work steps, etc.) is difficult to modify.

Type 3: Variable order robot. It is the same as Type 2, but its work order and other information are easy to modify.

Type 4: Repetitive robots. This type of robot can reproduce the actions originally taught by humans according to the information stored in the memory device. These teaching actions can be automatically repeated.

Type 5: Program-controlled robot. The human operator does not teach this robot manually, but provides the robot with a motion program to make it perform a given task.

Type 6: Intelligent robots. It can use sensor information to independently detect changes in their working environment or working conditions, and with the help of their self-decision-making ability, they can successfully perform corresponding tasks regardless of changes in the environmental conditions under which they perform tasks.

(2) RIA classification

The American Robot Association regards the last four types of machines in the JIRA classification as robots.

(3) AFRI classification

The French Industrial Robot Association divides robots into four types.

Type A: Category 1, manual or remote processing equipment.

Type B: Including Category 2 and Category 3, automatic processing equipment with pre-programmed work cycles.

Type C: Including Category 4 and Category 5, programmable and servo robots with point positions or continuous path trajectories, called the first generation robots.

Type D: The Category 6, which can obtain certain environmental data, is called the second-generation robot.

4. Classified by the degree of intelligence of the robot [9, 10, 12, 28]

(1) **General robots** do not have intelligence, but only have general programming capabilities and operating functions.

(2) **Intelligent robots** have different degrees of intelligence, which can be divided into:

   (a) Sensing robots can use sensory information (including vision, hearing, touch, proximity, force, infrared, ultrasound, and laser, etc.) to process sensory information to achieve control and operation.

   (b) Interactive robots, which conducts man-machine dialogue with the operator or programmer through the computer system to realize the control and operation of the robot.

   (c) Autonomous robots, which after being designed and manufactured, does not require human intervention and can automatically complete various anthropomorphic tasks in various environments.

5. Classified by the purpose of the robot [12]

(1) Industrial robots are used in industrial and agricultural production, mainly in the manufacturing sector for welding, painting, assembly, handling, inspection, and agricultural products processing and other operations.
(2) Exploration robots are used for space and ocean exploration, and can also be used for ground and underground adventure and exploration.
(3) Service robots. A kind of semi-autonomous or fully autonomous robot, the service work it is engaged in can make human beings survive better, and make equipment outside the manufacturing industry work better.
(4) Military robots. Used for military purposes and can be divided into aerial military robots, marine military robots and ground military robots.

6. Classification by robot mobility [12]

(1) Fixed robots. Fixed on a certain base, the whole robot (or manipulator) cannot move, only each joint can be moved.
(2) Mobile robot. The whole robot can move in a certain direction or any direction. This kind of robot can be divided into wheeled robots, crawler robots and walking robots, the latter of which are divided into one-legged, bipedal, four-legged, hexapod and eight-legged walking robots.

7. Other classifications of robots [4, 12]

(1) Manipulator imitates the movement of human upper limbs;
(2) Wheeled mobile robot imitates the movement of vehicles;
(3) The walking robot imitates the movement of human lower limbs;
(4) The underwater robot works underwater;
(5) The flying robot flies in the air;
(6) Sensing robots, especially vision robots, have sensors;
(7) Intelligent robot using artificial intelligence technology and having artificial intelligence technology;
(8) Robotic industrial automatic line. Robots are applied in batches on the production line.

1.4   Research Fields of Robotics

Robotics has an extremely wide range of research and application fields. These fields reflect a wide range of disciplines, involving many topics, such as robot architecture, mechanism, control, intelligence, sensing, robot assembly, robots in harsh environments, and robot languages. Robots have gained more and more common applications in various industries, agriculture, commerce, tourism, air and ocean, and national defense. The following are some of the more important research areas [8].

1. Robot vision [12, 16, 18, 27, 30, 37]
Machine vision or computer vision is to equip a computer system with a video input device so that it can "see" the surrounding things. Machine vision mainly uses computers to simulate human vision functions, extract information from images of objective things, process and understand them, and finally use them for actual detection, measurement and control.

Computer vision can generally be divided into low-level vision and high-level vision. Low-level vision mainly performs pre-processing functions, such as edge detection, moving target detection, texture analysis, and obtaining shapes, three-dimensional modeling, and surface colors through shadows. High-level vision is mainly to understand the observed image [32, 49].

Robot vision refers to a system that enables robots to have visual perception functions, and is one of the important areas of sensor-based intelligent robot systems. Robot vision can obtain a two-dimensional image of the environment through a vision sensor, analyze and interpret it through a vision processor, and then convert it into a symbol so that the robot can recognize an object and determine its location. Robot vision has been widely used in intelligent manufacturing, intelligent driving and navigation, intelligent transportation, intelligent agriculture and robot competitions.

2. Voice recognition technology [2, 29]
Speech recognition, also known as automatic speech recognition (ASR), is one of the important contents of natural language processing. Speech recognition is the use of a machine to convert speech signals into text information, and its ultimate goal is to make the machine understand human language. Speech recognition technology

refers to high technology that allows machines to convert speech signals into computer-readable text or commands through recognition and understanding. The essence of voice recognition is a pattern recognition based on speech characteristic parameters, that is, through learning, the system can classify the input speech according to a certain pattern, and then find the best matching result according to the judgment criterion.

Speech recognition technology has been widely used in robots, and well-known artificial intelligence and robotics companies from all over the world are scrambling to develop speech recognition products and put them on the market. Among the more influential companies and products are Apple's smart voice assistant Siri, Microsoft's social dialogue robot Xiaoice (Cortana), and China's Alibaba's personal voice assistant Xiaomi, and IFLYTEK's voice recognition SR301. Baidu's RavenH, Tencent's Dingdong smart screen, and a variety of voice recognition devices and chat robot products.

3. Sensors and perception systems

- Development of various new types of sensors, including vision, touch, hearing, proximity, force, telepresence, etc.
- Multi-sensing system and sensor fusion
- Sensor data integration
- Active vision and high-speed motion vision
- Modular sensor hardware
- Sensing technology under harsh working conditions
- Continuous language understanding and processing
- Sensing system software
- Virtual reality technology

4. Drive, modeling and motion control

- Ultra low inertia drive motor
- Direct drive and AC drive
- Modeling, control and performance evaluation of discrete event-driven systems
- Control mechanism (theory), including classic control, modern control and intelligent control
- Control system structure
- Control algorithm

- Multi-robot group coordinated control and group control
- Control system dynamics analysis
- Controller interface
- Online control and real-time control
- Autonomous operation and autonomous control
- Voice control and voice control

5. Automatic planning and scheduling

- Description of the environmental model
- Control the representation of knowledge
- Route plan
- Mission planning
- Planning in an unstructured environment
- Planning when there is uncertainty
- Planning and navigation of mobile robots in unknown environments
- Intelligent Algorithm
- Coordinated operation (movement) planning
- Assembly planning
- Planning based on sensory information
- Task negotiation and scheduling
- Scheduling of robots in manufacturing (processing) systems

6. Computer system

- The architecture of the intelligent robot control computer system
- General-purpose and special-purpose computer languages
- Robot planning and navigation based on computer vision [32]
- Standardized interface
- Neural computer and parallel processing
- Human-machine communication
- Multiple reality system (MAS)

7. Application areas [7, 31, 46]

- Application of robots in industry, agriculture and construction
- The application of robots in the service industry
- The application of robots in nuclear energy, high altitude and space, underwater and other dangerous environments
- Mining robot

- Military robot
- Education robot [1]
- Disaster rescue robot
- Rehabilitation robot
- Risk-removing robots and anti-riot robots
- Application of robots in CIMS and FMS

8. Others

- Microelectronics-mechanical system design and ultra-micro robots
- Collaborative design of products and their automatic processing

1.5　Summary of this Book

This book introduces the basic principles and applications of robotics including robot manipulators and mobile robots. It is a monograph on robotics and can be used as a reference for robotics research and a systematic robotics textbook. In addition to discussing general principles, some new methods and technologies are particularly elaborated, and a certain amount of space is used to describe the applications and development trends of robotics. This book contains the following specific contents:

(1) Chapter 1 briefly describes the origin and development of robotics, discusses the definition of robotics, analyzes the characteristics, structure and classification of robots, discusses the research fields of robotics, and analyzes the application fields of robotics, involving industrial robots, exploration robots, service robots and military robots.

　　The first part of this book discusses the mathematical foundations, main scientific and technological issues of the robot manipulator, involving five chapters from Chapters 2 to 6.

(2) Chapter 2 discusses the mathematical basis of robotics, including the position and posture transformation of any point in space, coordinate transformation, homogeneous coordinate transformation, object transformation and inverse transformation, and general rotation transformation. This basic mathematical knowledge provides powerful mathematical tools for

the following chapters to study robot kinematics, dynamics and control modeling.

(3) Chapter 3 describes the representation and solution of manipulator motion equations, including the motion equations of manipulator motion posture, direction angle, motion position and coordinates, as well as the linkage transformation matrix. For the solution of motion equations, the analytical and numerical solutions of inverse kinematics of the manipulator are discussed. In addition, the differential motion of the robot and its Jacobian matrix are also discussed.

(4) Chapter 4 deals with manipulator dynamics equations and dynamic characteristics, focusing on the analysis of two methods of manipulator dynamics equations, namely Lagrangian functional balance method and Newton-Euler dynamic balance method, and then analyzes the two-link manipulator On this basis, the steps to establish the Lagrangian equation are summarized, and the speed, kinetic energy and potential energy of a point on the manipulator link are calculated based on it, and then the dynamic equation of the three-link manipulator is derived. On this basis, the dynamic equations of the two-link manipulator and the velocity and acceleration equations of the three-link manipulator are deduced by examples. The dynamic characteristics of the manipulator are also reviewed.

(5) Chapter 5 studies the control principles and various control methods of the manipulator. These methods include robot position servo control, force and position hybrid control, variable structure control and so on. Manipulator control is based on manipulator kinematics and dynamics, combined with the basic principles of automatic control.

(6) Chapter 6 discusses the problem of manipulator planning. After explaining the role and tasks of robot planning, start with the robot planning in the building block world, and gradually carry out the discussion of robot planning. These planning methods include rule deduction method and logic algorithm method, planning with learning ability, planning based on expert systems, and planning based on machine learning.

The second part of this book discusses the key technological issues of mobile robots, including four chapters from Chapters 7 to 10.

(7) Chapter 7 introduces the architecture and dynamics model of mobile robots. Among them, three architectures of hierarchical, reactive and deliberate-reactive are discussed one by one for the architecture of mobile robots. The system structure of four-layer hierarchical intelligent robot is analyzed. For the dynamic model of the mobile robot, the dynamic model, stabilization and tracking problems of the wheeled mobile robot are studied, and examples of stabilization and tracking control are given.

(8) Chapter 8 discusses the localization and environmental modeling of mobile robots, including general methods of mobile robot localization and environmental modeling, simultaneous localization and modeling (SLAM) methods, vision-based topological environment modeling and localization methods, Cooperative localization and modeling methods of multiple mobile robots, and synchronous localization and modeling (SLAM) methods of mobile robots based on deep learning.

(9) Chapter 9 describes the navigation technology of mobile robots, first introduces the main methods and development trends of mobile robot navigation, and then discusses mobile robot global navigation based on approximate Voronoi diagrams, mobile robot local navigation strategies, composite navigation strategies, and feature point-based navigation strategy and robot path navigation based on ant colony algorithm. Finally, the application research of robot navigation based on machine learning and deep learning in autonomous path navigation of unmanned ships is reviewed.

(10) Chapter 10 explores the control of mobile robots. First, briefly introduce the intelligent control and intelligent control system, then introduce the adaptive fuzzy control of the robot and the control of the mobile robot based on neural network, and finally discuss the control of the mobile robot based on deep learning. The third part of this book summarizes the application of robots, including the two Chapters 11 and 12.

(11) Chapter 11 summarizes the application status of robots, and the application of robots in the fields of industry, exploration, service and military, and introduces the application of MATLAB in robot simulation.

(12) Chapter 12 discusses the current situation of robotics technology and the market, looks forward to the trend of robotics, and

introduces the robotics development plans of major economic powerful countries. In addition, the social problems caused by the application of robots and the challenges of cloning technology to intelligent robots are also discussed.

Each chapter contains related references, and there is an index of terminology at the end of the book, which helps readers to read robotics literature.

This book can be used by scientific and technical personnel engaged in robotics research, development and application, and can also be used as a teaching reference book for undergraduates and graduate students in colleges and vocational colleges.

1.6 Chapter Summary

As the beginning of this book, Section 1.1 of Chapter 1 discusses the origin of robots and the development of international robotics technology.

Section 1.2 introduces several main definitions of robots in the world, and then discusses the versatility and adaptability of robots. These characteristics are an important basis for the widespread application of robots. When discussing the composition of the robot system, a robot system is regarded as composed of four parts: the actuator, the environment, the task and the controller.

Section 1.3 analyzes the classification methods of robots and discusses the classification of robots according to the geometric structure, the control mode, the information input mode, the intelligence degree, the application purpose, and the mobility of the robot.

Section 1.4 briefly discusses the research fields of robotics, involving robot vision, robot speech recognition, sensing and perception systems, drive and control, automatic planning, computer systems, and application research.

Robots have been widely used in industrial production, sea and space exploration, service and military fields. Section 1.5 of this chapter introduces the application of industrial robots, exploration robots, service robots and military robots one by one.

Industrial robots are running in the automobile industry, electromechanical industry and other industrial sectors, contributing to human material production. Among them, welding robots and assembly robots are the two main application areas.

Mobile robots are becoming more and more widely used in the fields of exploration, service and military. In addition to special robots under severe working conditions, exploration robots are mainly space exploration robots and marine (underwater) exploration robots. Service robots have developed rapidly in recent years, and their number has greatly exceeded that of industrial robots, and is increasing year by year. Military robots are one of the focal points of competition among national power, economic power, technical power, and military power. Military robots are mostly ground-based military robots, and their technology is relatively mature.

References

[1] Adams, A. (1983). Current offering in robotics education. *Robotics Age*, 5(6):28–31.

[2] Amodei, D., Anubhai, R., Battenberg, E., *et al.* (2015). Deep Speech 2: End-to-End Speech Recognition in English and Mandarin. ArXiv preprint, ArXiv:1512.02595v1,2015.

[3] Analysis of the current situation of global industrial robots in 2020: General Industry has gradually become the main force in the new market. China Business Industry Research Institute, 2020-06-22 16:02, https://www.askci.com/news/chanye/20200622/1602311162338.shtml

[4] Aristidou, A. and Lasenby, J.F. (2011). A fast, iterative solver for the Inverse Kinematics problem. *Graphical Models*, 73:243–260.

[5] Asimov, I. (1950). I am Robot. Greenwich Cann Fawcett Grest Books, in nine short stories, 1970.

[6] Beni, G. and Hackwood, S. (editors). (1985). *Recent Advances in Robotics*. John Wiley & Sons., Inc.

[7] Bill Gates; Translated by Guo Kaisheng. (2007). Bill Gates predicted: Everyone will have robots in the future. 2007.02.01 09:17:47 http://people.techweb.com.cn/2007-02-01/149230.shtml.

[8] Bortz, A.B. (1983). Artificial intelligence and the nature of robotics. *Robotics Age*, 5(2):23–30.

[9] Cai, Z.X. (1987). Intelligent Control and Its Applications. IASTED 9th International Symposium on Robotics and Automation, May 1987.

[10] Cai, Z.X. (1988). *Principles and Applications of Robots*. Changsha: Central South University of Technology Press (in Chinese).

[11] Cai, Z.X. (2015). 40 years of robotics in China. *Science and Technology Review*, 33(21):13–22 (in Chinese).

[12] Cai, Z.X. and Xie, B. (2021). *Robotics*, 4th Edition. Beijing: Tsinghua University Press (in Chinese).

[13] Cai, Z.X. and Guo, F. (2013). Some problems in the development of industrial robots in China. *Robotics and Applications*, (3):9–12 (in Chinese).

[14] Cai, Z.X. and Weng, H. (2018). *Exploring the Kingdom of Robots.* Tsinghua University Press (in Chinese).

[15] Cai, Z.X. and Zhang, Z.J. (1988). Discussion on the development and application of robots. *Robot*, 10(3):61–63 (in Chinese).

[16] Cai, Z.X., Gu, M.Q. and Chen, B.F. (2013). Traffic Sign Recognition Algorithm Based on Multi-Modal Representation and Multi-Object Tracking. The 17th International Conference on Image Processing, Computer Vision, & Pattern Recognition, 1–7, Las Vegas, July 2013.

[17] Cai, Z.X., Liu, L.J., Cai, J.F. and Chen, B.F. (2020). *Artificial Intelligence and its Applications*, 6th Edition. Beijing: Tsinghua University Press (in Chinese).

[18] Cai, Z.X., Zheng, M.Y. and Zou, X.B. (2006). Real-time obstacle avoidance strategy for mobile robots based on lidar, *Journal of Central South University (Natural Science Edition)*, 37(2):324–329 (in Chinese).

[19] Cai, Z.X. (1986). On the definition of robots. *Electrical Automation*, (5):13, 22 (in Chinese).

[20] Cai, Z.X. and Chen, A.B. (2008). *AI Dictionary.* China, Beijing: Chemical Industry Press (in Chinese).

[21] Cai, Z.X. and Xie, B. (2021). *Fundamentals of Robotics*, 3rd Edition. Beijing: Machinery Industry Press (in Chinese).

[22] Capek, K. (1920). Rossum's Universal Robots (R.U.R), a Fantastic Melodrama.

[23] Coiffet, P. (1983). *Robot Technology Vol. 1: Modeling and Control,* Prentice-Hall, Inc.

[24] Coiffet, P. and Chirouze, M. (1983). *An Introduction to Robot Technology,* Hermes Publishing.

[25] Dorf, R.C. (1983). *Robotics and Automated Manufacturing.* Reston Publishing Company, Inc.

[26] Educational Systems: Robotics and Industrial Electronics. Heath Company, 1982.

[27] Fu, K.S., Gonzalez, R.C. and Lee, C.S.G. (1987). *Robotics: Control, Sensing, Vision, and Intelligence.* Wiley.

[28] Gevarter, W.B. (1984). *Artificial Intelligence Applications: Expert Systems, Computer Vision and Nature Language Processing.* NOYES Publications.

[29] Hannund, A., Case, C., Casper, J., *et al.* (2014). Deep Speech: Scaling up end-to-end speech recognition. ArXiv preprint, ArXiv: 1412.5567v2,2014.

[30] Hansen, P., Corke, P.I. and Boles, W. (2010). Wide-angle visual feature matching for outdoor localization. *The International Journal of Robotics Research*, 29(1-2):267–297.

[31] Hartley, J. (1983). *Robots at Work, A Practical Guide for Engineers and Managers*. IFS Publications.

[32] Hartley, R. and Zisserman, A. (2003). *A. Multiple View Geometry in Computer Vision*. Cambridge University Press.

[33] IFR. (2012). The continuing success story of industrial robots. 2012-11-11.http://www.msnbc.msn.com/id/23438322/ns/technology_ and_ science-innovation/t/japan-looks-robot-future/

[34] IFR. (2013). Executive Summary: World Robotics 2013 Industrial Robots. 2013-09-18. http://www.ifr.org/index.php?id=59&df= Executive_Summary_WR_2013.pdf

[35] Interpretation of the 2019 Global Industrial Robot Market Report. 2019-09-19 15:28:49 YI Intelligent Internet of Things, http://www.openpcba.com/web/contents/get?id=3887&tid=15

[36] John, J.C. (2018). *Introduction to Robotics: Mechanics and Control*, 4th Edition. Pearson.

[37] Ma, J.G. and Jia, Y.D. (2003). A mobile robot positioning method based on an omnidirectional camera. *Journal of Beijing Institute of Technology*, 23(3):317–321 (in Chinese).

[38] Miller, R.K. (1982). *Robotics in Industry: Applications for Foundries*, SFAI Institute.

[39] Niku, S.B. (2010). *Introduction to Robotics: Analysis, Control, Applications*. Wiley.

[40] Rathmill, K. (1986). *Robotic Assembly*. Springer-Verlag, New York, Inc., Secaucus, NJ.

[41] Rooks, B. compiled. (1983). *Developments in Robotics, 1983*. IFS (Publications) Ltd.

[42] Siciliano, B. and Khatib, O. (2008). Springer Handbook of Robotics. Springer Science & Business Media, Berlin/Heidelberg. https://doi.org/10.1007/978-3-540-30301-5

[43] Simons, G.L. (1980). *Robots in Industry*. NCC Publications Ltd.

[44] Song, J. (1987). Robots — An emerging technology with far-reaching influence. *Robots*, 1987(1):1–2 (in Chinese).

[45] Song, J. (editor in chief). (1991). *China Encyclopedia: Automatic Control and System Engineering*. Beijing-Shanghai: China Encyclopedia Press (in Chinese).

[46] Tanner, W.R. (1980). Can I Use a Robot? *Robotics Today*, Springer.

[47] The current situation and future development trend forecast of the global robotics industry segmentation market in 2019. 19-09-02 11:58 China Business Intelligence Network, https://baijiahao.baidu.com/ s?id=1643534568013791442&wfr=spider&for=pc

[48] Togai, M. (1984). Japan's Next Generation of Robot. *IEEE Computer*, March 1984, 19–25.

[49] Zhang, S., Huang, J., Li, H., *et al.* (2012). Automatic Image Annotation and Retrieval Using Group Sparsity. *IEEE Transactions on Systems Man & Cybernetics Part B Cybernetics A, Publication of the IEEE Systems Man & Cybernetics Society*, 42(3):838–849.

[50] Zhang, Z. and Cai, Z.X. (1986). Robotization—A new trend of automation. *Automation*, (6):2–3 (in Chinese).

[51] Zhou, L. (2019). Analysis of the current development status of the robot industry in various countries. July 13, 2019 09:15, Electronic enthusiasts network, http://www.elecfans.com/jiqiren/992424.html (in Chinese).

Part I

Robot Manipulator

Chapter 2

Mathematical Fundamentals

When discussing the composition of the robot system in the previous chapter, it was pointed out that the manipulator is the mechanical movement part of the robotic system. As an automated tool, the manipulator has the following characteristics: its actuator is a comprehensive rigid body used to ensure complex space movement, and it itself often needs to move as a unified body in the process of machining or assembly etc. Therefore, there is a need for an effective and convenient mathematical method to describe the displacement, velocity and acceleration of a single rigid body, as well as dynamic problems. This mathematical description method is not unique, and different people may use different methods. The matrix method will be used to describe the kinematics and dynamics of the robotic manipulator. This mathematical description is based on the homogeneous coordinates of the three-dimensional space points transformed by the fourth-order square matrix, which can link motion, transformation and mapping with matrix operations [2, 3, 6, 9, 14].

The study of the motion of the manipulator involves not only the manipulator itself, but also the relationship between objects and the relationship between the objects and the manipulator. The homogeneous coordinates and their transformations are used to express these relationships. Homogeneous coordinate transformation can express not only dynamic problems, but also manipulator control algorithms, computer vision, and computer graphics. Therefore, special attention is paid to this mathematical representation.

This chapter first introduces the pose and coordinate system description of the space point, then discusses the translation and rotation coordinate system mapping and the translation and rotation homogeneous coordinate transformation of the space point, then analyzes the transformation and transformation equations of the object, and finally studies the general rotation transformation of vectors.

2.1   Description of Pose and Coordinate System

When describing the relationship between objects (such as parts, tools, or manipulators), concepts such as position vectors, planes, and coordinate systems are used. First, establish these concepts and their representations [4, 13, 15, 18].

2.1.1   *Location description*

Once a coordinate system is established, a 3×1 position vector can be used to determine the position of any point in the space. For the rectangular coordinate system, the position of any point in space can be presented as a 3×1 column vector $^A\boldsymbol{p}$

$$^A\boldsymbol{p} = \begin{bmatrix} p_x \\ p_y \\ p_z \end{bmatrix} \tag{2.1}$$

Among them, p_x, p_y, p_z are the coordinate components of the point in the x, y, and z axis directions in the coordinate system $\{A\}$. The superscript of $^A\boldsymbol{p}$ represents the reference coordinate system $\{A\}$. $^A\boldsymbol{p}$ is called the position vector, as shown in Figure 2.1.

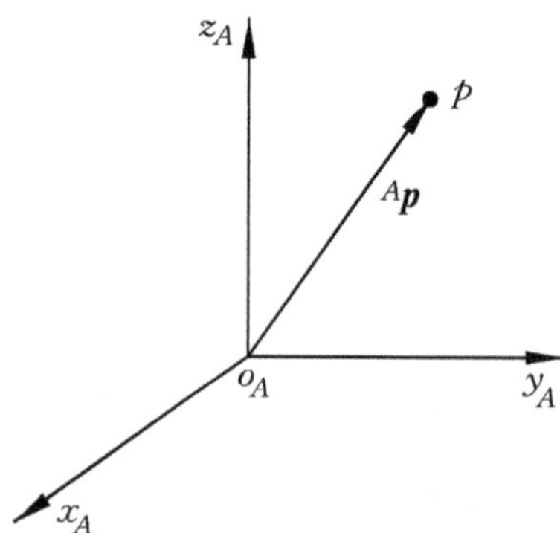

Figure 2.1.   Location representation.

2.1.2 *Orientation description*

In order to study the movement and operation of robotic manipulator, it is often necessary to not only indicate the position of a certain point in space, but also indicate the orientation of the object. The orientation of an object can be described by a coordinate system fixed to the object. In order to specify the orientation of a rigid body B in space, a rectangular coordinate system $\{B\}$ is set to be fixed to this rigid body. A 3×3 matrix $^A_B R$ composed of the three unit principal vectors x_B, y_B, z_B of the coordinate system $\{B\}$ relative to the direction cosine of the reference coordinate system $\{A\}$

$$^A_B R = \begin{bmatrix} ^A x_B & ^A y_B & ^A z_B \end{bmatrix} = \begin{bmatrix} r_{11} & r_{12} & r_{13} \\ r_{21} & r_{22} & r_{23} \\ r_{31} & r_{32} & r_{33} \end{bmatrix} \tag{2.2}$$

is used to indicate the orientation of the rigid body B relative to the coordinate system $\{A\}$. In the formula, $^A_B R$ is called the rotation matrix, the superscript A represents the reference coordinate system $\{A\}$, and the subscript B represents the coordinate system $\{B\}$ being described. There are nine elements in $^A_B R$ in total, but only three elements are independent. Since the three column vectors $^A x_B, ^A y_B$ and $^A z_B$ of $^A_B R$ are all unit vectors, and the pairs are perpendicular to each other, so its nine elements satisfy the following six constraints (orthogonal conditions):

$$^A x_B \cdot {}^A x_B = {}^A y_B \cdot {}^A y_B = {}^A z_B \cdot {}^A z_B = 1 \tag{2.3}$$

$$^A x_B \cdot {}^A y_B = {}^A y_B \cdot {}^A z_B = {}^A z_B \cdot {}^A x_B = 0 \tag{2.4}$$

It can be seen that the rotation matrix $^A_B R$ is orthogonal and satisfies the condition:

$$^A_B R^{-1} = {}^A_B R^T; \quad \left| {}^A_B R \right| = 1 \tag{2.5}$$

In the formula, the superscript T means transpose; $||$ is the determinant symbol.

Corresponding to the axis x, y or z for the rotation transformation with the angle of θ, the rotation matrices are:

$$R(x, \theta) = \begin{bmatrix} 1 & 0 & 0 \\ 0 & c\theta & -s\theta \\ 0 & s\theta & c\theta \end{bmatrix} \tag{2.6}$$

$$R(y, \theta) = \begin{bmatrix} c\theta & 0 & s\theta \\ 0 & 1 & 0 \\ -s\theta & 0 & c\theta \end{bmatrix} \tag{2.7}$$

$$R(z, \theta) = \begin{bmatrix} c\theta & -s\theta & 0 \\ s\theta & c\theta & 0 \\ 0 & 0 & 1 \end{bmatrix} \tag{2.8}$$

In the formula, s represents sin and c represents cos. This convention will be adopted in the future.

Figure 2.2 shows the orientation of an object (here, the gripper). This object is fixed to the coordinate system $\{B\}$ and moves relative to the reference coordinate system $\{A\}$.

2.1.3 *Pose description*

It has been discussed above that the position vector is used to describe the position of the point, and the rotation matrix is used to describe the orientation of the object. To fully describe the pose (position and posture) of a rigid body B in space, the object B is usually fixed to a certain coordinate system $\{B\}$. The origin of the coordinate of $\{B\}$ is generally selected on the feature point of the object B, such as the center of mass. Relative to the reference system $\{A\}$,

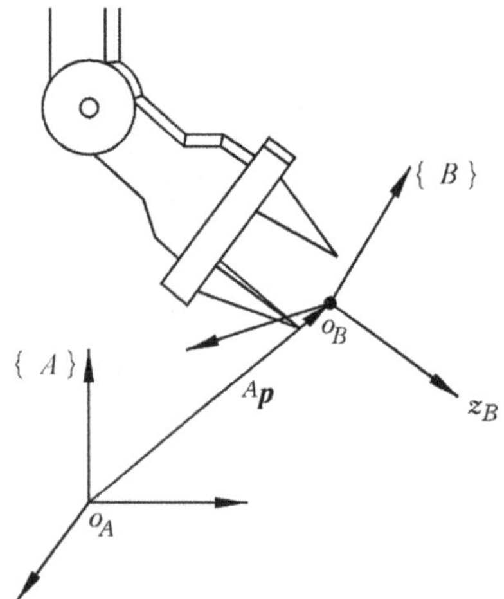

Figure 2.2.   Representation of orientation.

the origin position of the coordinate system $\{B\}$ and the orientation of the coordinate axis are described by the position vector ${}^A\boldsymbol{p}_{Bo}$ and the rotation matrix ${}^A_B R$ respectively. In this way, the pose of the rigid body B can be described by the coordinate system $\{B\}$, that is,

$$\{B\} = \{{}^A_B R \quad {}^A\boldsymbol{p}_{Bo}\} \tag{2.9}$$

When expressing the position, the rotation matrix (identity matrix) ${}^A_B R = I$ in equation (2.9); when expressing the orientation, the position vector ${}^A\boldsymbol{p}_{Bo} = \boldsymbol{o}$ in equation (2.9).

2.2 Mapping of Translation and Rotation Coordinate System

Any point in space is described differently in different coordinate systems. In order to clarify the relationship from one coordinate system description to another coordinate system description, it is necessary to discuss the mathematical problems of this transformation [1,8,16].

2.2.1 *Translation coordinate transformation*

Let the coordinate systems $\{B\}$ and $\{A\}$ have the same orientation, but the origin of the coordinate system $\{B\}$ does not coincide with the origin of the coordinate system $\{A\}$. Use the position vector ${}^A\boldsymbol{p}_{Bo}$ to describe its position relative to $\{A\}$, as shown in Figure 2.3. ${}^A\boldsymbol{p}_{Bo}$ is called the translation vector of $\{B\}$ relative to $\{A\}$. If the position of a point p in the coordinate system $\{B\}$ is ${}^B\boldsymbol{p}$, then its position vector ${}^A\boldsymbol{p}$ relative to the coordinate system $\{A\}$ can be obtained by adding the vectors, that is

$$^A\boldsymbol{p} = {}^B\boldsymbol{p} + {}^A\boldsymbol{p}_{Bo} \tag{2.10}$$

And call the above formula the coordinate translation equation.

2.2.2 *Rotation coordinate transformation*

Suppose that the coordinate system $\{B\}$ and $\{A\}$ have the same coordinate origin, but the orientation of the two is different, as shown in Figure 2.4. Use the rotation matrix ${}^A_B R$ to describe the orientation of $\{B\}$ relative to $\{A\}$. The description ${}^A\boldsymbol{p}$ and ${}^B\boldsymbol{p}$ of the same

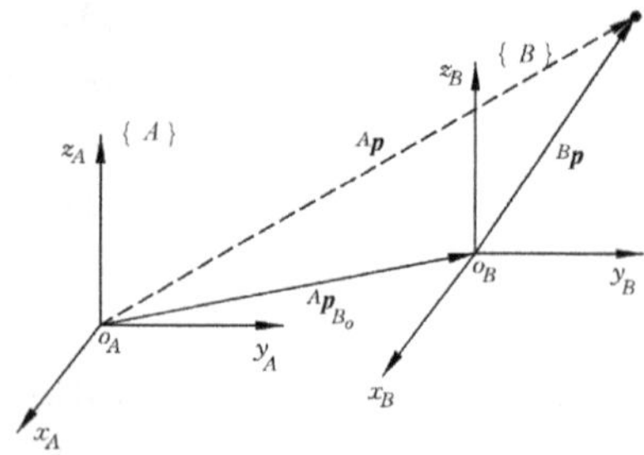

Figure 2.3. Translation transformation.

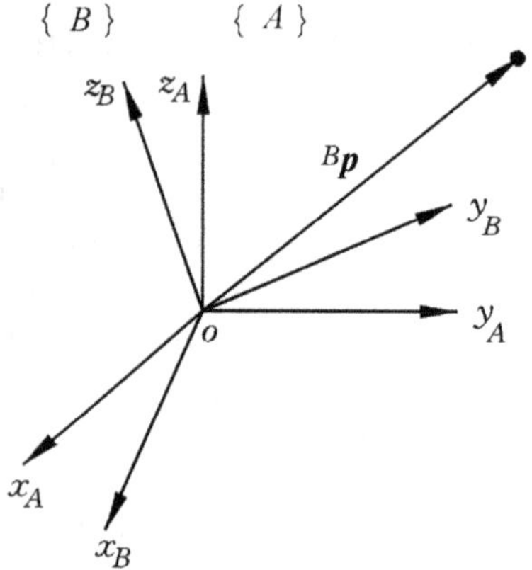

Figure 2.4. Rotation transformation.

point p in the two coordinate systems $\{A\}$ and $\{B\}$ has the following transformation relationship:

$$^A\boldsymbol{p} = {}^A_B R\,{}^B\boldsymbol{p} \tag{2.11}$$

call the above formula the coordinate rotation equation.

It can be similarly used ${}^A_B R$ to describe the position of the coordinate system $\{A\}$ relative to the coordinate system $\{B\}$. Both ${}^A_B R$ and ${}^B_A R$ are orthogonal matrices, and the two are inverse to each other. According to the property (2.5) of the orthogonal matrix, can get:

$$^B_A R = {}^A_B R^{-1} = {}^A_B R^T \tag{2.12}$$

For the most general case: the origin of the coordinate system $\{B\}$ does not coincide with the origin of the coordinate system $\{A\}$,

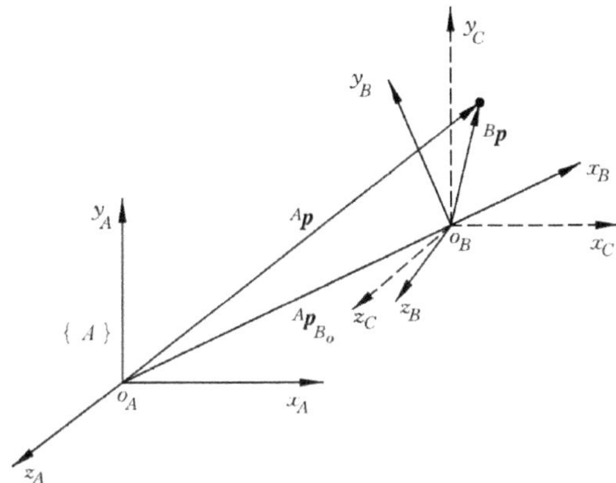

Figure 2.5. Compound transformation.

and the orientation of $\{B\}$ is not the same as that of $\{A\}$. Use the position vector $^Ap_{Bo}$ to describe the orientation of the coordinate origin of $\{B\}$ relative to; use the rotation matrix ^A_BR to describe the orientation of $\{B\}$ relative to $\{A\}$, as shown in Figure 2.5. The description ^Ap and ^Bp of any point p in the two coordinate systems and $\{B\}$ has the following transformation relationship:

$$^Ap = {}^A_BR\,{}^Bp + {}^Ap_{Bo} \tag{2.13}$$

The above formula can be regarded as a compound transformation of coordinate rotation and coordinate translation. In fact, a transition coordinate system $\{C\}$ is specified so that the origin of coordinate of $\{C\}$ coincides with the origin of $\{B\}$, and the orientation of $\{C\}$ is the same as that of $\{A\}$. According to formula (2.11), the transformation to the transition coordinate system can be obtained:

$$^Cp = {}^C_BR\,{}^Bp = {}^A_BR\,{}^Bp$$

Example 2.1. Knowing the initial pose of the coordinate system $\{B\}$ is coincided with $\{A\}$, first $\{B\}$ rotate $30°$ relative to axis z_A of the coordinate system $\{A\}$, then move 12 units along the axis x_A, of $\{A\}$, and move 6 units along the axis y_A of $\{A\}$. Find the position vector $^Ap_{Bo}$ and the rotation matrix ^A_BR. Assuming that the description of a point p in the coordinate system $\{B\}$ is $[3,7,0]^T$, find its description ^Ap in the coordinate system $\{A\}$.

According to formulas (2.8) and (2.1), we can get $^A_B R$ and $^A\boldsymbol{p}_{Bo}$ respectively:

$$^A_B R = R(z, 30^\circ) = \begin{bmatrix} c30^\circ & -s30^\circ & 0 \\ s30^\circ & c30^\circ & 0 \\ 0 & 0 & 1 \end{bmatrix}$$

$$= \begin{bmatrix} 0.866 & -0.5 & 0 \\ 0.5 & 0.866 & 0 \\ 0 & 0 & 1 \end{bmatrix}; \quad {}^A\boldsymbol{p}_{Bo} = \begin{bmatrix} 12 \\ 6 \\ 0 \end{bmatrix}$$

From formula (2.13), we have

$$^A\boldsymbol{p} = {}^A_B R {}^B\boldsymbol{p} + {}^A\boldsymbol{p}_{Bo} = \begin{bmatrix} -0.902 \\ 7.562 \\ 0 \end{bmatrix} + \begin{bmatrix} 12 \\ 6 \\ 0 \end{bmatrix} = \begin{bmatrix} 11.098 \\ 13.562 \\ 0 \end{bmatrix}$$

2.3   Homogeneous Coordinate Transformation of Translation and Rotation

Knowing the coordinates of a certain point in a rectangular coordinate system, the coordinates of the point in another rectangular coordinate system can be obtained by homogeneous coordinate transformation [11, 12].

2.3.1   *Homogeneous transformation*

The transformation formula (2.13) is non-homogeneous for the point $^B\boldsymbol{p}$, but it can be expressed as an equivalent homogeneous transformation:

$$\begin{bmatrix} ^A\boldsymbol{p} \\ 1 \end{bmatrix} = \begin{bmatrix} ^A_B R & {}^A\boldsymbol{p}_{Bo} \\ 0 & 1 \end{bmatrix} = \begin{bmatrix} ^B\boldsymbol{p} \\ 1 \end{bmatrix} \tag{2.14}$$

Among them, the 4×1 column vector represents the point in the three-dimensional space, which is called the homogeneous coordinate of the point, which is still recorded as $^A\boldsymbol{p}$ or $^B\boldsymbol{p}$. The above formula can be written in matrix form:

$$^A\boldsymbol{p} = {}^A_B T {}^B\boldsymbol{p} \tag{2.15}$$

In the formula, the homogeneous coordinate as $^A\boldsymbol{p}$ and $^B\boldsymbol{p}$ is a 4×1 column vector, which is different from the dimension in formula (2.13), and the fourth element 1 is added. The homogeneous transformation matrix $^A_B T$ is a 4×4 square matrix with the following form:

$$^A_B T = \begin{bmatrix} ^A_B R & ^A\boldsymbol{p}_{Bo} \\ 0 & 1 \end{bmatrix} \tag{2.16}$$

$^A_B T$ comprehensively shows translation transformation and rotation transformation. The transformation formulas (2.14) and (2.13) are equivalent. In essence, the formula (2.14) can be written as

$$^A\boldsymbol{p} = {}^A_B R\, {}^B\boldsymbol{p} + {}^A\boldsymbol{p}_{Bo}; \quad 1 = 1$$

Whether the position vector $^A\boldsymbol{p}$ and $^B\boldsymbol{p}$ is a 3×1 rectangular coordinate or a 4×1 homogeneous coordinate depends on the relationship between the contexts.

Example 2.2. Use the homogeneous transformation method to solve $^A\boldsymbol{p}$ in Example 2.1.

From the rotation matrix $^A_B R$ and position matrix $^A\boldsymbol{p}_{Bo}$ obtained in Example 2.1, the homogeneous transformation matrix can be obtained

$$^A_B T = \begin{bmatrix} ^A_B R & ^A\boldsymbol{p}_{Bo} \\ 0 & 1 \end{bmatrix} = \begin{bmatrix} 0.866 & -0.5 & 0 & 12 \\ 0.5 & 0.866 & 0 & 6 \\ 0 & 0 & 1 & 0 \\ 0 & 0 & 0 & 1 \end{bmatrix}$$

Substituting into the homogeneous transformation (2.15), we get:

$$^A\boldsymbol{p} = \begin{bmatrix} 0.866 & -0.5 & 0 & 12 \\ 0.5 & 0.866 & 0 & 6 \\ 0 & 0 & 1 & 0 \\ 0 & 0 & 0 & 1 \end{bmatrix} \begin{bmatrix} 3 \\ 7 \\ 0 \\ 1 \end{bmatrix} = \begin{bmatrix} 11.098 \\ 13.562 \\ 0 \\ 1 \end{bmatrix}$$

It is the position of the point p described by homogeneous coordinates.

So far, the Cartesian coordinate description and the homogeneous coordinate description of a point p in the available space are as follows:

$$p = \begin{bmatrix} x \\ y \\ z \end{bmatrix}$$

$$p = \begin{bmatrix} x \\ y \\ z \\ 1 \end{bmatrix} = \begin{bmatrix} wx \\ wy \\ wz \\ w \end{bmatrix}$$

In the formula, w is a non-zero constant and is a coordinate proportional coefficient.

The vector of the coordinate origin, that is, the zero vector is expressed as $[0,0,0,1]T$, which is undefined. A vector with the form $[a,b,c,0]T$ represents an infinite vector, which is used to represent the direction, that is, $[1, 0, 0, 0]$, $[0, 1, 0, 0]$, $[0, 0, 1, 0]$ indicate the direction of the axis x, y and z respectively.

Specifies the dot product of two vectors a and b

$$a \cdot b = a_x b_x + a_y b_y + a_z b_z \tag{2.17}$$

is a scalar, and the cross product (vector product) of two vectors is a vector perpendicular to the plane determined by the two multiplied vectors

$$a \times b = (a_y b_z - a_z b_y)i + (a_z b_x - a_x b_z)j + (a_x b_y - a_y b_x)k \tag{2.18}$$

Or use the following determinant to express:

$$a \times b = \begin{vmatrix} i & j & k \\ a_x & a_y & a_z \\ b_x & b_y & b_z \end{vmatrix} \tag{2.19}$$

2.3.2 *Translation homogeneous coordinate transformation*

A certain point in space is described by a vector $ai + bj + ck$. Among them, i, j, k are the unit vector on axis x, y, z. Using translational homogeneous transformation this point can be expressed as:

$$\mathbf{Trans}(a,b,c) = \begin{bmatrix} 1 & 0 & 0 & a \\ 0 & 1 & 0 & b \\ 0 & 0 & 1 & c \\ 0 & 0 & 0 & 1 \end{bmatrix} \tag{2.20}$$

Among them, Trans represents translation transformation.

The vector v obtained by performing translation transformation on the known vector $\boldsymbol{u} = [x,\, y,\, z]^{\mathrm{T}}$ is:

$$\begin{bmatrix} 1 & 0 & 0 & a \\ 0 & 1 & 0 & b \\ 0 & 0 & 1 & c \\ 0 & 0 & 0 & 1 \end{bmatrix} \begin{bmatrix} x \\ y \\ z \\ w \end{bmatrix} = \begin{bmatrix} x + aw \\ y + bw \\ z + cw \\ w \end{bmatrix} = \begin{bmatrix} x/w + a \\ y/w + b \\ z/w + c \\ 1 \end{bmatrix} \tag{2.21}$$

This transformation can be viewed as the sum of vector $(x/w)\boldsymbol{i} + (y/w)\boldsymbol{j} + (z/w)\boldsymbol{k}$ and vector $a\boldsymbol{i} + b\boldsymbol{j} + c\boldsymbol{k}$.

Multiplying each element of the transformation matrix by a non-zero constant does not change the characteristics of the transformation matrix.

Example 2.3. The vector $2\boldsymbol{i} + 3\boldsymbol{j} + 2\boldsymbol{k}$ is translation transformed by vector $4\boldsymbol{i} - 3\boldsymbol{j} + 7\boldsymbol{k}$

Consider the new point vector obtained by vector translation, consider the new point vector:

$$\begin{bmatrix} 1 & 0 & 0 & 4 \\ 0 & 1 & 0 & -3 \\ 0 & 0 & 1 & 7 \\ 0 & 0 & 0 & 1 \end{bmatrix} \begin{bmatrix} 2 \\ 3 \\ 2 \\ 1 \end{bmatrix} = \begin{bmatrix} 6 \\ 0 \\ 9 \\ 1 \end{bmatrix}$$

If the transformation matrix is multiplied by -5, and the translation transformation vector is multiplied by 2, then get:

$$\begin{bmatrix} -5 & 0 & 0 & -20 \\ 0 & -5 & 0 & 15 \\ 0 & 0 & -5 & -35 \\ 0 & 0 & 0 & -5 \end{bmatrix} \begin{bmatrix} 4 \\ 6 \\ 4 \\ 2 \end{bmatrix} = \begin{bmatrix} -60 \\ 0 \\ -90 \\ -10 \end{bmatrix}$$

It corresponds to the vector $[6,0,9,1]^{\mathrm{T}}$ and is the same as the point vector before multiplying by a constant.

2.3.3   *Rotation homogeneous coordinate transformation*

Corresponding to the axis x, y, z for the rotation transformation of the angle θ, then can get:

$$\mathbf{Rot}(x,\theta) = \begin{bmatrix} 1 & 0 & 0 & 0 \\ 0 & c\theta & -s\theta & 0 \\ 0 & s\theta & c\theta & 0 \\ 0 & 0 & 0 & 1 \end{bmatrix} \tag{2.22}$$

$$\mathbf{Rot}(y,\theta) = \begin{bmatrix} c\theta & 0 & s\theta & 0 \\ 0 & 1 & 0 & 0 \\ -s\theta & 0 & c\theta & 0 \\ 0 & 0 & 0 & 1 \end{bmatrix} \tag{2.23}$$

$$\mathbf{Rot}(z,\theta) = \begin{bmatrix} c\theta & -s\theta & 0 & 0 \\ s\theta & c\theta & 0 & 0 \\ 0 & 0 & 1 & 0 \\ 0 & 0 & 0 & 1 \end{bmatrix} \tag{2.24}$$

In the formula, Rot represents the rotation transformation. Below we illustrate this kind of rotation transformation.

Example 2.4. A known point $\boldsymbol{u} = 7\boldsymbol{i} + 3\boldsymbol{j} + 2\boldsymbol{k}$, by rotating it $90°$ around the axis z, can obtain:

$$\boldsymbol{v} = \begin{bmatrix} 0 & -1 & 0 & 0 \\ 1 & 0 & 0 & 0 \\ 0 & 0 & 1 & 0 \\ 0 & 0 & 0 & 1 \end{bmatrix} \begin{bmatrix} 7 \\ 3 \\ 2 \\ 1 \end{bmatrix} = \begin{bmatrix} -3 \\ 7 \\ 2 \\ 1 \end{bmatrix}$$

Figure 2.6(a) shows the position of the point vector in the coordinate system before and after the rotation transformation. It can be seen from the figure that the point $\boldsymbol{u}$ rotates $90°$ around axis z to the point v. If the point v rotates $90°$ around axis y axis, the point w is obtained. This transformation can also be seen from Figure 2.6(a), and can be obtained by equation (2.23):

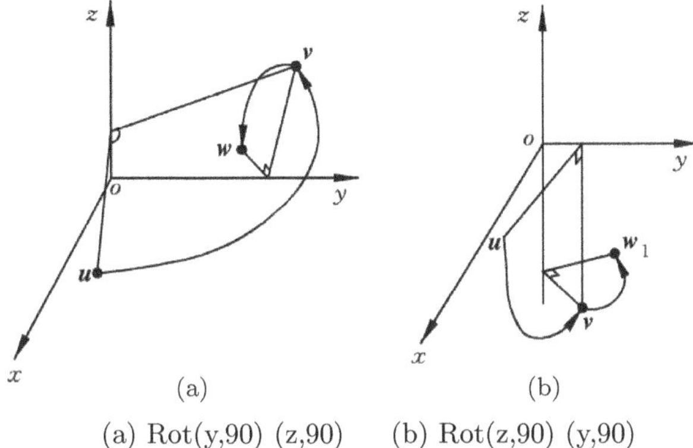

(a) Rot(y,90) (z,90)   (b) Rot(z,90) (y,90)

Figure 2.6.   Effect of rotation order on the transformation result.

$$\boldsymbol{w} = \begin{bmatrix} 0 & 0 & 1 & 0 \\ 0 & 1 & 0 & 0 \\ -1 & 0 & 0 & 0 \\ 0 & 0 & 0 & 1 \end{bmatrix} \begin{bmatrix} -3 \\ 7 \\ 2 \\ 1 \end{bmatrix} = \begin{bmatrix} 2 \\ 7 \\ 3 \\ 1 \end{bmatrix}$$

If the above two rotation transformations $\boldsymbol{v} = \mathbf{Rot}(z, 90)$ $\boldsymbol{u}$ and $\boldsymbol{w} = \mathbf{Rot}(y, 90)\boldsymbol{v}$ are combined, then the following formula can be obtained:

$$\boldsymbol{w} = \mathbf{Rot}(y, 90)\mathbf{Rot}(z, 90)\boldsymbol{u} \tag{2.25}$$

Because

$$\mathbf{Rot}(y, 90)\mathbf{Rot}(z, 90) = \begin{bmatrix} 0 & 0 & 1 & 0 \\ 1 & 0 & 0 & 0 \\ 0 & 1 & 0 & 0 \\ 0 & 0 & 0 & 1 \end{bmatrix} \tag{2.26}$$

So it can get:

$$\boldsymbol{w} = \begin{bmatrix} 0 & 0 & 1 & 0 \\ 1 & 0 & 0 & 0 \\ 0 & 1 & 0 & 0 \\ 0 & 0 & 0 & 1 \end{bmatrix} \begin{bmatrix} 7 \\ 3 \\ 2 \\ 1 \end{bmatrix} = \begin{bmatrix} 2 \\ 7 \\ 3 \\ 1 \end{bmatrix}$$

The result is the same as before.

If the order of rotation is changed, first let $\boldsymbol{u}$ rotate the axis y 90°, it will transform $\boldsymbol{u}$ to a different position $\boldsymbol{w}_1$ than $\boldsymbol{w}$, see Figure 2.6(b). The results $\boldsymbol{w}_1 \neq \boldsymbol{w}$ can also be derived from calculations. This result is inevitable, because matrix multiplication does not have commutative properties, that is, $\boldsymbol{AB} \neq \boldsymbol{BA}$. The movement interpretation of the left and right multiplications of the transformation matrix is different: the transformation order "from right to left" indicates that the movement is relative to a fixed coordinate system; the transformation order "from left to right" indicates that the motion is relative to the motion coordinate system.

Example 2.5. Consider the case where the rotation transformation is combined with the translation transformation. If the translation transformation $4i - 3j + 7k$ is performed on the basis of the rotation transformation in Figure 2.6(a), then according to equations (2.20) and (2.26), it can obtain:

$$\mathbf{Trans}(4, -3, 7)\,\mathbf{Rot}(y, 90)\,\mathbf{Rot}(z, 90) = \begin{bmatrix} 0 & 0 & 1 & 4 \\ 1 & 0 & 0 & -3 \\ 0 & 1 & 0 & 7 \\ 0 & 0 & 0 & 1 \end{bmatrix}$$

So there are:

$$t = \mathbf{Trans}(4, -3, 7)\,\mathbf{Rot}(y, 90)\,\mathbf{Rot}(z, 90)u = [6, 4, 10, 1]^{\mathrm{T}}$$

The result of this transformation is shown in Figure 2.7.

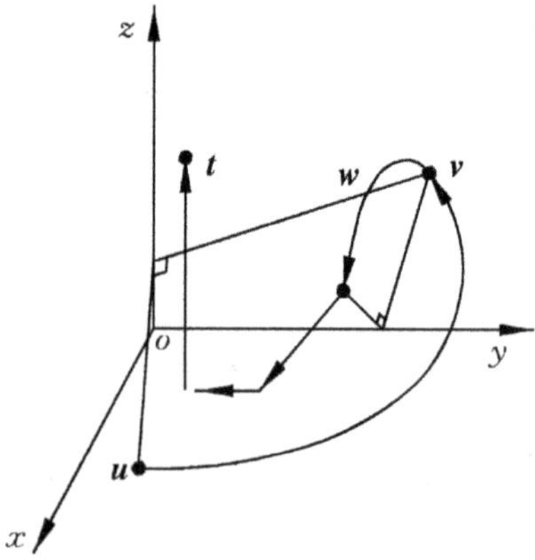

Figure 2.7.   Combination of translation transformation and rotation transformation.

2.4 Object Transformation and Transformation Equation

2.4.1 *Object location description*

The position and direction of an object in space can be described by the transformation method of describing a point in space. For example, the wedge-shaped object shown in Figure 2.8(a) can be represented by six points in the coordinate system where the object is fixed [5, 7, 17, 19].

If you first rotate the object $90°$ around the axis z, then rotate $90°$ around the axis y, and then translate 4 units along the axis x, then the transformation can be described by the following formula:

$$T = \mathbf{Trans}(4,0,0)\mathbf{Rot}(y,90)\mathbf{Rot}(z,90) = \begin{bmatrix} 0 & 0 & 1 & 4 \\ 1 & 0 & 0 & 0 \\ 0 & 1 & 0 & 0 \\ 0 & 0 & 0 & 1 \end{bmatrix}$$

This transformation matrix represents the rotation and translation operations on the coordinate system that coincides with the original reference coordinate system.

The six points of the above wedge-shaped object can be transformed as follows:

$$\begin{bmatrix} 0 & 0 & 1 & 4 \\ 1 & 0 & 0 & 0 \\ 0 & 1 & 0 & 0 \\ 0 & 0 & 0 & 1 \end{bmatrix} \begin{bmatrix} 1 & -1 & -1 & 1 & 1 & -1 \\ 0 & 0 & 0 & 0 & 4 & 4 \\ 0 & 0 & 2 & 2 & 0 & 0 \\ 1 & 1 & 1 & 1 & 1 & 1 \end{bmatrix} = \begin{bmatrix} 4 & 4 & 6 & 6 & 4 & 4 \\ 1 & -1 & -1 & 1 & 1 & -1 \\ 0 & 0 & 0 & 0 & 4 & 4 \\ 1 & 1 & 1 & 1 & 1 & 1 \end{bmatrix}$$

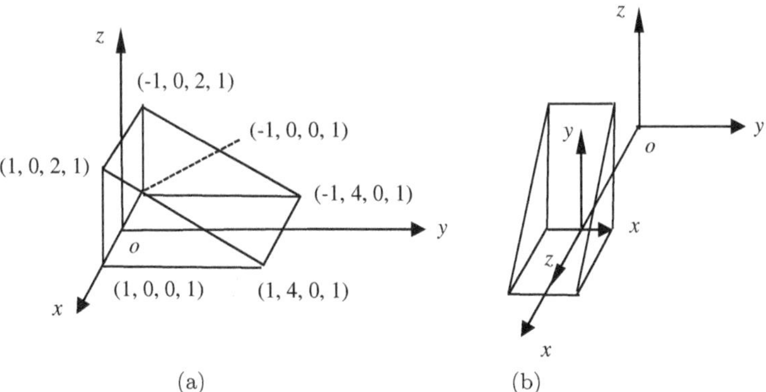

Figure 2.8. Transformation of wedge-shaped objects.

The transformation result is shown in Figure 2.8(b). It can be seen from the figure that the numerically described object has a definite relationship with the coordinate system that describes its position and direction.

2.4.2   *Inverse transformation of homogeneous transformation*

Given the coordinate system $\{A\}, \{B\}$ and $\{C\}$, if it is known that description of $\{B\}$ relative to $\{A\}$ is $^A_B T$, and description of $\{C\}$ relative to $\{B\}$ is $^B_C T$, then

$$^B\boldsymbol{p} = {}^B_C T\, {}^C\boldsymbol{p} \tag{2.27}$$

$$^A\boldsymbol{p} = {}^A_B T\, {}^B\boldsymbol{p} = {}^A_B T\, {}^B_C T\, {}^C\boldsymbol{p} \tag{2.28}$$

Define compound transformation

$$^A_C T = {}^A_B T\, {}^B_C T \tag{2.29}$$

to represent the description of $\{C\}$ relative to $\{A\}$. According to formula (2.6), we can get

$$
^A_C T = {}^A_B T\, {}^B_C T =
\begin{bmatrix} {}^A_B R & {}^A\boldsymbol{p}_{Bo} \\ 0 & 1 \end{bmatrix}
\begin{bmatrix} {}^B_C R & {}^B\boldsymbol{p}_{Co} \\ 0 & 1 \end{bmatrix}
$$

$$
= \begin{bmatrix} {}^A_B R\, {}^B_C R & {}^A_B R\, {}^B\boldsymbol{p}_{Co} + {}^A\boldsymbol{p}_{Bo} \\ 0 & 1 \end{bmatrix} \tag{2.30}
$$

From the description $^A_B T$ of the coordinate system $\{B\}$ relative to the coordinate system $\{A\}$, obtaining the description $^B_A T$ of the coordinate system $\{A\}$ relative to is an inverse problem of homogeneous transformation. One solution method is to inverse the 4×4 homogeneous transformation matrix $^A_B T$ directly; the other is to use the characteristics of the homogeneous transformation matrix to simplify the matrix inversion operation. The following discusses the transformation matrix inversion method firstly.

For a given $_B^A T$, find $_A^B T$, which is equivalent to given $_A^B T$ and $^A\boldsymbol{p}_{Bo}$, calculate $_A^B R$ and $^B\boldsymbol{p}_{Ao}$. Using the orthogonality of the rotation matrix, can get

$$_A^B R = {_B^A R}^{-1} = {_B^A R}^T \tag{2.31}$$

According to formula (2.13), find the description of the origin $^A\boldsymbol{p}_{Bo}$ in the coordinate system $\{B\}$

$$^B({^A\boldsymbol{p}_{Bo}}) = {_A^B R}\,{^A\boldsymbol{p}_{Bo}} + {^B\boldsymbol{p}_{Ao}} \tag{2.32}$$

$^B({^A\boldsymbol{p}_{Bo}})$ represent the description of the origin of $\{B\}$ relative to $\{B\}$, which is an $\boldsymbol{O}$ vector, so the above formula is 0, can get

$$^B\boldsymbol{p}_{Ao} = -{_A^B R}\,{^A\boldsymbol{p}_{Bo}} = -{_B^A R}^T\,{^A\boldsymbol{p}_{B0}} \tag{2.33}$$

Based on the above analysis, and according to formulas (2.31) and (2.33), we can get:

$$_A^B T = \begin{bmatrix} {_B^A R}^T & -{_B^A R}^T\,{^A\boldsymbol{p}_{Bo}} \\ 0 & 1 \end{bmatrix} \tag{2.34}$$

In the formula, $_A^B T = {_B^A T}^{-1}$. Equation (2.34) provides a convenient method for solving the inverse matrix of homogeneous transformation.

The following discusses the inversion method of the 4×4 homogeneous transformation matrix directly.

In fact, the inverse transformation is a transformation from the transformed coordinate system back to the original coordinate system, that is, the reference coordinate system describes the transformed coordinate system. For the object shown in Figure 2.8(b), its reference coordinate system is relative to the transformed coordinate system. The coordinate axes x, y, and z are $[0,0,1,0]^T$, $[1,0,0,0]^T$

and $[0,1,0,0]^{\mathrm{T}}$ respectively, and the origin is $[0,0,-4,1]^{\mathrm{T}}$. Thus, the available inverse transformation is:

$$
T^{-1} = \begin{bmatrix} 0 & 1 & 0 & 0 \\ 0 & 0 & 1 & 0 \\ 1 & 0 & 0 & -4 \\ 0 & 0 & 0 & 1 \end{bmatrix}
$$

Multiplying this inverse transformation by the transformation T to get the unit transformation can prove that this inverse transformation is indeed the inverse of the transformation T:

$$
T^{-1}T = \begin{bmatrix} 0 & 1 & 0 & 0 \\ 0 & 0 & 1 & 0 \\ 1 & 0 & 0 & -4 \\ 0 & 0 & 0 & 1 \end{bmatrix} \begin{bmatrix} 0 & 0 & 1 & 4 \\ 1 & 0 & 0 & 0 \\ 0 & 1 & 0 & 0 \\ 0 & 0 & 0 & 1 \end{bmatrix} = \begin{bmatrix} 1 & 0 & 0 & 0 \\ 0 & 1 & 0 & 0 \\ 0 & 0 & 1 & 0 \\ 0 & 0 & 0 & 1 \end{bmatrix}
$$

In general, the elements of the known transformation T

$$
T = \begin{bmatrix} n_x & o_x & a_x & p_x \\ n_y & o_y & a_y & p_y \\ n_z & o_z & a_z & p_z \\ 0 & 0 & 0 & 1 \end{bmatrix} \tag{2.35}
$$

Then its inverse transformation is

$$
\boldsymbol{T}^{-1} = \begin{bmatrix} n_x & n_y & n_z & -\boldsymbol{p}\cdot\boldsymbol{n} \\ o_x & o_y & o_z & -\boldsymbol{p}\cdot\boldsymbol{o} \\ a_x & a_y & a_z & -\boldsymbol{p}\cdot\boldsymbol{a} \\ 0 & 0 & 0 & 1 \end{bmatrix} \tag{2.36}
$$

In the formula, "$\bullet$" represents the dot product of the vector, $\boldsymbol{p}$, $\boldsymbol{n}$, $\boldsymbol{o}$, $\boldsymbol{a}$ are four column vectors, which are called the origin vector, normal vector, direction vector and proximity vector. In Chapter 3, these vectors will be further explained in conjunction with the gripping of the manipulator. It is not difficult to prove the correctness of this result by formula (2.36) right multiplication formula (2.35).

2.4.3 *Preliminary transformation equation*

In order to describe the operation of the robotic manipulator, it is necessary to establish the movement relationship between the links of the robotic manipulator and between the manipulator and the surrounding environment, that is, to construct the coordinate transformation relationship between various coordinate systems, so as to describe the relative relationship between the robotic manipulator and the environment. As shown in Figure 2.9(a), the working scene of a screw-holding manipulator, where $\{B\}$ represents the base coordinate system, $\{T\}$ is the tool system, $\{S\}$ is the workstation system, $\{G\}$ is the target system, and the corresponding homogeneous transformation can be used to describe the posture relationship between them.

$_S^B T$ indicates the pose of the workstation system $\{S\}$ relative to the base coordinate system $\{B\}$; $_G^S T$ indicates the pose of the target system $\{G\}$ relative to $\{S\}$; $_T^B T$ indicates the pose of the tool system $\{W\}$ relative to the base coordinate system $\{B\}$.

When operating an object, the pose $_T^G T$ of the tool system $\{T\}$ relative to the target system $\{G\}$ directly affects the operation effect, and is the goal of robotic manipulator control and planning. The relationship between $_T^G T$ and other transformations can be represented by a directed transformation graph, as shown in Figure 2.9(b), where the solid chain represents transformation what is known or can be obtained by simple measurement, and the dashed chain represents an unknown transformation. The description of the tool system $\{T\}$

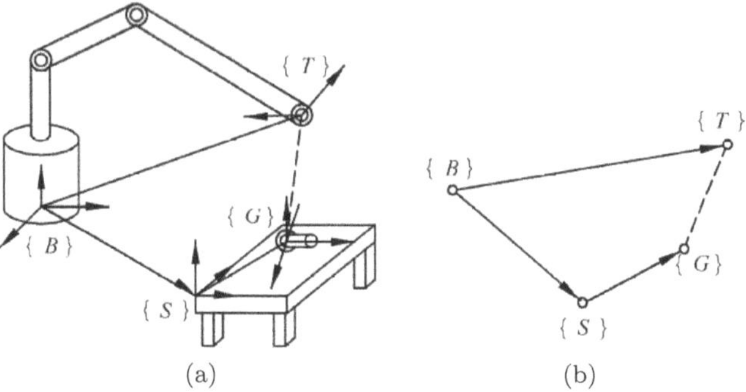

Figure 2.9. Transformation equations and its directed transformation diagram.

relative to the base coordinate system $\{B\}$ can be expressed by the product of the following transformation matrices:

$$^B_T T = {}^B_S T\, {}^S_G T\, {}^G_T T \tag{2.37}$$

After establishing such a matrix transformation equation, when only one of the above-mentioned matrix transformations is unknown, this unknown transformation can be expressed as the product of other known transformations. For the scene shown in Figure 2.9, if the pose $^T_G T$ of the target system $\{G\}$ relative to the tool system $\{T\}$ is required, then the inverse transformation $^B_T T^{-1}$ of $^B_T T$ can have a left multiplication and a right multiplication of $^T_G T$ on both sides of equation (2.37) at the same time, and get

$$^T_G T = {}^B_T T^{-1}\, {}^B_S T\, {}^S_G T \tag{2.38}$$

In this way, the originally unknown transformation $^T_G T$ is obtained through three known transformations.

2.5 General Rotation Transformation

The rotation transformation matrix that rotates around the axis x, y, and z has been studied previously. Now let us study the most general case, that is, study the rotation matrix when a certain vector (axis) $\boldsymbol{f}$ starts from the origin and rotates by an angle θ [10, 20].

2.5.1 *General rotation transformation formula*

Let $\boldsymbol{f}$ be the unit vector on the axis z of the coordinate system $\{C\}$, namely

$$C = \begin{bmatrix} n_x & o_x & a_x & 0 \\ n_y & o_y & a_y & 0 \\ n_z & o_z & a_z & 0 \\ 0 & 0 & 0 & 1 \end{bmatrix} \tag{2.39}$$

$$\boldsymbol{f} = a_x \boldsymbol{i} + a_y \boldsymbol{j} + a_z \boldsymbol{k} \tag{2.40}$$

Therefore, rotating around the vector $\boldsymbol{f}$ is equivalent to rotating around the axis z of the coordinate system $\{C\}$, namely:

$$Rot(\boldsymbol{f},\theta) = Rot(c_z,\theta) \tag{2.41}$$

If the coordinate system $\{T\}$ described by the reference coordinate is known, then another coordinate system $\{S\}$ described by the coordinate system $\{C\}$ can be obtained, because

$$T = CS \tag{2.42}$$

In the formula, S represents the position T relative to the coordinate system $\{C\}$. To solve S and get:

$$S = C^{-1}T \tag{2.43}$$

Rotating around $\boldsymbol{f}$ is equivalent to S rotating around the axis z of the coordinate system $\{C\}$:

$$Rot(\boldsymbol{f},\theta)T = C\,Rot(z,\theta)S$$

$$Rot(\boldsymbol{f},\theta)T = C\,Rot(z,\theta)C^{-1}T$$

So can get:

$$Rot(\boldsymbol{f},\theta) = C\,Rot(z,\theta)C^{-1} \tag{2.44}$$

Because $\boldsymbol{f}$ is the axis z of the coordinate system $\{C\}$, so it can find that by extending the formula (2.44), $Rot(z,\theta)C^{-1}$ is only a function of $\boldsymbol{f}$, because

$$
C\,Rot(z,\theta)C^{-1}
$$

$$
= \begin{bmatrix} n_x & o_x & a_x & 0 \\ n_y & o_y & a_y & 0 \\ n_z & o_z & a_z & 0 \\ 0 & 0 & 0 & 1 \end{bmatrix}
\begin{bmatrix} c\theta & -s\theta & 0 & 0 \\ s\theta & c\theta & 0 & 0 \\ 0 & 0 & 1 & 0 \\ 0 & 0 & 0 & 1 \end{bmatrix}
\begin{bmatrix} n_x & n_y & n_z & 0 \\ o_x & o_y & o_z & 0 \\ a_x & a_y & a_z & 0 \\ 0 & 0 & 0 & 1 \end{bmatrix}
$$

$$
= \begin{bmatrix} n_x & o_x & a_x & 0 \\ n_y & o_y & a_y & 0 \\ n_z & o_z & a_z & 0 \\ 0 & 0 & 0 & 1 \end{bmatrix}
$$

$$\begin{bmatrix} n_x c\theta - o_x c\theta & n_y c\theta - o_y s\theta & n_z c\theta - o_z s\theta & 0 \\ n_x s\theta + o_x c\theta & n_y s\theta + o_y c\theta & n_z s\theta + o_z c\theta & 0 \\ a_x & a_y & a_z & 0 \\ 0 & 0 & 0 & 1 \end{bmatrix}$$

$$= \begin{bmatrix} n_x n_x c\theta - n_x o_x s\theta + n_x o_x s\theta + o_x o_x c\theta + a_x a_x \\ n_y n_x c\theta - n_y o_x s\theta + n_x o_y s\theta + o_y o_x c\theta + a_y a_x \\ n_z n_x c\theta - n_z o_x s\theta + n_x o_z s\theta + o_z o_x c\theta + a_z a_x \\ 0 \end{bmatrix}$$

$$\begin{array}{c} n_x n_y c\theta - n_x o_y s\theta + n_y o_x s\theta + o_y o_x c\theta + a_x a_y \\ n_y n_y c\theta - n_y o_y s\theta + n_y o_y s\theta + o_y o_y c\theta + a_y a_y \\ n_z n_y c\theta - n_z o_y s\theta + n_y o_z s\theta + o_y o_z c\theta + a_z a_y \\ 0 \end{array}$$

$$\left.\begin{array}{cc} n_x n_z c\theta - n_x o_z s\theta + n_z o_x s\theta + o_z o_x c\theta + a_x a_z & 0 \\ n_y n_z c\theta - n_y o_z s\theta + n_z o_y s\theta + o_z o_y c\theta + a_y a_z & 0 \\ n_z n_z c\theta - n_z o_z s\theta + n_z o_z s\theta + o_z o_z c\theta + a_z a_z & 0 \\ 0 & 1 \end{array}\right] \qquad (2.45)$$

According to the properties of orthogonal vector dot product, vector multiplication, unit vector and similar matrix eigenvalues, and let $z = a$, $vers\theta = 1 - c\theta$, $\boldsymbol{f} = z$, simplify formula (2.45) (please calculate by reader yourself), can get:

$$Rot(\boldsymbol{f}, \theta)$$

$$= \begin{bmatrix} f_x f_x vers\theta + c\theta & f_y f_x vers\theta - f_z s\theta & f_z f_x vers\theta + f_y s\theta & 0 \\ f_x f_y vers\theta + f_z s\theta & f_y f_y vers\theta + c\theta & f_z f_y vers\theta - f_x s\theta & 0 \\ f_x f_z vers\theta - f_y s\theta & f_y f_z vers\theta + f_x s\theta & f_z f_z vers\theta + c\theta & 0 \\ 0 & 0 & 0 & 1 \end{bmatrix}$$

$$(2.46)$$

This is an important result.

From the above general rotation transformation formula, each basic rotation transformation can be obtained. For example, when $f_x = 1$, $f_y = 0$ and $f_z = 0$, that $Rot(\boldsymbol{f}, \theta)$ is $Rot(x, \theta)$. If these values are substituted into equation (2.46), can get:

$$Rot(x,\theta) = \begin{bmatrix} 1 & 0 & 0 & 0 \\ 0 & c\theta & -s\theta & 0 \\ 0 & s\theta & c\theta & 0 \\ 0 & 0 & 0 & 1 \end{bmatrix}$$

which is consistent with formula (2.22).

2.5.2 *Equivalent rotation angle and shaft*

Given any rotation transformation, the axis of the equivalent rotation angle θ can be obtained from equation (2.46). Known rotation transformation

$$R = \begin{bmatrix} n_x & o_x & a_x & 0 \\ n_y & o_y & a_y & 0 \\ n_z & o_z & a_z & 0 \\ 0 & 0 & 0 & 1 \end{bmatrix} \tag{2.47}$$

Let $R = Rot(\boldsymbol{f},\theta)$, i.e.,

$$\begin{bmatrix} n_x & o_x & a_x & 0 \\ n_y & o_y & a_y & 0 \\ n_z & o_z & a_z & 0 \\ 0 & 0 & 0 & 1 \end{bmatrix}$$

$$= \begin{bmatrix} f_x f_x vers\theta + c\theta & f_y f_x vers\theta - f_z s\theta & f_z f_x vers\theta + f_y s\theta & 0 \\ f_x f_y vers\theta + f_z s\theta & f_y f_y vers\theta + c\theta & f_z f_y vers\theta - f_x s\theta & 0 \\ f_x f_z vers\theta - f_y s\theta & f_y f_z vers\theta + f_x s\theta & f_z f_z vers\theta + c\theta & 0 \\ 0 & 0 & 0 & 1 \end{bmatrix} \tag{2.48}$$

Add the diagonal terms on both sides of the above formula separately and simplify it to:

$$n_x + o_y + a_z = \left(f_x^2 + f_y^2 + f_z^2 \right) vers\theta + 3c\theta = 1 + 2c\theta$$

And

$$c\theta = \frac{1}{2}\left(n_x + o_y + a_z - 1\right) \tag{2.49}$$

Subtracting the non-diagonal terms in equation (2.48) in pairs can get:

$$o_z - a_y = 2f_x s\theta$$
$$a_x - n_z = 2f_y s\theta \tag{2.50}$$
$$n_y - o_x = 2f_z s\theta$$

After squaring the lines of the above formula, add up and get:

$$\left(o_z - a_y\right)^2 + \left(a_x - n_y\right)^2 + \left(n_y - o_x\right)^2 = 4s^2\theta$$

And

$$s\theta = \pm\frac{1}{2}\sqrt{(o_z - a_y)^2 + (a_x - n_z)^2 + (n_y - o_x)^2} \tag{2.51}$$

The rotation is defined as a positive rotation around the vector f such that $0 \leq \theta \leq 180°$. At this time, the sign in formula (2.51) takes a positive sign. Therefore, the turning angle θ is uniquely determined as:

$$\tan\theta = \frac{\sqrt{(o_z - a_y)^2 + (a_x - n_z)^2 + (n_y - o_x)^2}}{n_x + o_y + a_z - 1} \tag{2.52}$$

The components of the vector f can be obtained by equation (2.50):

$$f_x = (o_z - a_y)/2s\theta$$
$$f_y = (a_x - n_z)/2s\theta \tag{2.53}$$
$$f_z = (n_y - n_x)/2s\theta$$

2.6 Chapter Summary

This chapter introduces the mathematical basis of robotic manipulator, including the representation of the position and posture of any point in space, the transformation of coordinates and homogeneous coordinates, the transformation and inverse transformation of objects, and the general rotation transformation.

For position description, a coordinate system needs to be established, and then a 3×1 position vector is used to determine the position of any point in the coordinate space, and it is represented by a 3×1 column vector, which is called a position vector. For the orientation of the object, it is also described by the coordinate system fixed to the object and represented by a 3×3 matrix. It also gives the rotation transformation matrix corresponding to the axes x, y and z for the rotation angle. On the basis of using a position vector to describe the position of a point and a rotation matrix to describe the orientation of an object, the pose of the object in space is represented by the position vector and the rotation matrix together.

After discussing translational and rotational coordinate transformation, further research on homogeneous coordinate transformation including translational homogeneous coordinate transformation and rotational homogeneous coordinate transformation. These transformation methods related to space points have established the foundation for the transformation and inverse transformation of space objects. In order to describe the operation of the manipulator, it is necessary to establish the movement relationship between the links of the manipulator and between the manipulator and the surrounding environment. To this end, the preliminary concept of the manipulator operation transformation equation is established, and the general matrix expression of the general rotation transformation and the equivalent rotation angle and rotation axis matrix expression are given.

The above conclusions provide mathematical tools for the study of robotic manipulator kinematics, dynamics, and control modeling. In recent years, when studying the rotational motion of rigid bodies in space, a new mathematical method has been adopted, that is, the use of exponential coordinates and exponential mapping and exponential product formulas to describe kinematic problems. Interested readers can refer to the relevant literature.

References

[1] Angeles, J. (2003). *Fundamentals of Robotic Mechanical Systems: Theory, Methods, and Algorithms*, 2nd Edition. New York: Springer.
[2] Asada, H. and Slotine, J.J.E. (1986). *Robot Analysis and Control*. John Wiley and Sons Inc.

[3] Brady, M. *et al.* (Eds.) (1983). *Robot Motion: Planning and Control.* MIT Press.

[4] Cai, Z.X. (2020). *Robotics*, Chapter 2 of 4th Edition. Beijing: Tsinghua University Press (in Chinese).

[5] Cai, Z.X. (2021). *Fundamentals of Robotics*, Chapter 2 of the 3rd Edition. Beijing: Machinery Industry Press (in Chinese).

[6] Cai, Z.X. (1988). *Robotics: Principles and Applications.* Changsha: Central South University of Technology Press (in Chinese).

[7] Cheng, L.L., Xu, J.X. and Wang, T. (2017). The transformation method between the robot coordinate system and the three-dimensional measurement coordinate system. Invention Patent, Publication/Announcement Number, CN106323286A, Public/ Announcement Day, 2017-01-11 (in Chinese).

[8] Cibicik, A. and Egeland, O. (2021). Kinematics and dynamics of flexible robotic manipulators using dual screws. *IEEE Transaction on Robotics*, 37(1):206–224.

[9] Craig, J. (2018). *Introduction to Robotics: Mechanics and Control*, Fourth Edition. Addison-Wesley Publishing Company.

[10] Fu, K.S., Gonzalez, R.C. and Lee, C.S.G. (1987). *Robotics: Control, Sensing, Vision, and Intelligence.* McGraw-Hill Book Company.

[11] Gao, X.F., Hao, L.N., Yang, H., *et al.* (2017). Kinematics solution of hybrid manipulator based on PSO algorithm. *7th IEEE Annual International Conference on CYBER Technology in Automation, Control, and Intelligent Systems* (CYBER), Honolulu, HI, July 31–Aug 04, 2017, pp. 199–204.

[12] Li, A.G., Wang, L. and Wu, D. (2010). Simultaneous robot-world and hand-eye calibration using dual-quaternions and Kronecker product. *International Journal of the Physical Sciences*, 5(10):1530–1536.

[13] Lu, Z.X., Xu, C.G., Pan, Q.X., *et al.* (2015). Automatic method for synchronizing workpiece frames in twin-robot nondestructive testing system. *Chinese Journal of Mechanical Engineering*, 28(4):860–868 (in Chinese).

[14] Paul, R.P. (1981). *Robot Manipulators: Mathematics, Programming and Control.* MIT Press, 1981.

[15] Qin, Z.-C. and Yang, L. (2020). Kinematics analysis and structure optimization of spray robot. *Heavy Machinery*, (3):98–102 (in Chinese).

[16] Reza, N.J. (2010). *Theory of Applied Robotics: Kinematics, Dynamics, and Control*, 2nd Edition. Springer.

[17] Wang, B.R., Fang, S.G. and Yan, D.M. (2012). Rigid-flexible coupling dynamics modeling of robot manipulators and modal analysis during swing. *China Mechanical Engineering*, 23(17):2092–2097 (in Chinese).

[18] Xu, K., Liu, Z.H., Zhao, B. *et al.* (2019). Composed continuum mechanism for compliant mechanical postural synergy: An anthropomorphic hand design example. *Mechanism and Machine Theory*, 132:108–122.

[19] Zhang, Y., Sreedharan, S., Kulkarni, A., *et al.* (2015). Plan explicability and predictability for robot task planning. arXiv: 1511.08158 [cs.AI], November 2015.

[20] Zsombor-Murray, P. and Gfrerrer, A. (2011). Mapping similarity between parallel and serial architecture kinematics. *2nd International Workshop on Fundamental Issues and Future Research Directions for Parallel Mechanisms and Manipulators*, MECCANICA, 46(1, SI): 183–194.

Chapter 3

Manipulator Kinematics

The work of the manipulator is directed by the controller, and the parameters of each joint corresponding to the driving end pose movement need to be calculated in real time. When the manipulator performs work tasks, its controller plans the pose sequence data according to the processing trajectory instructions, uses inverse kinematics algorithm to calculate the joint parameter sequence in real time, and drives the manipulator joints accordingly to make the end effector follow the predetermined pose sequence movement [2,6,8,10].

Manipulator kinematics or mechanism describes and studies the motion characteristics of manipulators from the perspective of geometry or mechanism, without considering the effects of the forces or moments that cause these motions. There are two basic problems in the kinematics of manipulators [4,7,11,16]:

(1) The expression problem of the motion equation of the manipulator, namely forward kinematics: For a given manipulator, the geometric parameters and joint variables of the connecting link are known, and the position and posture of the end effector of the manipulator relative to the reference coordinate system are desired. The manipulator programming language has the ability to specify work tasks according to Cartesian coordinates. The position of the object in the working space and the position of the robotic manipulator are all described by the position and posture of a certain coordinate system; this requires the establishment of the motion equation of the manipulator.

The expression problem of the equation of motion is forward kinematics, which belongs to problem analysis. Therefore, the problem of expressing the motion equation of the manipulator can also be called the manipulator motion analysis [17, 21].

(2) The problem of solving the motion equation of the manipulator, that is, inverse kinematics: Knowing the geometric parameters of the manipulator link given the expected position and posture (pose) of the robot end effector relative to the reference coordinate system, find the joint variables that the manipulator can achieve the expected posture. When the work task is described by the Cartesian coordinate system, these provisions must be transformed into a series of joint variables that can be driven by the arm. Determine the solution of each joint variable of the manipulator position and posture, this is the solution of the motion equation. The problem of solving manipulator motion equations is inverse kinematics, which belongs to problem synthesis. Therefore, the problem of solving manipulator motion equations can also be called manipulator motion synthesis [1, 9, 12, 14, 18].

At present, the contour motion of industrial robots is mainly controlled at a constant speed, and the two-level control of a computer and a motion control card is adopted to solve the problem of mapping between motion position and joint variables. If the requirement is extended to the calculation and control of contour speed changes, the Jacobian matrix can be used to further realize the linear transformation of the speed of each individual joint to the last link speed in the Cartesian coordinate system. Most industrial robots have six joints, which mean that the Jacobian matrix is a 6th-order square matrix.

In 1955, Denavit and Hartenberg proposed a general description method for robotic manipulators, which used connecting link parameters to describe the motion relationship of the mechanism. This method uses a 4×4 homogeneous transformation matrix to describe the spatial relationship between two adjacent links, and reduces the forward kinematics calculation problem to the calculation problem of the homogeneous transformation matrix. This matrix describes the manipulator transformation relationship of the end effector relative to the reference coordinate system. Inverse kinematics problems can be solved by multiple method, the most commonly used are algebraic methods, geometric methods, and iterative methods.

In the manipulator mechanism, there are currently four different reference coordinate system configuration conventions, each of which has its advantages intuitively. In addition to the original agreement between Denavit and Hartenberg, there have been Waldron and Paul versions, Craig versions, and Khalil and Dombre versions. The kinematics section of this book uses a Craig version based on the original Denavit-Hartenberg convention.

In the first two sections of this chapter, we will study the representation and solution method of the manipulator motion equation in turn, then use the PUMA560 robot as an example to analyze and synthesize the motion equation representation and solution of the manipulator, and finally discuss the differential motion of the manipulator and the Jacobian formula.

3.1 Representation of the Motion Equation of Manipulator

The manipulator is a kinematic chain formed by a series of links connected by joints. A series of rigid bodies on the joint chain are called links, and two adjacent links are connected by rotating joints or translation joints. To establish a coordinate system for each link of the manipulator, and use homogeneous transformation to describe the relative position and posture between these coordinate systems.

The six-link manipulator can have six degrees of freedom, and each link contains one degree of freedom, and can be positioned and oriented arbitrarily within its range of motion. According to the usual design of the robot, three degrees of freedom are used to specify the position, and the other three degrees of freedom are used to specify the posture. T_6 indicates the position and posture of the manipulator.

3.1.1 *Representation of manipulator movement posture and direction angle*

1. The direction of movement of the manipulator

A gripping hand of the manipulator can be represented by Figure 3.1. Put the origin of the described coordinate system at the center of the

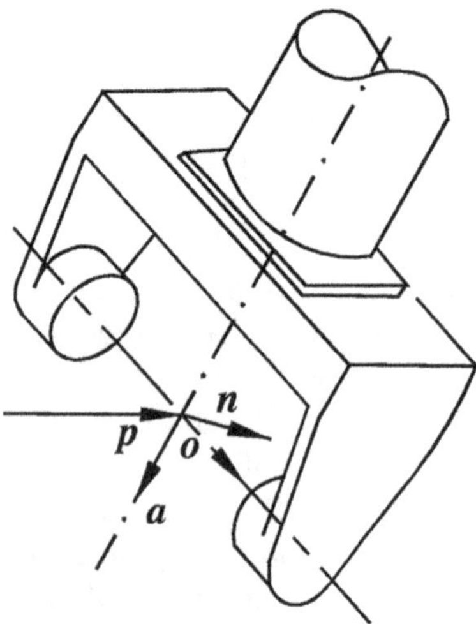

Figure 3.1.　Vectors o, a and p.

fingertip, and this origin is represented by the vector p. The direction of the three unit vectors describing the direction of the gripping hand is as follows: The direction of the vector z is in the direction of the gripping hand into the object, and is called the approach vector a; The direction of the vector y is from one fingertip to the other fingertip, at the specified gripping In the direction, it is called the direction vector o; the last vector is called the normal vector n, which forms a right-handed vector set together with the vectors o and a, and is specified by the intersection of the vectors $n = o \times a$. Therefore, transformation T_6 has the following elements:

$$T_6 = \begin{bmatrix} n_x & o_x & a_x & p_x \\ n_y & o_y & a_y & p_y \\ n_z & o_z & a_z & p_z \\ 0 & 0 & 0 & 1 \end{bmatrix} \tag{3.1}$$

The matrix T_6 of the six-link manipulator can be determined by specifying the values of its 16 elements. Among these 16 elements, only 12 elements have actual meaning. The bottom row consists of three zeros and one 1. The left column vector n is the intersection of the second column vector o and the third column vector a.

When there is no restriction on the value of $\boldsymbol{p}$, as long as the manipulator can reach the desired position, both vectors $\boldsymbol{o}$ and $\boldsymbol{a}$ are orthogonal unit vectors and are perpendicular to each other, namely: $\boldsymbol{o} \bullet \boldsymbol{o} = 1$, $\boldsymbol{a} \bullet \boldsymbol{a} = 1$, $\boldsymbol{o} \bullet \boldsymbol{a} = 0$. These constraints on the vectors $\boldsymbol{o}$ and $\boldsymbol{a}$ make it difficult to specify their components, unless the end effector is parallel to the coordinate system.

In addition to the above T matrix representation, the general rotation matrix discussed in Chapter 2 can also be used to specify the direction of the end of the manipulator as the angle of rotation around a certain axis $\boldsymbol{f}$, that is, $\mathrm{Rot}(f, \theta)$. Unfortunately, in order to achieve certain desired directions, there is no obvious intuitive feeling for this shaft.

2. Use Euler transformation to express motion posture

The motion posture of the manipulator is often specified by a rotation sequence x, y and z around an axis. This sequence of corners is called Euler's angle. Euler angles use a rotation angle ϕ around the axis z, then a rotation angle θ around the new axis y (y'), and finally a rotation angle ψ around the new axis z (z') to describe any possible posture, as shown in Figure 3.2.

In any rotation sequence, the rotation order is very important. This rotation sequence can be explained by the reverse order of rotation in the base system: first rotate around the axis z by angle ψ, then around the axis y by angle θ, and finally around the axis z by angle ϕ.

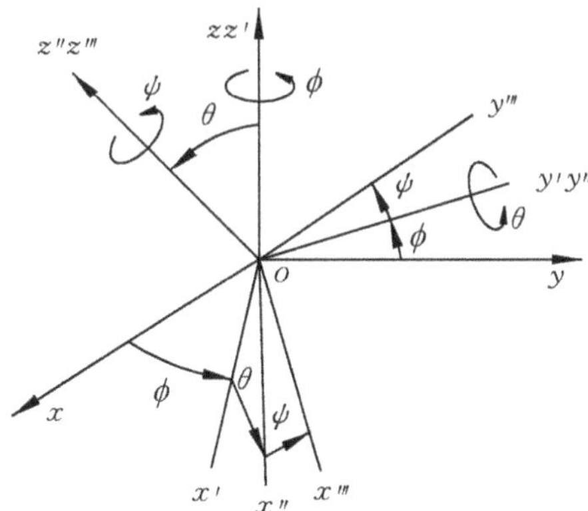

Figure 3.2. Definition of Euler angle.

Euler transformation Euler (ϕ, θ, ψ) can be obtained by multiplying 3 rotation matrices, namely:

$$\text{Euler}(\phi, \theta, \psi) = \text{Rot}(z, \phi)\text{Rot}(y, \theta)\text{Rot}(z, \psi)$$

$$\text{Euler}(\phi, \theta, \psi) = \begin{bmatrix} c\phi & -s\phi & 0 & 0 \\ s\phi & c\phi & 0 & 0 \\ 0 & 0 & 1 & 0 \\ 0 & 0 & 0 & 1 \end{bmatrix} \begin{bmatrix} c\theta & 0 & s\theta & 0 \\ 0 & 1 & 0 & 0 \\ -s\theta & 0 & c\theta & 0 \\ 0 & 0 & 0 & 1 \end{bmatrix}$$

$$\begin{bmatrix} c\psi & -s\psi & 0 & 0 \\ s\psi & c\psi & 0 & 0 \\ 0 & 0 & 1 & 0 \\ 0 & 0 & 0 & 1 \end{bmatrix}$$

$$= \begin{bmatrix} c\phi c\theta c\psi - s\phi s\psi & -c\phi c\theta s\psi - s\phi c\psi & c\phi s\theta & 0 \\ s\phi c\theta c\psi + c\phi s\psi & -s\phi c\theta s\psi + c\phi c\psi & s\phi s\theta & 0 \\ -s\theta c\psi & s\theta s\psi & c\theta & 0 \\ 0 & 0 & 0 & 1 \end{bmatrix}$$

$$(3.2)$$

3. Use RPY combination transformation to express movement posture

Another commonly used set of rotations is roll, pitch, and yaw.

If you imagine a ship sailing along the axis z, see Figure 3.3(a), then roll corresponds to the angle ϕ of rotation around the axis z, pitch corresponds to the angle θ of rotation around the axis y, and yaw corresponds to the angle ψ of rotation around the axis x. These rotations applicable to the end actuator of the manipulator are shown in Figure 3.3(b).

Regarding the rotation sequence, the following provisions can be made:

$$\text{RPY}(\phi, \theta, \psi) = \text{Rot}(z, \phi)\text{Rot}(y, \theta)\text{Rot}(x, \psi) \qquad (3.3)$$

In the formula, RPY represents the combined transformation of roll, pitch and yaw three rotations. In other words, first rotate the

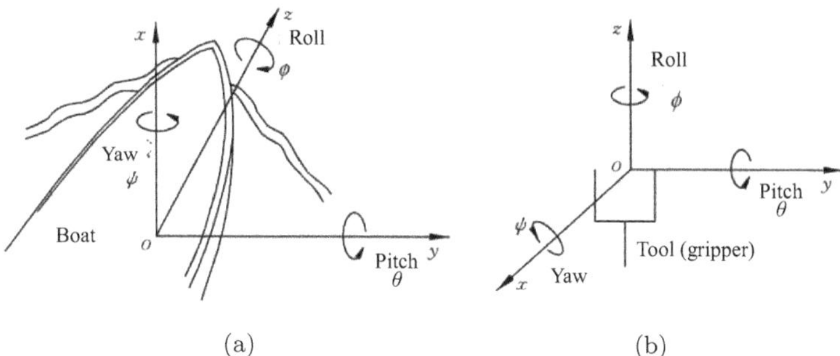

Figure 3.3. Use roll, pitch and yaw to represent the manipulator movement posture.

angle ψ around the axis x, then rotate the angle θ around the y axis, and finally rotate the angle ϕ around the axis z. This rotation transformation is calculated as follows:

$$\mathrm{RPY}(\phi, \theta, \psi) = \begin{bmatrix} c\phi & -s\phi & 0 & 0 \\ s\phi & c\phi & 0 & 0 \\ 0 & 0 & 1 & 0 \\ 0 & 0 & 0 & 1 \end{bmatrix} \begin{bmatrix} c\theta & 0 & s\theta & 0 \\ 0 & 1 & 0 & 0 \\ -s\theta & 0 & c\theta & 0 \\ 0 & 0 & 0 & 1 \end{bmatrix}$$

$$\begin{bmatrix} 1 & 0 & 0 & 0 \\ 0 & c\psi & -s\psi & 0 \\ 0 & s\psi & c\psi & 0 \\ 0 & 0 & 0 & 1 \end{bmatrix}$$

$$= \begin{bmatrix} c\phi c\theta & c\phi s\theta s\psi - s\phi c\psi & c\phi s\theta c\psi + s\phi s\psi & 0 \\ s\phi c\theta & s\phi s\theta s\psi + c\phi c\psi & s\phi s\theta c\psi - c\phi s\psi & 0 \\ -s\theta & c\theta s\psi & c\theta c\psi & 0 \\ 0 & 0 & 0 & 1 \end{bmatrix}$$

$$(3.4)$$

3.1.2 *Different coordinate system representation of translation transformation*

Once the motion posture of the manipulator is specified by a certain posture transformation, its position in the base system can be determined by multiplying a translation transformation corresponding to the vector $\boldsymbol{p}$ to the left:

$$T_6 = \begin{bmatrix} 1 & 0 & 0 & p_x \\ 0 & 1 & 0 & p_y \\ 0 & 0 & 1 & p_z \\ 0 & 0 & 0 & 1 \end{bmatrix} \text{ [A posture Transformation]} \qquad (3.5)$$

This translational transformation can be represented by different coordinates.

In addition to the Cartesian coordinates that have been discussed, cylindrical and spherical coordinates can also be used to express this translation.

1. Use cylindrical coordinates to represent the movement position

First, use cylindrical coordinates to indicate the position of the manipulator arm, which means its translation transformation. This corresponds to translation r along the axis x, then rotating α around the axis z, and finally translation z along the axis z, as shown in Figure 3.4(a),

That is

$$\text{Cyl}(z, \alpha, r) = \text{Trans}(0, 0, z)\text{Rot}(z, \alpha)\text{Trans}(r, 0, 0)$$

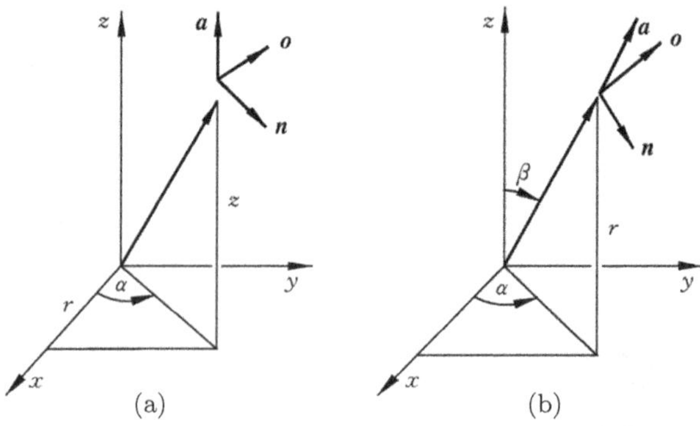

Figure 3.4.   Expressing position in cylindrical and spherical coordinates.

In the formula, Cyl represents the combined transformation of cylindrical coordinates. Calculate the above formula and simplify it to:

$$\mathbf{Cyl}(z,\alpha,r) = \begin{bmatrix} 1 & 0 & 0 & 0 \\ 0 & 1 & 0 & 0 \\ 0 & 0 & 1 & z \\ 0 & 0 & 0 & 1 \end{bmatrix} \begin{bmatrix} c\alpha & -s\alpha & 0 & 0 \\ s\alpha & c\alpha & 0 & 0 \\ 0 & 0 & 1 & 0 \\ 0 & 0 & 0 & 1 \end{bmatrix} \begin{bmatrix} 1 & 0 & 0 & r \\ 0 & 1 & 0 & 0 \\ 0 & 0 & 1 & 0 \\ 0 & 0 & 0 & 1 \end{bmatrix}$$

$$= \begin{bmatrix} c\alpha & -s\alpha & 0 & rc\alpha \\ s\alpha & c\alpha & 0 & rs\alpha \\ 0 & 0 & 1 & z \\ 0 & 0 & 0 & 1 \end{bmatrix} \tag{3.6}$$

If we multiply the above transformation formula with a posture transformation as shown in equation (3.5), the manipulator will rotate α around the axis z relative to the base system. If the posture of the robot end relative to the base system needs to be unchanged after transformation, then formula (3.6) should be rotated around the axis z by an angle $-\alpha$, which can be actually understood as: the cylindrical coordinate manipulator end coordinate system is transformed from the base coordinate system like this: start the end system coincides with the base system, and then the end system is relative to the x, z, and z axes of the base system, sequentially translates r, rotates α, and translates z, and finally rotates $-\alpha$ around the z axis of the transformed end coordinate system, that is:

$$\mathbf{Cyl}(z,\alpha,r) = \begin{bmatrix} c\alpha & -s\alpha & 0 & rc\alpha \\ s\alpha & c\alpha & 0 & rs\alpha \\ 0 & 0 & 1 & z \\ 0 & 0 & 0 & 1 \end{bmatrix} \begin{bmatrix} c(-\alpha) & -s(-\alpha) & 0 & 0 \\ s(-\alpha) & c(-\alpha) & 0 & 0 \\ 0 & 0 & 1 & 0 \\ 0 & 0 & 0 & 1 \end{bmatrix}$$

$$= \begin{bmatrix} 1 & 0 & 0 & rc\alpha \\ 0 & 1 & 0 & rs\alpha \\ 0 & 0 & 1 & z \\ 0 & 0 & 0 & 1 \end{bmatrix} \tag{3.7}$$

This is the form $\mathbf{Cyl}(z,\alpha,r)$ used to explain cylindrical coordinates.

2. Use spherical coordinates to represent the movement position

Now discuss the method of using spherical coordinates to represent the manipulator's motion position vector. This method corresponds to translation r along the axis z, then rotation around the axis y angle β, and finally around the axis z by angle α, as shown in Figure 3.4(b), which is:

$$\mathbf{Sph}(\alpha, \beta, r) = \mathbf{Rot}(z, \alpha)\mathbf{Rot}(y, \beta)\mathbf{Trans}(0, 0, r) \qquad (3.8)$$

In the formula, Sph represents the combined transformation of spherical coordinates. The calculation results of the above formula are as follows:

$$
Sph\,(\alpha, \beta, r) =
\begin{bmatrix}
c\alpha & -s\alpha & 0 & 0 \\
s\alpha & c\alpha & 0 & 0 \\
0 & 0 & 1 & 0 \\
0 & 0 & 0 & 1
\end{bmatrix}
\begin{bmatrix}
c\beta & 0 & s\beta & 0 \\
0 & 1 & 0 & 0 \\
-s\beta & 0 & c\beta & 0 \\
0 & 0 & 0 & 1
\end{bmatrix}
\begin{bmatrix}
1 & 0 & 0 & 0 \\
0 & 1 & 0 & 0 \\
0 & 0 & 1 & r \\
0 & 0 & 0 & 1
\end{bmatrix}
$$

$$
=
\begin{bmatrix}
c\alpha c\beta & -s\alpha & c\alpha s\beta & rc\alpha s\beta \\
s\alpha c\beta & c\alpha & s\alpha s\beta & rs\alpha s\beta \\
-s\beta & 0 & c\beta & rc\beta \\
0 & 0 & 0 & 1
\end{bmatrix}
\qquad (3.9)
$$

If you want the posture of the manipulator end coordinate system relative to the base system to remain unchanged after the transformation, you must use $\mathrm{Rot}(y,-\beta)$ and $\mathrm{Rot}(z,-\alpha)$ to have a right multiplication to formula (3.9), which can actually be understood as: the spherical coordinate manipulator end coordinate system is transformed from the base coordinate system like this: Start the end system coincides with the base system, and then the end system is translated r, rotated β, and translated α relative to the z, y, and z axes of the base system, and finally rotates $-\beta$, $-\alpha$ around the y axis and z axis of the transformed end coordinate system, which is:

$$\mathbf{Sph}(\alpha, \beta, r) = \mathbf{Rot}(z, \alpha)\mathbf{Rot}(y, \beta)\mathbf{Trans}(0, 0, r)$$

$$\mathbf{Rot}(y, -\beta)\mathbf{Rot}(z, -\alpha)$$

$$
= \begin{bmatrix} 1 & 0 & 0 & rc\alpha s\beta \\ 0 & 1 & 0 & rs\alpha s\beta \\ 0 & 0 & 1 & rc\beta \\ 0 & 0 & 0 & 1 \end{bmatrix} \tag{3.10}
$$

This is the form used to interpret spherical coordinates.

3.1.3 *Generalized linkage and generalized transformation matrix*

A coordinate system will be established for each link of the manipulator, and a homogeneous transformation will be used to describe the relative position and posture between these coordinate systems. The homogeneous transformation matrix of the end effector relative to the base coordinate system can be obtained by recursive mode, that is, the motion equation of the manipulator can be obtained.

1. Generalized link

The adjacent coordinate systems and their corresponding links can be represented by a homogeneous transformation matrix. To solve the transformation matrix required by the manipulator, a generalized link description is required for each link. After obtaining the corresponding generalized transformation matrix, it can be modified to suit each specific link.

The links are numbered starting from the fixed base of the manipulator, and the fixed base is generally referred to as link 0. The first movable link is link 1, and so on, the link at the end of the manipulator is link n. In order for the end effector to reach any position and posture in three-dimensional space, the manipulator needs at least six joints (corresponding to six degrees of freedom, three positions and three orientations).

The manipulator is composed of a series of links connected together. The various mechanical structures of links can be abstracted into two geometric elements and their parameters, that is, the angle between the common normal and the distance a_i and the two axes in the plane perpendicular to a_i; in addition, the connection relationship between adjacent links is also abstracted into two quantities, that is, the angle θ_i between the relative distance d_i

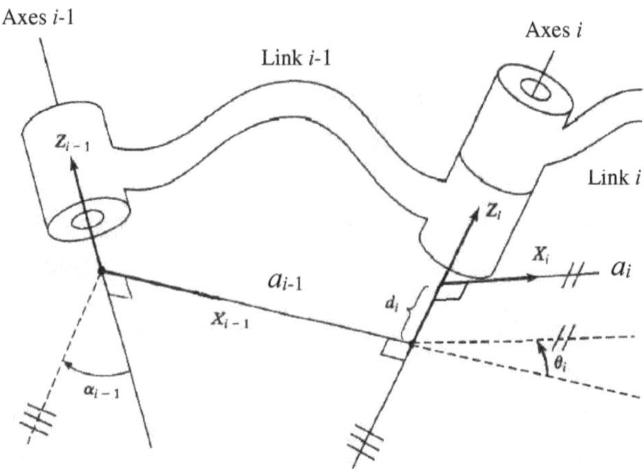

Figure 3.5. Schematic diagram of establishing the four parameters of the link and the coordinate system.

of the two connecting links and the male perpendicular of the two connecting links, as shown in Figure 3.5.

The establishment convention of Craig reference coordinate system is shown in Figure 3.5. Its characteristic is that the coordinate system z-axis and origin of each member are fixedly connected to the previous axis of the member. Except for the first and last link, each link has a normal at both ends of the axis, which is the common normal of the front and rear adjacent links. The distance between these two normals is d_i. It is called the length a_i of the link, the torsion angle α_i of the link, the distance d_i between the two links, and the angle θ_i between the two links.

There are two types of manipulator linkage joints: revolute joints and prismatic couplings. For revolute joints, θ_i is the joint variable, and for mobile joints, the distance d_i is the coupling (joint) variable. The origin of the coordinate system of the link i is at the intersection of the common normal of the axis $i-1$ and the axis i and the axis of the joint i. If the axes of two adjacent links intersect at a point, then the origin is at this intersection. If the two axes are parallel to each other, select the origin so that the distance d_{i+1} to the next link (the origin of the coordinate has been determined) is zero. The axis z_i of the link i and axis i is on a straight line, and x_i is on the common normal line of the axis i and axis $i-1$, and its direction is from i to

$i-1$, see Figure 3.5. When the two joint axes intersect, the direction of x_{i-1} and the intersection $z_{i-1} \times z_i$ of the two vectors are coaxial, in the same direction, or opposite, and the direction of x_{i-1} is always pointing from the axis $i-1$ to i along the common normal. When the two axes x_{i-1} and x_i are parallel and in the same direction, the θ_i of the ith revolute joint is zero.

When establishing the coordinate system of the manipulator link, first establish the coordinate axis z_i on the joint axis i of each link i. The positive direction of z_i can be selected from two directions, but all z axes should be as consistent as possible. The values of the four parameters a_i, α_i, θ_i and d_i, are all positive and negative except for $a_i \geq 0$, because α_i, θ_i are defined by rotation around the x_i and z_i axes, and their positive and negative are determined according to the right-hand rule for determining the direction of the rotation vector. d_i is the distance from x_{i-1} to x_i's vertical along the z_i axis. When the moving direction of this distance is consistent with the positive direction of z_i, its sign is taken as positive.

2. Generalized transformation matrix

After specifying the coordinate system for all the links, the relative relationship between the two adjacent link coordinate systems $i-1$ and i can be established by two rotations and two translations in the following order, as shown in Figure 3.6.

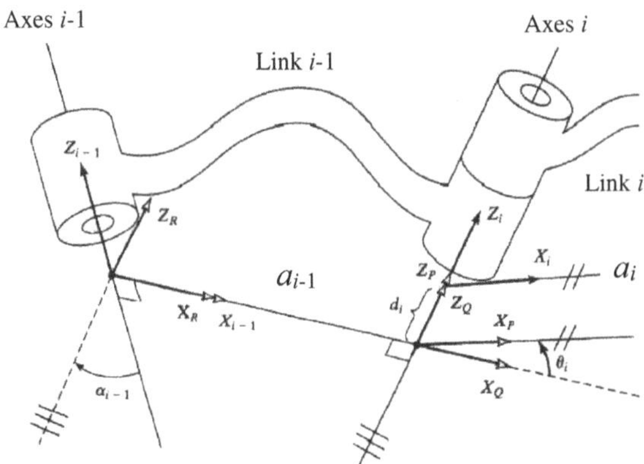

Figure 3.6. Schematic diagram of coordinate system transformation between adjacent ends of link.

(1) Rotate the α_{i-1} angle around the x_{i-1} axis to make z_{i-1} turn to z_R, which is consistent with the z_i direction, so that the coordinate system $\{i-1\}$ transitions to $\{R\}$.
(2) The coordinate system $\{R\}$ is translated a distance a_{i-1} along the x_{i-1} or x_R axis, and the coordinate system is moved to the i axis to make the coordinate system $\{R\}$ transition to $\{Q\}$.
(3) The coordinate system $\{Q\}$ rotates around z_Q or z_i axis to make $\{Q\}$ transition to $\{P\}$.
(4) The coordinate system $\{P\}$ is then translated a distance d_i along the axis z_i to make $\{P\}$ transition to coincide with the link i coordinate system $\{i\}$.

This relationship can be described by four homogeneous transformations that represent the relative position of the link i to link $i-1$. According to the chain rule of coordinate system transformation, the transformation matrix from coordinate system $\{i-1\}$ to coordinate system $\{i\}$ can be written as

$$^{i-1}_{i}T = {}^{i-1}_{R}T\,{}^{R}_{Q}T\,{}^{Q}_{P}T\,{}^{P}_{i}T \tag{3.11}$$

Each transformation in equation (3.11) has only one basic transformation (rotation or translation transformation) of link parameters. According to the settings of each intermediate coordinate system, equation (3.11) can be written as

$$^{i-1}_{i}T = Rot\,(x, \alpha_{i-1})\,Trans\,(a_{i-1}, 0, 0)\,Rot\,(z, \theta_i)\,Trans\,(0, 0, d_i) \tag{3.12}$$

Equation (3.12) can be calculated by multiplying four matrices, that is, the general transformation formula of $^{i-1}_{i}T$ is

$$^{i-1}_{i}T =
\begin{bmatrix}
c\theta_i & -s\theta_i & 0 & a_{i-1} \\
s\theta_i c\alpha_{i-1} & c\theta_i c\alpha_{i-1} & -s\alpha_{i-1} & -d_i s\alpha_{i-1} \\
s\theta_i s\alpha_{i-1} & c\theta_i s\alpha_{i-1} & c\alpha_{i-1} & d_i c\alpha_{i-1} \\
0 & 0 & 0 & 1
\end{bmatrix} \tag{3.13}$$

The relationship between the end of the manipulator and the base $^{0}_{6}T$ is:

$$^{0}_{6}T = {}^{0}_{1}T\,{}^{1}_{2}T\,{}^{2}_{3}T\,{}^{3}_{4}T\,{}^{4}_{5}T\,{}^{5}_{6}T$$

If the variables in the six joints of the robot are $\theta_1, \theta_2, d_3, \theta_4, \theta_5, \theta_6$, then the homogeneous matrix at the end relative to the base should also be a 4×4 matrix containing these six variables, namely:

$$^0_6 T(\theta_1, \theta_2, d_3, \theta_4, \theta_5, \theta_6) = {}^0_1 T(\theta_1) {}^1_2 T(\theta_2) {}^2_3 T(d_3) {}^3_4 T(\theta_4) {}^4_5 T(\theta_5) {}^5_6 T(\theta_6)$$
$$(3.14)$$

The above formula is the expression of the forward kinematics of the manipulator, that is, the value of each joint of the manipulator is known, and the pose of the end relative to the base is calculated.

If the robotic manipulator base has a fixed transformation relative to the workpiece reference system to Z, and the manipulator's tool end relative to the wrist end coordinate system $\{6\}$ also has a fixed transformation to E, then the transformation X of the manipulator tool end relative to the workpiece reference system is:

$$X = Z {}^0_6 T E$$

3.1.4 *Steps and examples for establishing a link coordinate system*

1. Summarize the steps to establish a link coordinate system
When establishing a coordinate system for each link in accordance with the above regulations, the corresponding link parameters can be summarized as follows:

a_i = Along the x_i axis, the distance from z_i to z_{i+1};
α_i = Rotate around the x_i axis from z_i to the angle of z_{i+1};
d_i = Along the z_i axis, the distance from x_{i-1} to x_i;
θ_i = Rotate the angle from x_{i-1} to x_i around the z_i axis.

Craig's law realizes the correspondence between the subscripts of the joint parameters and the joint axis. The only imperfection is that when calculating the homogeneous transformation matrix $^{i-1}_i T$ between two adjacent coordinate systems, the parameter is composed by the link parameters a_{i-1}, α_{i-1} whose subscript is $i-1$, and the joint parameters d_i, θ_i with subscript i, and subscripts are not completely unified.

For a robotic manipulator, the coordinate system of all links can be established in sequence according to the following steps:

(1) Find each joint axis and draw the extension line of these axes. In the following steps 2 to 5, only two adjacent axes (joint axis i and joint axis $i{+}1$) are considered.

(2) Find the common vertical line between the joint axis i and the joint axis $i{+}1$, and use the intersection point of the common vertical line and the joint axis i as the origin of the link coordinate system $\{i\}$ (when the joint axis i and the joint axis $i{+}1$ intersect, the intersection point is taken as the origin of the coordinate system $\{i\}$).

(3) Specify the direction of the z_i axis along the joint axis i.

(4) Specify the direction of the x_i axis along the common vertical line a_i, from the joint axis i to the joint axis $i{+}1$. If the joint axis i and the joint axis $i{+}1$ intersect, it is stipulated that the x_i axis is perpendicular to the plane where the two joint axes are located.

(5) Determine the y_i axis according to the right-hand rule.

(6) When the variable of the first joint is 0, the specified coordinate system $\{0\}$ coincides with the coordinate system $\{1\}$. For the coordinate system $\{n\}$, the origin and the direction of the x_n axis can be selected arbitrarily. But when selecting, usually try to make the link parameter be 0.

It is worth noting that the link coordinate system established according to the above method is not unique. First, when the z_i axis is selected to coincide with the joint axis i, there are two options for the direction of the z_i axis. In addition, when the joint axes intersect (at this time $a_i = 0$), since the x_i axis is perpendicular to the plane where the z_i axis and the $z_i{+}1$ axis are located, there are two options for the direction of the x_i axis. When the joint axis i is parallel to the joint axis $i{+}1$, the position of the origin of the coordinate system $\{i\}$ can be arbitrarily selected (usually the origin is selected to satisfy $d_i = 0$). In addition, when the joint is a translational joint, the selection of the coordinate system is also somewhat arbitrary.

The base is the 0 system, and the end is the n system: According to the establishment rules of the aforementioned coordinate system, there are countless ways to determine the x-axis of the 0 system and the n system. The general selection principle is to make more coefficients 0 and facilitate observation. There are generally two types of z-axis determination in the intermediate coordinate system, and when the z-axis intersects, there are also two types of x-axis. But as long as the definitions of the 0 series and the n series are fixed, no matter how diverse the intermediate definitions are, the final kinematics equation of the robotic manipulator should also be the same.

2. Example of establishing a link coordinate system

Example 3.1. Figure 3.7 shows a planar three-link manipulator. Because the three joints are all rotating joints, the manipulator is sometimes called RRR (or 3R) mechanism. To this end, the manipulator establishes a link coordinate system and writes its Denavit-Hartenberg parameters.

Solution: First define the reference coordinate system $\{0\}$, which is fixed on the base. When the variable value (θ_1) of the first joint is 0, the coordinate system $\{0\}$ coincides with the coordinate system $\{1\}$, so the established coordinate system $\{0\}$ is shown in Figure 3.5, and the z_0 axis coincides with the joint 1 axis.

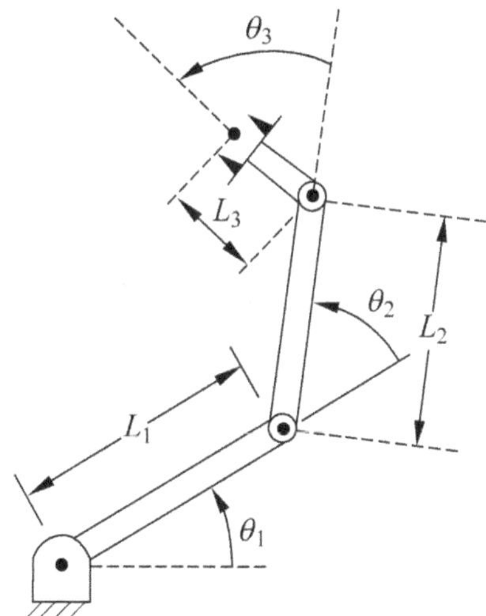

Figure 3.7.   A three-link plane manipulator.

Since the manipulator is located on a plane, all the joint axes of this manipulator are perpendicular to the plane where the manipulator is located (all the z-axes in the figure are perpendicular to the paper surface outwards, and are not shown for simplicity).

According to the previous regulations, the x_i axis is along the common perpendicular, and the z_i axis points to the z_{i+1} axis.

According to the right-hand rule, all y-axes can be determined. Each coordinate system is shown in Figure 3.8.

Find the corresponding link parameters below.

Because all joints are revolute joints, the joint variables are θ_1, θ_2 and θ_3 respectively. All the z-axes in Figure 3.5 are perpendicular to the paper surface outward and parallel to each other. According to the previous summary, the link torsion angle α_i represents the angle between adjacent z-axes, so all of α_i are zero.

Since all the x-axes are in the same plane, and the link offset d_i represents the distance between adjacent common perpendiculars, so all of d_i is zero.

According to regulations, a_i represents the distance moved from z_i to z_{i+1} along the x_i axis. Since the z_0 axis and z_1 axis coincide, so $a_0 = 0$. a_1 represents the distance between the z_1 axis and the z_2 axis, as shown in Figure 3.4,. $a_1 = L_1$, the same $a_2 = L_2$ can be obtained.

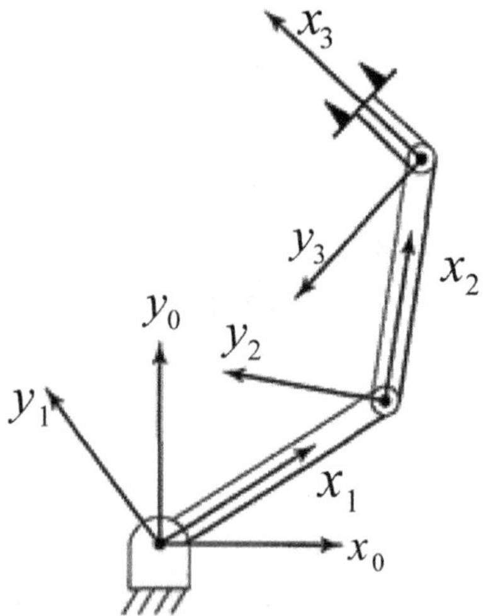

Figure 3.8.   Setting of the link coordinate system of a three-link manipulator.

Table 3.1. Parameter table of Denavit-Hartenberg corresponding to the three-link manipulator.

i	α_{i-1}	a_{i-1}	d_i	θ_i
1	0	0	0	θ_1
2	0	L_1	0	θ_2
3	0	L_2	0	θ_3

Therefore, the Denavit-Hartenberg parameters corresponding to the planar three-link manipulator are shown in Table 3.1.

3.2 Solving Kinematical Equation of Robot Manipulator

In the previous section, the forward kinematics of the manipulator was discussed. This section will study the more difficult inverse kinematics problem, that is, the solution of the manipulator motion equation: the expected position and posture of the tool coordinate system relative to the worktable coordinate system, find the joint variables of the manipulator that can reach the expected posture [3,13,15,19,20].

Most programming languages for manipulators use a Cartesian coordinate system to specify the end position of the manipulator. This designation can be used to solve the posture of the last link of the manipulator. However, before the manipulator can be driven to this posture, the position of all joints related to this position must be known.

3.2.1 *General problems solved by inverse kinematics*

1. The existence of solutions

Whether the inverse kinematics solution exists depends on whether the desired pose is in the working space of the manipulator. Simply put, the working space is the range that the end effector of the manipulator can reach. If the solution exists, the specified target point must be in the working space. If the desired pose of the end effector is in the working space of the manipulator, then there is at least a set of inverse kinematics solutions.

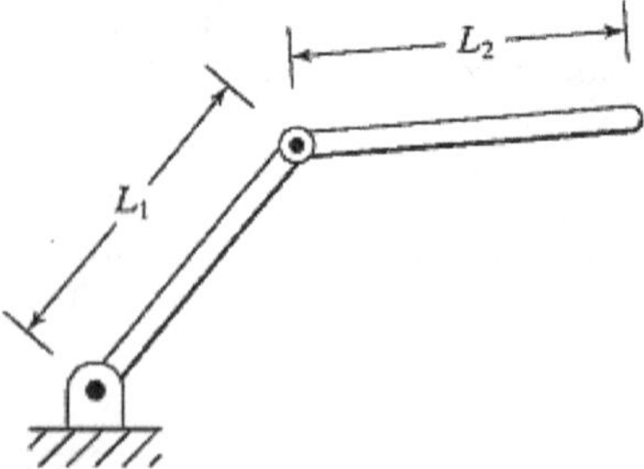

Figure 3.9.   Two-link manipulator with link lengths L_1 and L_2.

Now discuss the working space of the two-link manipulator shown in Figure 3.9. If $L_1 = L_2$, the reachable working space is a circle with a radius of $2L_1$. If $L_1 \neq L_2$, the reachable working space is a ring with outer diameter $L_1 + L_2$ and inner diameter $|L_1 - L_2|$. Inside the reachable working space, there are two possible solutions for the manipulator joint that reaches the target point; there is only one possible solution on the boundary of the working space.

The working space discussed here assumes that all joints can rotate 360 degrees, but this is rare in actual mechanisms. When the joint rotation cannot reach 360 degrees, the scope of the work space or the number of possible postures will be reduced accordingly.

When a manipulator has less than six degrees of freedom, it cannot achieve all poses in the three-dimensional space. Obviously, the planar manipulator shown in Figure 3.9 cannot extend out of the plane, so any target point whose z coordinate is not 0 is unreachable. In many practical situations, a manipulator with four or five degrees of freedom can operate beyond the plane, but it is obvious that such a manipulator cannot achieve all poses in the three-dimensional space.

2. Diversity problem

Another problem that may be encountered when solving inverse kinematics equations is the problem of multiple solutions. Figure 3.10 shows a three-link plane manipulator with an end effector. If the end effector of the manipulator needs to reach the posture shown in the figure, the link position in the figure is a set of possible inverse kinematics solutions. Note that when the first two links of the manipulator are in the dashed configuration in the figure, the pose of the end effector is exactly the same as the first configuration. That is

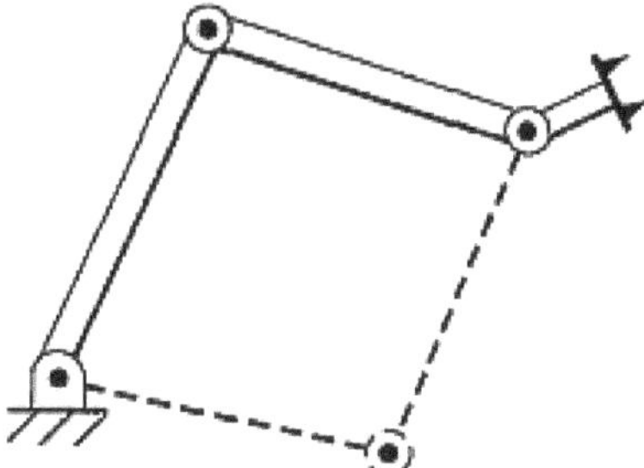

Figure 3.10. Three-link manipulator (the dotted line represents the second solution).

to say, for the planar three-link manipulator, there are two different solutions for its inverse kinematics.

The manipulator system can only select a set of solutions when performing manipulation. The selection criteria for the solutions are different for different applications. One of the more reasonable selection methods is the "shortest stroke solution," which is to make the manipulator move the shortest distance. For example, in Figure 3.11, if the end effector of the manipulator is at point A at the beginning, and it is hoped to move to point B, there are two possible configurations shown by the upper and lower dashed lines. In the absence of obstacles, according to the selection criteria of the shortest stroke solution, that is, to select the configuration that minimizes the movement of each joint, you can choose the configuration shown by the upper dashed line in Figure 3.11; but when there are obstacles in the environment In the case of objects, the "shortest stroke solution" may conflict. In this case, you may need to select the "longer stroke solution," that is, you need to follow the configuration shown by the dashed line in the lower part of Figure 3.11 to reach point B. Therefore, in order to enable the manipulator to reach the specified pose smoothly, it is usually desirable to be able to calculate all possible solutions when solving inverse kinematics.

The number of inverse kinematics solutions depends on the number of joints of the manipulator, as well as the link parameters and the range of joint motion. Generally speaking, the more the number of joints of the manipulator, the more the non-zero parameters of the link, and the more ways to reach a certain pose, that is, the more the number of inverse kinematics solutions.

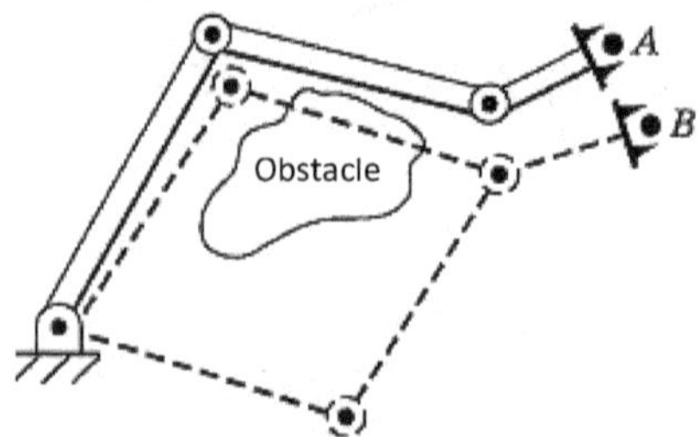

Figure 3.11. Multi-solution selection when there are obstacles in the environment.

3. Inverse kinematics solution method

As mentioned before, the solution of inverse kinematics of robotic manipulator is usually the solution of nonlinear equations. Unlike the solution of linear equations, there is no universal solution algorithm for nonlinear equations.

All the solving methods of inverse kinematics are divided into two categories: closed solution and numerical solution. Due to the iterative nature of the numerical solution, it is generally much slower than the corresponding closed solution. The numerical iterative solution of inverse kinematics equations has constituted a complete research field. Interested readers can refer to relevant references.

The following mainly discusses the closed solution method. In this book, "closed solution" refers to a solution based on an analytical form. The method of solving closed solutions can be divided into two categories: algebraic method and geometric method. Sometimes the difference between them is not obvious: algebraic description is introduced in any geometric method. Therefore, the two methods are similar, and their difference is only in the solution process.

If an algorithm can find all the joint variables to reach the required pose, then the manipulator is solvable. A new research result in inverse kinematics is that all serial type 6-DOF manipulators including revolute and translational joints are solvable. However, this solution is generally a numerical solution. For a 6-DOF manipulator, there is an analytical solution only under special circumstances. This kind of manipulator with analytical solution (closed solution) has the following characteristics: there are several orthogonal joint axes or many α_i are 0 or $\pm 90°$. Studies have shown that a sufficient condition for a manipulator with 6 rotary joints to have a closed solution is that the three adjacent joint axes intersect at one point. Almost all the 6-DOF manipulators designed today meet this condition. For example, the 4th, 5th, and 6th axes of the PUMA560 manipulator intersect, so most of them can be solved.

3.2.2 *Analytical solution of inverse kinematics*

In order to introduce the method of solving the inverse kinematics equation of the manipulator, this section first solves a simple planar three-link manipulator through two different methods.

1. Algebraic solution

Take the plane three-link manipulator introduced in Section 3.1.4 (Figure 3.7) as an example, its coordinate system setting is shown in Figure 3.8, and the link parameters are shown in Table 3.1.

According to the method introduced in Section 3.1, applying these link parameters can easily obtain the forward kinematics equation of this manipulator as:

$$
{}^{B}_{W}T = {}^{0}_{3}T = \begin{bmatrix} c_{123} & -s_{123} & 0 & L_1 c_1 + L_2 c_{12} \\ s_{123} & c_{123} & 0 & L_1 s_1 + L_2 s_{12} \\ 0 & 0 & 1 & 0 \\ 0 & 0 & 0 & 1 \end{bmatrix} \tag{3.15}
$$

In order to focus on the problem of inverse kinematics, it is assumed that the pose of the target point has been determined, that is, the transformation of the wrist coordinate system relative to the base coordinate system is known. The pose of the target point can be determined by the sum of three variables x, y, and ϕ, where $\boldsymbol{x}$, y are the Cartesian coordinates of the target point in the base coordinate system, and ϕ is the azimuth angle of the link 3 in the plane (relative to the x-axis positive direction of the base coordinate system), the transformation matrix of the target point with respect to the base coordinate system is as follows:

$$
{}^{B}_{W}T = \begin{bmatrix} c_\phi & -s_\phi & 0 & x \\ s_\phi & c_\phi & 0 & y \\ 0 & 0 & 1 & 0 \\ 0 & 0 & 0 & 1 \end{bmatrix} \tag{3.16}
$$

Among them, c_ϕ is the abbreviation of $\cos\phi$, and s_ϕ is the abbreviation of $\sin\phi$. Let equation (3.15) and equation (3.16) be equal, that is, the elements at corresponding positions are equal, four nonlinear equations can be obtained, and then find θ_1, θ_2 and θ_3:

$$c_\phi = c_{123} \tag{3.17}$$

$$s_\phi = s_{123} \tag{3.18}$$

$$x = L_1 c_1 + L_2 c_{12} \tag{3.19}$$

$$y = L_1 s_1 + L_2 s_{12} \tag{3.20}$$

Solve equations (3.17) through (3.20) algebraically. Squaring equation (3.19) and equation (3.20) at the same time, and then adding them together, get

$$x^2 + y^2 = L_1^2 + L_2^2 + 2L_1 L_2 c_2 \tag{3.21}$$

It can be solved by formula (3.21):

$$c_2 = \frac{x^2 + y^2 - L_1^2 - L_2^2}{2L_1 L_2} \tag{3.22}$$

The condition for the above formula to be solvable is that the value on the right side of formula (3.22) must be between -1 and 1. In this solution, the constraint condition can be used to check whether the solution exists. If the constraint conditions are not met, it means that the target point exceeds the reachable working space of the manipulator, and the manipulator cannot reach the target point, and its inverse kinematics has no solution.

Assuming that the target point is in the working space of the manipulator, the expression of s_2 is

$$s_2 = \pm\sqrt{1 - c_2^2} \tag{3.23}$$

According to formula (3.22) and formula (3.23) using the bivariate arctangent function to calculate, we can get

$$\theta_2 = \operatorname{atan2}(s_2, c_2) \tag{3.24}$$

Equation (3.24) has two sets of "positive" and "negative" solutions, corresponding to the two different sets of inverse kinematics solutions in this example.

After obtaining θ_2, θ_1 can be obtained according to equation (3.19) and equation (3.20). Write equation (3.19) and equation (3.20) as follows

$$x = k_1 c_1 - k_2 s_1 \tag{3.25}$$

$$y = k_1 s_1 + k_2 c_1 \tag{3.26}$$

Where

$$\begin{aligned} k_1 &= L_1 + L_2 c_2 \\ k_2 &= L_2 s_2 \end{aligned} \tag{3.27}$$

In order to solve the equation of this form, the following variable substitution can be carried out, let

$$r = \sqrt{k_1^2 + k_2^2} \tag{3.28}$$

and

$$\gamma = \operatorname{atan2}(k_2, k_1) \tag{3.29}$$

then

$$\begin{aligned} k_1 &= r \cos \gamma \\ k_2 &= r \sin \gamma \end{aligned} \tag{3.30}$$

Equation (3.25) and Equation (3.26) can be written as

$$\frac{x}{r} = \cos \gamma \cos \theta_1 - \sin \gamma \sin \theta_1 \tag{3.31}$$

$$\frac{y}{r} = \cos \gamma \sin \theta_1 + \sin \gamma \cos \theta_1 \tag{3.32}$$

That is

$$\cos(\gamma + \theta_1) = \frac{x}{r} \tag{3.33}$$

$$\sin(\gamma + \theta_1) = \frac{y}{r} \tag{3.34}$$

Using the bivariate arctangent function, can get

$$\gamma + \theta_1 = \operatorname{atan2}\left(\frac{y}{r}, \frac{x}{r}\right) = \operatorname{atan2}(y, x) \tag{3.35}$$

thereby

$$\theta_1 = \operatorname{atan2}(y, x) - \operatorname{atan2}(k_2, k_1) \tag{3.36}$$

It is worth noting that the selection of the symbol θ_2 will lead to the change of the symbol k_2, so it will affect the result of θ_1. The method of applying equations (3.28)–(3.30) for transformation solving often appears in solving inverse kinematics problems, that is, the method of solving equations of equation (3.25) or equation (3.26). If here $x = y = 0$, the value of equation (3.36) cannot be determined, and any value θ_1 can be taken at this time.

Finally, according to formula (3.17) and formula (3.18), θ_1, the sum of θ_2 and θ_3 can be solved:

$$\theta_1 + \theta_2 + \theta_3 = \text{atan2}\,(s_\phi, c_\phi) = \phi \tag{3.37}$$

Since θ_1 and θ_2 have been obtained before, we can solve for θ_3:

$$\theta_3 = \phi - \theta_1 - \theta_2 \tag{3.38}$$

So far, the inverse kinematics solution of the planar three-link manipulator has been completed by algebraic method. For the planar three-link manipulator, there are two possible inverse kinematics solutions, corresponding to the two values of equation (3.29).

The algebraic method is one of the basic methods for solving inverse kinematics. When solving the equation, the form of the solution has been determined. It can be seen that for many common algebra problems, there are often several fixed forms of transcendental equations, two of which have been encountered in the previous chapters.

2. Geometric solution

In the geometric method, in order to find the solution of the manipulator, it is necessary to decompose the spatial geometric parameters of the manipulator into plane geometric parameters. The geometric method is quite easy to solve the inverse kinematics for a manipulator with few degrees of freedom, or when the link parameters meet some specific values (such as when $\alpha_1 = 0$ or $\pm 90°$). For the planar three-link manipulator shown in Figure 3.7, if the end effector represented by the last link is not considered, the manipulator can be simplified to a planar two-link manipulator as shown in Figure 3.12. As long as the first two links can reach the designated position P, the end effector can reach the required posture. θ_1 and θ_2 can be solved directly by plane geometric relations.

In Figure 3.12, L_1, L_2, and the line connecting the origin of the coordinate system $\{0\}$ and the origin of the coordinate system $\{3\}$ form a triangle. In the connection OP in the figure, a group of dotted

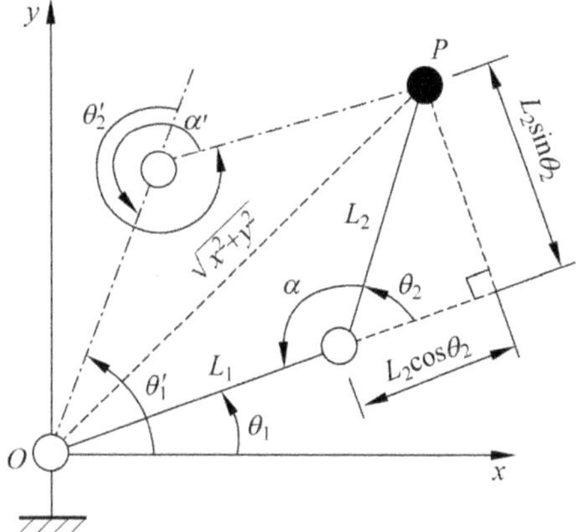

Figure 3.12.   The inverse kinematics solution of a planar two-link manipulator.

lines symmetrical to the positions of L_1 and L_2 represents another possible configuration of the triangle, and this group of configurations can also reach the position of the coordinate system {3}.

For the triangle represented by the solid line (the manipulator configuration in the lower part of Figure 3.12), according to the law of cosines, we can get

$$x^2 + y^2 = L_1^2 + L_2^2 - 2L_1L_2 \cos \alpha \tag{3.39}$$

That is

$$\alpha = \arccos \left(\frac{L_1^2 + L_2^2 - x^2 - y^2}{2L_1L_2} \right) \tag{3.40}$$

In order for the triangle to be established, the distance $\sqrt{x^2 + y^2}$ to the target point must be less than or equal to $L_1 + L_2$, the sum of the lengths of the two links. The above conditions can be calculated to check whether the solution exists. When the target point exceeds the working space of the manipulator, this condition is not satisfied, and there is no solution to inverse kinematics at this time.

After the angle α between the link L_1 and L_2 is obtained, θ_1 and θ_2 can be obtained through the plane geometric relationship:

$$\theta_2 = \pi - \alpha \tag{3.41}$$

$$\theta_1 = \arctan \left(\frac{y}{x} \right) - \arctan \left(\frac{L_2 \sin \theta_2}{L_1 + L_2 \cos \theta_2} \right) \tag{3.42}$$

As shown in Figure 3.12, when $\alpha' = -\alpha$, the manipulator has another set of symmetrical solutions:

$$\theta_2' = \pi + \alpha \tag{3.43}$$

$$\theta_1' = \arctan\left(\frac{y}{x}\right) + \arctan\left(\frac{L_2 \sin\theta_2}{L_1 + L_2 \cos\theta_2}\right) \tag{3.44}$$

The angles in the plane can be added directly so the sum of the angles of the three links is the azimuth angle of the last link:

$$\theta_1 + \theta_2 + \theta_3 = \phi \tag{3.45}$$

It can be solved by the above formula:

$$\theta_3 = \phi - \theta_1 - \theta_2 \tag{3.46}$$

So far, all the solutions of the inverse kinematics of this manipulator have been obtained by geometric solution.

In addition to the manipulator inverse kinematics algebraic solution and geometric solution discussed above, other analytical solutions are also available. Such as Euler transformation solution, RPY transformation solution and spherical transformation solution. Due to space limitations, it will not be introduced here, and interested readers can refer to the literature [6–8, 15].

3.2.3　*Numerical solution of inverse kinematics*

The inverse kinematics solution of the manipulator has a great relationship with the structure of the manipulator. It is necessary to select the appropriate inverse kinematics algorithm according to different manipulator structures. According to Pieper's criterion, if the three adjacent joint axes of the manipulator intersect at one point or the three axes are parallel, the algebraic method can be used to solve the problem. However, if the structure of the manipulator does not meet this criterion, the analytical solution cannot be obtained. At this time, only the numerical solution method can be used to obtain the inverse kinematics solution. In addition, the inverse kinematics of redundant manipulators (the degree of freedom in the joint space is greater than the degree of freedom required in the task space) has infinitely many solutions, and the inverse kinematics problems can usually only be solved by numerical solutions.

There are many types of inverse kinematics numerical solutions. This section mainly introduces three numerical solutions with more applications. For other solutions, please refer to related references.

1. Degradation method under cyclic coordinates

Cyclic-Coordinate Descent (CCD) is a heuristic direct search algorithm. Each iteration of this method contains n steps, corresponding to the nth joint to the first joint respectively. In the i-th step, only the i-th joint can be changed, and the other joints are fixed. By minimizing the objective function, in each step, the optimal joint change value is calculated, so that the entire manipulator is gradually iteratively optimized from the end effector to the first joint until the end effector's pose reaches the desired pose or less than a certain error.

Suppose the target position and posture of the end effector of the manipulator are P_d and $R_d = [f_1|f_2|f_3]$ respectively, where $f_j (j = 1,2,3)$ are the unit axial vectors of the x, y, and z coordinate axes, respectively. The current position and posture of the end effector of the manipulator are $P_h(q)$ and c respectively, where $q = [q_1, q_2, \ldots, q_n]^T$ is the nx1 dimensional joint variable. The system-related errors are defined as follows.

The position error is defined as:

$$\delta P(q) = (P_d - P_h(q)) \cdot (P_d - P_h(q)) \tag{3.47}$$

That is, the Euclidean distance between the current end effector position of the manipulator and the target position.

The attitude error is defined as:

$$\delta O(q) = \sum_{j=1}^{3} w_j(f_j \cdot h_j(q) - 1)^2 \tag{3.48}$$

Where w_j is the weight of the attitude error of the x, y, and z coordinate axes. According to the orthogonal condition of the rotation matrix, when the current posture of the end effector of the manipulator is consistent with the target posture, $f_j \cdot h_j(q) = 1$, the posture error of the corresponding direction is zero.

Therefore, the total error is defined as:

$$e(q) = w_p \cdot \delta P(q) + w_o \cdot \delta O(q) \tag{3.49}$$

Where w_p and w_o are the weights of position error and heading error respectively, and the value is any positive real number. By defining the pose error of the end effector of the manipulator, the inverse kinematics problem is transformed into finding a set of solutions: $q = [q_1, q_2, \ldots, q_n]^T$, so that the total system error $e(q) < \varepsilon, (\varepsilon \to 0)$. Combining the above definition of the pose error of the end effector with the Rodrigue rotation equation, the optimization objective function of the i-th rotation joint can be written as:

$$g(\phi_i) = k_1(1 - cos\phi_i) + k_2 cos\phi_i + k_3 sin\phi_i \qquad (3.50)$$

Among them, ϕ_i is the amount of joint parameter change, and its value should be such that when the above formula takes the maximum value, the total error is the smallest. Where:

$$\begin{cases} k_1 = w_p(P_{id}(q) \cdot z_i)(P_{ih}(q) \cdot z_i) + w_o \sum_{j=1}^{3} (f_j \cdot z_i)(h_j(q) \cdot z_i) \\ k_2 = w_p(P_{id}(q) \cdot P_{ih}(q)) + w_o \sum_{j=1}^{3} (f_j \cdot h_j(q)) \\ k_3 = z_i \cdot [w_p(P_{id}(q) \times P_{ih}(q))] + w_o \sum_{j=1}^{3} (h_j(q) \times f_j) \end{cases}$$

Among them, z_i is the unit axial vector of the i-th joint rotation axis; $P_{id}(q) = P_d - P_i(q)$, $P_{ih}(q) = P_h - P_i(q)$, $P_i(q)$ is the position vector of the origin of the i-th coordinate system. The formula (3.50) is a one-variable linear function about the rotation of the i-th joint. Obviously, the condition for the maximum value of this formula is:

$$\begin{cases} \dfrac{dg(\phi_i)}{d\phi_i} = (k_1 - k_2)sin\phi_i + k_3 cos\phi_i = 0 \\ \dfrac{d^2 g(\phi_i)}{d\phi_i^2} = (k_1 - k_2)cos\phi_i - k_3 sin\phi_i < 0 \end{cases}$$

The solution can be:

$$\begin{cases} tan\phi_i > \dfrac{k_1 - k_2}{k_3} \\ \phi_i = arctan(\dfrac{k_3}{k_1 - k_2}) \end{cases} \qquad (3.51)$$

For translation joints, there are:

$$\phi_i = (P_{id}(q) - P_{ih}(q)) \cdot z_i \qquad (3.52)$$

According to the specific type of joint, the optimal adjustment value of the i-th joint variable in the iteration process is calculated by the formula (3.51) or (3.52) in each iteration process.

The process of using the cyclic coordinate descent method to iteratively solve the inverse kinematics problem is summarized as follows:

(1) Input the forward kinematics model of the manipulator, the joint limit, the expected pose of the end effector and the maximum error ε of the end, and give the initial value of the manipulator joint;

(2) Use the forward kinematics model to calculate the position and orientation of each joint, and calculate the total error of the current end effector pose based on formula (3.49). If the total error value is less than ε, stop the iteration, otherwise go to the next step;

(3) From the last joint of the manipulator to the first joint, according to the specific type of the joint, if the current joint is a rotary joint, the corresponding angle is rotated according to formula (3.51), and if it is a translation joint, the corresponding distance is translated according to formula (3.52), and then go to the second step.

2. Forward and backward reaching solution

Forward and Backward Reaching Inverse Kinematics (FABRIK) is also a heuristic inverse kinematics algorithm. Unlike the cyclic coordinate descent method, which only iterates from the end effector to the base mark, FABRIK will iterate once from the end effector to the base mark, and then iterate from the base coordinate to the end effector. Therefore, FABRIK overcomes the problem that the weight of joints close to the end effector is too large in the cyclic coordinate descent method.

Consistent with the previous article, use $P_i(i = 1, 2, \ldots, n)$ to represent the position of the i-th joint, P_1 is the position of the first joint, and P_n is the position of the n-th joint, which is also considered the position of the end effector. Considering that the manipulator is a tandem robotic manipulator with a single end effector, and P_d as the target position, the specific execution process of the FABRIK algorithm is as follows:

First calculate the Euclidean distance between the joints of the current manipulator, $d_i = |P_{i+1} - P_i|(i = 1, 2, \ldots, n)$. Thereafter, the distance *dist* between the position P_1 of joint 1 and the target position P_d is calculated. If $dist < \sum_1^{n-1} d_i$, then the target is

   Robotics: From Manipulator to Mobilebot

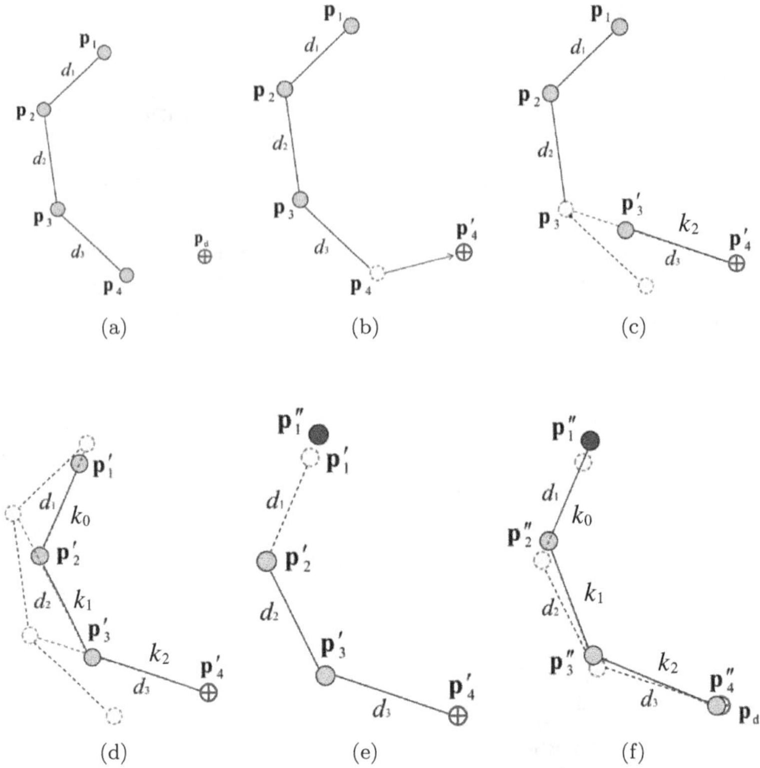

Figure 3.13.   FABRIK algorithm iteration example.

in the reachable range, otherwise, the target point is unreachable. Figure 3.13 is the process of iterative solution of inverse kinematics by a 4-DOF manipulator using FABRIK algorithm. The operation process of the algorithm is explained below in conjunction with Figure 3.13.

For the situation where the target point is reachable, an iterative process consists of two stages. The initial state of the manipulator is shown in Figure 3.13(a). In the first stage, the algorithm gradually estimates the position of each joint from the end effector forward. First, let the new position P_n of the end effector be the target position and mark it as P'_n shown in Figure 3.13(b). Thereafter, as shown in Figure 3.13 (c), let l_{n-1} be the connection line between P'_n and P_{n-1}, and the position on the connection line with a distance d_{n-1} of is the position of the new $(n-1)$th joint, denoted by P'_{n-1}. After that, follow the same method to determine the positions of the new joints one by one until the position P_1' of joint 1 is set. It can be seen from Figure 3.13(d) that the position of joint 1 after iteration has changed and is not at the original position. Because joint 1 is usually the position of the base coordinate system, its position will not change, so in the second stage, as shown in Figure 3.13(e), set the new position P_1'' and overlap with position P_1 of joint 1, and then in the same way set the connection between the joints and the new joint position are set according to the distance between the joints, until the position P_n'' of the new end effector is calculated, and an iterative calculation process is completed.

For the situation where the target is not reachable, the same iterative process is adopted, but because the end effector cannot reach the target position, it is necessary to set a certain error range as a condition to stop the iteration or set the maximum number of iterations to ensure that the algorithm stops.

What is discussed above is the single end effector and unconstrained situation where the FABRIK algorithm is applied. Generally, the end effector of the manipulator still needs to meet the target orientation. In the FABRIK algorithm, the target orientation is regarded as a posture constraint, and it is added to the iterative process, so that each joint of the manipulator can meet the requirements of joint restriction and orientation. In addition, the FABRIK algorithm can also be applied to the case of multi-end effector manipulators. For details, please refer to relevant references.

3. Genetic algorithm optimization solution

In the case that the algebraic solution does not exist or the robot is a redundant manipulator, the inverse kinematics problem of the manipulator can be solved based on the optimization method, and the forward kinematics model of the manipulator is combined with optimization algorithms such as genetic algorithm, particle swarm algorithm, and firefly algorithm. In combination, the optimal solution is approached through successive iterative optimization.

The following is based on the genetic algorithm to give the basic idea of the optimization method to solve the inverse kinematics problem:

(1) Encode each joint of the manipulator as variables;
(2) Initialize the population in the entire manipulator joint space;

(3) Decode the individual, calculate the joint variables and bring it into the forward kinematics model of the manipulator, and calculate the individual end matrix;

(4) Bring the individual solution and individual terminal matrix into the fitness function for calculation to obtain the individual fitness function value;

(5) Check whether the conditions for stopping evolution are met, and if the stopping conditions are met, the operation will be stopped;

(6) Perform selection, mutation, and crossover operations based on the individual's fitness function value to generate the next-generation population, and go to step 3.

Through the use of genetic algorithms, the inverse kinematics problem is transformed into an optimization problem of optimizing the joint configuration of the manipulator to find the best pose. This method has a certain versatility, is simple and easy to use, and can guide the structure optimization direction of the manipulator by setting a suitable fitness function.

3.3　Analysis and Comprehensive Examples of Manipulator Movement

The equation of motion expressed in Cartesian coordinates can be solved by equation (3.13) and so on. The elements of the right-hand formula of these matrices are either zero, constant, or a function of the sixth to sixth joint variables. The matrix equality means that the corresponding elements are respectively equal, and 12 equations can be obtained from each matrix equation, and each equation corresponds to each component of the four vectors. The following will take the PUMA560 as an example to solve the kinematics equations with joint angles as variables [5, 10, 16, 21].

3.3.1　*Examples of forward kinematics of manipulator*

PUMA560 is an articulated manipulator whose six joints are all revolving joints. The first three joints determine the position of the wrist reference point, and the rear three joints determine the position of the wrist. Like most industrial robots, the last three joint axes intersect at one point. This point is selected as the reference point of the wrist and also as the origin of the linkage coordinate system $\{4\}$, $\{5\}$ and $\{6\}$. The axis of joint 1 is in the vertical direction, and the axis of joints 2 and 3 are in the horizontal direction and parallel, with a distance of. The axes of joints 1 and 2 intersect vertically, and the axes of joints 3 and 4 intersect vertically with a distance of. The coordinate system of each link is shown in Figure 3.14, and the corresponding link parameters are listed in Table 3.2. Among them, $= 431.8$mm, $= 20.32$mm, $= 149.09$mm, $= 433.07$mm.

According to equation (3.13) and the link parameters shown in Table 3.2, the transformation matrix of each link can be obtained as

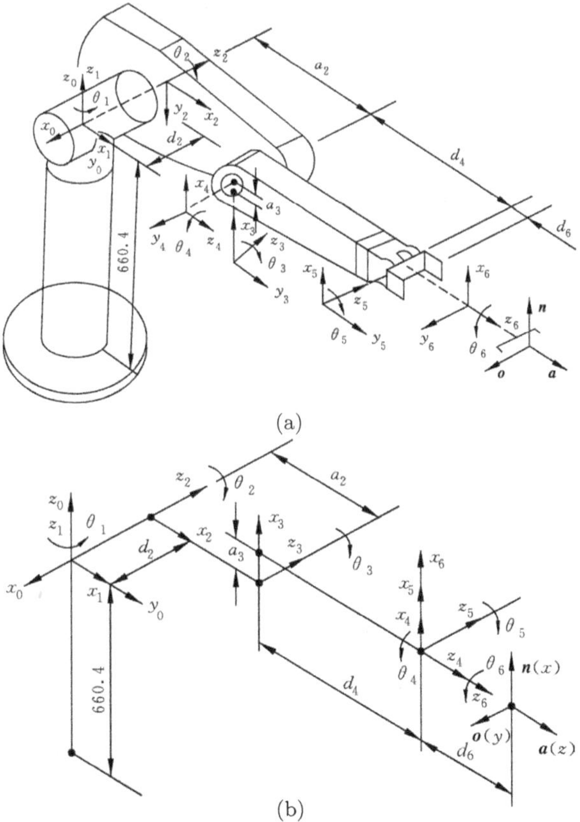

Figure 3.14.   The linkage coordinate system of PUMA560 manipulator.

Table 3.2.   Link parameters of PUMA 560 manipulator.

Link i	Variable θ_i	α_{i-1}	a_{i-1}	d_i	Variable range
1	$\theta_1(90°)$	$0°$	0	0	$-160° \sim 160°$
2	$\theta_2(0°)$	$-90°$	0	d_2	$-225° \sim 45°$
3	$\theta_3(-90°)$	$0°$	a_2	0	$-45° \sim 225°$
4	$\theta_4(0°)$	$-90°$	a_3	d_4	$-110° \sim 170°$
5	$\theta_5(0°)$	$90°$	0	0	$-100° \sim 100°$
6	$\theta_6(0°)$	$-90°$	0	0	$-266° \sim 266°$

follows:

$$
{}^0T_1 = \begin{bmatrix} c\theta_1 & -s\theta_1 & 0 & 0 \\ s\theta_1 & c\theta_1 & 0 & 0 \\ 0 & 0 & 1 & 0 \\ 0 & 0 & 0 & 1 \end{bmatrix} \qquad
{}^1T_2 = \begin{bmatrix} c\theta_2 & -s\theta_2 & 0 & 0 \\ 0 & 0 & 1 & d_2 \\ -s\theta_2 & -c\theta_2 & 0 & 0 \\ 0 & 0 & 0 & 1 \end{bmatrix}
$$

$$
{}^2T_3 = \begin{bmatrix} c\theta_3 & -s\theta_3 & 0 & a_2 \\ s\theta_3 & c\theta_3 & 0 & 0 \\ 0 & 0 & 1 & 0 \\ 0 & 0 & 0 & 1 \end{bmatrix} \qquad
{}^3T_4 = \begin{bmatrix} c\theta_4 & -s\theta_4 & 0 & a_3 \\ 0 & 0 & 1 & d_4 \\ -s\theta_4 & -c\theta_4 & 0 & 0 \\ 0 & 0 & 0 & 1 \end{bmatrix}
$$

$$
{}^4T_5 = \begin{bmatrix} c\theta_5 & -s\theta_5 & 0 & 0 \\ 0 & 0 & -1 & 0 \\ s\theta_5 & c\theta_5 & 0 & 0 \\ 0 & 0 & 0 & 1 \end{bmatrix} \qquad
{}^5T_6 = \begin{bmatrix} c\theta_6 & -s\theta_6 & 0 & 0 \\ 0 & 0 & 1 & 0 \\ -s\theta_6 & -c\theta_6 & 0 & 0 \\ 0 & 0 & 0 & 1 \end{bmatrix}
$$

Multiply each link transformation matrix to get the kinematic equation of PUMA560

$$
{}^0_6T = {}^0_1T(\theta_1)\,{}^1_2T(\theta_2)\,{}^2_3T(\theta_3)\,{}^3_4T(\theta_4)\,{}^4_5T(\theta_5)\,{}^5_6T(\theta_6) \tag{3.53}
$$

This equation is a function of joint variables $\theta_1, \theta_2, \ldots, \theta_6$. To solve this equation of motion, you need to calculate some intermediate results (these intermediate results help solve the inverse kinematics problem in the next section):

$$
{}^4_6T = {}^4_5T\,{}^5_6T = \begin{bmatrix} c_5c_6 & -c_5s_6 & -s_5 & 0 \\ s_6 & c_6 & 0 & 0 \\ s_5c_6 & -s_5s_6 & c_5 & 0 \\ 0 & 0 & 0 & 1 \end{bmatrix} \tag{3.54}
$$

Since the joints 2 and 3 of PUMA560 are parallel to each other, multiply and

$$\,^3_6T = \,^3_4T\,^4_6T = \begin{bmatrix} c_4c_5c_6 - s_4s_6 & -c_4c_5s_6 - s_4c_6 & -c_4s_5 & a_3 \\ s_5c_6 & -s_5s_6 & c_5 & d_4 \\ -s_4c_5c_6 - c_4s_6 & s_4c_5s_6 - c_4c_6 & s_4s_5 & 0 \\ 0 & 0 & 0 & 1 \end{bmatrix}$$

(3.55)

Since the joints 2 and 3 of PUMA560 are parallel to each other, multiply $\,^1_2T(\theta_2)$ and $\,^2_3T(\theta_3)$

$$\,^1_3T = \,^1_2T\,^2_3T = \begin{bmatrix} c_{23} & -s_{23} & 0 & a_2c_2 \\ 0 & 0 & 1 & d_2 \\ -s_{23} & -c_{23} & 0 & -a_2s_2 \\ 0 & 0 & 0 & 1 \end{bmatrix}$$

(3.56)

where, $c_{23} = \cos(\theta_2+\theta_3) = c_2c_3 - s_2s_3$; $s_{23} = \sin(\theta_2+\theta_3) = c_2s_3 + s_2c_3$. It can be seen that when the two rotating joints are parallel, a simpler expression can be obtained by using the formula of the sum of angles.

Then multiply formula (3.56) and formula (3.55), we can get:

$$\,^1_6T = \,^1_3T\,^3_6T = \begin{bmatrix} ^1n_x & ^1o_x & ^1a_x & ^1p_x \\ ^1n_y & ^1o_y & ^1a_y & ^1p_y \\ ^1n_z & ^1o_z & ^1a_z & ^1p_z \\ 0 & 0 & 0 & 1 \end{bmatrix}$$

Where

$$\begin{aligned}
^1n_x &= c_{23}(c_4c_5c_6 - s_4s_6) - s_{23}s_5c_6 \\
^1n_y &= -s_4c_5c_6 - c_4s_6 \\
^1n_z &= -s_{23}(c_4c_5c_6 - s_4s_6) - c_{23}s_5c_6 \\
^1o_x &= -c_{23}(c_4c_5s_6 + s_4c_6) + s_{23}s_5s_6 \\
^1o_y &= s_4c_5s_6 - c_4c_6 \\
^1o_z &= s_{23}(c_4c_5s_6 + s_4c_6) + c_{23}s_5s_6 \\
^1a_x &= -c_{23}c_4s_5 - s_{23}c_5 \\
^1a_y &= s_4s_5 \\
^1a_z &= s_{23}c_4s_5 - c_{23}c_5 \\
^1p_x &= a_2c_2 + a_3c_{23} - d_4s_{23} \\
^1p_y &= d_2 \\
^1p_z &= -a_3s_{23} - a_2s_2 - d_4c_{23}
\end{aligned}$$

(3.57)

Finally, the product of the six link coordinate transformation matrices can be obtained, that is, the forward kinematics equation of the PUMA560 robot is:

$$
{}^{0}_{6}T = {}^{0}_{1}T\,{}^{1}_{6}T =
\begin{bmatrix}
n_x & o_x & a_x & p_x \\
n_y & o_y & a_y & p_y \\
n_z & o_z & a_z & p_z \\
0 & 0 & 0 & 1
\end{bmatrix}
$$

where

$$
\left.
\begin{aligned}
n_x &= c_1\left[c_{23}(c_4c_5c_6 - s_4s_6) - s_{23}s_5c_6\right] + s_1(s_4c_5c_6 + c_4s_6) \\
n_y &= s_1\left[c_{23}(c_4c_5c_6 - s_4s_6) - s_{23}s_5c_6\right] - c_1(s_4c_5c_6 + c_4s_6) \\
n_z &= -s_{23}(c_4c_5c_6 - s_4s_6) - c_{23}s_5c_6 \\
o_x &= c_1\left[c_{23}(-c_4c_5s_6 - s_4c_6) + s_{23}s_5s_6\right] + s_1(c_4c_6 - s_4c_5s_6) \\
o_y &= s_1\left[c_{23}(-c_4c_5s_6 - s_4c_6) + s_{23}s_5s_6\right] - c_1(c_4c_6 - s_4c_5c_6) \\
o_z &= -s_{23}(-c_4c_5s_6 - s_4c_6) + c_{23}s_5s_6 \\
a_x &= -c_1(c_{23}c_4s_5 + s_{23}c_5) - s_1s_4s_5 \\
a_y &= -s_1(c_{23}c_4s_5 + s_{23}c_5) + c_1s_4s_5 \\
a_z &= s_{23}c_4s_5 - c_{23}c_5 \\
p_x &= c_1\left[a_2c_2 + a_3c_{23} - d_4s_{23}\right] - d_2s_1 \\
p_y &= s_1\left[a_2c_2 + a_3c_{23} - d_4s_{23}\right] + d_2c_1 \\
p_z &= -a_3s_{23} - a_2s_2 - d_4c_{23}
\end{aligned}
\right\}
\tag{3.58}
$$

Equation (3.58) represents the arm transformation matrix of the PUMA560 manipulator, which describes the pose of the end link coordinate system {6} relative to the base coordinate system {0}, which is the basic equation for all kinematics analysis of the PUMA 560 manipulator.

In order to check the correctness of the obtained ${}^{0}_{6}T$, calculate the value of the arm transformation matrix ${}^{0}_{6}T$ when $\theta_1 = 90°, \theta_2 = 0°, \theta_3 = -90°, \theta_4 = \theta_5 = \theta_6 = 0°$. The calculation result is:

$$
{}^{0}_{6}T =
\begin{bmatrix}
0 & 1 & 0 & -d_2 \\
0 & 0 & 1 & a_2 + d_4 \\
1 & 0 & 0 & a_3 \\
0 & 0 & 0 & 1
\end{bmatrix}
$$

It is exactly the same as the situation shown in Figure 3.14.

3.3.2 *Examples of inverse kinematics of the manipulator*

There are many methods for solving or synthesizing the motion equations of the manipulator, and several of them have been introduced in the previous section. It can solve the equation of motion represented by Cartesian coordinates. The elements of the right formula of the matrix are either zero, constant, or a function of the sixth to sixth joint variables. The matrix equality means that the corresponding elements are respectively equal, and 12 equations can be obtained from each matrix equation, and each equation corresponds to each component of the four vectors.

The inverse kinematics solution of the manipulator can be described as: Find all possible solutions of $\theta_1, \theta_2, \ldots, \theta_n$ with a known value of ${}^0_nT$. This is usually a non-linear problem. For the PUMA560 robot, considering the equation given in equation (3.58), the exact description of its inverse kinematics solution is as follows: known 16 values of ${}^0_nT$(4 of which are meaningless), solving for the six joint angles $\theta_1, \theta_2, \ldots, \theta_6$ in equation (3.58).

As an example of algebraic solution for a 6-degree-of-freedom manipulator, the kinematics equation of the PUMA560 derived above is solved. It is worth noting that the following solution is not applicable to the inverse kinematics solution of all manipulators, but for most general-purpose manipulators, such a solution is the most commonly used.

Write PUMA560's equation of motion (3.58) as:

$$
{}^0_6T =
\begin{bmatrix}
n_x & o_x & a_x & p_x \\
n_y & o_y & a_y & p_y \\
n_z & o_z & a_z & p_z \\
0 & 0 & 0 & 1
\end{bmatrix}
= {}^0_1T(\theta_1){}^1_2T(\theta_2){}^2_3T(\theta_3){}^3_4T(\theta_4){}^4_5T(\theta_5){}^5_6T(\theta_6)
$$

$$(3.59)$$

If the posture of the end link has been given, that is, $\boldsymbol{n}$, $\boldsymbol{o}$, $\boldsymbol{a}$ and $\boldsymbol{p}$ are known, the value of joint variables $\theta_1, \theta_2, \ldots, \theta_6$ is called the inverse kinematic solution. By using the unknown link inverse transformation and multiplying both sides of the equation (3.59) to the left, the joint variables can be separated and solved. Specific steps are as follows:

1. Seek θ_1

Use the inverse transformation to multiply both sides of the equation (3.59) to the left,

$$^0_1T^{-1}(\theta_1)^0_6T = {}^1_2T(\theta_2)^2_3T(\theta_3)^3_4T(\theta_4)^4_5T(\theta_5)^5_6T(\theta_6) \tag{3.60}$$

That is

$$\begin{bmatrix} c_1 & s_1 & 0 & 0 \\ -s_1 & c_1 & 0 & 0 \\ 0 & 0 & 1 & 0 \\ 0 & 0 & 0 & 1 \end{bmatrix} \begin{bmatrix} n_x & o_x & a_x & p_x \\ n_y & o_y & a_y & p_y \\ n_z & o_z & a_z & p_z \\ 0 & 0 & 0 & 1 \end{bmatrix} = {}^1_6T \tag{3.61}$$

Let the elements (2,4) at both ends of the matrix equation (3.61) correspond to be equal, we can get

$$-s_1 p_x + c_1 p_y = d_2 \tag{3.62}$$

Use triangle substitution:

$$p_x = \rho \cos \phi; p_y = \rho \sin \phi \tag{3.63}$$

In the formula, $\rho = \sqrt{p_x^2 + p_y^2}; \phi = \text{atan2}(p_y, p_x)$. Substituting the substitution formula (3.63) into the formula (3.62), the solution of:

$$\left. \begin{aligned} \sin(\phi - \theta_1) &= d_2/\rho; \quad \cos(\phi - \theta_1) = \pm\sqrt{1 - (d_2/\rho)^2} \\ \phi - \theta_1 &= \text{atan2}\left[\frac{d_2}{\rho}, \pm\sqrt{1 - \left(\frac{d_2}{\rho}\right)^2} \right] \\ \theta_1 &= \text{atan2}(p_y, p_x) - \text{atan2}(d_2, \pm\sqrt{p_x^2 + p_y^2 - d_2^2}) \end{aligned} \right\} \tag{3.64}$$

In the formula, the positive and negative signs correspond to the two possible solutions of.

2. Seek θ_3

After selecting a solution of θ_1, make the elements (1,4) and (3,4) at both ends of the matrix equation (3.61) correspond to the same, that is, two equations are obtained:

$$\left. \begin{aligned} c_1 p_x + s_1 p_y &= a_3 c_{23} - d_4 s_{23} + a_2 c_2 \\ -p_z &= a_3 s_{23} + d_4 c_{23} + a_2 s_2 \end{aligned} \right\} \tag{3.65}$$

The sum of squares of formulas (3.66) and (3.65) is:

$$a_3 c_3 - d_4 s_3 = k \tag{3.66}$$

Where $k = \dfrac{p_x^2 + p_y^2 + p_z^2 - a_2^2 - a_3^2 - d_2^2 - d_4^2}{2a_2}$

θ_2 in Equation (3.66) has been eliminated, and equations (3.66) and (3.62) have the same form, so they can be solved by trigonometric substitution:

$$\theta_3 = \text{atan2}(a_3, d_4) - \text{atan2}(k, \pm\sqrt{a_3^2 + d_4^2 - k^2}) \qquad (3.67)$$

In the formula, the positive and negative signs correspond to two possible solutions for θ_3.

3. Seek θ_2

To solve for θ_2, multiply the inverse transformation ${}^0_3T^{-1}$ on both sides of the matrix equation (3.59),

$$ {}^0_3T^{-1}(\theta_1, \theta_2, \theta_3){}^0_6T = {}^3_4T(\theta_4){}^4_5T(\theta_5){}^5_6T(\theta_6) \qquad (3.68) $$

That is

$$
\begin{bmatrix}
c_1 c_{23} & s_1 c_{23} & -s_{23} & -a_2 c_3 \\
-c_1 s_{23} & -s_1 s_{23} & -c_{23} & a_2 s_3 \\
-s_1 & c_1 & 0 & -d_2 \\
0 & 0 & 0 & 1
\end{bmatrix}
\begin{bmatrix}
n_x & o_x & a_x & p_x \\
n_y & o_y & a_y & p_y \\
n_z & o_z & a_z & p_z \\
0 & 0 & 0 & 1
\end{bmatrix}
= {}^3_6T \quad (3.69)
$$

In the formula, the transformation ${}^3_6T$ is given by formula (3.55). Let the elements (1,4) and (2,4) on both sides of the matrix equation (3.69) correspond to each other and be equal, we can get:

$$
\left.
\begin{aligned}
c_1 c_{23} p_x + s_1 c_{23} p_y - s_{23} p_z - a_2 c_3 &= a_3 \\
-c_1 s_{23} p_x - s_1 s_{23} p_y - c_{23} p_z + a_2 s_3 &= d_4
\end{aligned}
\right\} \qquad (3.70)
$$

Solve simultaneously and get s_{23} and c_{23}:

$$
\left.
\begin{aligned}
s_{23} &= \frac{(-a_3 - a_2 c_3)p_z + (c_1 p_x + s_1 p_y)(a_2 s_3 - d_4)}{p_z^2 + (c_1 p_x + s_1 p_y)^2} \\
c_{23} &= \frac{(-d_4 + a_2 s_3)p_z - (c_1 p_x + s_1 p_y)(-a_2 c_3 - a_3)}{p_z^2 + (c_1 p_x + s_1 p_y)^2}
\end{aligned}
\right\} \qquad (3.71)
$$

The denominator of and expression are equal and positive. then

$$\theta_{23} = \theta_2 + \theta_3 = \text{atan2}\,[-(a_3 + a_2 c_3)p_z + (c_1 p_x + s_1 p_y)(a_2 s_3 - d_4),$$
$$(-d_4 + a_2 s_3)p_z + (c_1 p_x + s_1 p_y)(a_2 c_3 + a_3)] \qquad (3.72)$$

According to the four possible combinations of solutions of θ_1 and θ_3, the corresponding four possible values can be obtained from equation (3.72), and then four possible solutions of θ_{23} can be obtained:

$$\theta_2 = \theta_{23} - \theta_3 \tag{3.73}$$

4. Seek θ_4

Because the left side of equation (3.69) is known, let the elements (1,3) and (3,3) on both sides correspond to the same, then we can get:

$$\begin{cases} a_x c_1 c_{23} + a_y s_1 c_{23} - a_z s_{23} = -c_4 s_5 \\ -a_x s_1 + a_y c_1 = s_4 s_5 \end{cases} \tag{3.74}$$

As long as $s_5 \neq 0$, you can find θ_4

$$\theta_4 = \mathrm{atan2}(-a_x s_1 + a_y c_1, -a_x c_1 c_{23} - a_y s_1 c_{23} + a_z s_{23}) \tag{3.75}$$

When $s_5 = 0$, the robot is in a singular position. At this time, the joint axes 4 and 6 overlap, and only the sum or difference can be solved. The singularity can be judged by whether the two variables of atan 2 in formula (3.75) are close to zero. If they are all close to zero, it is a singular configuration. In the singular configuration, the value of θ_4 can be selected arbitrarily, and then the corresponding value θ_6 can be calculated.

5. Seek θ_5

According to the obtained θ_4, and then solve the θ_5, multiply both ends of the equation (3.63) to the left with the inverse transform ${}_4^0T^{-1}(\theta_1, \theta_2, \theta_3, \theta_4)$ at the same time, there is

$$ {}_4^0T^{-1}(\theta_1, \theta_2, \theta_3, \theta_4) {}_6^0T = {}_5^4T(\theta_5){}_6^5T(\theta_6) \tag{3.76}$$

The $\theta_1, \theta_2, \theta_3, \theta_4$ in the left side of the factor (3.76) has been solved, and the inverse transformation ${}_4^0T^{-1}(\theta_1, \theta_2, \theta_3, \theta_4)$ is:

$$\begin{bmatrix} c_1 c_{23} c_4 + s_1 s_4 & s_1 c_{23} c_4 - c_1 s_4 & -s_{23} c_4 & -a_2 c_3 c_4 + d_2 s_4 - a_3 c_4 \\ -c_1 c_{23} s_4 + s_1 c_4 & -s_1 c_{23} s_4 - c_1 c_4 & s_{23} s_4 & a_2 c_3 s_4 + d_2 c_4 + a_3 s_4 \\ -c_1 s_{23} & -s_1 s_{23} & -c_{23} & a_2 s_3 - d_4 \\ 0 & 0 & 0 & 1 \end{bmatrix}$$

The right side ${}_6^4T(\theta_5, \theta_6) = {}_5^4T(\theta_5){}_6^5T(\theta_6)$ of the equation (3.80) is given by equation (3.48). Let the elements (1,3) and (3,3) on both sides of formula (3.54) correspond to the same, we can get

$$\left.\begin{array}{r}a_x(c_1c_{23}c_4 + s_1s_4) + a_y(s_1c_{23}c_4 - c_1s_4) - a_z(s_{23}c_4) = -s_5 \\ a_x(-c_1s_{23}) + a_y(-s_1s_{23}) + a_z(-c_{23}) = c_5\end{array}\right\} \quad (3.77)$$

The closed solution of θ_5 can be obtained:

$$\theta_5 = \text{atan2}(s_5, c_5) \qquad (3.78)$$

6. Seek θ_6

Multiplying both ends of the formula (3.59) to the left with inverse transformation ${}_5^0T^{-1}(\theta_1, \theta_2, \theta_3, \theta_4, \theta_5)$ at the same time, we can get

$$_5^0T^{-1}(\theta_1, \theta_2, \ldots, \theta_5)_6^0T = {}_6^5T(\theta_6) \qquad (3.79)$$

Let the elements (1,1) and (3,1) on both sides of the matrix equation (3.79) correspond to each other and be equal:

$$\left.\begin{array}{r}-n_x(c_1c_{23}s_4 - s_1c_4) - n_y(s_1c_{23}s_4 + c_1c_4) + n_z(s_{23}s_4) = s_6 \\ n_x[(c_1c_{23}c_4 + s_1s_4)c_5 - c_1s_{23}s_5] + n_y[(s_1c_{23}c_4 - c_1s_4)c_5 - s_1s_{23}s_5] \\ -n_z(s_{23}c_4c_5 + c_{23}s_5) = c_6\end{array}\right\}$$

$$(3.80)$$

Thus the closed solution of θ_6 can be obtained:

$$\theta_6 = \text{atan2}(s_6, c_6) \qquad (3.81)$$

Since the signs $\pm$ appear in equations (3.64) and (3.67), these equations may have 4 sets of solutions. As shown in Figure 3.15, there are 4 possible solutions for the PUMA560 manipulator to reach the same target posture, and they have exactly the same posture for the hand. For each set of solutions shown in the figure, another 4 sets of symmetrical solutions can be obtained by "flipping" the wrist joint of the manipulator arm. These 4 sets of solutions can be obtained by the following inversion formula:

$$\left.\begin{array}{l}\theta_4' = \theta_4 + 180° \\ \theta_5' = -\theta_5 \\ \theta_6' = \theta_6 + 180°\end{array}\right\} \qquad (3.82)$$

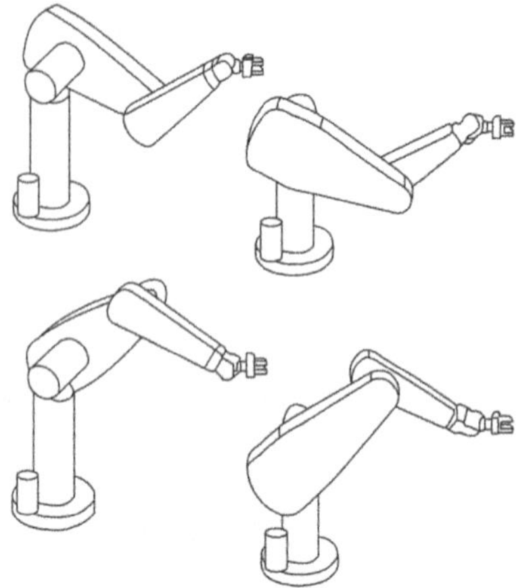

Figure 3.15.   PUMA 560 4 sets of solutions.

There may be 8 solutions for the inverse motion solution of PUMA560. Therefore, there are 8 sets of different solutions for the PUMA560 manipulator to reach a certain target pose. However, due to structural limitations, for example, each joint variable cannot move within the full 360° range, and some solutions cannot be realized. When there are multiple solutions for the manipulator, the most satisfactory set of solutions should be selected to meet the work requirements of the manipulator.

3.4   Jacobian Formula of Manipulator [4, 10, 15]

When the manipulator is operated and controlled, the position and posture of the manipulator are often involved in small changes. These changes can be represented by small changes in the homogeneous transformation matrix describing the position of the manipulator. Mathematically, this small change can be expressed by differential change. The differential relationship in the movement of the manipulator is very important. For example, when using a camera to observe the end effector of a manipulator, it is necessary to transform a differential change in one coordinate system into a differential change in another coordinate system. For example, the coordinate system of the camera is established on. Another application of differential change is when the differential change of a pair is known, the corresponding change of the coordinate of each joint needs to be requested. Differential relations are also very important for studying the dynamics of manipulators.

3.4.1 *Differential motion of the manipulator*

Given a transformation whose element is a function of a certain variable, then the differential transformation of this variable is such a transformation that its element is the derivative of the original transformation element. A method has been developed to make the differential transformation of the coordinate system equivalent to the transformation of the base system. This method can be extended to any two coordinate systems to make their differential motions equal.

The transformation of the manipulator includes translation transformation, rotation transformation, scale transformation and projection transformation. Here, the discussion is limited to translation transformation and rotation transformation. In this way, the derivative term can be expressed as differential translation and differential rotation.

1. Differential translation and differential rotation

Either a given coordinate system or a base coordinate system can be used to express differential translation and rotation.

Given the coordinate system T, $T + dT$ can be expressed as:

$$T + dT = \mathbf{Trans}(d_x, d_y, d_z)\mathbf{Rot}(f, d\theta)T$$

In the formula, $\mathbf{Trans}(d_x, d_y, d_z)$ represents the transformation of the differential translation d_x, d_y, d_z in the base system; $\mathbf{Rot}(f, d\theta)$ represents the transformation of the differential rotation $d\theta$ around the vector f in the base system. The expression dT can be obtained from the above formula:

$$dT = (\mathbf{Trans}(d_x, d_y, d_z)\mathbf{Rot}(f, d\theta) - I)T \qquad (3.83)$$

Similarly, the differential translation and rotation of a given coordinate system $dT = (\mathbf{Trans}(d_x, d_y, d_z)\mathbf{Rot}(f, d\theta) - I)T$ can also be used to express the differential change:

$$T + dT = T\mathbf{Trans}(d_x, d_y, d_z)\mathbf{Rot}(f, d\theta)$$

In the formula, $\mathbf{Trans}(d_x, d_y, d_z)$ represents the differential translation transformation to the coordinate system T; $\mathbf{Rot}(f, d\theta)$

represents the differential rotation $d\theta$ around the vector $\boldsymbol{f}$ in the coordinate system T. At this time:

$$dT = T(\mathbf{Trans}(d_x, d_y, d_z)\mathbf{Rot}(f, d\theta) - I) \qquad (3.84)$$

There is a common term in formulas (3.83) and (3.84). When the differential motion is performed on the base system, it is specified as Δ; and when the motion is performed on the coordinate system T, it is recorded as $^T\Delta$. Therefore, when the base system is subjected to a differential change, $dT = \Delta T$; and when the coordinate system T is subjected to a differential change, $dT = T^T\Delta$.

The homogeneous transformation representing the differential translation is:

$$\mathbf{Trans}(d_x, d_y, d_z) = \begin{bmatrix} 1 & 0 & 0 & d_x \\ 0 & 1 & 0 & d_y \\ 0 & 0 & 1 & d_z \\ 0 & 0 & 0 & 1 \end{bmatrix}$$

At this time, the variable of **Trans** is the differential vector $\boldsymbol{d}$ represented by the differential change $d_x i + d_y j + d_z k$. When discussing the general rotation transformation in Chapter 2, there are the following formulas:

$\mathbf{Rot}(f, \theta)$

$$= \begin{bmatrix} f_x f_x vers\theta + c\theta & f_y f_x vers\theta - f_z s\theta & f_z f_x vers\theta + f_y s\theta & 0 \\ f_x f_y vers\theta + f_z s\theta & f_y f_y vers\theta + c\theta & f_z f_y vers\theta - f_x s\theta & 0 \\ f_x f_z vers\theta - f_y s\theta & f_y f_z vers\theta + f_x s\theta & f_z f_z vers\theta + c\theta & 0 \\ 0 & 0 & 0 & 1 \end{bmatrix}$$

See formula (2.46). For differential change $d\theta$, the corresponding sine function, cosine function and orthogonal function are:

$$\lim_{\theta \to 0} \sin\theta = d\theta, \quad \lim_{\theta \to 0} \cos\theta = 1, \quad \lim_{\theta \to 0} vers\theta = 0$$

Substituting them into equation (2.46), the differential rotation homogeneous transformation can be expressed as:

$$Rot(\mathbf{f}, d\theta) = \begin{bmatrix} 1 & -f_z d\theta & f_y d\theta & 0 \\ f_z d\theta & 1 & -f_x d\theta & 0 \\ -f_y d\theta & f_x d\theta & 1 & 0 \\ 0 & 0 & 0 & 1 \end{bmatrix}$$

Substituting in $\Delta = Trans(d_x, d_y, d_z)Rot(f, d\theta) - I$, you can get:

$$\Delta = \begin{bmatrix} 1 & 0 & 0 & d_x \\ 0 & 1 & 0 & d_y \\ 0 & 0 & 1 & d_z \\ 0 & 0 & 0 & 1 \end{bmatrix} \begin{bmatrix} 1 & -f_z d\theta & f_y d\theta & 0 \\ f_z d\theta & 1 & -f_x d\theta & 0 \\ -f_y d\theta & f_x d\theta & 1 & 0 \\ 0 & 0 & 0 & 1 \end{bmatrix} - \begin{bmatrix} 1 & 0 & 0 & 0 \\ 0 & 1 & 0 & 0 \\ 0 & 0 & 1 & 0 \\ 0 & 0 & 0 & 1 \end{bmatrix}$$

Simplified:

$$\Delta = \begin{bmatrix} 0 & -f_z d\theta & f_y d\theta & d_x \\ f_z d\theta & 0 & -f_x d\theta & d_y \\ -f_y d\theta & f_x d\theta & 0 & d_z \\ 0 & 0 & 0 & 0 \end{bmatrix} \tag{3.85}$$

The differential rotation $d\theta$ around the vector f is equivalent to the differential rotations $\delta_x, \delta_y, \delta_z$ around the three axes x, y, and z, namely $f_x d\theta = \delta_x, f_y d\theta = \delta_y, f_z d\theta = \delta_z$. Substituting the formula (3.75) to get:

$$\Delta = \begin{bmatrix} 0 & -\delta_z & \delta_y & d_x \\ \delta_z & 0 & -\delta_x & d_y \\ -\delta_y & \delta_x & 0 & d_z \\ 0 & 0 & 0 & 0 \end{bmatrix} \tag{3.86}$$

The similarly available expression $^T\Delta$ is:

$$^T\Delta = \begin{bmatrix} 0 & -^T\delta_z & ^T\delta_y & ^Td_x \\ ^T\delta_z & 0 & -^T\delta_x & ^Td_y \\ -^T\delta_y & ^T\delta_x & 0 & ^Td_z \\ 0 & 0 & 0 & 0 \end{bmatrix} \tag{3.87}$$

Therefore, the differential translation and rotation transformation Δ can be regarded as composed of the differential translation vector d and the differential rotation vector δ, which are $d = d_x i + d_y j + d_z k, \delta = \delta_x i + \delta_y j + \delta_z k$ respectively. We use the column vector D to contain the above two vectors, which are called the differential motion vectors of the rigid body or coordinate system:

$$D = \begin{bmatrix} d_x \\ d_y \\ d_z \\ \delta_x \\ \delta_y \\ \delta_z \end{bmatrix} \quad \text{or} \quad D = \begin{bmatrix} d \\ \delta \end{bmatrix} \qquad (3.88)$$

Similarly, there are the following types:

$$^T\boldsymbol{d} = {}^T d_x \boldsymbol{i} + {}^T d_y \boldsymbol{j} + {}^T d_z \boldsymbol{k}$$

$$^T\boldsymbol{\delta} = {}^T \delta_x \boldsymbol{i} + {}^T \delta_y \boldsymbol{j} + {}^T \delta_z \boldsymbol{k}$$

$$^T D = \begin{bmatrix} {}^T d_x \\ {}^T d_y \\ {}^T d_z \\ {}^T \delta_x \\ {}^T \delta_y \\ {}^T \delta_z \end{bmatrix} \quad \text{or} \quad {}^T D = \begin{bmatrix} {}^T d \\ {}^T \delta \end{bmatrix} \qquad (3.89)$$

Example 3.2. The given coordinate system {A} and its differential translation and differential rotation to the base system are:

$$A = \begin{bmatrix} 0 & 0 & 1 & 10 \\ 1 & 0 & 0 & 5 \\ 0 & 1 & 0 & 0 \\ 0 & 0 & 0 & 1 \end{bmatrix}$$

$$\boldsymbol{d} = 1\boldsymbol{i} + 0\boldsymbol{j} + 0.5\boldsymbol{k}$$

$$\boldsymbol{\delta} = 0\boldsymbol{i} + 0.1\boldsymbol{j} + 0\boldsymbol{k}$$

Try to find the differential transformation d$\boldsymbol{A}$.

Solution: First, according to formula (3.86), the following formula can be obtained:

$$\Delta = \begin{bmatrix} 0 & 0 & 0.1 & 1 \\ 0 & 0 & 0 & 0 \\ -0.1 & 0 & 0 & 0.5 \\ 0 & 0 & 0 & 0 \end{bmatrix}$$

Then according to $\mathrm{d}\boldsymbol{T} = \Delta\boldsymbol{T}$, $\mathrm{d}\boldsymbol{A} = \Delta\boldsymbol{A}$, namely:

$$dA = \begin{bmatrix} 0 & 0 & 0.1 & 1 \\ 0 & 0 & 0 & 0 \\ -0.1 & 0 & 0 & 0.5 \\ 0 & 0 & 0 & 0 \end{bmatrix} \begin{bmatrix} 0 & 0 & 1 & 10 \\ 1 & 0 & 0 & 5 \\ 0 & 1 & 0 & 0 \\ 0 & 0 & 0 & 1 \end{bmatrix}$$

$$= \begin{bmatrix} 0 & 0.1 & 0 & 1 \\ 0 & 0 & 0 & 0 \\ 0 & 0 & -0.1 & -0.5 \\ 0 & 0 & 0 & 0 \end{bmatrix}$$

This differential change of the coordinate system $\{A\}$ is shown in Figure 3.16.

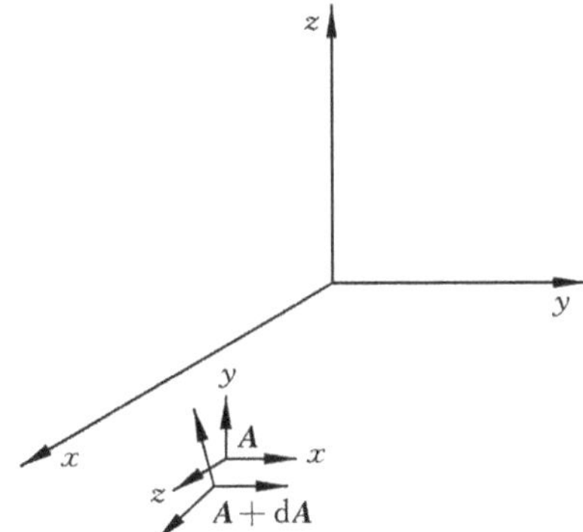

Figure 3.16.   Differential change of coordinate system {A}.

2. Equivalent transformation of differential motion

To obtain the Jacobian matrix of the manipulator, small changes in position and posture in one coordinate system need to be transformed into equivalent expressions in another coordinate system.

According to $\mathrm{d}\boldsymbol{T} = \Delta\boldsymbol{T}$ and $\mathrm{d}\boldsymbol{T} = \boldsymbol{T}^T\Delta$, when the 2 coordinate systems are equivalent, $\Delta\boldsymbol{T} = \boldsymbol{T}^T\Delta$, after transformation, we get:

$$T^{-1}\Delta T = {}^T\Delta \tag{3.90}$$

From the formula (3.86):

$$\Delta T = \begin{bmatrix} 0 & -\delta_z & \delta_y & d_x \\ \delta_z & 0 & -\delta_x & d_y \\ -\delta_y & \delta_x & 0 & d_z \\ 0 & 0 & 0 & 0 \end{bmatrix} \begin{bmatrix} n_x & o_x & a_x & p_x \\ n_y & o_y & a_y & p_y \\ n_z & o_z & a_z & p_z \\ 0 & 0 & 0 & 1 \end{bmatrix}$$

$$\Delta T = \begin{bmatrix} -\delta_z n_y + \delta_y n_z & -\delta_z o_y + \delta_y o_z & -\delta_z a_y + \delta_y a_z & -\delta_z p_y + \delta_y p_z + d_x \\ \delta_z n_x + \delta_x n_z & \delta_z o_x - \delta_x o_z & \delta_z a_x - \delta_x a_z & \delta_z p_x - \delta_x p_z + d_y \\ -\delta_y n_x + \delta_x n_y & -\delta_y o_x + \delta_x o_y & -\delta_y a_x + \delta_x a_y & -\delta_y p_x + \delta_x p_y + d_z \\ 0 & 0 & 0 & 0 \end{bmatrix}$$

$$\tag{3.91}$$

It is equivalent to the following formula:

$$\Delta T = \begin{bmatrix} (\delta \times n)_x & (\delta \times o)_x & (\delta \times a)_x & (\delta \times p + d)_x \\ (\delta \times n)_y & (\delta \times o)_y & (\delta \times a)_y & (\delta \times p + d)_y \\ (\delta \times n)_z & (\delta \times o)_z & (\delta \times a)_z & (\delta \times p + d)_z \\ 0 & 0 & 0 & 0 \end{bmatrix}$$

Multiply the upper formula with T^{-1} left to get:

$$T^{-1}\Delta T = \begin{bmatrix} n_x & n_y & n_z & -p \cdot n \\ o_x & o_y & o_z & -p \cdot o \\ a_x & a_y & a_z & -p \cdot a \\ 0 & 0 & 0 & 1 \end{bmatrix}$$

$$\times \begin{bmatrix} (\delta \times n)_x & (\delta \times o)_x & (\delta \times a)_x & (\delta \times p + d)_x \\ (\delta \times n)_y & (\delta \times o)_y & (\delta \times a)_y & (\delta \times p + d)_y \\ (\delta \times n)_z & (\delta \times o)_z & (\delta \times a)_z & (\delta \times p + d)_z \\ 0 & 0 & 0 & 0 \end{bmatrix}$$

$$T^{-1}\Delta T = \begin{bmatrix} n \cdot (\delta \times n) & n \cdot (\delta \times o) & n \cdot (\delta \times a) & n \cdot (\delta \times p + d) \\ o \cdot (\delta \times n) & o \cdot (\delta \times o) & o \cdot (\delta \times a) & o \cdot (\delta \times p + d) \\ a \cdot (\delta \times n) & a \cdot (\delta \times o) & a \cdot (\delta \times a) & a \cdot (\delta \times p + d) \\ 0 & 0 & 0 & 0 \end{bmatrix}$$

Applying the two properties $a \cdot (b \times c) = b \cdot (c \times a)$ 及 $a \cdot (a \times c) = 0$ of three-vector multiplication, and according to formula (3.90), the above formula can be transformed into:

$$^{T}\Delta = \begin{bmatrix} 0 & -\delta \cdot (n \times o) & \delta \cdot (a \times n) & \delta \cdot (p \times n) + d \cdot n \\ \delta \cdot (n \times o) & 0 & -\delta \cdot (o \times a) & \delta \cdot (p \times o) + d \cdot o \\ -\delta \cdot (a \times n) & \delta \cdot (o \times a) & 0 & \delta \cdot (p \times a) + d \cdot a \\ 0 & 0 & 0 & 0 \end{bmatrix}$$

Simplified:

$$^{T}\Delta = \begin{bmatrix} 0 & -\delta \cdot a & \delta \cdot o & \delta \cdot (p \times n) + d \cdot n \\ \delta \cdot a & 0 & -\delta \cdot n & \delta \cdot (p \times o) + d \cdot o \\ -\delta \cdot o & \delta \cdot n & 0 & \delta \cdot (p \times a) + d \cdot a \\ 0 & 0 & 0 & 0 \end{bmatrix} \tag{3.92}$$

Since $^{T}\Delta$ has been defined by equation (3.87), so let the elements of equations (3.87) and (3.92) are equal respectively, the following equations can be obtained:

$$\left. \begin{aligned} ^{T}d_x &= \delta \cdot (p \times n) + d \cdot n \\ ^{T}d_y &= \delta \cdot (p \times o) + d \cdot o \\ ^{T}d_z &= \delta \cdot (p \times a) + d \cdot a \end{aligned} \right\} \tag{3.93}$$

$$^{T}\delta_x = \delta \cdot n, \quad ^{T}\delta_y = \delta \cdot o, \quad ^{T}\delta_z = \delta \cdot a \tag{3.94}$$

Where $\boldsymbol{n}$, $\boldsymbol{o}$, $\boldsymbol{a}$, $\boldsymbol{p}$ are respectively the column vectors of the differential coordinate transformation $\boldsymbol{T}$. From the above two formulas, the relationship between differential motion $^{T}\boldsymbol{D}$ and $\boldsymbol{D}$ can be obtained as follows:

$$
\begin{bmatrix} {}^T d_x \\ {}^T d_y \\ {}^T d_z \\ {}^T \delta_x \\ {}^T \delta_y \\ {}^T \delta_z \end{bmatrix} =
\begin{bmatrix}
n_x & n_y & n_z & (p \times n)_x & (p \times n)_y & (p \times n)_z \\
o_x & o_y & o_z & (p \times o)_x & (p \times o)_y & (p \times o)_z \\
a_x & a_y & a_z & (p \times a)_x & (p \times a)_y & (p \times a)_z \\
0 & 0 & 0 & n_x & n_y & n_z \\
0 & 0 & 0 & o_x & o_y & o_z \\
0 & 0 & 0 & a_x & a_y & a_z
\end{bmatrix}
\begin{bmatrix} d_x \\ d_y \\ d_z \\ \delta_x \\ \delta_y \\ \delta_z \end{bmatrix}
\tag{3.95}
$$

Using the property $a \cdot (b \times c) = c \cdot (a \times b)$ of three-vector multiplication, we can further write equations (3.93) and (3.94) as:

$$
\left.\begin{aligned}
{}^T d_x &= n \cdot ((\delta \times p) + d) \\
{}^T d_y &= o \cdot ((\delta \times p) + d) \\
{}^T d_z &= a \cdot ((\delta \times p) + d)
\end{aligned}\right\}
\tag{3.96}
$$

$$
\left.\begin{aligned}
{}^T \delta_x &= n \cdot \delta \\
{}^T \delta_y &= o \cdot \delta \\
{}^T \delta_z &= a \cdot \delta
\end{aligned}\right\}
\tag{3.97}
$$

Using the above two formulas, the differential change of the base coordinate system can be transformed into the differential change of the coordinate system T very conveniently. The formula (3.95) can be abbreviated as:

$$
\begin{bmatrix} {}^T d \\ {}^T \delta \end{bmatrix} =
\begin{bmatrix} R^T & -R^T S(p) \\ 0 & R^T \end{bmatrix}
\begin{bmatrix} d \\ \delta \end{bmatrix}
\tag{3.98}
$$

Where R is the rotation matrix,

$$
R = \begin{bmatrix} n_x & o_x & a_x \\ n_y & o_y & a_y \\ n_z & o_z & a_z \end{bmatrix}
\tag{3.99}
$$

For any three-dimensional vector $p = [p_x, p_y, p_z]^T$, its antisymmetric matrix $S(p)$ is defined as

$$
S(p) = \begin{bmatrix} 0 & -p_x & p_y \\ p_z & 0 & -p_x \\ -p_y & p_x & 0 \end{bmatrix}
\tag{3.100}
$$

Example 3.3. Given the coordinate system $\{A\}$ and its differential translation d and differential rotation δ relative to the base coordinate system, the same as in Example 3.2. Try to find the equivalent differential translation and differential rotation of the coordinate system $\{A\}$.

Solution: Because

$$n = 0i + 1j + 0k$$

$$o = 0i + 0j + 1k$$

$$a = 1i + 0j + 0k$$

$$p = 10i + 5j + 0k$$

as well as

$$\delta \times p = \begin{bmatrix} i & j & k \\ 0 & 0.1 & 0 \\ 10 & 5 & 0 \end{bmatrix}$$

which is $\delta \times p = 0i + 0j - 1k$. After adding d, $\delta \times p + d = 1i + 0j - 0.5k$.

According to equations (3.95) and (3.94), the equivalent differential translation and differential rotation can be obtained as:

$$^Ad = 0i - 0.5j + 1k, \quad ^A\delta = 0.1i + 0j + 0k$$

According $dT = T^T\Delta$ to compute $dA = A^A\Delta$, to check whether the obtained differential motion is correct. According to formula (3.87):

$$^A\Delta = \begin{bmatrix} 0 & 0 & 0 & 0 \\ 0 & 0 & -0.1 & 0.5 \\ 0 & 0.1 & 0 & 1 \\ 0 & 0 & 0 & 0 \end{bmatrix}$$

$$dA = \begin{bmatrix} 0 & 0 & 1 & 10 \\ 1 & 0 & 0 & 5 \\ 0 & 1 & 0 & 0 \\ 0 & 0 & 0 & 1 \end{bmatrix} \begin{bmatrix} 0 & 0 & 0 & 0 \\ 0 & 0 & -0.1 & -0.5 \\ 0 & 0.1 & 0 & 1 \\ 0 & 0 & 0 & 0 \end{bmatrix}$$

which is:

$$dA = \begin{bmatrix} 0 & 0.1 & 0 & 0 \\ 0 & 0 & 0 & 0 \\ 0 & 0 & -0.1 & -0.5 \\ 0 & 0 & 0 & 0 \end{bmatrix}$$

The results obtained are consistent with Example 3.2. It can be seen that the obtained differential translation and differential rotation of $\{A\}$ are correct.

3. The differential relationship in the transformation

Equations (3.95) to (3.97) can be used to transform the differential motion between any two coordinate systems. Among them, by formulas (3.96) and (3.97), the elements of $T\Delta$ can be determined according to the differential coordinate transformation T and the differential rotation transformation Δ. If you want to find the differential vector Δ from each differential vector of $^{T}\Delta$, you can multiply the left by T and the right by T^{-1} from the formula (3.90) to obtain the following transformation expression:

$$\Delta = T^{T}\Delta T^{-1}$$

Or transform to

$$\Delta = (T^{-1})^{-1T}\Delta(T^{-1}) \tag{3.101}$$

There are two coordinate systems $\{A\}$ and $\{B\}$, the latter is defined relative to $\{A\}$. Then, either the coordinate system $\{A\}$ or the coordinate system $\{B\}$ can be used to represent the differential motion. The conversion chart in Figure 3.17 shows this situation.

Figure 3.17 shows the differential change relationship between $\{A\}$ and $\{B\}$ coordinate systems. According to Figure 3.17, $\Delta AB = AB^{B}\Delta$, the solution to Δ is:

$$\Delta = AB^{B}\Delta B^{-1}A^{-1}$$

Or transform into:

$$\Delta = (A^{-1}B^{-1})^{-1B}\Delta(B^{-1}A^{-1}) \tag{3.102}$$

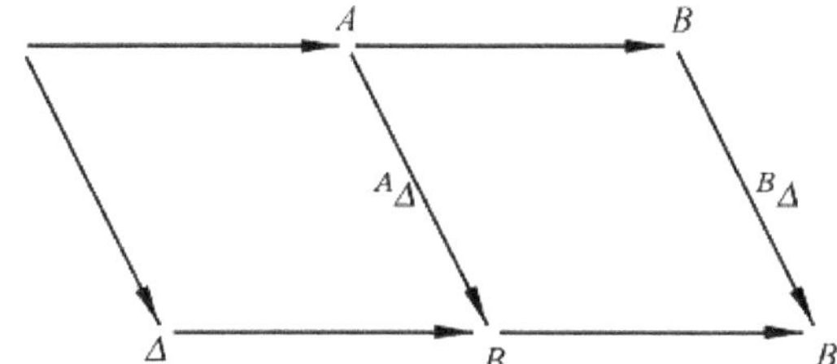

Figure 3.17. Differential transformation diagram between two coordinate systems.

This formula expresses the relationship between the differential motion in the coordinate system $\{\boldsymbol{B}\}$ and the differential motion in the base coordinate system. It has the general form of equation (3.101); and $(B^{-1}A^{-1})$ corresponds to $\boldsymbol{T}$ in equation (3.101).

Similarly, according to Figure 3.17, we can get $A^A\Delta B = AB^B\Delta$, or $^A\Delta B = B^B\Delta$. To solve $^A\Delta$ and get $^A\Delta = B^B\Delta B^{-1}$, namely

$$^A\Delta = (B^{-1})^{-1}{}^B\Delta(B^{-1}) \tag{3.103}$$

This formula expresses the relationship between the differential motion in the coordinate system $\{\boldsymbol{A}\}$ and the differential motion in the coordinate system $\{\boldsymbol{B}\}$. Among them, B^{-1} corresponds to $\boldsymbol{T}$ in equations (3.96) and (3.97). Here, $\boldsymbol{T}$ is no longer a coordinate system matrix, but a differential coordinate transformation matrix. It can be obtained directly from Figure 3.17, that is, starting from the arrow of the known differential change transformation, backtracking to the path taken by the equivalent differential change to be found. For the first case above, the path from the arrow of $^B\Delta$ to the arrow of Δ is $\boldsymbol{B}^{-1}\boldsymbol{A}^{-1}$, and for the second case, from $^B\Delta$ to $^A\Delta$, the path traverses is $\boldsymbol{B}^{-1}$.

Example 3.4. A camera is installed on the link 5 of the manipulator. This connection is determined by the following formula:

$$^T CAM = \begin{bmatrix} 0 & 0 & -1 & 5 \\ 0 & -1 & 0 & 0 \\ -1 & 0 & 0 & 10 \\ 0 & 0 & 0 & 1 \end{bmatrix}$$

The current position of the last link of the manipulator is described
by the following formula:

$$
A_6 = \begin{bmatrix} 0 & -1 & 0 & 0 \\ 1 & 0 & 0 & 0 \\ 0 & 0 & 1 & 8 \\ 0 & 0 & 0 & 1 \end{bmatrix}
$$

The observed target object is $^{CAM}\boldsymbol{O}$. To guide the end of the manipulator to the target object, the differential change in the coordi-
nate system CAM that needs to be known is: $^{CAM}\boldsymbol{d} = -1\boldsymbol{i} + 0\boldsymbol{j} +
0\boldsymbol{k}$, $^{CAM}\boldsymbol{\delta} = 0\boldsymbol{i}+0\boldsymbol{j}+0.1\boldsymbol{k}$, try to find the required differential change
in the coordinate system.

Solution: The above situation can be described by Figure 3.18(a)
and the following equation

$$
T_5 A_6 EX = T_5 \; CAM \; O
$$

In the formula, T_5 describes the relationship between the link 5 and
the base coordinate system; A_6 describes the link 6 in the coordinate
system of the link 5; $\boldsymbol{E}$ is an unknown transformation describing the
object to the end; $\boldsymbol{O}$ describes the object in the camera coordinate
system.

The transformation diagram is shown in Figure 3.18(b). From the
figure, the relative differential coordinate transformation $\boldsymbol{T}$ related

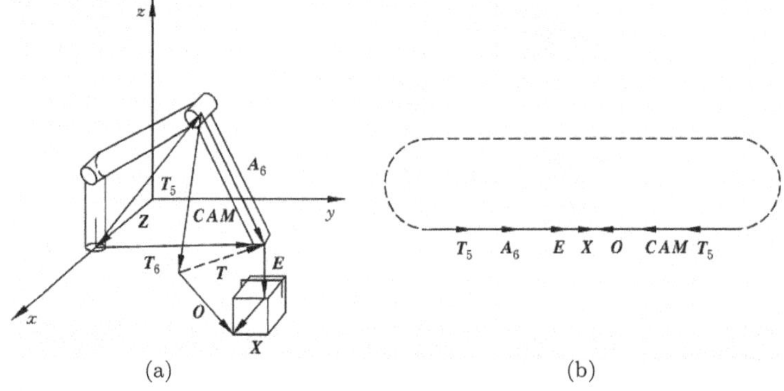

Figure 3.18.　The posture and differential transformation diagram of Example 3.5.

to $^{T_6}\Delta$ to $^{CAM}\Delta$ can be obtained as:

$$T = CAM^{-1}T^{-1}T_5A_6 = CAM^{-1}A_6$$

Because

$$CAM^{-1} = \begin{bmatrix} 0 & 0 & -1 & 10 \\ 0 & -1 & 0 & 0 \\ -1 & 0 & 0 & 5 \\ 0 & 0 & 0 & 1 \end{bmatrix}$$

Then we can get this differential coordinate transformation:

$$T = \begin{bmatrix} 0 & 0 & -1 & 10 \\ 0 & -1 & 0 & 0 \\ -1 & 0 & 0 & 5 \\ 0 & 0 & 0 & 1 \end{bmatrix} \begin{bmatrix} 0 & -1 & 0 & 0 \\ 1 & 0 & 0 & 0 \\ 0 & 0 & 1 & 8 \\ 0 & 0 & 0 & 1 \end{bmatrix} = \begin{bmatrix} 0 & 0 & -1 & 2 \\ -1 & 0 & 0 & 0 \\ 0 & 1 & 0 & 5 \\ 0 & 0 & 0 & 1 \end{bmatrix}$$

also because

$$\delta \times p = \begin{bmatrix} i & j & k \\ 0 & 0 & 0.1 \\ 2 & 0 & 5 \end{bmatrix} = 0i + 0.2j + 0k$$

Finally, according to equations (3.93) and (3.94), the differential change in the coordinate system T_6 can be obtained as follows:

$$^{T_6}d = -0.2i + 0j + 1k$$
$$^{T_6}\delta = 0i + 0.1j + 0k$$

3.4.2 *Definition and solution of Jacobian matrix*

The differential motion of the manipulator is analyzed above. On this basis, the linear mapping relationship between the manipulator operating space velocity and the joint space velocity, namely the Jacobian matrix, will be studied.

1. Definition of Jacobian matrix

The linear transformation between the operating speed of the manipulator and the joint speed is defined as the Jacobian matrix of the manipulator, which can be regarded as the transmission ratio of the movement speed from the joint space to the operating space.

Let the manipulator's motion equation

$$x = x(q) \tag{3.104}$$

It represents the displacement relationship between the operating space x and the joint space q. Derivation of both sides of equation (3.104) with respect to time, the differential relationship between q and x is obtained

$$\dot{x} = J(q)\dot{q} \tag{3.105}$$

In the formula, $\dot{x}$ is called the generalized speed of the end in the operating space, referred to as the operating speed; $\dot{q}$ is the joint speed; $J(q)$ is the $6 \times n$ partial derivative matrix, called as the Jacobian matrix of the manipulator. Its element i-th row and j-th column is

$$J_{ij}(q) = \frac{\partial x_i(q)}{\partial q_j}, \quad i = 1, 2, \ldots, 6; \quad j = 1, 2, \ldots, n \tag{3.106}$$

It can be seen from equation (3.105) that for a given $q \in R^n$, Jacobian $J(q)$ is a linear transformation from the joint space velocity $\dot{q}$ to the operating space velocity $\dot{x}$.

The generalized velocity $\dot{x}$ of a rigid body or coordinate system is a 6-dimensional column vector composed of linear velocity V and angular velocity ω:

$$\dot{x} = \begin{bmatrix} V \\ \omega \end{bmatrix} = \lim_{\Delta t \to 0} \frac{1}{\Delta t} \begin{bmatrix} d \\ \delta \end{bmatrix} \tag{3.107}$$

According to formula (3.105), we can get:

$$\dot{x} = J(q)\dot{q} \tag{3.108}$$

From the formula (3.108) we have:

$$D = \begin{bmatrix} d \\ \delta \end{bmatrix} = \lim_{\Delta t \to 0} \dot{x}\Delta t$$

Substituting formula (3.121) into the above formula can be obtained:

$$D = \lim_{\Delta t \to 0} J(q)\dot{q}\Delta t$$

i.e.

$$D = J(q)dq \qquad (3.109)$$

For a manipulator with n joints, its Jacobian $J(q)$ is an $6 \times n$ order matrix. The first three rows represent the transfer ratio of the linear velocity v of the gripping hand, the last three rows represent the transfer ratio of the angular velocity v of the gripping hand, the last three rows represent the transfer ratio of the angular velocity ω of the gripping hand, and each column represents the corresponding joint velocity $\dot{q}_i$ for the gripping hand. In this way, the Jacobian can be divided into:

$$\begin{bmatrix} v \\ \omega \end{bmatrix} = \begin{bmatrix} J_{l1} & J_{l2} & \cdots & J_{\mathrm{l}n} \\ J_{a1} & J_{a2} & \cdots & J_{an} \end{bmatrix} \begin{bmatrix} \dot{q}_1 \\ \dot{q}_2 \\ \vdots \\ \dot{q}_n \end{bmatrix} \qquad (3.110)$$

Therefore, the linear velocity v and angular velocity ω of the gripping hand can be expressed as a linear function of the speed $\dot{q}$ of each joint:

$$\left. \begin{aligned} v &= J_{l1}\dot{q}_1 + J_{l2}\dot{q}_2 + \cdots + J_{\mathrm{l}n}\dot{q}_n \\ \omega &= J_{a1}\dot{q}_1 + J_{a2}\dot{q}_2 + \cdots + J_{an}\dot{q}_n \end{aligned} \right\} \qquad (3.111)$$

In the formula, J_{li} and J_{ai} respectively represent the linear velocity and angular velocity of the gripping of the hand caused by the unit joint velocity of the joint.

2. Solution of Jacobian matrix

The equations (3.88), (3.89), (3.95), (3.98), (3.108), and (3.109) discussed above are the basic formulas for calculating the Jacobian matrix, and these formulas can be used for calculation. Two methods of directly constructing the Jacobian matrix are introduced below.

(1) Vector product method

The vector product method for solving the Jacobian matrix of the manipulator is based on the concept of the motion coordinate system and was proposed by Whitney, J.C. Figure 3.19 shows the transmission of joint speed. The linear speed v and angular speed ω of the end grip are related to the joint speed.

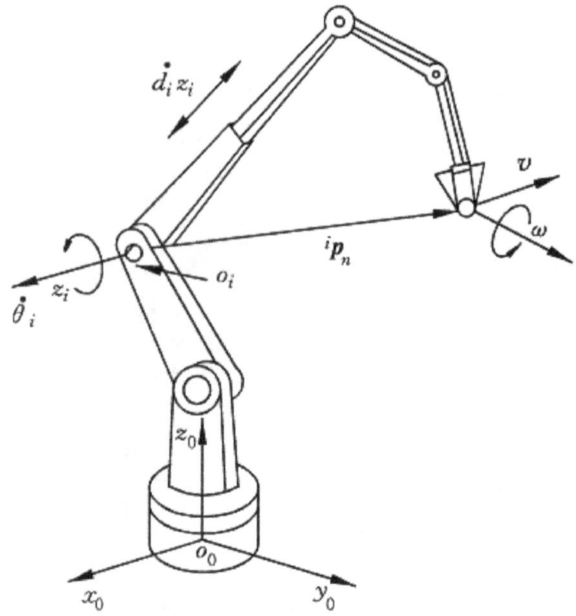

Figure 3.19.   Transmission of joint speed.

For translational joint i, there are

$$\begin{bmatrix} v \\ \omega \end{bmatrix} = \begin{bmatrix} z_i \\ 0 \end{bmatrix} \dot{q}_i, \quad J_i = \begin{bmatrix} z_i \\ 0 \end{bmatrix} \tag{3.112}$$

For revolute joint i, there are

$$\begin{bmatrix} v \\ \omega \end{bmatrix} = \begin{bmatrix} z_i \times {}^i p_n^0 \\ z_i \end{bmatrix} \dot{q}_i, \quad J_i = \begin{bmatrix} z_i \times {}^i p_n^0 \\ z_i \end{bmatrix} = \begin{bmatrix} z_i \times ({}^0_i R\,{}^i p_n) \\ z_i \end{bmatrix} \tag{3.113}$$

In the formula, ${}^i p_n^0$ is the representation of the position vector of the hand-clamping coordinate origin relative to the coordinate system $\{i\}$ in the base coordinate system $\{o\}$, namely

$$ {}^i p_n^0 = {}^0_i R\,{}^i p_n \tag{3.114} $$

Rather, z_i is the representation of the z-axis unit vector of the coordinate system i in the base coordinate system o.

(2) Differential transformation method

For the revolute joint i, the differential rotation of the link i relative to the link $(i-1)$ around the z_i axis of the coordinate system $\{i\}$, the differential motion vector is:

$$d = \begin{bmatrix} 0 \\ 0 \\ 0 \end{bmatrix}, \qquad \delta = \begin{bmatrix} 0 \\ 0 \\ 1 \end{bmatrix} d\theta_i \tag{3.115}$$

Using equation (3.95), the corresponding differential motion vector of the gripping hand is:

$$\begin{bmatrix} {}^T d_x \\ {}^T d_y \\ {}^T d_z \\ {}^T \delta_x \\ {}^T \delta_y \\ {}^T \delta_y \end{bmatrix} = \begin{bmatrix} (p \times n)_z \\ (p \times o)_z \\ (p \times a)_z \\ n_z \\ o_z \\ a_z \end{bmatrix} d\theta_i \tag{3.116}$$

For a translation joint, the clink i makes a differential movement $\mathrm{d}\ di$ relative to the link $(i-1)$ along the z_i axis, and the differential motion vector is:

$$d = \begin{bmatrix} 0 \\ 0 \\ 1 \end{bmatrix} \mathrm{d}d_i, \quad \delta = \begin{bmatrix} 0 \\ 0 \\ 0 \end{bmatrix} \tag{3.117}$$

And the differential motion vector of the gripped hand is:

$$\begin{bmatrix} {}^T d_x \\ {}^T d_y \\ {}^T d_z \\ {}^T \delta_x \\ {}^T \delta_y \\ {}^T \delta_y \end{bmatrix} = \begin{bmatrix} n_z \\ o_z \\ a_z \\ 0 \\ 0 \\ 0 \end{bmatrix} \mathrm{d}d_i \tag{3.118}$$

Therefore, the Jacobian matrix $J(q)$ of i-th column can be obtained as follows:

For revolute joint i:

$$^T J_{li} = \begin{bmatrix} (p \times n)_z \\ (p \times o)_z \\ (p \times a)_z \end{bmatrix}, \qquad ^T J_{ai} = \begin{bmatrix} n_z \\ o_z \\ a_z \end{bmatrix} \qquad (3.119)$$

For translational joint i:

$$^T J_{li} = \begin{bmatrix} n_z \\ o_z \\ a_z \end{bmatrix}, \qquad ^T J_{ai} = \begin{bmatrix} 0 \\ 0 \\ 0 \end{bmatrix} \qquad (3.120)$$

In the formula, n, o, a and p are the four column vectors of $^i_n T$.

The above method of seeking Jacobian $^T J(q)$ is constructive. As long as the transformation $^{i-1}_i T_i$ of each link is known, the Jacobian can be automatically generated without the need to solve equations and other procedures. The automatic generation steps are as follows:

(1) Calculate the transformation $^0 T_1, {}^1 T_2, \ldots, {}^{n-1} T_n$ of each link.
(2) Calculate the transformation from each link to the end link (see Figure 3.20):

$$^{n-1}_n T = {}^{n-1}_n T, \quad {}^{n-2}_n T = {}^{n-2}_{n-1} T^{n-1}_n T, \quad \ldots, \quad {}^{i-1}_n$$
$$T = {}^{i-1}_i T^i_n T, \quad \ldots, \quad {}^0_n T = {}^0_1 T^1_n T$$

(3) Calculate the elements of each column of $J(q)$, the $^T J_i$ of i-th column is determined by $^i_n T$. Calculate $^T J_{li}$, $^T J_{ai}$ according to formulas (3.117) and (3.118). The relationship between $^T J_i$ and $^i_n T$ is shown in Figure 3.20.

3.4.3   *Example of Jacobian matrix calculation for manipulator*

The following still takes the PUMA560 type manipulator as an example to illustrate the method of calculating the differential motion and Jacobian matrix of the specific manipulator.

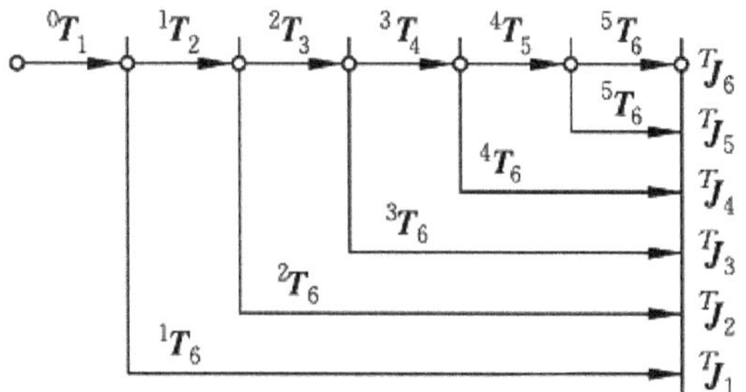

Figure 3.20. The relationship between $^T J_i$ and $^i_n T$.

The six joints of PUMA560 are all revolute joints, and its Jacobian matrix contains six columns. According to formula (3.119), the elements of each column can be calculated. Two methods are used for calculation.

1. Differential transformation method

The corresponding transformation matrix is $^1_6 T$ corresponded to first column $^T J_1(\boldsymbol{q})$ of $^T \boldsymbol{J}(\boldsymbol{q})$, and formula (3.57) lists the elements of $^1_6 T$, which can be obtained by formula (3.119)

$$^T J_1(\boldsymbol{q}) = \begin{bmatrix} ^T J_{1x} \\ ^T J_{1y} \\ ^T J_{1z} \\ -s_{23}(c_4 c_5 c_6 - s_4 s_6) - c_{23} s_5 c_6 \\ s_{23}(c_4 c_5 s_6 + s_4 c_6) + c_{23} s_5 s_6 \\ s_{23} c_4 s_5 - c_{23} c_5 \end{bmatrix} \tag{3.121}$$

Where

$$^T J_{1x} = -d_2 [c_{23}(c_4 c_5 c_6 - s_4 s_6) - s_{23} s_5 c_6]$$
$$- (a_2 c_2 + a_3 c_{23} - d_4 s_{23})(s_4 c_5 c_6 + c_4 s_6)$$
$$^T J_{1y} = -d_2 [-c_{23}(c_4 c_5 s_6 + s_4 c_6) + s_{23} s_5 s_6]$$
$$+ (a_2 c_2 + a_3 c_{23} - d_4 s_{23})(s_4 c_5 s_6 - c_4 c_6)$$
$$^T J_{1z} = d_2 (c_{23} c_4 s_5 + s_{23} c_5) + (a_2 c_2 + a_3 c_{23} - d_4 s_{23})(s_4 s_5)$$

In the same way, use the transformation matrix ^2_6T to get the second column of $^T\boldsymbol{J}(\boldsymbol{q})$:

$$^T J_2(\boldsymbol{q}) = \begin{bmatrix} ^T J_{2x} \\ ^T J_{2y} \\ ^T J_{2z} \\ -s_4 c_5 c_6 - c_4 s_6 \\ s_4 c_5 s_6 - c_4 c_6 \\ s_4 s_5 \end{bmatrix} \tag{3.122}$$

Where

$$^T J_{2x} = a_3 s_5 c_6 - d_4(c_4 c_5 c_6 - s_4 s_6)$$
$$+ a_2 \left[s_3(c_4 c_5 c_6 - s_4 s_6) + c_3 s_5 c_6 \right]$$
$$^T J_{2y} = -a_3 s_5 s_6 - d_4(-c_4 c_5 s_6 - s_4 c_6)$$
$$+ a_2 \left[s_3(-c_4 c_5 s_6 - s_4 c_6) + c_3 s_5 s_6 \right]$$
$$^T J_{2z} = a_3 c_6 + d_4 c_4 s_5 + a_2(-s_3 c_4 s_5 + c_3 c_6)$$

Also available,

$$^T J_3(\boldsymbol{q}) = \begin{bmatrix} -d_4(c_4 c_5 c_6 - s_4 s_6) + a_3(s_5 c_6) \\ d_4(c_4 c_5 s_6 + s_4 c_6) - a_3(s_5 s_6) \\ d_4 c_4 s_5 + a_3 c_6 \\ -s_4 c_5 c_6 - c_4 s_6 \\ s_4 c_5 s_6 - c_4 c_6 \\ s_4 s_5 \end{bmatrix} \tag{3.123}$$

$$^T J_4(\boldsymbol{q}) = \begin{bmatrix} 0 \\ 0 \\ 0 \\ s_5 c_6 \\ -s_5 s_6 \\ c_5 \end{bmatrix} \tag{3.124}$$

$$^T J_5(q) = \begin{bmatrix} 0 \\ 0 \\ 0 \\ -s_6 \\ -c_6 \\ 0 \end{bmatrix} \tag{3.125}$$

$$^T J_6(q) = \begin{bmatrix} 0 \\ 0 \\ 0 \\ 0 \\ 0 \\ 1 \end{bmatrix} \tag{3.126}$$

2. Vector product method

The six joints of PUMA 560 are all revolute joints, so its Jacobian matrix has the following form:

$$\boldsymbol{J}(\boldsymbol{q}) = \begin{bmatrix} z_1 \times {}^1p_6^0 & z_2 \times {}^2p_6^0 & \cdots & z_6 \times {}^6p_6^0 \\ z_1 & z_2 & \cdots & z_6 \end{bmatrix} \tag{3.127}$$

From Figure 3.20 and the listed link transformation matrices $^0_1T, ^1_2T, \ldots, ^5_6T$ (see section 3.3.1), you can calculate the intermediate terms, and then find the columns of $J(q)$, that is $J_1(q), J_2(q), \ldots, J_6(q)$, to obtain $J(q)$. The specific calculation process is omitted here.

3.5 Chapter Summary

The kinematics of the manipulator studied in this chapter involves the representation, solution and example of the manipulator motion equation, as well as the analysis and calculation of the Jacobian matrix of the manipulator. These contents are an important basis for the study of manipulator dynamics and control.

The first section studies the expression of the motion equation of the manipulator, and describes the positional relationship of the coordinate system where the end effector of the manipulator is relative to the base coordinate system through the homogeneous transformation matrix. On the basis of the coordinate system layout and the definition of link parameters in the previous chapter, the general form of coordinate system transformation between adjacent links is derived, and then these independent transformations are connected to obtain the position and gesture of link n relative to link 0. Use the transformation matrix to represent the movement direction of the manipulator, and use the rotation angle (i.e. Euler angle) transformation sequence to represent the movement posture, or use the roll, pitch and yaw angles to represent the movement posture of the manipulator. Once the motion posture of the manipulator is determined by a certain posture transformation matrix, its position in the base system can be determined by multiplying a translation transformation corresponding to the vector p to the left. This translation transformation can be represented by Cartesian coordinates, cylindrical coordinates or spherical coordinates. In order to further discuss the motion equation of the manipulator, the transformation matrix of the generalized link (including the revolute joint link and the prismatic joint link) is also given and analyzed, and the general link transformation matrix and the directed transformation graph of the manipulator are obtained.

The second section studies the solution of manipulator motion equations. First, it analyzes the solvability and versatility of inverse kinematics. Then, taking a planar three-link manipulator as an example, it introduces two main solving methods of inverse kinematics: analytical solution and the numerical solution method, in which the analytical solution method includes algebraic solution method, geometric solution method, and obtains the solution formula of each joint position; the numerical solution method involves the degradation method under cyclic coordinates, the forward and backward reach solution method and the genetic algorithm optimization solution method to obtain the position of each joint.

The third section gives an example of the representation (analysis) and solution (synthesis) of the motion equation of the manipulator. According to the equations obtained in the first and second sections, combined with the actual link parameters of the PUMA560 manipulator, the transformation matrix of each link and the transformation matrix of the manipulator are obtained, that is, the motion equation of the manipulator. When solving the motion equation of the manipulator, according to the pose of the end effector of the manipulator and the link parameters of the manipulator, all the joint variables are obtained one by one, and the solution of the motion equation of the PUMA560 is completed.

The fourth section studies the small changes in the position and posture of the manipulator. First, the differential motion of the manipulator (including differential translational motion and differential rotational motion) is discussed, and the differential motion vector of a rigid body (or coordinate system) is obtained. Then the problem of equivalent transformation of the differential motion of the manipulator is discussed, which lays the foundation for the derivation of the Jacobian matrix of the manipulator. In addition, the differential relationship in the equivalent transformation formula is also analyzed. This section also cites three examples, which will help the understanding of differential motion and its equivalent transformation. On the basis of the above analysis and research, the problem of linear mapping between manipulator operation space velocity and joint space velocity, namely the Jacobian matrix problem, is studied. This part of the research involves the definition and calculation of the Jacobian matrix, and takes the PUMA560 robot as an example to illustrate the differential motion of the specific manipulator and the method for deriving the Jacobian matrix.

References

[1] Ahmad, M., Kumar, N., and Kumari, R. (2019). A hybrid genetic algorithm approach to solve inverse kinematics of a mechanical manipulator. *International Journal of Scientific and Technology Research*, 8(9):1777–1782.

[2] Angeles, J. (2003). *Fundamentals of Robotic Mechanical Systems: Theory, Methods, and Algorithms*, 2nd Edition. New York: Springer.

[3] Aristidou, A. and Lasenby, J. (2011). Fabrik: A fast, iterative solver for the Inverse Kinematics problem. *Graphical Models*, 73(5):243–260.

[4] Asada, H. and Slotine, J.J.E. (1983). *Robot Motion: Planning and Control*. MIT Press.

[5] Cai, Z.X. (1986). Computer simulation of Stanford Robotic Manipulator. The Second National Conference on Computer Simulation (in Chinese).

[6] Cai, Z.X. (2021). *Fundamentals of Robotics*, Chapter 3, 3rd edition. Beijing: China Machine Press (in Chinese).

[7] Cai, Z.X. (2021). *Robotics*, Chapter 3, 4th Edition. Beijing: Tsinghua University Press (in Chinese).

[8] Cai, Z.X. (1988). *Robotics: Principles and Applications*, Chapter 3. Changsha: Central South University of Technology Press (in Chinese).

[9] Chapelle, F. and Bidaud, P. (2001). A closed form for inverse kinematics approximation of general 6R manipulators using genetic programming. *Proceedings of the IEEE International Conference on Robotics and Automation*, 3364–3369.

[10] Craig, J.J. (2018). *Introduction to Robotics: Mechanics and Control*, 4th Edition. Pearson Education, Inc.

[11] Dereli, S. and Koker, R. (2019). A Meta-heuristic proposal for inverse kinematics solution of 7-DOF serial robotic manipulator: Quantum behaved particle swarm algorithm. *Artificial Intelligence Review*, 53(2):949–964.

[12] Dereli, S. (2018). IW-PSO approach to the inverse kinematics problem solution of a 7-DOF serial robot manipulator. *International Journal of Natural and Engineering Sciences*, 36(1):77–85.

[13] Lansley, A., Vamplew, P., Smith, P., and Foale, C. (2016). Caliko: An inverse kinematics software library implementation of the FABRIK algorithm. *Journal of Open Research Software*, 4:e36.

[14] Momani, S., Abo-Hammour, Z.S., and Alsmadi, O.M. (2015). Solution of inverse kinematics problem using genetic algorithms. *Applied Mathematics and Information Sciences*, 10(1):1–9.

[15] Paul, R.P. (1981). *Robot Manipulators: Mathematics, Programming and Control*. MIT Press.

[16] Reza, N.J. (2010). *Theory of Applied Robotics: Kinematics, Dynamics, and Control*, 2nd Edition. Springer.

[17] Rokbani, N., Casals, A., and Alimi, A.M. (2015). *IK-FA, A New Heuristic Inverse Kinematics Solver Using Firefly Algorithm*. Cham: Springer International Publishing, pp. 369–395.

[18] Serkan, D. and Rait, K. (2019). Calculation of the inverse kinematics solution of the 7-DOF redundant robot manipulator by the firefly algorithm and statistical analysis of the results in terms of speed and accuracy. *Inverse Problems in Science and Engineering*, DOI:10.1080/17415977.2019.1602124.

[19] Starke, S., Hendrich, N., and Zhang, J. (2019). Memetic evolution for generic full-body inverse kinematics in robotics and animation. *IEEE Transactions on Evolutionary Computation*, 23(3):406–420.

[20] Stevenson, C.N. (1883). The Kinematic Analysis of Industrial Robots. TR-EE 83-50, Nov. 1883, School of EE, Purdue University.

[21] Wang, L.C.T. and Chen, C.C. (1991). A combined optimization method for solving the inverse kinematics problems of mechanical manipulators. *IEEE Transactions on Robotics & Automation*, 7(4):489–499.

Chapter 4

Manipulator Dynamics

The operating robot (manipulator) is an active mechanical device, in principle, each degree of freedom can have a separate driving. From a control point of view, the manipulator system represents a redundant, multi-variable and intrinsically non-linear automatic control system, and it is also a complex dynamic coupling system [4, 22]. Each control task itself is a dynamic task. Therefore, studying the dynamics of the robotic manipulator is to further discuss the control problem [1, 2, 7, 23].

To analyze the dynamic mathematical model of manipulator operation, the following two theories are mainly used [12, 15, 21]:

(1) Basic theory of dynamics, including Newton-Euler equation.
(2) Lagrange mechanics, especially the second-order Lagrange equation.

In addition, there are applications of Gaussian principle, Appel equation, spinor duality method and Kane method to analyze dynamic problems [22].

The first method is the dynamic balance method of force. When using this method, it is necessary to obtain the acceleration from kinematics and eliminate the internal forces. For more complex systems, this analysis method is very complicated and troublesome. Therefore, this chapter only discusses some relatively simple examples. The second method is the Lagrangian functional balance method, which only requires speed and not internal force. Therefore, this is a straightforward and convenient method. In this book, we mainly use this method to analyze and solve the dynamics of the manipulator. Of particular interest is finding symbolic solutions to dynamics problems, because it helps in-depth understanding of manipulator control problems.

There are two opposite problems in studying dynamics. One is to know the force or moment of each joint of the manipulator, find the displacement, velocity and acceleration of each joint, and find the motion trajectory. The second is to know the motion trajectory of the manipulator, that is, the displacement, velocity and acceleration of each joint, and find the driving force or torque required by each joint. The former is called the forward problem of dynamics, and the latter is called the inverse problem of dynamics. The dynamic equation of a general manipulator is represented by six nonlinear differential simultaneous equations. In fact, except for some relatively simple cases, it is impossible to obtain general solutions to these equations. The dynamic equations will be obtained in matrix form and simplified to obtain the information needed for control. In actual control, it is often necessary to make certain assumptions about the dynamic equations and simplify processing.

4.1　Dynamic Equations of Rigid Bodies [8, 15, 19]

The Lagrangian function L is defined as the difference between the kinetic energy K and the potential energy P of the system, namely

$$L = K - P \tag{4.1}$$

Among them, K and P can be expressed in any convenient coordinate system.

The system dynamics equation, namely the Lagrangian equation is as follows:

$$\boldsymbol{F}_i = \frac{d}{dt}\frac{\partial L}{\partial \dot{q}_i} - \frac{\partial L}{\partial q_i}, \quad i = 1, 2, \cdots n \tag{4.2}$$

Where q_i is the coordinate representing the kinetic energy and potential energy, $\dot{q}_i$ is the corresponding speed, and $\boldsymbol{F}_i$ is the force or moment acting on the i-th coordinate. Whether $\boldsymbol{F}_i$ is a force or a moment is determined by $\boldsymbol{q}_i$ that is the linear coordinate or the angular coordinate. These forces, moments and coordinates are called

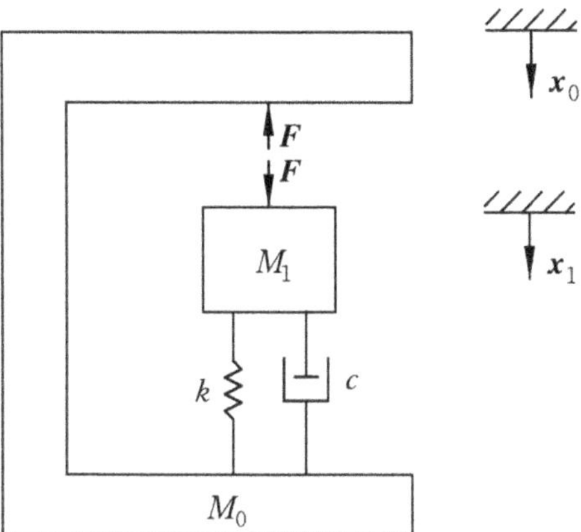

Figure 4.1. Kinetic energy and potential energy of general objects.

generalized forces, generalized moments and generalized coordinates, and n is the number of links.

4.1.1 *Kinetic energy and potential energy of a rigid body* [11, 20]

In the theoretical mechanics or physical mechanics, the kinetic energy and potential energy of a general object in translation as shown in Figure 4.1 have been calculated, as follows:

$$K = \frac{1}{2}M_1\dot{x}_1^2 + \frac{1}{2}M_0\dot{x}_0^2$$

$$P = \frac{1}{2}k(x_1 - x_0)^2 - M_1gx_1 - M_0gx_0$$

$$D = \frac{1}{2}c(\dot{x}_1 - \dot{x}_0)^2$$

$$W = \boldsymbol{F}\boldsymbol{x}_1 - \boldsymbol{F}\boldsymbol{x}_0$$

In the formula, K, P, D and W respectively represent the kinetic energy and potential energy of the object, the energy consumed and the work done by the external force; M_0 and M_1 is the mass of the bracket and the moving object; $\boldsymbol{x}_0$ and $\boldsymbol{x}_1$ is the motion coordinate: $\boldsymbol{g}$ is the acceleration of gravity: k is the spring Hooke coefficient: c is the friction coefficient: $\boldsymbol{F}$ is the applied force.

For this problem, there are two situations:

(1) $x_0 = 0$, x_1 is the generalized coordinate

$$\frac{d}{dt}\left(\frac{\partial K}{\partial \dot{x}_1}\right) - \frac{\partial K}{\partial x_1} + \frac{\partial D}{\partial \dot{x}_1} + \frac{\partial P}{\partial x_1} = \frac{\partial W}{\partial x_1}$$

Among them, the first term of the left-hand formula is the change of kinetic energy with speed (or angular velocity) and time; the second term is the change of kinetic energy with position (or angle); the third term is the change of energy consumption with speed; the fourth term is potential energy change with location. The right type is the actual applied force or moment. Substituting the corresponding expressions, and simplifying, we can get:

$$\frac{d}{dt}(M_1\dot{x}_1) - 0 + c_1\dot{x}_1 + kx_1 - M_1g = F$$

Expressed as a general form:

$$M_1\ddot{x}_1 + c_1\dot{x}_1 + dx_1 = F + M_1g$$

That is the kinetic equation of when $x_0 = 0$. Among them, the three terms on the left represent the acceleration, resistance and elastic force of the object, and the two terms on the right represent the applied force and gravity.

(2) $x_0 = 0$, x_0 and x_1 are generalized coordinates, at this time the following formula:

$$M_1\ddot{x}_1 + c(\dot{x}_1 - \dot{x}_0) + k(x_1 - x_0) - M_1g = F$$
$$M_0\ddot{x}_0 + c(\dot{x}_1 - \dot{x}_0) - k(x_1 - x_0) - M_0g = -F$$

Or express them as a matrix form:

$$\begin{bmatrix} M_1 & 0 \\ 0 & M_0 \end{bmatrix}\begin{bmatrix} \ddot{x}_1 \\ \ddot{x}_0 \end{bmatrix} + \begin{bmatrix} c & -c \\ -c & c \end{bmatrix}\begin{bmatrix} \dot{x}_1 \\ \dot{x}_0 \end{bmatrix} + \begin{bmatrix} k & -k \\ -k & k \end{bmatrix}\begin{bmatrix} x_1 \\ x_0 \end{bmatrix} = \begin{bmatrix} F \\ -F \end{bmatrix}$$

Let's consider the kinetic energy and potential energy of the two-link manipulator (see Figure 4.2). This kind of movement mechanism has an open kinematic chain, which has many similarities with compound pendulum movement. In the figure, m_1 and m_2 is the mass of link 1 and link 2, and expressed by the point

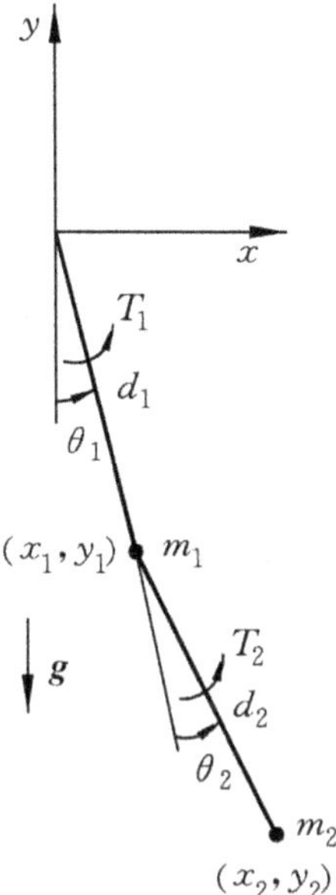

Figure 4.2.   Two-link manipulator (1).

mass of the end of the link; d_1 and d_2 are the length of the two links respectively, θ_1 and θ_2 are generalized coordinates; g is the acceleration of gravity.

First calculate the kinetic energy K_1 and potential energy P_1 of link 1. Because

$K_1 = \dfrac{1}{2}m_1 v_1^2, v_1 = d_1\dot{\theta}_1, P_1 = m_1 g h_1, h_1 = -d_1\cos\theta_1,$ so have:

$$K_1 = \frac{1}{2}m_1 d_1^2 \dot{\theta}_1^2$$

$$P_1 = -m_1 g d_1 \cos\theta_1$$

Then find the kinetic energy K_2 and potential energy P_2 of link 2.

$$K_2 = \frac{1}{2}m_2 v_2^2, \quad P_2 = mgy_2$$

Where

$$v_2^2 = \dot{x}_2^2 + \dot{y}_2^2$$
$$x_2 = d_1\sin\theta_1 + d_2\sin(\theta_1 + \theta_2)$$
$$y_2 = -d_1\cos\theta_1 - d_2\cos(\theta_1 + \theta_2)$$

$$\dot{x}_2 = d_1 \cos\theta_1 \dot{\theta}_1 + d_2 \cos(\theta_1 + \theta_2)(\dot{\theta}_1 + \dot{\theta}_2)$$

$$\dot{y}_2 = d_1 \sin\theta_1 \dot{\theta}_1 + d_2 \sin(\theta_1 + \theta_2)(\dot{\theta}_1 + \dot{\theta}_2)$$

So it can be obtained:

$$v_2^2 = d_1^2 \dot{\theta}_1^2 + d_2^2(\dot{\theta}_1^2 + 2\dot{\theta}_1\dot{\theta}_2 + \dot{\theta}_2^2) + 2d_1 d_2 \cos\theta_2(\dot{\theta}_1^2 + \dot{\theta}_1\dot{\theta}_2)$$

as well as

$$K_2 = \frac{1}{2}m_2 d_1^2 \dot{\theta}_1^2 + \frac{1}{2}m_2 d_2^2(\dot{\theta}_1 + \dot{\theta}_2)^2 + m_2 d_1 d_2 \cos\theta_2(\dot{\theta}_1^2 + \dot{\theta}_1\dot{\theta}_2)$$

$$P_2 = -m_2 g d_1 \cos\theta_1 - m_2 g d_2 \cos(\theta_1 + \theta_2)$$

In this way, the total kinetic energy and total position energy of the two-link manipulator system are:

$$K = K_1 + K_2$$
$$= \frac{1}{2}(m_1 + m_2)d_1^2 \dot{\theta}_1^2 + \frac{1}{2}m_2 d_2^2(\dot{\theta}_1 + \dot{\theta}_2)^2$$
$$+ m_2 d_1 d_2 \cos\theta_2(\dot{\theta}_1^2 + \dot{\theta}_1\dot{\theta}_2) \tag{4.3}$$

$$P = P_1 + P_2$$
$$= -(m_1 + m_2)g d_1 \cos\theta_1 - m_2 g d_2 \cos(\theta_1 + \theta_2) \tag{4.4}$$

4.1.2 *Lagrange equation and Newton-Euler equation* [6, 16, 17, 19]

1. Lagrangian functional balance method

The Lagrangian function L of the two-link manipulator system can be obtained according to equations (4.1), (4.3) and (4.4):

$$L = K - P$$
$$= \frac{1}{2}(m_1 + m_2)d_1^2 \dot{\theta}_1^2 + \frac{1}{2}m_2 d_2^2(\dot{\theta}_1^2 + 2\dot{\theta}_1\dot{\theta}_2 + \dot{\theta}_2^2)$$
$$+ m_2 d_1 d_2 \cos\theta_2(\dot{\theta}_1^2 + \dot{\theta}_1\dot{\theta}_2) + (m_1 + m_2)g d_1 \cos\theta_1$$
$$+ m_2 g d_2 \cos(\theta_1 + \theta_2) \tag{4.5}$$

Find the partial derivative and derivative of L:

$$\frac{\partial L}{\partial \theta_1} = -(m_1 + m_2)gd_1 \sin \theta_1 - m_2 g d_2 \sin(\theta_1 + \theta_2)$$

$$\frac{\partial L}{\partial \theta_2} = -m_2 d_1 d_2 \sin \theta_2 (\dot{\theta}_1^2 + \dot{\theta}_1 \dot{\theta}_2) - m_2 g d_2 \sin(\theta_1 + \theta_2)$$

$$\frac{\partial L}{\partial \dot{\theta}_1} = (m_1 + m_2)\, d_1^2 \dot{\theta}_1 + m_2 d_2^2 \dot{\theta}_1 + m_2 d_2^2 \dot{\theta}_2 + 2m_2 d_1 d_2 \cos \theta_2 \dot{\theta}_1$$

$$+ m_2 d_1 d_2 \cos \theta_2 \dot{\theta}_2$$

$$\frac{\partial L}{\partial \dot{\theta}_2} = m_2 d_2^2 \dot{\theta}_1 + m_2 d_2^2 \dot{\theta}_2 + m_2 d_1 d_2 \cos \theta_2 \dot{\theta}_1$$

As well as

$$\frac{d}{dt}\frac{\partial L}{\partial \dot{\theta}_1} = [(m_1 + m_2)d_1^2 + m_2 d_2^2 + 2m_2 d_1 d_2 \cos \theta_2]\ddot{\theta}_1$$

$$+ (m_2 d_2^2 + m_2 d_1 d_2 \cos \theta_2)\ddot{\theta}_2 - 2m_2 d_1 d_2 \sin \theta_2 \dot{\theta}_1 \dot{\theta}_2$$

$$- m_2 d_1 d_2 \sin \theta_2 \dot{\theta}_2^2$$

$$\frac{d}{dt}\frac{\partial L}{\partial \dot{\theta}_2} = m_2 d_2^2 \ddot{\theta}_1 + m_2 d_2^2 \ddot{\theta}_2 + m_2 d_1 d_2 \cos \theta_2 \ddot{\theta}_1 - m_2 d_1 d_2 \sin \theta_2 \dot{\theta}_1 \dot{\theta}_2$$

Substituting the corresponding derivatives and partial derivatives into equation (4.2), the dynamic equation of torque T_1 and T_2 can be obtained:

$$T_1 = \frac{d}{dt}\frac{\partial L}{\partial \dot{\theta}_1} - \frac{\partial L}{\partial \theta_1}$$

$$= [(m_1 + m_2)d_1^2 + m_2 d_2^2 + 2m_2 d_1 d_2 \cos \theta_2]\ddot{\theta}_1$$

$$+ (m_2 d_2^2 + m_2 d_1 d_2 \cos \theta_2)\ddot{\theta}_2$$

$$- 2m_2 d_1 d_2 \sin \theta_2 \dot{\theta}_1 \dot{\theta}_2 - m_2 d_1 d_2 \sin \theta_2 \dot{\theta}_2^2$$

$$+ (m_1 + m_2)gd_1 \sin \theta_1 + m_2 g d_2 \sin(\theta_1 + \theta_2) \qquad (4.6)$$

$$T_2 = \frac{d}{dt}\frac{\partial L}{\partial \dot{\theta}_2} - \frac{\partial L}{\partial \theta_2}$$

$$= (m_2 d_2^2 + m_2 d_1 d_2 \cos \theta_2)\ddot{\theta}_1 + m_2 d_2^2 \ddot{\theta}_2$$

$$+ m_2 d_1 d_2 \sin \theta_2 \dot{\theta}_1^2 + m_2 g d_2 \sin(\theta_1 + \theta_2) \qquad (4.7)$$

The general form and matrix form of equations (4.6) and (4.7) are as follows:

$$T_1 = D_{11}\ddot{\theta}_1 + D_{12}\ddot{\theta}_2 + D_{111}\dot{\theta}_1^2 + D_{122}\dot{\theta}_2^2$$
$$+ D_{112}\dot{\theta}_1\dot{\theta}_2 + D_{121}\dot{\theta}_2\dot{\theta}_1 + D_1 \tag{4.8}$$

$$T_2 = D_{21}\ddot{\theta}_1 + D_{22}\ddot{\theta}_2 + D_{211}\dot{\theta}_1^2 + D_{222}\dot{\theta}_2^2$$
$$+ D_{212}\dot{\theta}_1\dot{\theta}_2 + D_{221}\dot{\theta}_2\dot{\theta}_1 + D_2 \tag{4.9}$$

$$\begin{bmatrix} T_1 \\ T_2 \end{bmatrix} = \begin{bmatrix} D_{11} & D_{12} \\ D_{21} & D_{22} \end{bmatrix} \begin{bmatrix} \ddot{\theta}_1 \\ \ddot{\theta}_2 \end{bmatrix} + \begin{bmatrix} D_{111} & D_{122} \\ D_{211} & D_{222} \end{bmatrix} \begin{bmatrix} \dot{\theta}_1^2 \\ \dot{\theta}_2^2 \end{bmatrix}$$

$$+ \begin{bmatrix} D_{112} & D_{121} \\ D_{212} & D_{221} \end{bmatrix} \begin{bmatrix} \dot{\theta}_1\dot{\theta}_2 \\ \dot{\theta}_2\dot{\theta}_1 \end{bmatrix} + \begin{bmatrix} D_1 \\ D_2 \end{bmatrix} \tag{4.10}$$

Where D_{ii} is called the effective inertia of joint i, because the acceleration of joint i will produce an inertial force $D_{ii}\ddot{\theta}_i$ on joint i; D_{ij} is called coupling inertia between joints i and j, because the accelerations $\ddot{\theta}_i$ and $\ddot{\theta}_j$ of joint i and j will produce an inertial force equal $D_{ij}\ddot{\theta}_i$ or $D_{ij}\ddot{\theta}_j$ at joint j or i respectively; $D_{ijk}\dot{\theta}_1^2$ is the centripetal force generated on joint i by the speed $\dot{\theta}_j$ of joint j; $D_{ijk}\dot{\theta}_j\dot{\theta}_k + D_{ikj}\dot{\theta}_k\dot{\theta}_j$ is the Coriolis force acting on joint i caused by the speed $\dot{\theta}_j$ and $\dot{\theta}_k$ of joint j and k; D_i represents the gravity at joint i.

Comparing equations (4.6), (4.7) with (4.8), (4.9), the coefficients of this system can be obtained as follows:

Effective inertia

$$D_{11} = (m_1 + m_2)d_1^2 + m_2d_2^2 + 2m_2d_1d_2\cos\theta_2$$
$$D_{22} = m_2d_2^2$$

Coupled inertia

$$D_{12} = m_2d_2^2 + m_2d_1d_2\cos\theta_2 = m_2(d_2^2 + d_1d_2\cos\theta_2)$$

Centripetal acceleration coefficient

$$D_{111} = 0$$
$$D_{122} = -m_2d_1d_2\sin\theta_2$$
$$D_{211} = m_2d_1d_2\sin\theta_2$$
$$D_{222} = 0$$

Coriolis acceleration coefficient

$$D_{112} = D_{121} = -m_2 d_1 d_2 \sin\theta_2$$

$$D_{212} = D_{221} = 0$$

Gravity term

$$D_1 = (m_1 + m_2)gd_1 \sin\theta_1 + m_2 g d_2 \sin(\theta_1 + \theta_2)$$

$$D_2 = m_2 g d_2 \sin(\theta_1 + \theta_2)$$

The following assigns some numbers to the above example to estimate the values T_1 and T_2 of this two-link manipulator under static and fixed gravity load. The calculation conditions are as follows:

(1) Joint 2 is locked, that is, maintain a constant speed $(\ddot{\theta}_2) = 0$, that is, $\dot{\theta}_2$ is a constant value;
(2) Joint 2 is unconstrained, namely $T_2 = 0$.
   Under the first condition, equations (4.8) and (4.9) are simplified to

$$T_1 = D_{11}\ddot{\theta}_1 = I_1\ddot{\theta}_1, T_2 = D_{12}\ddot{\theta}_1$$

Under the second condition, solution:

$$T_2 = D_{12}\ddot{\theta}_1 + D_{22}\ddot{\theta}_2 = 0, T_1 = D_{11}\ddot{\theta}_1 + D_{12}\ddot{\theta}_2$$

Solve it and get

$$\ddot{\theta}_2 = -\frac{D_{12}}{D_{22}}\ddot{\theta}_1$$

$$T_1 = \left(D_{11} - \frac{D_{12}^2}{D_{22}}\right)\ddot{\theta}_1 = I_i\ddot{\theta}_1$$

Take $d_1 = d_2 = 1$, $m_1 = 2$, calculate $m_2 = 1.4$ and 100 (respectively represent the three different situations of the manipulator under no load, full ground load and load in outer space; for the latter, a large load is allowed due to weightlessness) of each coefficient value under three different values. Table 4.1 shows these coefficient values and their relationship with location θ_2. Among them, $m_1 = m_2 = 1$, for no load, $m_1 = 2, m_2 = 4$ for full ground load, and $m_1 = 2, m_2 = 100$ for external space load.

Table 4.1.　Coefficient values of two-link manipulators under different loads.

Load	θ_2	$\cos\theta_2$	D_{11}	D_{12}	D_{22}	I_1	I_f
No load	0°	1	6	2	1	6	2
	90°	0	4	1	1	4	3
	180°	−1	2	0	1	2	2
	270°	0	4	1	1	4	3
Full ground load	0°	1	18	8	4	18	2
	90°	0	10	4	4	10	6
	180°	−1	2	0	4	2	2
	270°	0	10	4	4	10	6
Load in outer space	0°	1	402	200	100	402	2
	90°	0	202	100	100	202	102
	180°	−1	2	0	100	2	2
	270°	0	202	100	100	202	102

The two rightmost columns in Table 4.1 are the effective moments of inertia on joint 1. Under no-load, when θ_2 changes, the effective inertia value of joint 1 changes within the range of 3:1 (when joint 2 is locked) or 3:2 (when joint 2 is free). It can also be seen from Table 4.1 that under full ground load, the effective inertia of joint 1 varies with θ_2 within the range of 9:1, and the effective inertia value is increased to 3 times in compared with the no-load condition. In the case of an outer space load of 100, the effective inertia has a larger variation range, up to 201:1. These changes in inertia will have a significant impact on the control of the manipulator.

2. Newton-Euler dynamic balance method

In order to compare with the Lagrangian method and see which method is relatively simple, use the Newton-Euler dynamic balance method to find the dynamic equation of the same two-link system mentioned above. Its general form is:

$$\frac{\partial W}{\partial q_i} = \frac{d}{dt}\frac{\partial K}{\partial \dot{q}_i} - \frac{\partial K}{\partial q_i} + \frac{\partial D}{\partial \dot{q}_i} + \frac{\partial P}{\partial q_i}, \quad i = 1, 2, \cdots n \qquad (4.11)$$

The meanings of W, K, D, P, and q_i in the formula are the same as the Lagrangian method; i is the code of the link, and n is the number of links.

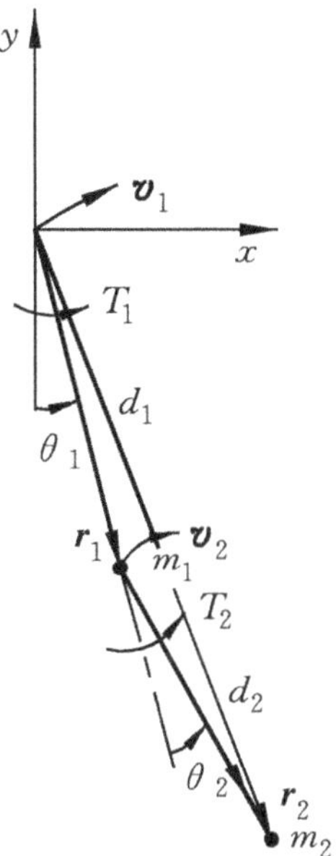

Figure 4.3.   Two-link manipulator (2).

The position vector r_1 and r_2 of mass m_1 and m_2 (see Figure 4.3) are:

$$r_1 = r_0 + (d_1 \cos \theta_1)i + (d_1 \sin \theta_1)j$$
$$= (d_1 \cos \theta_1)i + (d_1 \sin \theta_1)j$$
$$r_2 = r_1 + [d_2 \cos(\theta_1 + \theta_2)]i + [d_2 \sin(\theta_1 + \theta_2)]j$$
$$= [d_1 \cos \theta_1 + d_2 \cos(\theta_1 + \theta_2)]i + [d_1 \sin \theta_1 + d_2 \sin(\theta_1 + \theta_2)]j$$

Speed vector v_1 and v_2:

$$v_1 = \frac{dr_1}{dt} = [-\dot{\theta}_1 d_1 \sin \theta_1]i + [\dot{\theta}_1 d_1 \cos \theta_1]j$$

$$v_2 = \frac{dr_2}{dt} = [-\dot{\theta}_1 d_1 \sin \theta_1 - (\dot{\theta}_1 + \dot{\theta}_2)d_2 \sin(\theta_1 + \theta_2)]i$$
$$= [\dot{\theta}_1 d_1 \cos \theta_1 - (\dot{\theta}_1 + \dot{\theta}_2)d_2 \cos(\theta_1 + \theta_2)]j$$

Then find the square of the speed, the calculation result is:

$$v_1^2 = d_1^2 \dot{\theta}_1^2$$
$$v_2^2 = d_1^2 \dot{\theta}_1^2 + d_2^2(\dot{\theta}_1^2 + 2\dot{\theta}_1\dot{\theta}_2 + \dot{\theta}_2^2) + 2d_1 d_2(\dot{\theta}_1^2 + \dot{\theta}_1\dot{\theta}_2) \cos \theta_2$$

Then the system kinetic energy can be obtained:

$$
K = \frac{1}{2}m_1 v_1^2 + \frac{1}{2}m_2 v_2^2
$$

$$
= \frac{1}{2}(m_1 + m_2)d_1^2\dot{\theta}_1^2 + \frac{1}{2}m_2 d_2^2(\dot{\theta}_1^2 + 2\dot{\theta}_1\dot{\theta} + \dot{\theta}_2^2)
$$

$$
+ m_2 d_1 d_2(\dot{\theta}_1^2 + \dot{\theta}_1\dot{\theta}_2)\cos\theta_2
$$

The potential energy of the system decreases as r increases (position decreases). Calculate with the coordinate origin as the reference point:

$$
P = -m_1 g r_1 - m_2 g r_2
$$

$$
= -(m_1 + m_2)g d_1 \cos\theta_1 - m_2 g d_2 \cos(\theta_1 + \theta_2)
$$

System energy consumption

$$
D = \frac{1}{2}C_1\dot{\theta}_1^2 + \frac{1}{2}C_2\dot{\theta}_2^2
$$

Work done by external torque

$$
D = \frac{1}{2}C_1\dot{\theta}_1^2 + \frac{1}{2}C_2\dot{\theta}_2^2
$$

So far, four scalar equations about K, P, D, and W have been obtained. With these four equations, the dynamic equations of the system can be calculated according to equation (4.11). To this end, first find the relevant derivative and partial derivative.

As $q_i = \theta_1$,

$$
\frac{\partial K}{\partial \dot{\theta}_1} = (m_1 + m_2)d_1^2\dot{\theta}_1 + m_2 d_2^2(\dot{\theta}_1 + \dot{\theta}_2) + m_2 d_1 d_2(2\dot{\theta}_1 + \dot{\theta}_2)\cos\theta_2
$$

$$
\frac{d}{dt}\frac{\partial K}{\partial \dot{\theta}_1} = (m_1 + m_2)d_1^2\ddot{\theta}_1 + m_2 d_2^2(\ddot{\theta}_1 + \ddot{\theta}_2) + m_2 d_1 d_2(2\ddot{\theta}_1 + \ddot{\theta}_2)\cos\theta_2
$$

$$
- m_2 d_1 d_2(2\dot{\theta}_1 + \dot{\theta}_2)\dot{\theta}_2\sin\theta_2
$$

$$\frac{\partial K}{\partial \theta_1} = 0$$

$$\frac{\partial D}{\partial \dot{\theta}_1} = C_1 \dot{\theta}_1$$

$$\frac{\partial P}{\partial \theta_1}(m_1 + m_2)gd_1 \sin \theta_1 + m_2 d_2 g \sin(\theta_1 + \theta_2)$$

$$\frac{\partial W}{\partial \theta_1} = T_1$$

Substituting the obtained derivatives of the above list into equation (4.11), after merging and sorting, we can get:

$$
\begin{aligned}
T_1 = {} & [(m_1 + m_2)d_1^2 + m_2 d_2^2 + 2m_2 d_1 d_2 \cos \theta_2]\ddot{\theta}_1 \\
& + [m_2 d_2^2 + m_2 d_1 d_2 \cos \theta_2]\ddot{\theta}_2 + c_1 \dot{\theta}_1 - (2m_2 d_1 d_2 \sin \theta_2)\dot{\theta}_1 \dot{\theta}_2 \\
& - (m_2 d_1 d_2 \sin \theta_2)\dot{\theta}_2^2 + [(m_1 + m_2)gd_1 \sin \theta_1 \\
& + m_2 d_2 g \sin(\theta_1 + \theta_2)]
\end{aligned}
\tag{4.12}
$$

As $q_i = \theta_2$,

$$\frac{\partial K}{\partial \dot{\theta}_2} = m_2 d_2^2(\dot{\theta}_1 + \dot{\theta}_2) + m_2 d_1 d_2 \dot{\theta}_1 \cos_2$$

$$\frac{d}{dt}\frac{\partial K}{\partial \dot{\theta}_2} = m_2 d_2^2(\ddot{\theta}_1 + \ddot{\theta}_2) + m_2 d_1 d_2 \ddot{\theta}_1 \cos \theta_2 - m_2 d_1 d_2 \dot{\theta}_1 \ddot{\theta}_2 \sin \theta_2$$

$$\frac{\partial K}{\partial \dot{\theta}_2} = -m_2 d_2^2(\dot{\theta}_1^2 + \dot{\theta}_1 \dot{\theta}_2) \sin \theta_2$$

$$\frac{\partial D}{\partial \dot{\theta}_2} = C_2 \dot{\theta}_2$$

$$\frac{\partial P}{\partial \dot{\theta}_2} = m_2 g d_2 \sin(\theta_1 + \theta_2)$$

$$\frac{\partial W}{\partial \theta_2} = T_2$$

Substituting the above formulas into (4.11), and simplifying it to get:

$$
\begin{aligned}
T_2 = {} & (m_2 d_2^2 + m_2 d_1 d_2 \cos \theta_2)\ddot{\theta}_1 + m_2 d_2^2 \ddot{\theta}_2 + m_2 d_1 d_2 \sin \theta_2 \dot{\theta}_1^2 \\
& + c_2 \dot{\theta}_2 + m_2 g d_2 \sin(\theta_1 + \theta_2)
\end{aligned}
\tag{4.13}
$$

Equations (4.12) and (4.13) can also be written in general forms like equations (4.8) and (4.9).

Comparing equations (4.6), equations (4.7) with equations (4.12), and equations (4.13), it can be seen that if friction loss is not considered (take $c_1 = c_2 = 0$), then equations (4.6) and (4.12) are completely consistent, and equation (4.7) is exactly the same as equation (4.13). In equations (4.6) and (4.7), the energy consumed by friction is not considered, while equations (4.12) and (4.13) consider this loss. Therefore, this difference appears between the two results sought.

4.2   Calculation and Simplification of Manipulator Dynamic Equations [13, 14, 18, 19]

Based on the analysis of the simple two-link manipulator system, we then analyze any manipulator described by a set of A-transforms and find its dynamic equations. The derivation process is divided into five steps:

(1) Calculate the speed of any point on any link;
(2) Calculate the kinetic energy of each link and the total kinetic energy of the manipulator;
(3) Calculate the potential energy of each link and the total potential energy of the manipulator;
(4) Establish the Lagrangian function of the manipulator system;
(5) Derivation of the Lagrangian function to obtain the dynamic equation.

Figure 4.4 shows the structure of a four-link manipulator. Let us start with this example to find the speed, the mass point, the kinetic energy and potential energy of the manipulator, and the Lagrangian operator at a certain point (such as point P) on a certain link (such as link 3) of the manipulator, and then find the dynamic equation of the system. Then, from the special to the general, the general expressions of the velocity, kinetic energy, potential energy and dynamics equations of any manipulator are derived.

4.2.1   *Calculation of particle velocity*

The position of point P on link 3 in Figure 4.4 is:

$$^0\boldsymbol{r}_p = T_3{}^3\boldsymbol{r}_p$$

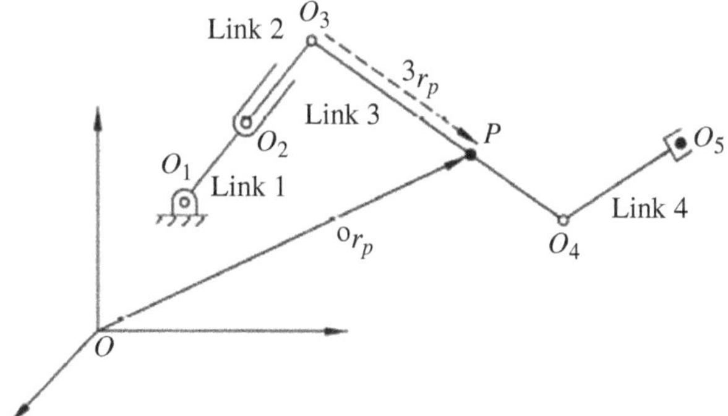

Figure 4.4. Four-link manipulator.

Where $^0\boldsymbol{r}_p$ is the position vector in the general (base) coordinate system; $^3\boldsymbol{r}_p$ is the position vector in the local (relative to joint O_3) coordinate system; T_3 is the transformation matrix, including rotation transformation and translation transformation.

For a point on any link i, its position is:

$$^0\boldsymbol{r} = T_i\,^i\boldsymbol{r} \tag{4.14}$$

The speed of point P is:

$$^0\boldsymbol{v}_p = \frac{d}{dt}(^0\boldsymbol{r}_p) = \frac{d}{dt}(T_3\,^3\boldsymbol{r}_p) = \dot{T}_3\,^3\boldsymbol{r}_p$$

In the formula, $\dot{T}_3 = \frac{dT_3}{dt} = \sum_{j=1}^3 \frac{\partial T_3}{\partial q_i}\dot{q}_j$, so there are:

$$^0\boldsymbol{v}_p = \left(\sum_{j=1}^3 \frac{\partial T_3}{\partial q_j}\dot{q}_i\right)(^3\boldsymbol{r}_p)$$

The speed at any point on the link i is:

$$\boldsymbol{v} = \frac{d\boldsymbol{r}}{dt} = \left(\sum_{j=1}^i \frac{\partial T_i}{\partial q_j}\dot{q}_j\right)\,^i\boldsymbol{r} \tag{4.15}$$

Acceleration at point P

$$^0\boldsymbol{a}_p = \frac{d}{dt}(^0\boldsymbol{v}_p) = \frac{d}{dt}(\dot{T}_3\,^3\boldsymbol{r}_p) = \ddot{T}_3\,^3\boldsymbol{r}_p = \frac{d}{dt}\left(\sum_{j=1}^{3}\frac{\partial T_3}{\partial q_i}\dot{q}_i\right)(^3\boldsymbol{r}_p)$$

$$= \left(\sum_{j=1}^{3}\frac{\partial T_3}{\partial q_i}\frac{d}{dt}\dot{q}_i\right)(^3\boldsymbol{r}_p) + \left(\sum_{k=1}^{3}\sum_{j=1}^{3}\frac{\partial^2 T_3}{\partial q_j \partial q_k}\dot{q}_k\dot{q}_j\right)(^3\boldsymbol{r}_p)$$

$$= \left(\sum_{j=1}^{3}\frac{\partial T_3}{\partial q_i}\ddot{q}_i\right)(^3\boldsymbol{r}_p) + \left(\sum_{k=1}^{3}\sum_{j=1}^{3}\frac{\partial^2 T_3}{\partial q_j \partial q_k}\dot{q}_k\dot{q}_j\right)(^3\boldsymbol{r}_p)$$

Speed squared

$$(^0\boldsymbol{v}_p)^2 = (^0\boldsymbol{v}_p)\cdot(^0\boldsymbol{v}_p) = \mathrm{Trace}[(^0\boldsymbol{v}_p)\cdot(^0\boldsymbol{v}_p)^T]$$

$$= \mathrm{Trace}\left[\sum_{j=1}^{3}\frac{\partial T_3}{\partial q_j}\dot{q}_j(^3\boldsymbol{r}_p)\cdot\sum_{k=1}^{3}\left(\frac{\partial T_3}{\partial q_k}\dot{q}_k\right)(^3\boldsymbol{r}_p)^T\right]$$

$$= \mathrm{Trace}\left[\sum_{j=1}^{3}\sum_{k=1}^{3}\frac{\partial T_3}{\partial q_j}(^3\boldsymbol{r}_p)(^3\boldsymbol{r}_p)^T\frac{\partial T_3^T}{\partial q_k}\dot{q}_j\dot{q}_k\right]$$

For any point in any manipulator, the speed squared is:

$$\boldsymbol{v}^2 = \left(\frac{dr}{dt}\right)^2 = \mathrm{Trace}\left[\sum_{j=1}^{i}\frac{\partial T_i}{\partial q_j}\dot{q}_j\,^i\boldsymbol{r}\sum_{k=1}^{i}\left(\frac{\partial T_i}{\partial q_k}\dot{q}_k\,^i\boldsymbol{r}\right)^T\right]$$

$$= \mathrm{Trace}\left[\sum_{j=1}^{i}\sum_{k=1}^{i}\frac{\partial T_i}{\partial q_k}\,^i\boldsymbol{r}\,^i\boldsymbol{r}^T\left(\frac{\partial T_i}{\partial q_k}\right)^T\dot{q}_k\dot{q}_k\right] \tag{4.16}$$

In the formula, Trace represents the trace of the matrix. For a square matrix of order n, its trace is the sum of the elements on its main diagonal.

4.2.2 *Calculation of kinetic energy and potential energy of particle*

Let the mass of any point P on the link 3 be dm, then its kinetic energy is:

$$
\begin{aligned}
dK_3 &= \frac{1}{2}v_p^2 dm \\
&= \frac{1}{2}\text{Trace}\left[\sum_{j=1}^{3}\sum_{k=1}^{3}\frac{\partial T_3}{\partial q_i}{}^3 r_p({}^3 r_p)^T\left(\frac{\partial T_3}{\partial q_k}\right)^T \dot{q}_i\dot{q}_k\right]dm \\
&= \frac{1}{2}\text{Trace}\left[\sum_{j=1}^{3}\sum_{k=1}^{3}\frac{\partial T_3}{\partial q_i}({}^3 r_p dm\,{}^3 r_p^T)^T\left(\frac{\partial T_3}{\partial q_k}\right)^T \dot{q}_i\dot{q}_k\right]
\end{aligned}
$$

The kinetic energy of the mass point of the position vector ${}^i\boldsymbol{r}$ on any manipulator link i is as follows:

$$
\begin{aligned}
dK_i &= \frac{1}{2}\text{Trace}\left[\sum_{j=1}^{i}\sum_{k=1}^{i}\frac{\partial T_i}{\partial q_j}{}^j\boldsymbol{r}^i\boldsymbol{r}^T\frac{\partial T_i^T}{\partial q_k}\dot{q}_j\dot{q}_k\right]dm \\
&= \frac{1}{2}\text{Trace}\left[\sum_{j=1}^{i}\sum_{k=1}^{i}\frac{\partial T_i}{\partial q_j}({}^i\boldsymbol{r}dm\,{}^i\boldsymbol{r}^T)^T\frac{\partial T_i^T}{\partial q_k}\dot{q}_j\dot{q}_k\right]
\end{aligned}
$$

For link 3 integral dK_3, the kinetic energy of link 3 is:

$$
\begin{aligned}
K_3 = \int_{\text{Link 3}} dK_3 = \frac{1}{2}\text{Trace}\Bigg[\sum_{j=1}^{3}\sum_{k=1}^{3}\frac{\partial T_3}{\partial q_j}\left(\int_{\text{Link 3}}{}^3 r_p{}^3 r_p^T dm\right) \\
\times \left(\frac{\partial T_3}{\partial q_k}\right)^T \dot{q}_j\dot{q}_k\Bigg]
\end{aligned}
$$

In the formula, the integral $\int {}^3\boldsymbol{r}_p{}^3\boldsymbol{r}_p^T dm$ is called the pseudo-inertia matrix of the link and is recorded as:

$$
J_3 = \int_{\text{Link 3}}{}^3\boldsymbol{r}_p{}^3\boldsymbol{r}_p^T dm
$$

So that,

$$K_3 = \frac{1}{2}\text{Trace}\left[\sum_{j=1}^{3}\sum_{k=1}^{3}\frac{\partial T_3}{\partial q_j}J_3\left(\frac{\partial T_3}{\partial q_k}\right)^T \dot{q}_j\dot{q}_k\right]$$

The kinetic energy of any link i on any manipulator is:

$$K_i = \int_{\text{Link }i} dK_i$$

$$= \frac{1}{2}\text{Trace}\left[\sum_{j=1}^{i}\sum_{k=1}^{i}\frac{\partial T_i}{\partial q_j}I_i\left(\frac{\partial T_i}{\partial q_k}\right)\dot{q}_j\dot{q}_k\right] \tag{4.17}$$

Where I_i is the pseudo-inertia matrix, and its general expression is:

$$I_i = \int_{\text{Link }i} {}^i r\,{}^i r^T dm = \int_i {}^i r\,{}^i r^T dm$$

$$= \begin{bmatrix} \int_i {}^i x^2 dm & \int_i {}^i x^i y dm & \int_i {}^i x^i z dm & \int_i {}^i x dm \\ \int_i {}^i x^i y dm & \int_i {}^i y^2 dm & \int_i {}^i y^i z dm & \int_i {}^i y dm \\ \int_i {}^i x^i z dm & \int_i {}^i y^i z dm & \int_i {}^i z^2 dm & \int_i {}^i z dm \\ \int_i {}^i x dm & \int_i {}^i y dm & \int_i {}^i z dm & \int_i dm \end{bmatrix}$$

According to theoretical mechanics or physics, the moment of inertia, vector product, and first-order moment of an object are:

$$I_{xx} = \int (y^2 + z^2)dm, \quad I_{yy} = \int (x^2 + z^2)dm,$$

$$I_{zz} = \int (x^2 + y^2)dm$$

$$I_{xy} = I_{yx} = \int xy dm, \quad I_{xz} = I_{zx} = \int xz dm,$$

$$I_{yz} = I_{zy} = \int yz dm$$

$$mx = \int x dm, \quad my = \int y dm, \quad mz = \int z dm$$

If you make

$$\int x^2 dm = -\frac{1}{2}\int (y^2 + z^2)dm + \frac{1}{2}\int (x^2 + z^2)dm + \frac{1}{2}\int (x^2 + y^2)dm$$

$$= (-I_{xx} + I_{yy} + I_{zz})/2$$

$$\int y^2 dm = +\frac{1}{2}\int (y^2 + z^2)dm - \frac{1}{2}\int (x^2 + z^2)dm + \frac{1}{2}\int (x^2 + y^2)dm$$

$$= (+I_{xx} - I_{yy} + I_{zz})/2$$

$$\int z^2 dm = +\frac{1}{2}\int (y^2 + z^2)dm + \frac{1}{2}\int (x^2 + z^2)dm - \frac{1}{2}\int (x^2 + y^2)dm$$

$$= (+I_{xx} + I_{yy} - I_{zz})/2$$

So I_i can be expressed as:

$$I_i = \begin{bmatrix} \frac{-I_{ixx}+I_{iyy}+I_{izz}}{2} & I_{ixy} & I_{ixz} & m_i\bar{x}_i \\ I_{ixy} & \frac{I_{ixx}-I_{iyy}+I_{izz}}{2} & I_{iyz} & m_i\bar{y}_i \\ I_{ixz} & I_{iyz} & \frac{I_{ixx}+I_{iyy}-I_{izz}}{2} & m_i\bar{z}_i \\ m_i\bar{x}_i & m_i\bar{y}_i & m_i\bar{z}_i & m_i \end{bmatrix} \tag{4.18}$$

The general functions of a manipulator with two links are:

$$K = \sum_{i=1}^{n} K_i = \frac{1}{2}\sum_{i=1}^{n} \text{Trace}\left[\sum_{j=1}^{n}\sum_{k=1}^{i} \frac{\partial T_i}{\partial q_j} I_i \frac{\partial T_i^T}{\partial q_k} \dot{q}_i \dot{q}_k\right] \tag{4.19}$$

In addition, the transmission kinetic energy of the link i is:

$$K_{ai} = \frac{1}{2}I_{ai}\dot{q}_i^2$$

Where I_{ai} is the equivalent moment of inertia of the transmission device, for the translation joint, I_a is the equivalent mass; $\dot{q}_i$ is the speed of the joint i.

The total kinetic energy of the transmission of all joints is:

$$K_a = \frac{1}{2}\sum_{i=1}^{n} I_{ai}\dot{q}_i^2$$

So the total kinetic energy of the manipulator system (including the transmission device) is:

$$K_t = K + K_a$$

$$= \frac{1}{2}\sum_{i=1}^{6}\sum_{j=1}^{i}\sum_{k=1}^{i} \text{Trace}\left(\frac{\partial T_i}{\partial q_i}I_i\frac{\partial T_i^T}{\partial q_k}\right)\dot{q}_j\dot{q}_k + \frac{1}{2}\sum_{i=1}^{6} I_{ai}\dot{q}_i^2 \quad (4.20)$$

Let's calculate the potential energy of the manipulator again. As well know, the potential energy of an object with mass m at height h is:

$$P = mgh$$

The potential energy of the mass point $\mathbf{d}m$ at the upper position ^ir of the link i is:

$$\mathbf{d}P_i = -\mathbf{d}m\mathbf{g}^{T0}r = -\mathbf{g}^T T_i{}^i r \mathbf{d}m$$

In the formula, $\mathbf{g}^T = [g_x, g_y, g_z, 1]$.

$$P_i = \int_{\text{Link}\,i} \mathbf{d}P_i = -\int_{\text{Link}\,i} \mathbf{g}^T T_i{}^i r \mathbf{d}m = -\mathbf{g}^T T_i \int_{\text{Link}\,i} {}^i r \mathbf{d}m$$

$$= -\mathbf{g}^T T_i m_i{}^i r_i = -m_i \mathbf{g}^T T_i{}^i r_i$$

Among them, m_i is the mass of the link i: ^ir_i is the position of the center of gravity of the link i relative to its front-end joint coordinate system.

Since the gravity effect P_{ai} of the transmission device is generally very small and can be ignored, the total position energy of the manipulator system is:

$$P = \sum_{i=1}^{n}(P_i - P_{ai}) \approx \sum_{i=1}^{n} P_i$$

$$= -\sum_{i=1}^{n} m_i \mathbf{g}^T T_i{}^i r_i \quad (4.21)$$

4.2.3 *Derivation of manipulator dynamics equation* [5, 9, 10]

According to formula (4.1), find the Lagrangian function

$$L = K_t - P$$

$$= \frac{1}{2} \sum_{i=1}^{n} \sum_{j=1}^{i} \sum_{k=1}^{i} \text{Trace} \left(\frac{\partial T_i}{\partial q_i} I_i \frac{\partial T_i^T}{\partial q_k} \right) \dot{q}_j \dot{q}_k$$

$$+ \frac{1}{2} \sum_{i=1}^{n} I_{ai} \dot{q}_i^2 + \sum_{i=1}^{n} m_i \boldsymbol{g}^T T_i^{\,i} r_i, \quad n = 1, 2, \ldots \quad (4.22)$$

Then find the kinetic equation according to equation (4.2). Derivative first

$$\frac{\partial L}{\partial \dot{q}_p} = \frac{1}{2} \sum_{i=1}^{n} \sum_{k=1}^{i} \text{Trace} \left(\frac{\partial T_i}{\partial q_p} I_i \frac{\partial T_i^T}{\partial q_k} \right) \dot{q}_k$$

$$+ \frac{1}{2} \sum_{i=1}^{n} \sum_{j=1}^{i} \text{Trace} \left(\frac{\partial T_i}{\partial q_i} I_i \frac{\partial T_i^T}{\partial q_p} \right) \dot{q}_j + I_{ap} \dot{q}_p$$

$$p = 1, 2, \cdots n$$

According to formula (4.18), I_i is a symmetric matrix, that is $I_i^T = I_i$, so the following formula holds:

$$\text{Trace} \left(\frac{\partial T_i}{\partial q_j} I_i \frac{\partial T_i^T}{\partial q_k} \right) = \text{Trace} \left(\frac{\partial T_i}{\partial q_k} I_i^T \frac{\partial T_i^T}{\partial q_j} \right) = \text{Trace} \left(\frac{\partial T_i}{\partial q_k} I_i \frac{\partial T_i^T}{\partial q_j} \right)$$

$$\frac{\partial L}{\partial \dot{q}_p} = \sum_{i=1}^{n} \sum_{k=1}^{i} \text{Trace} \left(\frac{\partial T_i}{\partial q_k} I_i \frac{\partial T_i^T}{\partial q_p} \right) \dot{q}_k + I_{ap} \dot{q}_p$$

When $p > i$, the variable q_p of the rear link has no effect on the front links, that is $\partial T_i / \partial q_p = 0, p > i,$. This can be obtained:

$$\frac{\partial L}{\partial \dot{q}_p} = \sum_{i=p}^{n} \sum_{k=1}^{i} \text{Trace} \left(\frac{\partial T_i}{\partial q_k} I_i \frac{\partial T_i^T}{\partial q_p} \right) + \dot{q}_k + I_{ap} \dot{q}_p$$

Because

$$\frac{d}{dt} \left(\frac{\partial T_i}{\partial q_j} \right) = \sum_{k=1}^{i} \frac{\partial}{\partial q_k} \left(\frac{\partial T_i}{\partial q_i} \right) \dot{q}_k$$

So we have

$$\frac{d}{dt}\frac{\partial L}{\partial \dot{q}_p} = \sum_{i=p}^{n}\sum_{k=1}^{i}\operatorname{Trace}\left(\frac{\partial T_i}{\partial q_k}I_i\frac{\partial T_i^T}{\partial q_p}\right)\ddot{q}_k + I_{ap}\ddot{q}_p$$

$$+ \sum_{i=p}^{n}\sum_{j=1}^{i}\sum_{k=1}^{i}\operatorname{Trace}\left(\frac{\partial^2 T_i}{\partial q_j \partial q_k}I_i\frac{\partial T_i^T}{\partial q_i}\right)\dot{q}_j\dot{q}_k$$

$$+ \sum_{i=p}^{n}\sum_{j=1}^{i}\sum_{k=1}^{i}\operatorname{Trace}\left(\frac{\partial^2 T_i}{\partial q_p \partial q_k}I_i\frac{\partial T_i^T}{\partial q_i}\right)\dot{q}_j\dot{q}_k$$

$$= \sum_{i=p}^{n}\sum_{k=1}^{i}\operatorname{Trace}\left(\frac{\partial T_i}{\partial q_k}I_i\frac{\partial T_i^T}{\partial q_p}\right)\ddot{q}_k + I_{ap}\ddot{q}_p$$

$$+ 2\sum_{i=p}^{n}\sum_{j=1}^{i}\sum_{k=1}^{i}Trace\left(\frac{\partial^2 T_i}{\partial q_j \partial q_k}I_i\frac{\partial T_i^T}{\partial q_k}\right)\dot{q}_j\dot{q}_k$$

Find the $\partial L/\partial q_p$ item:

$$\frac{\partial L}{\partial q_p} = \frac{1}{2}\sum_{i=p}^{n}\sum_{j=1}^{i}\sum_{k=1}^{i}\operatorname{Trace}\left(\frac{\partial^2 T_i}{\partial q_j \partial q_k}I_i\frac{\partial T_i^T}{\partial q_k}\right)\dot{q}_j\dot{q}_k$$

$$+ \frac{1}{2}\sum_{i=p}^{n}\sum_{i=1}^{i}\sum_{k=1}^{i}\operatorname{Trace}\left(\frac{\partial^2 T_i}{\partial q_k \partial q_p}I_i\frac{\partial T_i^t}{\partial q_j}\right)\dot{q}_j\dot{q}_k + \sum_{i=p}^{n}m_i\boldsymbol{g}^T\frac{\partial T_i}{\partial q_p}{}^i r_i$$

$$= \sum_{i=p}^{n}\sum_{j=1}^{i}\sum_{k=1}^{i}\operatorname{Trace}\left(\frac{\partial^2 T_i}{\partial q_p \partial q_j}I_i\frac{\partial T_i^T}{\partial q_k}\right)\dot{q}_j\dot{q}_k + \sum_{i=p}^{n}m_i\boldsymbol{g}^T\frac{\partial T_i}{\partial q_p}{}^i r_i$$

In the above two equations, the dummy elements j and k of the second sum are exchanged, and then combined with the first sum to obtain a simplified form. Substituting the above two formulas into the right formula of (4.2), we get:

$$\frac{d}{dt}\frac{\partial L}{\partial \dot{q}_p} - \frac{\partial L}{\partial q_p} = \sum_{i=p}^{n}\sum_{k=1}^{i}\operatorname{Trace}\left(\frac{\partial T_i}{\partial q_k}I_i\frac{\partial T_i^T}{\partial q_p}\right)\ddot{q}_k + I_{ap}\ddot{q}_p$$

$$+ \sum_{i=p}^{n}\sum_{j=1}^{i}\sum_{k=1}^{i}\operatorname{Trace}\left(\frac{\partial^2 T_i}{\partial q_j \partial q_k}I_i\frac{\partial T_i^T}{\partial q_p}\right)\dot{q}_j\dot{q}_k$$

$$- \sum_{i=p}^{n}m_i\boldsymbol{g}^T\frac{\partial T_i}{\partial q_p}{}^i r_i$$

By exchanging the dummy elements in the sum formulas listed above, replacing i for p, replacing j for i, and replacing m for j, the dynamic equation of the manipulator system with a links can be obtained as follows:

$$T_i = \sum_{j=i}^{n} \sum_{k=1}^{j} \mathrm{Trace}\left(\frac{\partial T_j}{\partial q_k} I_j \frac{\partial T_j^T}{\partial q_i} \right) \ddot{q}_k + I_{ai}\ddot{q}_i$$

$$+ \sum_{j=1}^{n} \sum_{k=1}^{j} \sum_{m=1}^{j} \mathrm{Trace}\left(\frac{\partial^2 T_i}{\partial q_k \partial q_m} I_j \frac{\partial T_j^T}{\partial q_i} \right) \dot{q}_k \dot{q}_m - \sum_{j=1}^{n} m_j \boldsymbol{g}^T \frac{\partial T_{i}}{\partial q_i}{}^i r_i$$

$$(4.23)$$

These equations are independent of the order of summation. We write the formula (4.23) in the following form:

$$T_i = \sum_{j=1}^{n} D_{ij}\ddot{q}_j + I_{ai}\ddot{q}_i + \sum_{j=1}^{6} \sum_{k=1}^{6} D_{ijk}\dot{q}_j \dot{q}_k + D_i \qquad (4.24)$$

In the formula, take $n = 6$, and

$$D_{ij} = \sum_{p=\max i,j}^{6} \mathrm{Trace}\left(\frac{\partial T_p}{\partial q_j} I_p \frac{\partial T_p^T}{\partial q_i} \right) \qquad (4.25)$$

$$D_{ijk} = \sum_{p=\max i,j,k}^{6} \mathrm{Trace}\left(\frac{\partial^2 T_p}{\partial q_j \partial q_k} I_i \frac{\partial T_p^T}{\partial q_i} \right) \qquad (4.26)$$

$$D_i = \sum_{p=i}^{6} -m_p \boldsymbol{g}^T \frac{\partial T_p}{\partial q_i}{}^p r_p \qquad (4.27)$$

The above equations are the same as the inertia term and gravity term in §4.1.2. These items are particularly important in manipulator control, because they directly affect the stability and positioning accuracy of the manipulator system. Centripetal force and Coriolis force are important only when the manipulator moves at high speed. At this time, the error generated by them is not large. The inertia of the transmission device often has a considerable value, and the relative importance of the structural dependence of the effective inertia and the coupling inertia term is reduced.

4.2.4 *Simplification of manipulator dynamics equation* [4, 12]

The calculation of inertia term and gravity term in Section 4.2.3 must be simplified to facilitate the actual calculation.

1. Simplification of inertia term D_{ij}

When discussing the Jacobian matrix in subsection 3.3.2, we got the partial derivative $\partial T_6/\partial q_i = T_6^{T_6}\Delta_i$, which is actually a special case of $p = 6$. It can be generalized to the general form:

$$\frac{\partial T_p}{\partial q_i} = T_p^{T_p}\Delta_i \tag{4.28}$$

Where, $^{T_p}\Delta_i = (A_i A_{i+1} \cdots A_p)^{-1}{}^{i-1}\Delta_i(A_i A_{i+1} \cdots A_p)$, and the differential coordinate transformation is

$$^{i-1}T_p = (A_i A_{i+1} \cdots A_p)$$

For rotary joints, according to equation (3.93), the differential translation vector and differential rotation vector can be obtained as follows:

$$\left.\begin{array}{l} ^{p}d_{ix} = -{}^{i-1}n_{px}{}^{j-1}p_{py} + {}^{i-1}n_{py}{}^{i-1}p_{px} \\[4pt] ^{p}d_{jy} = -{}^{i-1}o_{px}{}^{i-1}p_{py} + {}^{i-1}o_{py}{}^{i-1}p_{px} \\[4pt] ^{p}d_{iz} = -{}^{i-1}a_{px}{}^{i-1}p_{py} + {}^{i-1}a_{py}{}^{i-1}p_{px} \end{array}\right\} \tag{4.29}$$

$$^{p}\boldsymbol{\delta}_i = {}^{i-1}n_{pz}\boldsymbol{i} + {}^{i-1}o_{pz}\boldsymbol{j} + {}^{i-1}a_{pz}\boldsymbol{k} \tag{4.30}$$

The following abbreviations are used in the above formula: write $^{T_p}\boldsymbol{d}_i$ as $^{p}\boldsymbol{d}_i$, write $^{T_{i-1}}n$ as $^{i-1}n_p$, and so on.

For prismatic (translation) joints, according to equation (3.94), the vectors can be obtained as:

$$^{p}\boldsymbol{d}_i = {}^{i-1}n_{pz}\boldsymbol{i} + {}^{i-1}o_{pz}\boldsymbol{j} + {}^{i-1}a_{pz}\boldsymbol{k}$$

$$^{p}\boldsymbol{\delta}_i = 0\boldsymbol{i} + 0\boldsymbol{j} + 0\boldsymbol{k}$$

Substituting formula (4.28) into formula (4.25), we get:

$$D_{ij} = \sum_{p=\max i,j}^{6} \text{Trace}(T_p{}^{p}\Delta_j I_p{}^{p}\Delta_i^T T_p^T)$$

Expand the middle three terms of the above formula:

$$D_{ij} = \sum_{p=\max i,j}^{6} \text{Trace} \left(T_p \begin{bmatrix} 0 & -{}^p\delta_{jz} & {}^p\delta_{jy} & {}^pd_{jx} \\ {}^p\delta_{jz} & 0 & -{}^p\delta_{jx} & {}^pd_{jy} \\ -{}^p\delta_{jy} & {}^p\delta_{jx} & 0 & {}^pd_{jz} \\ 0 & 0 & 0 & 0 \end{bmatrix} \right.$$

$$\times \begin{bmatrix} \frac{-I_{xx}+I_{yy}+I_{zz}}{2} & I_{xy} & I_{xz} & m_i\bar{x}_i \\ I_{xy} & \frac{-I_{xx}-I_{yy}+I_{zz}}{2} & I_{yz} & m_i\bar{y}_i \\ I_{xz} & I_{yz} & \frac{I_{xx}+I_{yy}-I_{zz}}{2} & m_i\bar{z}_i \\ m_i\bar{x}_i & m_i\bar{y}_i & m_i\bar{z}_i & m_i \end{bmatrix}$$

$$\times \left. \begin{bmatrix} 0 & {}^p\delta_{ix} & -{}^p\delta_{iy} & 0 \\ -{}^p\delta_{iz} & 0 & {}^p\delta_{ix} & 0 \\ {}^p\delta_{iy} & -{}^p\delta_{ix} & 0 & 0 \\ {}^p\delta_{ix} & {}^p\delta_{iy} & {}^pd_{iz} & 0 \end{bmatrix} T_p^T \right)$$

The middle three terms are obtained by transposing formula (3.79), (4.18) and formula (3.79). Each element in the bottom row and right column of the matrix obtained by multiplying them is zero. When they multiply T_p to the left and T_p^T to the right, only the rotation part T_p of the transformation is used. Under this operation, the trace of the matrix is an invariant. Therefore, as long as the traces of the three items in the above expression are traced. Its simplified vector form is:

$$D_{ij} = \sum_{p=\max i,y}^{6} m_p[{}^p\boldsymbol{\delta}_i{}^T k_p{}^p\boldsymbol{\delta}_j + {}^p\boldsymbol{d}_i{}^p\boldsymbol{d}_j + {}^p\bar{\boldsymbol{r}}_p({}^p\boldsymbol{d}_i \times {}^p\boldsymbol{\delta}_j + {}^p\boldsymbol{d}_j \times {}^p\boldsymbol{\delta}_j)]$$

$$\tag{4.31}$$

Where

$$k_p = \begin{bmatrix} k_{pxx}^2 & -k_{pxy}^2 & -k_{pxz}^2 \\ -k_{pxy}^2 & k_{pyy}^2 & -k_{pyz}^2 \\ -k_{pxz}^2 & -k_{pyz}^2 & k_{pzz}^2 \end{bmatrix}$$

As well as

$$m_p k_{pxx}^2 = I_{pxx}, \quad m_p k_{pyy}^2 = I_{pyy}, \quad m_p k_{pzz}^2 = I_{pzz},$$

$$m_p k_{pxy}^2 = I_{pxy}, \quad m_p k_{pyz}^2 = I_{pyz}, \quad m_p k_{pxz}^2 = I_{pxz}$$

If the non-diagonal inertia terms in the above formula are set to 0, it is a normal hypothesis, then the formula (4.29) is further simplified as:

$$D_{ij} = \sum_{p=\max i,j}^{6} m_p \Big\{ \big[{}^p\delta_{ix} k_{pxx}^2 {}^p\delta_{jx} + {}^p\delta_{iy} k_{pyy}^2 {}^p\delta_{jy} + {}^p\delta_{iz} k_{pzz}^2 {}^p\delta_{jz}\big]$$

$$+ [{}^p\boldsymbol{d}_i \cdot {}^p\boldsymbol{d}_j] + [{}^p\bar{\boldsymbol{r}}_p \cdot (\boldsymbol{d}_i \times {}^p\boldsymbol{\delta}_j + {}^p\boldsymbol{d}_j \times {}^p\boldsymbol{\delta}_i)] \Big\} \qquad (4.32)$$

It can be seen from formula (4.32) that each element of the D_{ij} sum is composed of three groups of terms. The first group of items ${}^p\delta_{ix} k_{pxx}^2 \cdots$ represents the distribution effect of mass m_p on the link p. The second group of items represents the mass distribution of the link p, which is recorded as the effective moment arm ${}^p\boldsymbol{d}_i \cdot {}^p\boldsymbol{d}_j$. The last group of items is generated because the center of mass of the link p is not at the origin of the coordinate system of the link p. When the centroids of the links are far apart, the items in the second part above will play a major role, and the influence of the first group of items and the third group of items can be ignored.

2. Simplification of inertia term D_{ij}

In formula (4.32), when $i = j$, D_{ij} can be further simplified to D_{ii} as follows:

$$D_{ii} = \sum_{p=i}^{6} m_p \Big\{ [{}^p\delta_{ix}^2 k_{pxx}^2 + {}^p\delta_{iy}^2 k_{pyy}^2 + {}^p\delta_{iz}^2 k_{pzz}^2]$$

$$+ [{}^p\boldsymbol{d}_i \cdot \boldsymbol{d}_i] + [{}^{2p}\bar{\boldsymbol{r}}_p \cdot ({}^p\boldsymbol{d}_i \times {}^p\boldsymbol{\delta}_i)] \Big\} \qquad (4.33)$$

If it is a revolute joint, then substituting equations (4.29) and (4.30) into the above equations can be obtained:

$$D_{ii} = \sum_{p=i}^{6} m_p \Big\{ [n_{px}^2 k_{pxx}^2 + o_{py}^2 k_{pyy}^2 + a_{pz}^2 k_{pzz}^2] + [\bar{\boldsymbol{p}}_p \cdot \bar{\boldsymbol{p}}_p]$$

$$+ [{}^{2p}\bar{\boldsymbol{r}}_p \cdot [(\bar{\boldsymbol{p}}_p \cdot \boldsymbol{n}_p)\boldsymbol{i} + (\bar{\boldsymbol{p}}_p \cdot \boldsymbol{o}_p)\boldsymbol{j} + (\bar{\boldsymbol{p}}_p \cdot \boldsymbol{a}_p)\boldsymbol{k}]] \Big\} \qquad (4.34)$$

Where, $\boldsymbol{n}_p$, $\boldsymbol{o}_p$, $\boldsymbol{a}_p$ and p_p are vectors of ${}^{(i-1)}T_p$, and

$$\bar{\boldsymbol{p}} = p_x \boldsymbol{i} + p_y \boldsymbol{j} + 0\boldsymbol{k}$$

The relevant corresponding items in formula (4.32) and formula (4.33) can be made equal:

$$^p\delta_{ix}^2 k_{pxx}^2 + {}^p\delta_{iy}^2 k_{pyy}^2 + {}^p\delta_{iz}^2 k_{pzz}^2 = n_{px}^2 k_{pxx}^2 + o_{py}^2 k_{pyy}^2 + a_{pz}^2 k_{pzz}^2$$

$$^p\boldsymbol{d}_i \cdot {}^p\boldsymbol{d}_i = \bar{\boldsymbol{p}}_p \cdot \bar{\boldsymbol{p}}_p$$

$$^p\boldsymbol{d}_i \times {}^p\delta_i = (\bar{p}_p \cdot n_p)i + (\bar{p}_p \cdot o_p)j + (\bar{p}_p \cdot a_p)k$$

Just like formula (4.22), each element of the sum formula D_{ii} is also composed of three terms. If it is a prismatic joint, $^p\delta_i = 0$, $^p d_i \cdot {}^p d_i = 1$, then

$$D_{ii} = \sum_{p=i}^{6} m_p \tag{4.35}$$

3. Simplification of the gravity term D_i

Substituting formula (4.28) into formula (4.27), we get:

$$D_i = \sum_{p=i}^{6} -m_p \boldsymbol{g}^T T_p{}^p\Delta_i{}^p\bar{r}_p$$

Separate T_p into $T_{i-1}{}^{i-1}T_p$, and rear multiply $^p\Delta_i$ by $^{i-1}T_p^{-1 i-1}T_p$, get:

$$D_i = \sum_{p=i}^{6} -m_p \boldsymbol{g}^T T_{i-1}{}^{i-1}T_p{}^p\Delta_i^{i-1}T_p^{-1 i-1}T_p{}^p\bar{r}_p \tag{4.36}$$

When $^{i-1}\Delta_i = {}^{i-1}T_p^{-1}$, $^i r_p = {}^iT_p{}^p\bar{r}_p$, D_i can be further simplified as:

$$D_i = -\boldsymbol{g}^T T_{i-1}{}^{i-1}\Delta_i \sum_{p=i}^{6} m_p{}^{i-1}\bar{r}_p \tag{4.37}$$

Definition $^{i-1}\boldsymbol{g} = -\boldsymbol{g}^T T_{i-1}{}^{i-1}\Delta_i$, then:

$$^{i-1}\boldsymbol{g} = -[g_x g_y g_z 0] \begin{bmatrix} n_x & o_x & a_x & p_x \\ n_y & o_y & a_y & p_y \\ n_z & o_z & a_z & p_z \\ 0 & 0 & 0 & 1 \end{bmatrix} \begin{bmatrix} 0 & -\delta_x & \delta_y & d_x \\ \delta_z & 0 & -\delta_x & d_y \\ -\delta_y & \delta_x & 0 & d_z \\ 0 & 0 & 0 & 0 \end{bmatrix}$$

Corresponds to the revolute joint i, $^{i-1}\Delta_i$ corresponds to the rotation around the axis z. Therefore, the above formula can be simplified as:

$$^{i-1}\boldsymbol{g} = -[g_x g_y g_z 0] \begin{bmatrix} n_x & o_x & a_x & p_x \\ n_y & o_y & a_y & p_y \\ n_z & o_z & a_z & p_z \\ 0 & 0 & 0 & 1 \end{bmatrix} \begin{bmatrix} 0 & -1 & 0 & 0 \\ 1 & 0 & 0 & 0 \\ 0 & 0 & 0 & 0 \\ 0 & 0 & 0 & 0 \end{bmatrix}$$

$$= [-\boldsymbol{g} \cdot \boldsymbol{o}, \boldsymbol{g} \cdot \boldsymbol{n}, 0, 0] \tag{4.38}$$

For prismatic joints, $^{i-1}\Delta_i$ corresponds to the translation along the axis z, and then has the following formula:

$$^{i-1}\boldsymbol{g} = -[g_x g_y g_z 0] \begin{bmatrix} n_x & o_x & a_x & p_x \\ n_y & o_y & a_y & p_y \\ n_z & o_z & a_z & p_z \\ 0 & 0 & 0 & 1 \end{bmatrix} \begin{bmatrix} 0 & 0 & 0 & 0 \\ 0 & 0 & 0 & 0 \\ 0 & 0 & 0 & 1 \\ 0 & 0 & 0 & 0 \end{bmatrix}$$

$$= [0, 0, -\boldsymbol{g} \cdot \boldsymbol{a}] \tag{4.39}$$

Thus, D_i can be written as:

$$D_i = {}^{i-1}\boldsymbol{g} \sum_{p=i}^{6} m_p {}^{i-1}\bar{\boldsymbol{r}}_p \tag{4.40}$$

4.3   Examples of Manipulator Dynamics Equations

4.3.1   *The dynamic equation of the two-link manipulator* [9, 19]

In the first section of this chapter, the dynamic equations of the two-link manipulator have been discussed, and see Figure 4.2 and Figure 4.3 and the formulas of related parties. Here, only the effective inertia term, coupled inertia term and gravity term calculation of the two-link manipulator are discussed.

First, specify the coordinate system of the manipulator, as shown in Figure 4.5, and calculate the matrix A and the matrix T. Table 4.2 shows the parameters of each link.

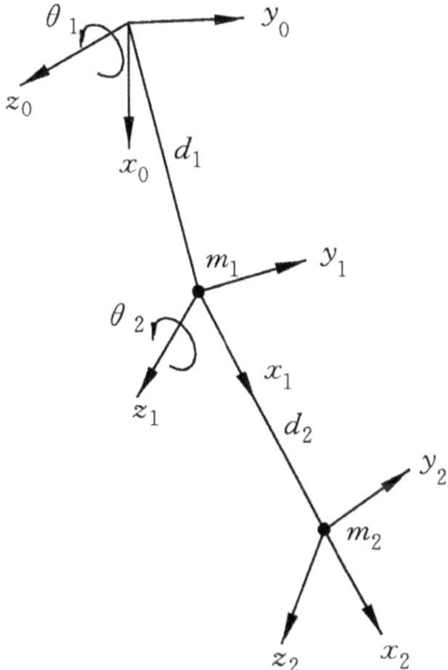

Figure 4.5.   Coordinate system of two-link manipulator.

Table 4.2.   Two-link manipulator link parameters.

Link	Variable	α	a	d	$\cos\alpha$	$\sin\alpha$
1	θ_1	$0°$	d_1	0	1	0
2	θ_2	$0°$	d_2	0	1	0

The matrix A and matrix T are as follows:

$$
A_1 = {}^0T_1 = \begin{bmatrix} c_1 & -s_1 & 0 & d_1c_1 \\ s_1 & c_1 & 0 & d_1s_1 \\ 0 & 0 & 1 & 0 \\ 0 & 0 & 0 & 1 \end{bmatrix}
$$

$$
A_2 = {}^1T_2 = \begin{bmatrix} c_2 & -s_2 & 0 & d_2c_2 \\ s_2 & c_2 & 0 & d_2s_2 \\ 0 & 0 & 1 & 0 \\ 0 & 0 & 0 & 1 \end{bmatrix}
$$

$$^0T_2 = \begin{bmatrix} c_{12} & -s_{12} & 0 & d_1c_1 + d_2c_{12} \\ s_{12} & c_{12} & 0 & d_1s_1 + d_2s_{12} \\ 0 & 0 & 1 & 0 \\ 0 & 0 & 0 & 1 \end{bmatrix}$$

Since the two joints are of the rotary type, $\boldsymbol{d}$ and $\boldsymbol{\delta}$ can be calculated according to equations (4.29) and (4.30).

Based on $^0T_1$, have the following formula:

$$^1\boldsymbol{d}_1 = 0\boldsymbol{i} + d_1\boldsymbol{j} + 0\boldsymbol{k}, \quad ^1\boldsymbol{\delta}_1 = 0\boldsymbol{i} + 0\boldsymbol{j} + 1\boldsymbol{k}$$

Based on $^1T_2$, the following formula can be obtained:

$$^2\boldsymbol{d}_1 = 0\boldsymbol{i} + d_2\boldsymbol{j} + 0\boldsymbol{k}, \quad ^2\boldsymbol{\delta}_2 = 0\boldsymbol{i} + 0\boldsymbol{j} + 1\boldsymbol{k}$$

Based on $^0T_2$, the following formula can be obtained:

$$^2\boldsymbol{d}_1 = s_2d_1\boldsymbol{i} + (c_2d_1 + d_2)\boldsymbol{j} + 0\boldsymbol{k}, \quad ^2\boldsymbol{\delta}_1 = 0\boldsymbol{i} + 0\boldsymbol{j} + 1\boldsymbol{k}$$

For this simple manipulator, all moments of inertia are zero, just like $^i\boldsymbol{r}_i$ and $^2\boldsymbol{r}_2$ are zero. Therefore, from equation (4.34), we can immediately get:

$$D_{11} = \sum_{p=1}^{2} m_p \left\{ [n_{px}^2 k_{pxx}^2] + [\bar{\boldsymbol{p}}_p \cdot \bar{\boldsymbol{p}}_p] \right\}$$

$$= m_1(p_{1x}^2 + p_{1y}^2) + m_2(p_{2x}^2 + p_{2y}^2)$$

$$= m_1 d_1^2 + m_2(d_1^2 + d_2^2 + 2c_2d_1d_2)$$

$$= (m_1 + m_2)d_1^2 + m_2d_2^2 + 2m_2d_1d_2c_2$$

$$D_{22} = \sum_{p=2}^{2} m_p \left\{ [n_{px}^2 k_{pxx}^2] + [\bar{\boldsymbol{p}}_p \cdot \bar{\boldsymbol{p}}_p] \right\}$$

$$= m_2(^1p_{2x}^2 + ^1p_{2y}^2) = m_2d_2^2$$

According to equation (4.32), find D_{12}:

$$D_{12} = \sum_{p=\max 1,2}^{2} m_p \left\{ [^p\boldsymbol{d}_1 \cdot {}^p\boldsymbol{d}_2] \right\} = m_2(^2\boldsymbol{d}_1 \cdot {}^2\boldsymbol{d}_2)$$

$$= m_2(c_2d_1 + d_2)d_2 = m_2(c_2d_1d_2 + d_2^2)$$

Finally, the gravitational terms D_1 and D_2 are calculated, and for this purpose, ^2g and ^1g are calculated first.

Because of $^{i-1}\boldsymbol{g} = [-\boldsymbol{g}\cdot\boldsymbol{o} \quad \boldsymbol{g}\cdot\boldsymbol{n} \quad 0 \quad 0]$, the following formulas can be obtained:

$$\boldsymbol{g} = [g \quad 0 \quad 0 \quad 0]$$

$$^0\boldsymbol{g} = [0 \quad g \quad 0 \quad 0]$$

$$^1\boldsymbol{g} = [gs_1 \quad gc_1 \quad 0 \quad 0]$$

Then find the centroid vector $^i\boldsymbol{r}_p$

$$^2\bar{\boldsymbol{r}}_2 = \begin{bmatrix} 0 \\ 0 \\ 0 \\ 1 \end{bmatrix} \quad ^1\bar{\boldsymbol{r}}_2 = \begin{bmatrix} c_2 d_2 \\ s_2 d_2 \\ 0 \\ 1 \end{bmatrix} \quad ^0\bar{\boldsymbol{r}}_2 = \begin{bmatrix} c_1 d_1 + c_{12} d_2 \\ s_1 d_1 + s_{12} d_2 \\ 0 \\ 1 \end{bmatrix}$$

$$^1\bar{\boldsymbol{r}}_1 = \begin{bmatrix} 0 \\ 0 \\ 0 \\ 1 \end{bmatrix} \quad ^0\bar{\boldsymbol{r}}_1 = \begin{bmatrix} c_1 d_1 \\ s_1 d_1 \\ 0 \\ 1 \end{bmatrix}$$

Then we can obtain D_1 and D_2 according to equation (4.37):

$$D_1 = m_1\,^0\boldsymbol{g}\,^0\bar{\boldsymbol{r}}_1 + m_2\,^0\boldsymbol{g}\,^0\bar{\boldsymbol{r}}_2 = m_1 g s_1 d_1 + m_2 g (s_1 d_1 + s_{12} d_2)$$

$$= (m_1 + m_2) g d_1 s_1 + m_2 g d_2 s_{12}$$

$$D_2 = m_2\,^1\boldsymbol{g}\,^1\bar{\boldsymbol{r}}_2 = m_2 g (s_1 c_2 + c_1 s_2) = m_2 g d_2 s_{12}$$

The items obtained above can be compared with D_{11}, D_{22}, D_1 and D_2 in §4.1.2 to verify the correctness of the calculation results.

The calculation examples of dynamic equations of five-link and six-link manipulators have been introduced in some literatures, such as the dynamic equations of Stanford manipulators, and the dynamic equations of PUMA560 industrial manipulator. It will not list here.

4.3.2 *Velocity and acceleration equations of three-link manipulator* [3, 9]

In the control of robotic manipulator, it is often necessary to know the speed and acceleration of the end of each link, or the control system is required to provide a certain driving torque or force to

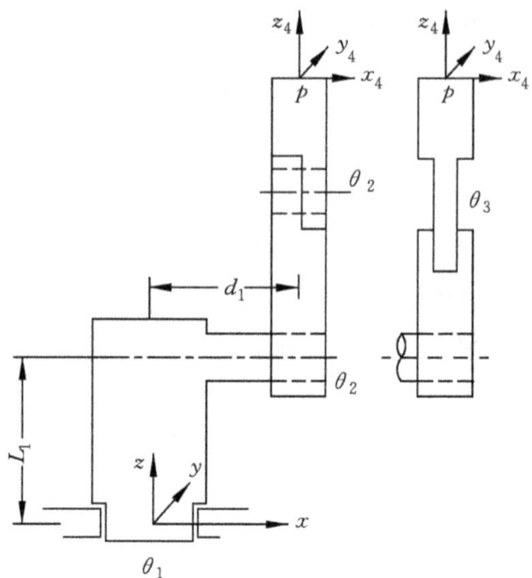

Figure 4.6.　Three-link manipulator device.

ensure the determined speed and acceleration of each link of the manipulator. Therefore, it is necessary to exemplify how to establish the speed and acceleration equations of the manipulator.

Figure 4.6 shows the structure and coordinate system of a three-link manipulator. The velocity and acceleration equations will be established below.

1. Position equation

The relative position equation of the end of the manipulator device 1 to the origin of the base coordinate system is:

$$\begin{bmatrix} x \\ y \\ z \\ 1 \end{bmatrix} = \phi_1 T_{12} \phi_2 T_{23} \phi_3 T_{34} \begin{bmatrix} x_4 \\ y_4 \\ z_4 \\ 1 \end{bmatrix} = T_3 \begin{bmatrix} x_4 \\ y_4 \\ z_4 \\ 1 \end{bmatrix}$$

Where

$$\phi_1 = Rot(z_1, \theta_1) = \begin{bmatrix} c_1 & -s_1 & 0 & 0 \\ s_1 & c_1 & 0 & 0 \\ 0 & 0 & 1 & 0 \\ 0 & 0 & 0 & 1 \end{bmatrix}$$

$$T_{12} = Trans(d_1, 0, L_1)Rot(y, 90) = \begin{bmatrix} 0 & 0 & 1 & d_1 \\ 0 & 1 & 0 & 0 \\ -1 & 0 & 0 & L_1 \\ 0 & 0 & 0 & 1 \end{bmatrix}$$

$$\phi_2 = Rot(z_2, \theta_2) = \begin{bmatrix} c_2 & -s_2 & 0 & 0 \\ s_2 & c_2 & 0 & 0 \\ 0 & 0 & 1 & 0 \\ 0 & 0 & 0 & 1 \end{bmatrix}$$

$$T_{23} = Trans(-L_2, 0, 0) = \begin{bmatrix} 1 & 0 & 0 & -L_2 \\ 0 & 1 & 0 & 0 \\ 0 & 0 & 1 & 0 \\ 0 & 0 & 0 & 1 \end{bmatrix}$$

$$\phi_3 = Rot(z_2, \theta_2) = \begin{bmatrix} c_3 & -s_3 & 0 & 0 \\ s_3 & c_3 & 0 & 0 \\ 0 & 0 & 1 & 0 \\ 0 & 0 & 0 & 1 \end{bmatrix}$$

$$T_{34} = Trans(-L_3, 0, 0)Rot(y, 90) = \begin{bmatrix} 0 & 0 & -1 & -L_3 \\ 0 & 1 & 0 & 0 \\ 1 & 0 & 0 & 0 \\ 0 & 0 & 0 & 1 \end{bmatrix}$$

So there is the following formula:

$$T_3 = \begin{bmatrix} 0 & -s_1 & c_1 & d_1c_1 \\ 0 & c_1 & s_1 & d_1s_1 \\ -1 & 0 & 0 & L_1 \\ 0 & 0 & 0 & 1 \end{bmatrix} \begin{bmatrix} c_2 & -s_2 & 0 & -L_2c_2 \\ s_2 & c_2 & 0 & -L_2s_2 \\ 0 & 0 & 1 & 0 \\ 0 & 0 & 0 & 1 \end{bmatrix}$$

$$\times \begin{bmatrix} 0 & -s_3 & -c_3 & -L_3c_3 \\ 1 & c_3 & -s_2 & -L_3s_3 \\ 0 & 0 & 0 & 0 \\ 0 & 0 & 0 & 1 \end{bmatrix}$$

$$
= \begin{bmatrix} c_1 & s_1 s_2 s_3 - s_1 c_2 c_3 & s_1 s_2 c_3 + s_1 c_2 s_3 \\ s_1 & -c_1 s_2 s_3 + c_1 c_2 c_3 & c_1 s_2 c_3 + c_1 c_2 c_3 \\ 0 & c_2 s_3 + s_2 c_3 & c_2 c_3 + s_2 s_3 \\ 0 & 0 & 0 \end{bmatrix}
$$

$$
\begin{bmatrix} L_3 s_1 (s_2 c_3 + c_2 s_3) + L_2 s_1 s_2 + d_1 c_1 \\ -L_3 c_1 (s_2 c_3 + c_2 s_3) - L_2 c_1 s_2 + d_1 s_1 \\ L_3 (c_2 c_3 - s_2 s_3) + L_2 c_2 + L_1 \\ 1 \end{bmatrix}
$$

$$
= \begin{bmatrix} c_1 & -s_1 c_{23} & s_1 s_{23} & d_1 c_1 + s_1 (L_2 s_2 + L_2 s_{23}) \\ s_1 & c_1 c_{23} & c_1 c_{23} & d_1 s_1 - c_1 (L_2 s_2 + L_2 s_{23}) \\ 0 & s_{23} & -c_{23} & L_1 + L_2 c_2 + L_3 c_{23} \\ 0 & 0 & 0 & 1 \end{bmatrix}
$$

2. Velocity equation

The velocity equation of the end of the manipulator (tool) to the origin of the base coordinate system is:

$$
\begin{bmatrix} \dot{x} \\ \dot{y} \\ \dot{z} \\ 0 \end{bmatrix} = \frac{d}{dt} \begin{bmatrix} x \\ y \\ z \\ 1 \end{bmatrix} = [\omega_1 \theta_1 \phi_1 T_{12} \phi_2 T_{23} \phi_{23} T_{34} + \omega_2 \phi_1 T_{12} \theta_2 \phi_2 T_{23} \phi_3 T_{34}
$$

$$
+ \omega_3 \phi_1 T_{12} \phi_2 T_{23} \theta_3 \phi_3 T_{34}] \begin{bmatrix} x_4 \\ y_4 \\ z_4 \\ 1 \end{bmatrix}
$$

In the formula, ω_1, ω_2 and ω_3 are the rotational angular velocities of the axes z_1, z_2 and z_3 respectively; θ_1, θ_2 and θ_3 are the rotation derivation calculation matrices, and

$$
\theta_1 = \theta_2 = \theta_3 = \begin{bmatrix} 0 & -1 & 0 & 0 \\ 1 & 0 & 0 & 0 \\ 0 & 0 & 0 & 0 \\ 0 & 0 & 0 & 0 \end{bmatrix}
$$

So there are:

$$\dot{T}_{31} = \omega_1 \theta_1 \phi_1 T_{12} \phi_2 T_{23} \phi_3 T_{34}$$

$$= \ldots$$

$$= \omega_1 \begin{bmatrix} -s_1 & -c_1 & 0 & 0 \\ c_1 & -s_1 & 0 & 0 \\ 0 & 0 & 0 & 0 \\ 0 & 0 & 0 & 0 \end{bmatrix} \begin{bmatrix} 0 & 0 & 1 & d_1 \\ 0 & 1 & 0 & 0 \\ -1 & 0 & 0 & L_1 \\ 0 & 0 & 0 & 1 \end{bmatrix}$$

$$\times \begin{bmatrix} c_2 & -s_2 & 0 & -L_2 c_2 \\ 0 & 1 & 0 & -L_2 s_2 \\ 0 & 0 & 1 & 0 \\ 0 & 0 & 0 & 1 \end{bmatrix} \begin{bmatrix} 0 & -s_3 & -c_3 & -L_3 c_3 \\ 0 & c_3 & -s_3 & -L_3 s_3 \\ 1 & 0 & 0 & 0 \\ 0 & 0 & 0 & 1 \end{bmatrix}$$

$$= \omega_1 \begin{bmatrix} -s_1 & -c_1 c_{23} & c_1 s_{23} & L_3 c_1 s_{23} + L_2 c_1 s_2 - d_1 s_1 \\ c_1 & -s_1 c_{23} & s_1 s_{23} & L_2 s_2 s_{23} + L_2 s_1 s_2 + d_1 c_1 \\ 0 & 0 & 0 & 0 \\ 0 & 0 & 0 & 0 \end{bmatrix}$$

$$\dot{T}_{32} = \omega_2 \phi_1 T_{12} \theta_2 \phi_2 T_{23} \phi_3 T_{34}$$

$$= \ldots$$

$$= \omega_2 \begin{bmatrix} 0 & s_1 & c_1 & d_1 c_1 \\ 0 & -c_1 & s_1 & d_2 s_1 \\ -1 & 0 & 0 & L_1 \\ 0 & 0 & 0 & 1 \end{bmatrix} \begin{bmatrix} -s_2 & -c_2 & 0 & 0 \\ c_2 & -s_2 & 0 & 0 \\ 0 & 0 & 0 & 0 \\ 0 & 0 & 0 & 0 \end{bmatrix}$$

$$\times \begin{bmatrix} 1 & - & 0 & -L_2 \\ 0 & 1 & 0 & 0 \\ 0 & 0 & 1 & 0 \\ 0 & 0 & 0 & 1 \end{bmatrix} \begin{bmatrix} 0 & -s_3 & -c_3 & -L_3 c_3 \\ 0 & c_3 & -s_3 & -L_3 s_3 \\ 1 & 0 & 0 & 0 \\ 0 & 0 & 0 & 1 \end{bmatrix}$$

$$
= \omega_2
\begin{bmatrix}
0 & -s_1c_2s_3 - s_1s_2s_3 & -s_1c_2c_3 + s_1s_2s_3 \\
0 & c_1c_2s_3 + c_1s_2c_3 & c_1c_2c_3 - c_1s_2s_3 \\
0 & -s_2s_3 + c_2c_3 & -s_2c_3 - c_2s_3 \\
0 & 0 & 0
\end{bmatrix}
$$

$$
\begin{matrix}
L_3(-s_1c_2c_3 + s_1s_2s_3) - L_2s_1c_2 \\
L_3(c_1c_2c_3 - c_1c_2s_3) + L_2c_1c_2 \\
-L_3(s_2c_3 + c_2s_3) - L_2s_2 \\
0
\end{matrix}
$$

$$
= \omega_2
\begin{bmatrix}
0 & -s_1s_{23} & -s_1c_{23} & -L_3s_1c_{23} - L_2s_1c_2 \\
0 & c_1s_{23} & c_1c_{23} & L_3c_1c_{23} + L_2c_1c_2 \\
0 & c_{23} & -s_{23} & -L_3s_{23} - L_2s_2 \\
0 & 0 & 0 & 0
\end{bmatrix}
$$

$$
\dot{T}_{33} = \omega_3\phi_1 T_{12}\phi_2 T_{23}\phi_3 T_{34}
$$

$$
= \cdots
$$

$$
= \omega_3
\begin{bmatrix}
0 & s_1 & c_1 & d_1c_1 \\
0 & -c_1 & s_1 & d_1s_1 \\
-1 & 0 & 0 & L_1 \\
0 & 0 & 0 & 1
\end{bmatrix}
\begin{bmatrix}
c_2 & -s_2 & 0 & -L_2c_2 \\
s_2 & c_2 & 0 & -L_2s_2 \\
0 & 0 & 1 & 0 \\
0 & 0 & 0 & 1
\end{bmatrix}
$$

$$
\times
\begin{bmatrix}
0 & -1 & 0 & 0 \\
1 & 0 & 0 & 0 \\
0 & 0 & 0 & 0 \\
0 & 0 & 0 & 0
\end{bmatrix}
\begin{bmatrix}
0 & -s_3 & -c_3 & -L_3c_3 \\
0 & c_3 & -s_3 & -L_3s_3 \\
1 & 0 & 0 & 0 \\
0 & 0 & 0 & 1
\end{bmatrix}
$$

$$
= \omega_3
\begin{bmatrix}
0 & \begin{matrix}-s_1c_2s_3 \\ -s_1s_2c_3\end{matrix} & \begin{matrix}-s_1c_2c_3 \\ +s_1s_2s_3\end{matrix} & \begin{matrix}L_3(-s_1c_2c_3 + s_1s_2s_3) \\ L_3(c_1c_2c_3 - c_1s_2s_3) \\ -L_3(s_2c_3 + c_2s_3) \\ 0\end{matrix} \\
0 & c_1c_2s_3 + c_1s_2c_3 & c_1c_2c_3 - c_1s_2s_3 & \\
0 & -s_2s_3 + c_2c_3 & -s_2c_3 - c_2s_3 & \\
0 & 0 & 0 &
\end{bmatrix}
$$

$$
= \omega_3
\begin{bmatrix}
0 & -s_1s_{23} & -s_1c_{23} & -L_3s_1c_{23} \\
0 & c_1s_{23} & c_1c_{23} & L_3c_1c_{23} \\
0 & c_{23} & -s_{23} & -L_3s_{23} \\
0 & 0 & 0 & 0
\end{bmatrix}
$$

Therefore, the available velocity equation is:

$$
\begin{bmatrix} \dot{x} \\ \dot{y} \\ \dot{z} \\ 0 \end{bmatrix} = \omega_1 \begin{bmatrix} -s_1 & -c_1 c_{23} & c_1 s_{23} & L_3 c_1 s_{23} + L_2 c_1 s_2 - d_1 s_1 \\ c_1 & -s_1 c_{23} & s_1 s_{23} & L_3 s_1 s_{23} + L_2 s_1 s_2 + d_1 c_1 \\ 0 & 0 & 0 & 0 \\ 0 & 0 & 0 & 0 \end{bmatrix} \begin{bmatrix} x_4 \\ y_4 \\ z_4 \\ 1 \end{bmatrix}
$$

$$
+ \omega_2 \begin{bmatrix} 0 & -s_1 s_{23} & -s_1 c_{23} & -L_3 s_1 c_{23} - L_2 s_1 c_2 \\ 0 & c_1 s_{23} & c_1 c_{23} & L_3 c_1 c_{23} + L_2 c_1 c_2 \\ 0 & c_{23} & -s_{23} & -L_3 s_{23} - L_2 s_2 \\ 0 & 0 & 0 & 0 \end{bmatrix} \begin{bmatrix} x_4 \\ y_4 \\ z_4 \\ 1 \end{bmatrix}
$$

$$
+ \omega_3 \begin{bmatrix} 0 & -s_1 s_{23} & -s_1 c_{23} & -L_3 s_1 c_{23} \\ 0 & c_1 s_{23} & c_1 c_{23} & L_3 c_1 c_{23} \\ 0 & c_{23} & -s_{23} & -L_3 s_{23} \\ 0 & 0 & 0 & 0 \end{bmatrix} \begin{bmatrix} x_4 \\ y_4 \\ z_4 \\ 1 \end{bmatrix}
$$

3. Acceleration equation [4–6]

In the formula, and are the angular accelerations of rotation around the axis and respectively.

$$
\begin{bmatrix} \ddot{x} \\ \ddot{y} \\ \ddot{z} \\ 0 \end{bmatrix} = \frac{d}{dt} \begin{bmatrix} \dot{x} \\ \dot{y} \\ \dot{z} \\ 0 \end{bmatrix}
$$

$$
\begin{aligned}
= [&(\omega_1{}^2 \theta_1 \theta_1 \phi_1 T_{12} \phi_2 T_{23} \phi_3 T_{34} + \omega_1 \omega_2 \theta_1 \phi_1 T_{12} \phi_2 T_{23} \phi_3 T_{34} \\
&+ \omega_1 \omega_3 \theta_1 \phi_1 T_{12} \phi_2 T_{23} \theta_3 \phi_3 T_{34} + \alpha_1 \theta_1 T_{12} \phi_2 T_{23} \phi_3 T_{34}) \\
&+ (\omega_2 \omega_1 \theta_1 \phi_1 T_{12} \theta_2 \phi_2 T_{23} \phi_3 T_{34} + \omega_2{}^2 \phi_1 T_{12} \theta_2 \theta_2 \phi_2 T_{23} \phi_3 T_{23} \\
&+ \omega_2 \omega_3 \phi_1 T_{12} \theta_2 \phi_2 T_{23} \theta_3 \phi_3 T_{34} + \alpha_2 \phi_1 T_{12} \theta_2 \phi_2 T_{12} \phi_3 T_{34}) \\
&+ (\omega_3 \omega_1 \theta_1 \phi_1 T_{12} \phi_2 T_{23} \theta_3 \phi_3 T_{34} + \omega_3 \omega_2 \phi_1 T_{12} \theta_2 \phi_2 T_{23} \theta_3 \phi_3 T_{34} \\
&+ \omega_3{}^2 \phi_1 T_{12} \phi_2 T_{23} \theta_3 \theta_3 \phi_3 T_{34} + \alpha_3 \phi_1 T_{12} \phi_2 T_{23} \theta_3 \phi_3 T_{34})] T_4 \\
= &(\ddot{T}_{31} + \ddot{T}_{32} + \ddot{T}_{33}) T_4 = \ddot{T}_3 T_4
\end{aligned}
$$

In the formula, $T_4 = [x_4 \quad y_4 \quad z_4 \quad 1]^T$, α_1, α_2 and α_3 are the angular accelerations of rotation around the axis z_1, z_2 and z_3 respectively.

$$\ddot{T}_{31} = \omega_1{}^2\theta_1\theta_1\phi_1 T_{12}\phi_2 T_{23}\phi_3 T_{34} + \omega_1\omega_2\theta_1\phi_1 T_{12}\theta_2\phi_2 T_{23}\phi_3 T_{34}$$
$$+ \omega_1\omega_2\theta_1\phi_1 T_{12}\phi_2 T_{23}\theta_3\phi_3 T_{34} + \alpha_1\theta_1 T_{12}\phi_2 T_{23}\phi_3 T_{34}$$
$$= \omega_1\theta_1\dot{T}_{31} + \omega_1\theta_1\dot{T}_{32} + \omega_1\theta_1\dot{T}_{33} + \alpha_1/\omega_1\dot{T}_{31}$$
$$= \omega_1\theta_1(\dot{T}_{31} + \dot{T}_{32} + \dot{T}_{33}) + \alpha_1/\omega_1\dot{T}_{31}$$
$$= \ldots$$

$$= \begin{bmatrix} -\omega_1{}^2 c_1 - \alpha_1 s_1 & \omega_1{}^2 s_1 c_{23} - \omega_1(\omega_2 + \omega_3) c_1 s_{23} - \alpha_1 c_1 c_{23} \\[2mm] -\omega_1{}^2 s_1 + \alpha_1 c_1 & -\omega_1{}^2 c_1 c_{23} - \omega_1(\omega_2 + \omega_3) s_1 s_{23} - \alpha_1 s_1 c_{23} \\[2mm] 0 & 0 \\[2mm] 0 & 0 \end{bmatrix}$$

$$\begin{array}{c} -\omega_1{}^2 s_1 s_{23} - \omega_1(\omega_2 + \omega_3) c_1 c_{23} + \alpha_1 c_1 s_{23} \\[2mm] \omega_1{}^2 c_1 s_{23} - \omega_1(\omega_2 + \omega_3) s_1 c_{23} - \alpha_1 s_1 s_{23} \\[2mm] 0 \\[2mm] 0 \end{array}$$

$$\begin{array}{c} -\omega_1{}^2 L_3 s_1 s_{23} - \omega_1(\omega_2 + \omega_3) L_3 c_1 c_{23} - \omega_1\omega_2 L_2 c_1 c_2 \\ + \omega_1{}^2 L_2 s_1 s_2 + \alpha_1 L_3 c_1 s_{23} + \alpha_1 L_2 c_1 s_2 - \alpha_1 d_1 s_1 + d_1\omega_1{}^2 c_1 \\ \omega_2{}^2 L_3 c_1 s_{23} - \omega_1(\omega_2 + \omega_3) L_3 s_1 c_{23} - \omega_1\omega_2 L_2 s_1 c_2 + \omega_1{}^2 L_2 c_1 s_2 \\ + \alpha_1 L_3 s_1 s_{23} + \alpha_1 L_2 s_1 s_2 + \alpha_1 d_1 c_1 + d_2\omega_1{}^2 s_1 \\ 0 \\ 0 \end{array}$$

$$\ddot{T}_{32} = \omega_2\omega_1\theta_1\phi_1 T_{12}\theta_2\phi_2 T_{23}\phi_3 T_{34} + \omega_2{}^2\phi_1 T_{12}\theta_2\theta_2\phi_2 T_{23}\phi_3 T_{34}$$
$$+ \omega_2\omega_3\phi_1 T_{12}\theta_2\phi_2 T_{23}\theta_3\phi_3 T_{34} + \alpha_2\theta_1 T_{12}\theta_2\phi_2 T_{23}\phi_3 T_{34}$$
$$= \omega_1\theta_1\dot{T}_{32} + \alpha_2/\omega_2\dot{T}_{32} + \omega_2{}^2[\phi_1 T_{12}\theta_2][\theta_2][\phi_2 T_{23}][\phi_3 T_{34}]$$
$$+ \omega_2\omega_3[\phi_1 T_{12}\theta_2][\phi_2 T_{23}][\theta_3][\phi_3 T_{34}]$$
$$= \ldots$$

$$
= \begin{bmatrix}
0 & -\omega_1\omega_2 c_1 s_{23} - \alpha_2 s_1 s_{23} - \omega_2(\omega_2+\omega_3)s_1 c_{23} \\[6pt]
0 & -\omega_1\omega_2 s_1 s_{23} + \alpha_2 c_1 s_{23} + \omega_2(\omega_2+\omega_3)c_1 c_{23} \\[6pt]
0 & \alpha_2 c_{23} - \omega_2(\omega_2+\omega_3)s_{23} \\[6pt]
0 & 0
\end{bmatrix}
$$

$$
\begin{aligned}
&-\omega_1\omega_2 c_1 c_{23} - \alpha_2 s_1 c_{23} + \omega_2(\omega_2+\omega_3)s_1 s_{23} \\
&-\omega_1\omega_2 s_1 c_{23} + \alpha_2 c_1 c_{23} - \omega_2(\omega_2+\omega_3)c_1 s_{23} \\
&\quad -\alpha_2 s_{23} - \omega_2(\omega_2+\omega_3)c_{23} \\
&\quad\quad 0
\end{aligned}
$$

$$
\left.\begin{aligned}
&-\omega_1\omega_2(L_3 c_1 c_{23} + L_2 c_1 c_2) - \alpha_2(L_3 s_1 c_{23} + L_2 s_1 c_2) \\
&\quad + \omega_2{}^2 L_2 s_1 s_2 + \omega_2(\omega_2+\omega_3)L_3 s_1 s_{23} \\
&-\omega_1\omega_2(L_3 s_1 c_{23} + L_2 s_1 c_2) + \alpha_2(L_3 c_1 c_{23} + L_2 c_1 c_2) \\
&\quad - \omega_2{}^2 L_2 c_1 s_2 - \omega_2(\omega_2+\omega_3)L_3 c_1 s_{23} \\
&-\alpha_2(L_3 s_{23} + L_2 s_2) - \omega_2{}^2 L_2 c_2 - \omega_2(\omega_2+\omega_3)L_3 c_{23} \\
&\quad\quad 0
\end{aligned}\right]
$$

$$
\begin{aligned}
\ddot{T}_{33} &= \omega_1\theta_1\dot{T}_{33} + \omega_3\omega_2\phi_1 T_{12}\theta_2\phi_2 T_{23}\theta_3\phi_3 T_{34} \\
&\quad + \omega_3{}^2\phi_1 T_{12}\phi_2 T_{23}\theta_3\theta_3\phi_3 T_{34} + \alpha_3/\omega_3 T_{33} \\
&= \cdots
\end{aligned}
$$

$$
= \begin{bmatrix}
0 & -(\omega_3\omega_1 c_1 + \alpha_3 s_1)s_{23} - \omega_3(\omega_2-\omega_3)s_1 c_{23} \\
0 & -(\omega_3\omega_1 s_1 - \alpha_3 c_1)s_{23} + \omega_3(\omega_2-\omega_3)c_1 c_{23} \\
0 & -\omega_3(\omega_2+\omega_3)s_{23} + \alpha_3 c_{23} \\
0 & 0
\end{bmatrix}
$$

$$
\begin{aligned}
&-(\omega_3\omega_1 c_1 + \alpha_3 s_1)c_{23} + \omega_3(\omega_2-\omega_3)s_1 s_{23} \\
&-(\omega_3\omega_1 s_1 - \alpha_3 c_1)c_{23} - \omega_3 + \omega_3(\omega_2-\omega_3)c_1 s_{23} \\
&\quad -\omega_3(\omega_2-\omega_3)c_{23} - \alpha_3 s_{23} \\
&\quad\quad 0
\end{aligned}
$$

$$
\left.\begin{aligned}
&-L_3(\omega_3\omega_1 c_1 + \alpha_3 s_1)c_{23} + L_3\omega_3(\omega_2-\omega_3)s_1 s_{23} \\
&-L_3(\omega_3\omega_1 s_1 - \alpha_3 c_1)c_{23} - L_3\omega_3(\omega_2-\omega_3)c_1 s_{23} \\
&\quad -L_3\omega_3(\omega_2+\omega_3)c_{23} - L_3\alpha_3 s_{23} \\
&\quad\quad 0
\end{aligned}\right]
$$

 Robotics: From Manipulator to Mobilebot

Then the acceleration equation can be obtained:

$$
\begin{bmatrix} \ddot{x} \\ \ddot{y} \\ \ddot{z} \\ 0 \end{bmatrix} = (\ddot{T}_{31} + \ddot{T}_{32} + \ddot{T}_{33}) \begin{bmatrix} x_4 \\ y_4 \\ z_4 \\ 1 \end{bmatrix} = \ddot{T}_3 \begin{bmatrix} x_4 \\ y_4 \\ z_4 \\ 1 \end{bmatrix}
$$

Where

$$
\ddot{T}_3 = \begin{bmatrix} -\omega_1{}^2 c_1 - \alpha_1 s_1 \\ -\omega_1{}^2 s_1 + \alpha_1 c_1 \\ 0 \\ 0 \end{bmatrix}
$$

$$
(\omega_1^2 - \omega_2^2 + \omega_3^2 - 2\omega_2\omega_3)s_1 c_{23} - 2\omega_1(\omega_2 + \omega_3)c_1 s_{23}
$$
$$
-\alpha_1 c_1 c_{23} - (\alpha_2 + \alpha_3)s_1 s_{23}
$$
$$
-(\omega_1^2 - \omega_2^2 + \omega_3^2 - 2\omega_2\omega_3)c_1 c_{23} - 2\omega_1(\omega_2 + \omega_3)s_1 s_{23}
$$
$$
-\alpha_1 s_1 c_{23} + (\alpha_2 + \alpha_3)c_1 s_{23} - (\omega_2 + \omega_3)^2 s_{23} + (\alpha_2 + \alpha_3)c_{23}
$$
$$
0
$$

$$
-(\omega_1^2 - \omega_2^2 + \omega_3^2 - 2\omega_2\omega_3)s_1 s_{23} - 2\omega_1(\omega_2 + \omega_3)c_1 s_{23}
$$
$$
-(\alpha_2 + \alpha_3)s_1 c_{23} + \alpha_1 c_1 s_{23}
$$
$$
(\omega_1^2 - \omega_2^2 + \omega_3^2 - 2\omega_2\omega_3)c_1 s_{23} - 2\omega_1(\omega_2 + \omega_3)s_1 c_{23}
$$
$$
+(\alpha_2 + \alpha_3)c_1 c_{23} + \alpha_1 s_1 s_{23} - (\omega_2 + \omega_3)^2 c_{23} - (\alpha_2 + \alpha_3)s_{23}
$$
$$
0
$$

$$
-L_3(\omega_1^2 - \omega_2^2 + \omega_3^2 - 2\omega_2\omega_3)s_1 s_{23} - 2\omega_1(\omega_2 + \omega_3)L_3 c_1 c_{23}
$$
$$
-L_3(\alpha_2 + \alpha_3)s_1 c_{23}
$$
$$
+L_3\alpha_1 c_1 s_{23} - 2L_2\omega_1\omega_2 c_1 c_2 - L_2\alpha_1 s_1 c_2
$$
$$
+L_2(\omega_1^2 + \omega_2^2)s_1 s_2 + L_2\alpha_1 c_1 s_2 + d_1\omega_1^2 c_1 - d_1\alpha_1 s_1
$$
$$
L_3(\omega_1^2 - \omega_2^2 + \omega_3^2 - 2\omega_2\omega_3)c_1 s_{23} - 2\omega_1(\omega_2 + \omega_3)L_3 s_1 c_{23}
$$
$$
+L_3(\alpha_2 + \alpha_3)c_1 c_{23} + L_3\alpha_1 s_1 s_{23} - 2L_2\omega_1\omega_2 s_1 c_2
$$
$$
-L_2\alpha_1 c_1 c_2 + L_2(\omega_1^2 + \omega_2^2)c_1 s_2 + L_2\alpha_1 s_1 s_2 - d_1\omega_1^2 s_1 + d_1\alpha_1 c_1
$$
$$
-L_3(\omega_1^2 - \omega_2^2 + \omega_3^2 - 2\omega_2\omega_3)c_{23} - L_3(\alpha_2 + \alpha_3)s_{23}
$$
$$
-L_2\alpha_2 s_2 - L_2\omega_2^2 c_2
$$
$$
0
$$

4.4 Chapter Summary

The study of robot dynamics is of special significance for fast-moving robots and their control. Section 1 of this chapter studies the problem of rigid body dynamics, focusing on the analysis of two methods of solving the dynamic equations of robot manipulators, namely the Lagrangian functional balance method and the Newton-Euler dynamic balance method.

Section 2 summarizes the steps to establish the Lagrangian equation based on the analysis of the two-link manipulator, and calculates the speed, kinetic energy and potential energy of a point on the manipulator link, and then derives the four-link manipulator dynamic equation and its simplified calculation formula.

In this chapter, after obtaining the general expression of the dynamic equation of the robot manipulator, the dynamic equation of the two-link manipulator and the velocity and acceleration equation of the three-link manipulator are analyzed and calculated in Section 3 for example. The calculations are quite complicated and must be very careful.

References

[1] Angeles, J. (2003). *Fundamentals of Robotic Mechanical Systems: Theory, Methods, and Algorithms*, 2nd Edition. New York: Springer.

[2] Bajd, T., Mihelj, M., Lenarcic, J., *et al.* (2010). *Robotics (Intelligent Systems, Control and Automation: Science and Engineering).* Springer.

[3] Cai, Z.X. (1986). Computer simulation of Stanford Robotic Manipulator. *The Second National Conference on Computer Simulation* (in Chinese).

[4] Cai, Z.X. (1988). *Robotics: Principles and Applications.* Changsha: Central South University of Technology Press (in Chinese).

[5] Cai, Z.X. and Xie, B. (2021). *Fundamentals of Robotics*, Chapter 4, 3rd Edition. Beijing: Mechanical Industry Press (in Chinese).

[6] Cai, Z.X. and Xie, B. (2021). *Robotics*, Chapter 4, 4th Edition. Beijing: Tsinghua University Press, 2022 (in Chinese).

[7] Cai, Z.X., *et al.* (2016). *Key Techniques of Navigation Control for Mobile Robots under Unknown Environment.* Beijing: Science Press.

[8] Craig, J.J. (2005). *Introduction to Robotics: Mechanics and Control*, 3rd Edition. Pearson Education, Inc.

[9] Craig, J.J. (2018). *Introduction to Robotics: Mechanics and Control*, 4th Edition. Pearson Education, Inc.

[10] Durham, W. (2013). *Aircraft Flight Dynamics and Control*. New York: John Wiley & Sons.

[11] Featherstone, R. (2014). *Rigid Body Dynamics Algorithms*. New York: Springer.

[12] Featherstone, R. and Orin, D. (2016). Dynamics. *Handbook of Robotics*, 2nd Edition, Siciliano, B. and Khatib, O. (Eds.). Springer, pp. 37–66.

[13] Fu, K.S., Gonzalez, R.C. and Lee, C.S.G. (1987). *Robotics: Control, Sensing, Vision, and Intelligence*. McGraw-Hill Book Company.

[14] Gudino-Lau, J. and Arteaga, M. (2005). Dynamic model and simulation of cooperative robots: A case study. *Robotica*, 23(5):615–624.

[15] Hamiton, J.F., Cipra, R.J., *et al.* (1984). *Design and Analysis of Robot Manipulators*. EE-597T, Purdue University.

[16] Huo, W. (2005). *Robot Dynamics and Control*. Beijing: Higher Education Press (in Chinese).

[17] Lynch, K.M. and Park, F.C. (2017). *Modern Robotics: Mechanics, Planning, and Control*. Cambridge University Press.

[18] Marchese, A.D., Tedrake, R. and Rus, D.L. (2016). Dynamics and trajectory optimization for a soft spatial fluidic elastomer manipulator. *International Journal of Robotics Research*, 35(8):1000–1019.

[19] Paul, R.P. (1981). *Robot Manipulators: Mathematics, Programming and Control*. MIT Press.

[20] Reza, N.J. (2010). *Theory of Applied Robotics: Kinematics, Dynamics, and Control*, 2nd Edition. Springer.

[21] Silver, W.M. (1982). On the equivalence of Lagrangian and Newton-Euler dynamics for manipulators. *International Journal of Robotics Research*, 1(2):60–70.

[22] Vukobratovvic, M. and Poykonjak, V. (1982). *Fundamentals of Robotics 1, Dynamics of Manipulation Robots: Theory and Applications*. Springer-Verlag.

[23] Xiong, Y.L., Li, W.L., Chen, W.B., *et al.* (2017). *Robotics Modeling, Control and Vision*. Wuhan: Huazhong University of Science and Technology Press (in Chinese).

Chapter 5

Manipulator Control

Starting from this chapter, we will discuss the control of the manipulator in order to design and select a reliable and applicable manipulator controller, and make the manipulator move according to the specified trajectory to meet the control requirements.

This chapter will start with by discussing the basic knowledge of manipulator control and transmission, and then introduce and analyze the position control, force control, force/position hybrid control, resolved motion control, adaptive control, and intelligent control of the manipulator. Among them, some control methods are more traditional, while others are more novel and need to be developed and improved further.

5.1 Overview of Manipulator Control and Transmission

As mentioned earlier, from a control point of view, the robotic manipulator system represents a redundant, multi-variable and essentially nonlinear control system, while at the same time a complex coupled dynamic system. Each control task itself is a dynamic task. In actual research, the manipulator control system is often simplified into a number of low-level subsystems to describe.

Table 5.1.　Classification and analysis methods of manipulator control.

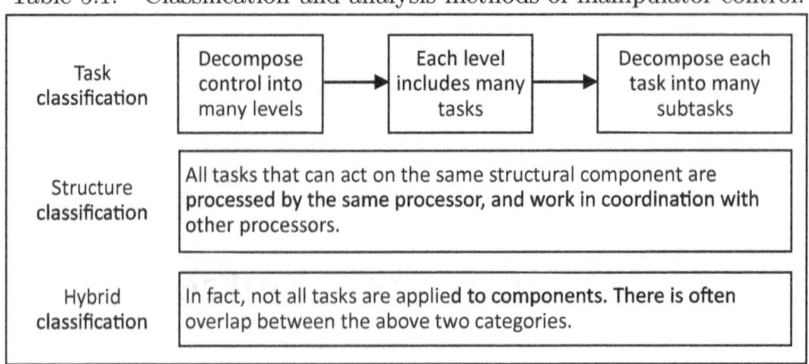

Task classification	Decompose control into many levels	→	Each level includes many tasks	→	Decompose each task into many subtasks
Structure classification	All tasks that can act on the same structural component are processed by the same processor, and work in coordination with other processors.
Hybrid classification	In fact, not all tasks are applied to components. There is often overlap between the above two categories.

5.1.1　*Classification, variables and levels of manipulator control* [1, 13, 14, 44]

1. Controller classification

Manipulator controllers have a variety of structural forms, including non-servo control, servo control, position and speed feedback control, force (torque) control, sensor-based control, nonlinear control, resolved acceleration control, sliding mode control, and optimal control, adaptive control, hierarchical control and various intelligent controls, etc.

Table 5.1 shows the main methods of classification and analysis of manipulator control systems.

The following discussion does not involve structural details but is related to control principles.

2. Main control variables

The control variables of each joint of the manipulator are shown in Figure 5.1. If you want to teach the manipulator to grasp the workpiece A, you must know the state of the end effector of the manipulator (such as the gripper) relative to A at any time, including the position, posture, and open/close state. The position of workpiece A is given by a set of coordinate axes of the worktable where it is located. This set of coordinate axes is called task axis (R_0). The state of the end effector is represented by many values or parameters of this set of coordinate axes, and these parameters are the components of the control vector X. The task is to find out for the change of the control vector X over time, that is $X(t)$, it represents the real-time position of the end-effector in space. Only when the joints from θ_1

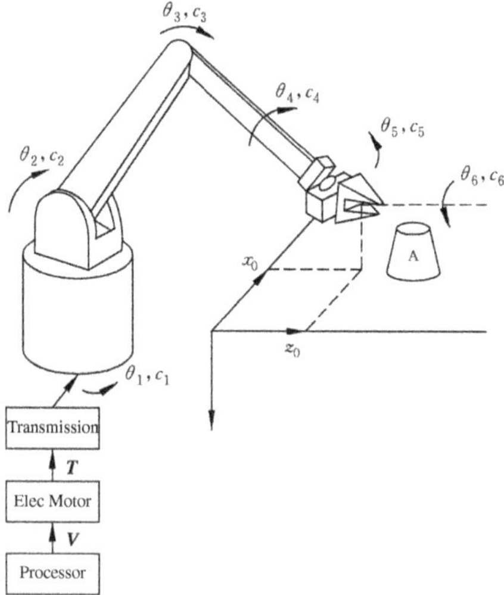

Figure 5.1.   The control variables of each joint of the manipulator.

to θ_6 is moved, then $\boldsymbol{X}$ is changed. Use the vector $\boldsymbol{\Theta}(t)$ to represent the joint variables the DC motor servo from θ_1 to θ_6.

Therefore, the control of a manipulator is essentially the control to the following two-way equations:

$$V(t) \leftrightarrow T(t) \leftrightarrow C(t) \leftrightarrow \Theta(t) \leftrightarrow X(t) \tag{5.1}$$

5.1.2   *Principle and transfer function of DC control system* [1, 19, 21, 27, 28]

The mathematical model of the DC motor servo control system is discussed below to prepare the necessary knowledge for studying the position controller of the manipulator.

Figure 5.2 shows the working principle diagram of a DC motor servo drive with a reduction gear and a rotating load. In the figure, the parameters of the servo motor are specified as follows:

r_f, l_f — Excitation loop resistance and inductance;
i_f, V_f — Excitation circuit current and voltage;
R_m, L_m — Resistance and inductance of armature loop;
i_m, V_m — The current and voltage of the armature circuit;
θ_m, ω_m — Angular displacement and speed of armature (rotor);
J_m, f_m — Moment of inertia and viscous friction coefficient of
        motor rotor;

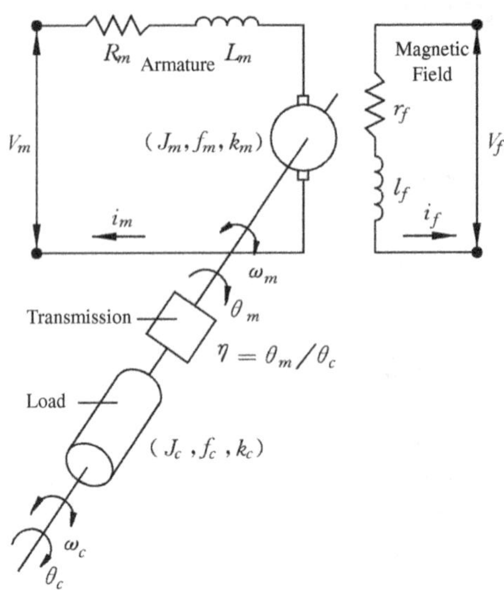

Figure 5.2.   Principle of DC motor servo drive.

T_m, k_m — Motor torque and torque constant;
k_e — The electric potential constant of the motor;
θ_c, ω_c — Load angular displacement and speed;
$\eta = \theta_m/\theta_c$ — Reduction ratio;
J_c, f_c — Load moment of inertia and load viscous friction
coefficient;
k_c — Load return coefficient.

These parameters are used to calculate the transfer function of the servo motor.

First, calculate the transfer function of the field control motor. The following equations can be established:

$$V_f = r_f i_f + l_f \frac{di_f}{dt} \tag{5.2}$$

$$T_m = k_m i_f \tag{5.3}$$

$$T_m = J\frac{d^2\theta_m}{dt^2} + F\frac{d\theta_m}{dt} + K\theta_m \tag{5.4}$$

In the formula, $J = J_m + J_c/\eta^2, F = f_m + f_c/\eta^2, K = k_c/\eta^2$ respectively represent the total moment of inertia of the drive system to the drive shaft, the total viscous friction coefficient and the total

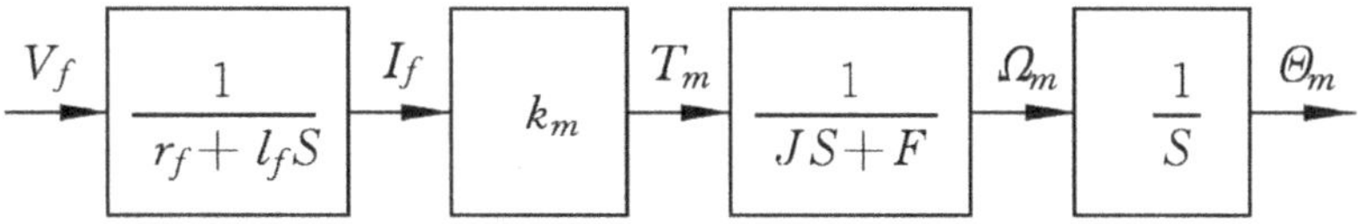

$$V_f \xrightarrow{\quad} \boxed{\dfrac{1}{r_f + l_f S}} \xrightarrow{I_f} \boxed{k_m} \xrightarrow{T_m} \boxed{\dfrac{1}{JS+F}} \xrightarrow{\Omega_m} \boxed{\dfrac{1}{S}} \xrightarrow{\Theta_m}$$

Figure 5.3.   Open-loop block diagram of excitation control DC motor with load.

feedback coefficient. Quoting the Laplace transform, the above three equations become:

$$V_f(S) = (r_f + l_f S)I_f(S) \tag{5.5}$$

$$T_m(S) = k_m I_f(S) \tag{5.6}$$

$$T_m(S) = (JS^2 + FS + K)\Theta_m(S) \tag{5.7}$$

The equivalent block diagram is shown in Figure 5.3.

According to equations (5.5) to (5.7), the open-loop transfer function of the motor can be obtained as follows:

$$\frac{\Theta_m(S)}{V_f(S)} = \frac{k_m}{(r_f + l_f S)(JS^2 + FS + K)} \tag{5.8}$$

In fact, it is often assumed that $K = 0$, so there are:

$$\frac{\Theta_m(S)}{V_f(S)} = \frac{k_m}{S(r_f + l_f S)(JS + F)}$$

$$= \frac{k_m}{r_f F} \cdot \frac{1}{S\left(1 + \frac{l_f}{r_f}S\right)\left(1 + \frac{J}{F}S\right)}$$

$$= \frac{k_0}{S(1 + \tau_e S)(1 + \tau_m S)} \tag{5.9}$$

In the formula, τ_e is the electrical time constant and τ_m is the mechanical time constant. Compared with τ_m, τ_e can be ignored, so

$$\frac{\Theta_m(S)}{V_f(S)} = \frac{k_0}{S(1 + \tau_m S)} \tag{5.10}$$

Because of $\omega_m = d\theta_m/dt$, the above formula becomes:

$$\frac{\Omega_m(S)}{V_f(S)} = \frac{k_0}{1 + \tau_m S} \tag{5.11}$$

Let's calculate the transfer function of the armature-controlled DC motor. At this time, the equation becomes:

$$V_m = R_m i_m + L_m \frac{di_m}{dt} + k_e \omega_m \tag{5.12}$$

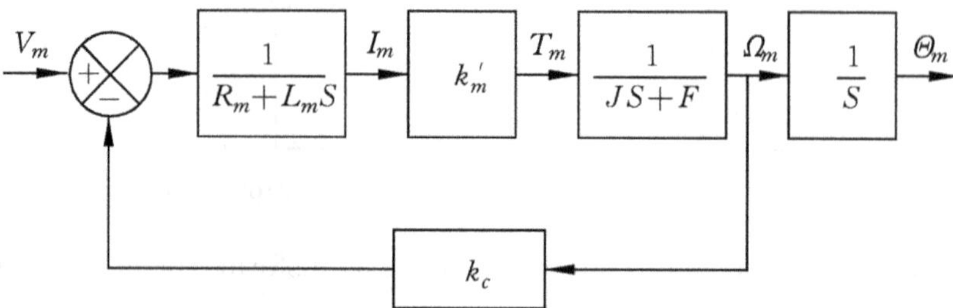

Figure 5.4. Block diagram of the armature control DC motor transmission device.

$$T_m = k'_m i_m \tag{5.13}$$

$$T_m = J\frac{d^2\theta_m}{dt^2} + F\frac{d\theta_m}{dt} + K\theta_m \tag{5.4}$$

In the formula, k_e is to consider the coefficient of the back electromotive force generated when the motor rotates, and this electric potential is proportional to the angular speed of the motor.

Using the same method described above, the following relations can be obtained:

$$\frac{\Theta_m(S)}{V_m(S)} = \frac{k'_m}{JL_m S^3 + (JR_m + FL_m)S^2 + (L_m K + R_m F + k'_m k_e)S + kR_m} \tag{5.14}$$

Taking into account $K \approx 0$ in fact, so the above formula becomes:

$$\frac{\Theta_m(S)}{V_m(S)} = \frac{k'_m}{S[(R_m + L_m S)(F + JS) + k_e k'_m]} \tag{5.15}$$

That is the required transfer function of the armature control DC motor. Figure 5.4 is its block diagram.

5.1.3   *Speed adjustment of DC motor*

Figure 5.5 shows a closed-loop position control structure diagram of an excitation-controlled DC motor. It is required that the required output position θ_o in the figure is equal to the input θ_i of the system. For this reason, θ_i is set by the input potentiometer, and θ_o is measured by the feedback potentiometer, and the two are compared, and the difference is amplified and sent to the excitation winding.

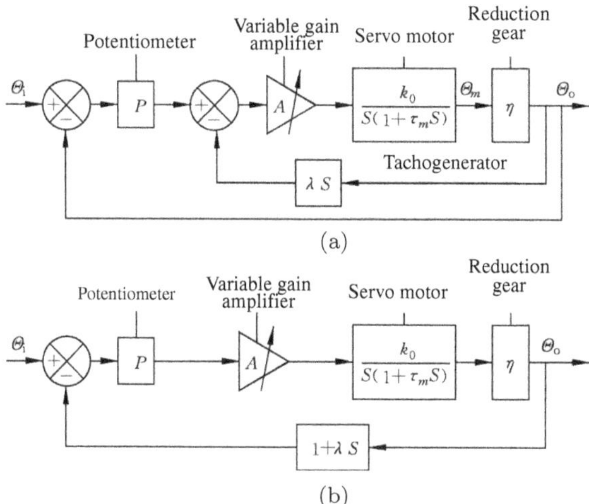

Figure 5.5. DC motor control principle diagram with speed feedback.

From the viewpoint of stability and accuracy, to obtain satisfactory servo drive performance, a compensation network must be introduced into the servo circuit. More precisely, compensation $e(t) = \theta_i(t) - \theta_o(t)$ related to the error signal must be introduced. Among them, $\theta_i(t)$ and $\theta_o(t)$ are the input and output displacements respectively. There are mainly the following four types of compensation:

Proportional compensation: proportional to $e(t)$.
Differential compensation: proportional to $de(t)/dt$ of the derivative of $e(t)$.
Integral compensation: proportional to $\int_0^t e(t)dt$ of the integral of $e(t)$.
Speed measurement compensation: proportional to the differential of the output position.

The first three types of compensation mentioned above belong to the feedforward control, while the speed measurement compensation belongs to the feedback control. In the actual system, at least two types of compensation should be used in combination, such as proportional-differential compensation (PD), proportional-integral compensation (PI) and proportional-integral-derivative compensation (PID).

When proportional-differential compensation is used, the output signal of the compensation component is:

$$e'(S) = (k + \lambda_d S)e(S) \tag{5.16}$$

When proportional-integral compensation is used, the output signal of the compensation component is:

$$e'(S) = \left(k + \frac{\lambda_i}{S}\right)e(S) \tag{5.17}$$

When proportional-derivative-integral compensation is used, the output signal of the compensation component is:

$$e'(S) = \left(k + \lambda_d S + \frac{\lambda_i}{S}\right)e(S) \tag{5.18}$$

When a tachogenerator is used to realize speed feedback, the compensation signal is:

$$e'(S) = e(S) - \lambda_t S \Theta_o(S) = \Theta_i(S) - (1 + \lambda_t S)\Theta_o(S) \tag{5.19}$$

In the above four equations, k, λ_d, λ_i and λ_t are proportional compensation coefficient, differential compensation coefficient, integral compensation coefficient and speed feedback coefficient. Figure 5.10 shows the control principle structure diagram of a DC motor with speed measurement feedback. In the Figure 5.5(a) and (b) are equivalent.

5.2   Position Control of the Manipulator

As a cascade link manipulator, its dynamic characteristics generally have a high degree of nonlinearity. To control such a manipulator driven by a motor, it is very important to use appropriate mathematical equations to express its motion. This kind of mathematical expression is a mathematical model, or model for short. The computer that controls the movement of the manipulator uses this mathematical model to predict and control the movement process to be carried out [32, 37, 38, 47, 52].

When designing the model, the following two assumptions were put forward:

(1) Each segment of the manipulator is an ideal rigid body, so all joints are ideal, and there is no friction and clearance.
(2) There is only one degree of freedom between two adjacent links, which are either fully rotating or fully translating.

5.2.1 *General structure of manipulator position control*

1. Basic control structure of manipulator

The operation of the manipulator is often to control the position and posture of the end tool of the manipulator to achieve point-to-point control (PTP control, such as handling and spot-welding robots) or continuous path control (CP control, such as arc welding and painting robots). Therefore, realizing the position control of the manipulator is the most basic control task of the manipulator. Manipulator position control is sometimes called pose control or trajectory control. For some operations, such as assembly, grinding, etc., only position control is not enough, and force control is also required.

The position control structure of the manipulator mainly has two forms, namely the joint space control structure and the rectangular coordinate space control structure, as shown in Figure 5.6(a) and (b) respectively.

In Figure 5.6(b), $q_d = [q_{d1}, q_{d2}, \ldots, q_{dn}]^T$ is the desired joint position vector, $\dot{q}_d$ and $\ddot{q}_d$ is the desired joint velocity vector and acceleration vector, q and $\dot{q}$ is the actual joint position vector and

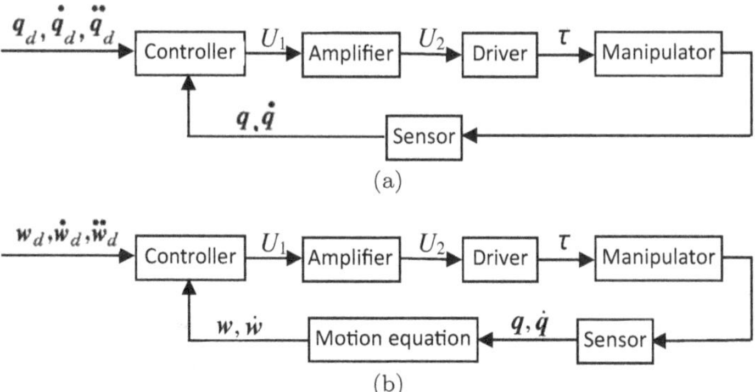

Figure 5.6. Basic structure of manipulator position control.

velocity vector. $\boldsymbol{\tau} = [\tau_1, \tau_2, \ldots, \tau_n]^T$ is the joint drive torque vector, and $\boldsymbol{U}_1, \boldsymbol{U}_2$ is the corresponding control vector.

In Figure 5.6(b), $w_d = [\boldsymbol{p}_d^T, \ \boldsymbol{\phi}_d^T]^T$ is the desired tool pose, where $\boldsymbol{p}_d = [x_d, \ y_d, \ z_d]$ is the desired tool position, and $\boldsymbol{\phi}_d$ is the desired tool pose. $\dot{\boldsymbol{w}}_d = [\boldsymbol{v}_d^T, \ \boldsymbol{\omega}_d^T]^T$, where $\boldsymbol{v}_d = [v_{d_x}, v_{d_y}, v_{d_z}]^T$ is the desired tool linear velocity, $\boldsymbol{w}_d = [\omega_{d_x}, \omega_{d_y}, \omega_{d_z}]^T$ is the desired tool angular velocity, $\ddot{\boldsymbol{w}}_d$ is the desired tool acceleration, $\boldsymbol{w}$ and $\dot{\boldsymbol{w}}$ is the actual tool pose and tool velocity. Industrial robots in operation generally adopt the control structure shown in Figure 5.6(a). The desired trajectory of the control structure is the position, velocity and acceleration of the joint, so it is easy to realize the servo control of the joint. The main problem of this control structure is: because it is often required to move the trajectory of the end of the manipulator in the Cartesian coordinate space, in order to achieve trajectory tracking, the desired trajectory of the end of the manipulator needs to be transformed into the desired trajectory expressed in the joint space by inverse kinematics trajectory.

2. Servo control structure of PUMA manipulator

Manipulator controllers are generally implemented by computers. The control structure of the computer has many forms, the common ones are centralized control, decentralized control and hierarchical control. The PUMA manipulator adopts a two-level hierarchical control structure.

The manipulator control system takes the manipulator as the control object, and its design method and parameter selection can still refer to the general computer control system. However, the design method of continuous system is still used more often, that is, the manipulator control system is first designed as a continuous system, and then the designed control law is discretized, and finally realized by a computer. For some design methods (such as the design method of self-tuning control), the direct discretization design method is adopted, that is, the manipulator control object model is discretized first, and then the discrete controller is directly designed, and then realized by the computer.

Most of the existing industrial robots adopt independent joint PID control. The control structure of the PUMA manipulator is a typical example. However, the independent joint PID control does not consider the nonlinearity of the controlled object (manipulator) and the coupling between the joints, which limits the improvement of control accuracy and speed. In addition to the independent joint PID control introduced in this section, some new control methods will be discussed in subsequent sections.

5.2.2 *Structure and model of single joint position controller* [13, 27–29]

Using conventional technology, it is possible to design the linear feedback controller of the manipulator by independently controlling each link or joint. The influence of gravity and the interaction between the joints can be eliminated by pre-calculated feedforward. In order to reduce the computational workload, the compensation signal is often approximate, or simplified calculation formulas are used.

1. Manipulator position control system structure

The number of joints of industrial manipulators available on the market is $3 \sim 7$. The most typical industrial manipulator has 6 joints, 6 degrees of freedom, and grippers (usually called hands or end effectors). Cincinnati-Milacron T3, Unimason's PUMA650 and Stanford manipulators are industrial manipulators with six joints, which are driven by hydraulic, pneumatic or electric drives. Among them, the Stanford manipulator has feedback control, and its joint control block diagram is shown in Figure 5.7. It can be seen from the figure that it has an optical encoder to form position and speed feedback together with the tachogenerator. This industrial manipulator is a positioning device, each of its joints has a position control system.

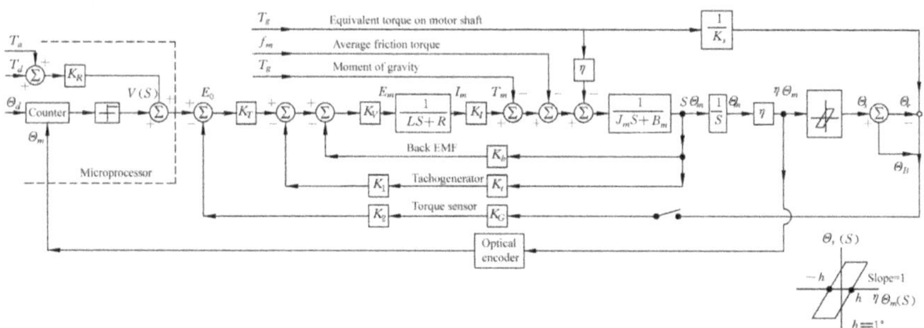

Figure 5.7. Block diagram of the position control system of the Stanford manipulator.

If there is no path constraint, the controller only needs to know that the gripper must pass through all the designated turning points on the path. The input of the control system is the Cartesian coordinates of the required turning points on the path. These coordinate points may be input in two ways, namely:

(1) Enter the system in digital form;
(2) Supply the system in teaching mode, and then perform coordinate transformation, that is, calculate the corresponding joint coordinates $[q_1, \ldots, q_6]$ of each designated turning point in the Cartesian coordinate system. The calculation method is related to the coordinate point signal input method.

For digital input mode, perform digital calculation $\boldsymbol{f}^{-1}[q_1, \ldots, q_6]$; for teaching input mode, perform analog calculation. Among them, $\boldsymbol{f}^{-1}[q_1, \ldots, q_6]$ is the inverse function of $\boldsymbol{f}[q_1, \ldots, q_6]$, and $\boldsymbol{f}[q_1, \ldots, q_6]$ is a vector function with six coordinate values. Finally, the manipulator's joint coordinate points are positioned and controlled point by point. If the manipulator is allowed to move only one joint in turn and lock the other joints, then each joint controller is very simple. If multiple joints move at the same time, the interaction of the forces between the joints will produce coupling, making the control system complicated.

2. Transfer function of single joint controller

Think of the manipulator as a rigid structure. Figure 5.8 shows a schematic diagram of a single joint motor gear-load combined device.

In the figure, J_a — the moment of inertia of the drive motor of a joint; J_m — the moment of inertia of the gripping load of a joint of the manipulator at the transmission end; J_l — the moment of inertia of the manipulator link; B_m — the damping coefficient of the

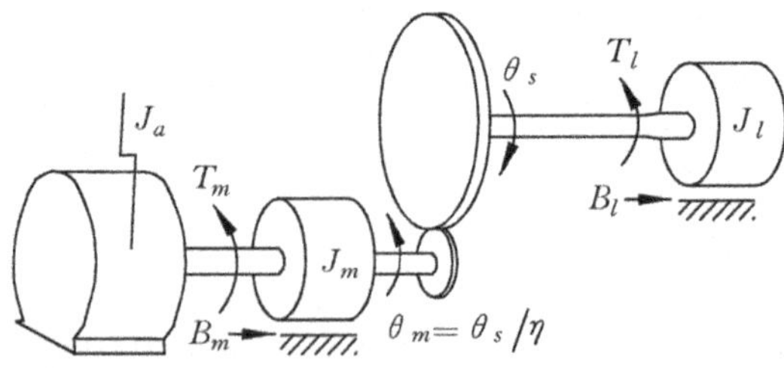

Figure 5.8.   Schematic diagram of a joint motor-gear-load combined device.

transmission end; B_l — load end damping coefficient; θ_m — drive end angular displacement; θ_s — load end angular displacement; N_m, N_s — number of gear teeth on the drive shaft and load; r_m, r_s — drive shaft and load shaft Gear pitch radius; $\eta = r_m/r_s = N_m/N_s$ — reduction gear transmission ratio.

Let $\boldsymbol{F}$ be the force acting on the meshing point of the gear transmitted from the motor to the load, then $T_1' = Fr_m$ is the equivalent load torque converted to the motor shaft, and

$$T_1 = Fr_s \tag{5.20}$$

And because of $\theta_m = 2\pi/N_m, \theta_s = 2\pi/N_s$, then

$$\theta_s = \theta_m N_m/N_s = \eta\theta_m \tag{5.21}$$

The relationship between the angular velocity and angular acceleration of the transmission side and the load side is as follows:

$$\dot{\theta}_s = \eta\dot{\theta}_m, \ddot{\theta}_s = \eta\ddot{\theta}_m$$

The load moment T_1 is used to overcome the effect of the inertia $J_1\ddot{\theta}_s$ of the link and the damping effect $B_1\dot{\theta}_s$, namely

$$T_1 = J_1\ddot{\theta}_s + B_1\dot{\theta}_s$$

Or rewritten as:

$$T_1 - B_1\dot{\theta}_s = J_1\ddot{\theta}_s \tag{5.22}$$

On the drive shaft side, according to the same principle the following can be obtained:

$$T_m = T_1' - B_m\dot{\theta}_m = (J_a - J_m)\ddot{\theta}_m \tag{5.23}$$

And the following two formulas are established:

$$T_1' = \eta^2(J_1\ddot{\theta}_m + B_1\dot{\theta}_m) \tag{5.24}$$

$$T_m = (J_a + J_m + \eta^2 J_1)\ddot{\theta}_m + (B_m + \eta^2 B_1)\dot{\theta}_m \tag{5.25}$$

Or

$$T_m = J\ddot{\theta}_m + B\dot{\theta}_m \tag{5.26}$$

Where $J = J_{eff} = J_a + J_m + \eta^2 J_1$ is the equivalent moment of inertia on the drive shaft; $B = B_{eff} = B_m + \eta^2 B_1$ is the equivalent damping coefficient on the drive shaft.

In the previous section, the transfer function of the armature control DC motor was established, see equation (5.14) and equation (5.15). Therefore, the transfer function similar to equation (5.15) is obtained as follows:

$$\frac{\Theta_m(S)}{V_m(S)} = \frac{K_I}{S[L_m J S^2 + (R_m J + L_m B)S + (R_m B + k_e K_I)]} \quad (5.27)$$

K_I and B in the formula are equivalent to and F in the formula (5.15).

Because

$$e(t) = \theta_d(t) - \theta_s(t) \quad (5.28)$$

$$\theta_s(t) = \eta \theta_m(t) \quad (5.29)$$

$$V_m(t) = K_\theta[\theta_d(t) - \theta_s(t)] \quad (5.30)$$

Its Laplace transform is:

$$E(S) = \Theta_d(S) - \Theta_s(S) \quad (5.31)$$

$$\Theta_s(S) = \eta \Theta_m(S) \quad (5.32)$$

$$V_m(S) = K_\theta[\Theta_d(S) - \Theta_s(S)] \quad (5.33)$$

Where K_θ is the transformation coefficient.

Figure 5.9(a) shows the block diagram of this position controller. From equation (5.27) to equation (5.33), the open-loop transfer function can be obtained as:

$$\frac{\Theta_s(S)}{E(S)} = \frac{\eta K_\theta K_I}{S[L_m J S^2 + (R_m J + L_m B)S + (R_m B + k_e K_I)]} \quad (5.34)$$

Since in fact $\omega L_m \ll R_m$, the terms L_m contained in the above formula can be ignored, and the above formula is simplified to:

$$\frac{\Theta_s(S)}{E(S)} = \frac{\eta K_\theta K_I}{S(R_m J S + R_m B + k_e K_I)} \quad (5.35)$$

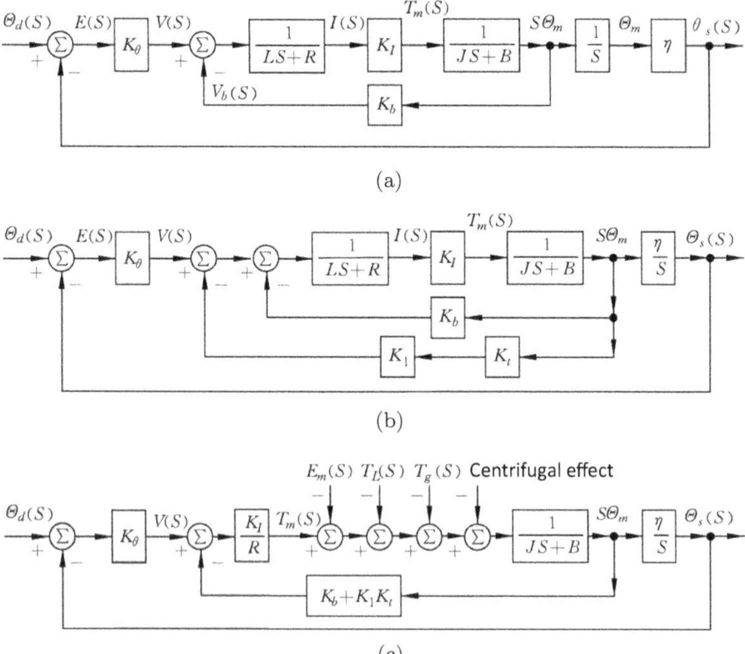

Figure 5.9.   Structure diagram of robot position controller.

Then find the closed-loop transfer function:

$$
\frac{\Theta_s(S)}{\Theta_d(S)} = \frac{\Theta_s(S)/E(S)}{1 + \Theta_s(S)/E(S)}
$$

$$
= \frac{\eta K_\theta K_I}{R_m J} \cdot \frac{1}{S^2 + (R_m B + K_I k_e)S/(R_m J) + K_\theta K/(R_m J)}
\tag{5.36}
$$

The obtained formula (5.36) is the closed-loop transfer function of the second-order system. In theory, it is always stable. To improve the response speed, it is usually necessary to increase the gain of the system (such as increase K_θ) and introduce some damping into the system by the negative feedback of the motor drive shaft speed to strengthen the effect of the back EMF. To do this, you can use a tachogenerator, or calculate the difference in angular displacement of the drive shaft within a certain time interval. Figure 5.15(b) shows the position control system with speed feedback. In the figure, K_t is the transfer coefficient of the tachogenerator; K_1 is the gain of the speed feedback signal amplifier. Because the feedback voltage of the motor armature circuit has changed from $k_e\theta_m(t)$ to $k_e\theta_m(t) + K_1 K_t\theta_m(t) = (k_e + K_1 K_t)\theta_m(t)$, therefore its open-loop transfer function and closed-loop transfer function have also become:

$$\frac{\Theta_s(S)}{E(S)} = \frac{\eta K_\theta}{S} \cdot \frac{S\theta_m(S)}{K_\theta E(S)}$$

$$= \frac{\eta K_\theta K_I}{R_m J S^2 + [R_m B + K_I(k_e + K_1 K_t)]S} \tag{5.37}$$

$$\frac{\Theta_s(S)}{\Theta_d(S)} = \frac{\Theta_s(S)/E(S)}{1 + \Theta_s(S)/E(S)}$$

$$= \frac{\eta K_\theta K_I}{R_m J S^2 + [R_m B + K_I(k_e + K_1 K_t)]S + \eta K_\theta K_I}$$

$$\tag{5.38}$$

For a specific robotic manipulator, its characteristic parameters and other values η, K_I, K_t, k_e, R_m, J and B are provided by the component manufacturer or determined through experiments. For example, the combined device of joint 1 and joint 2 of the Stanford manipulator includes U9M4T and U12M4T DC motors and 030/105 tachogenerators. The relevant parameters are shown in Table 5.2.

The effective moment of inertia of each joint of the Stanford-JPL manipulator is shown in Table 5.3.

It is worth noting that the conversion constant K_θ and amplifier gain K_1 must be determined according to the corresponding resonant frequency and damping coefficient of the manipulator structure.

The motor must overcome the average friction torque f_m, applied load torque T_1, gravitational torque T_g and centrifugal torque T_c of the motor-speed measuring unit. These physical quantities represent the effect of the actual additional load on the robotic manipulator. Insert these effects into the position controller block diagram in Figure 5.15(b) at the point where the motor generates the relevant torque, and then the control block diagram shown in Figure 5.15(c) can be obtained. In the figure, $F_m(S)$, $T_L(S)$ and $T_g(S)$ are the Laplace transform variables of f_m, T_l, and T_g respectively.

Table 5.2. Motor-tachometer unit parameter values.

Parameter / Type	U9M4T	U12M4T
$K_I(\text{oz·in/A})$	6.1	14.4
$J_a(\text{oz·in·S}^2/\text{rad})$	0.008	0.033
$B_m(\text{oz·in·S/rad})$	0.01146	0.04297
$k_e(\text{V·S/rad})$	0.04297	0.10123
$L_m(\mu\text{H})$	100	100
$R_m(\Omega)$	1.025	0.91
$K_t(\text{V·S/rad})$	0.02149	0.05062
$f_m(\text{pz·in})$	6.0	6.0
η	0.01	0.01

Table 5.3. Stanford-JPL manipulator effective inertia.

Joint number	Minimum value (with no load) $(\text{kg} \cdot \text{m}^2)$	Maximum value (with no load) $(\text{kg} \cdot \text{m}^2)$	Maximum (when fully loaded) $(\text{kg} \cdot \text{m}^2)$
1	1.417	6.176	9.570
2	3.590	6.950	10.300
3	7.257	7.257	9.057
4	0.108	0.123	0.234
5	0.114	0.114	0.225
6	0.040	0.040	0.040

5.2.3 *Coupling and compensation of multi-joint position controllers* [3, 27–30]

If the other joints of the manipulator are locked and one joint is moved in turn, this working method is obviously inefficient. This kind of work process makes the time to perform the prescribed tasks too long and is therefore uneconomical. However, if more than one joint is required to move at the same time, the forces and moments between the moving joints will interact, and the aforementioned position controller cannot be appropriately applied to each joint. Therefore, to overcome this interaction, additional compensation must be added. To determine this compensation, it is necessary to analyze the dynamic characteristics of the manipulator.

1. Lagrangian formula for dynamic equations

The dynamic equation represents the dynamic characteristics of a system. The general form of dynamic equations and Lagrangian equations are discussed in Chapter 4 of this book, as follows:

$$T_i = \frac{d}{dt}\frac{\partial L}{\partial \dot{q}_i} - \frac{\partial L}{\partial q_i}, \quad i = 1, 2, \ldots, n \tag{4.2}$$

$$T_i = \sum_{i=1}^{6} D_{ij}\ddot{q}_j + J_{ai}\ddot{q}_i + \sum_{j=1}^{6}\sum_{k=1}^{6} D_{ijk}\dot{q}_j\dot{q}_k + D_i \tag{4.24}$$

In the formula, $n = 6$, D_{ij}, D_{ijk} and D_i are represented by formula (4.25), formula (4.26) and formula (4.27) respectively.

Lagrange equation (4.26) and equation (4.27) is an important method to calculate the dynamic equation of the robotic manipulator system, and it is used to discuss and calculate the problems related to compensation.

2. Coupling and compensation between joints

It can be seen from equation (4.24) that the force or moment T_i required by each joint is composed of five parts. In the formula, the first term represents the effect of the inertia of all joints. In the case of a single joint movement, all other joints are locked, and the inertia of each joint is concentrated together. In the case of multiple joints moving at the same time, there is the effect of coupling inertia between the joints. These torque terms $\sum_{j=1}^{6} D_{ij}\ddot{q}_j$ must be input to the controller input of joint i through feedforward to compensate for the interaction between the joints, as shown in Figure 5.10. The second term in equation (4.24) represents the moment of inertia of the joint i transmission device with the equivalent moment of inertia J on the transmission shaft; it has been discussed in the single joint controller. The last term of equation (4.24) is obtained from the acceleration of gravity, which is also compensated by the feedforward term τ_a. This is an estimated gravitational moment signal and is calculated by the following formula

$$\tau_a = (R_m/KK_R)\bar{\tau}_g \tag{5.39}$$

Where $\bar{\tau}_g$ is the estimated value of the gravitational moment τ_g. Use D_i as the best estimate for the controller i. According to formula (4.27), the value $\bar{\tau}_g$ of joint i can be set.

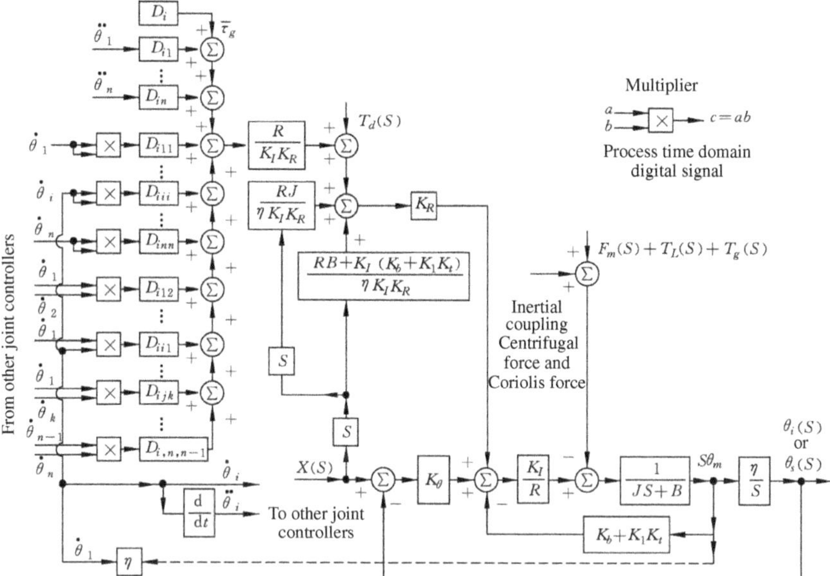

Figure 5.10. The i-th joint full controller with n joints.

The third and fourth terms in formula (4.24) represent the effects of centripetal force and Coriolis force, respectively. These torque terms must also be fed forward to the controller of joint i to compensate for the actual interaction between the joints, which is also shown in Figure 5.10. Figure 5.10 shows the complete block diagram of the joint i $(i = 1, 2, \ldots, n)$ controller of the industrial robotic manipulator. To realize these n controllers, the value D_{ij}, D_{ijk} and D_i of each feedforward component of a specific manipulator must be calculated.

3. Calculation of coupled inertia compensation

The calculation of D_{ij} is very complicated and time-consuming. In order to illustrate the difficulty of calculation, the formula (4.24) is extended as follows:

$$
\begin{aligned}
T_i =\ & D_{i1}\ddot{q}_1 + D_{i2}\ddot{q}_2 + \cdots + D_{i6}\ddot{q}_6 + J_{ai}\ddot{q}_i \\
& + D_{i11}\dot{q}_1^2 + D_{i22}\dot{q}_2^2 + \cdots \dot{q}_{i66}^2\dot{q}_6^2 \\
& + D_{i12}\dot{q}_1\dot{q}_2 + D_{i13}\dot{q}_1\dot{q}_3 + \cdots D_{i16}\dot{q}_1\dot{q}_6 \\
& + \cdots \\
& + D_{i45}\dot{q}_4\dot{q}_5 + \cdots D_{i56}\dot{q}_5\dot{q}_6 + D_i
\end{aligned}
\tag{5.40}
$$

For $i = 1$, $D_{i1} = D_{11}$ in the above formula. Let $\theta_i = q_i$, $i = 1, 2, \ldots 6$, then the expression of D_{11} is as follows:

$$D_{11} = m_1 k_{122}^2$$

$$+ m_2[k_{211}^2 s^2\theta_2 + k_{233}^2 c^2\theta_2 + r_2(2\bar{y}_2 + r_2)]$$

$$+ m_3[k_{322}^2 s^2\theta_2 + k_{333}^2 c^2\theta_2 + r_3(2\bar{z}_2 + r_3)s^2\theta_2 + r_2^2]$$

$$+ m_4\left\{ \frac{1}{2}k_{411}^2[s^2\theta_2(2s^2\theta_4 - 1) + s^2\theta_4] + \frac{1}{2}k_{422}^2(1 + c^2\theta_2 + s^2\theta_4) \right.$$

$$+ \frac{1}{2}k_{433}^2[s^2\theta_2(1 - 2s^2\theta_4) - s^2\theta_4] + r_3^2 s^2\theta_2 + r_2^2 - 2\bar{y}_4 r_3 s^2\theta_2$$

$$\left. + 2\bar{z}_4(r_2 s\theta_4 + r_3 s\theta_2 c\theta_2 c\theta_4) \right\}$$

$$+ m_5\left\{ \frac{1}{2}(-k_{511}^2 + k_{533}^2 + k_{533}^2)[(s\theta_2 s\theta_5 - c\theta_2 s\theta_4 c\theta_5)^2 \right.$$

$$+ c^2\theta_4 c^2\theta_5] + \frac{1}{2}(k_{511}^2 - k_{522}^2 - k_{533}^2)(s^2\theta_4 + c^2\theta_2 c^2\theta_4)$$

$$+ \frac{1}{2}(k_{511}^2 + k_{522}^2 - k_{533}^2)[(s\theta_2 c\theta_5 + c\theta_2 s\theta_4 s\theta_5)^2 + c^2\theta_4 c^2\theta_5]$$

$$\left. + r_3^2 s^2\theta_2 + r_2^2 + 2\bar{z}_5[r_3(s^2\theta_2 c\theta_5 + s\theta_2 s\theta_4 s\theta_5) - r_2 c\theta_4 s\theta_5] \right\}$$

$$+ m_6\left\{ \frac{1}{2}(-k_{611}^2 + k_{622}^2 + k_{633}^2)[(s\theta_2 s\theta_5 c\theta_6 - c\theta_2 s\theta_4 c\theta_5 c\theta_6 \right.$$

$$- c\theta_s c\theta_4 s\theta_6)^2 + (c\theta_4 c\theta_5 c\theta_6 - s\theta_4 s\theta_6)^2]$$

$$+ \frac{1}{2}(k_{611}^2 - k_{622}^2 + k_{633}^2)[(c\theta_2 s\theta_4 c\theta_5 s\theta_6 + s\theta_2 s\theta_5 s\theta_6$$

$$- c\theta_2 c\theta_4 c\theta_6)^2 + (c\theta_4 c\theta_5 s\theta_6 + s\theta_4 c\theta_6)^2]$$

$$+ \frac{1}{2}(k_{611}^2 + k_{622}^2 - k_{633}^2)[(c\theta_2 s\theta_4 s\theta_5 + s\theta_2 c\theta_5)^2 + c^2\theta_4 s^2\theta_5]$$

$$+ [r_6 c\theta_2 s\theta_4 s\theta_5 + (r_6 c\theta_5 + r_3)s\theta_2 + (r_6 c\theta_4 s\theta_5 - r_2)^2$$

$$+ 2z_6[r_6(s^2\theta_2 c^2\theta_5 + c^2\theta_4 s^2\theta_5 + c^2\theta_2 s^2\theta_5 + 2s\theta_2 c\theta_2 s\theta_4 s\theta_5 c\theta_5)$$

$$\left. + r_3(s\theta_2 c\theta_2 s\theta_4 s\theta_5 + s^2\theta_2 c\theta_5) - r_2 c\theta_4 s\theta_5] \right\}$$

It is not difficult to see that the calculation of D_{i1} is not a simple task. Especially when the manipulator is moving, if its position and attitude parameters change, then the calculation task is more difficult. Therefore, we are trying to find new ways to simplify this calculation. There are 3 simplification methods, namely geometric/digital method, hybrid method and differential transformation method.

Bejezy's geometric/digital method involves the characteristics of rotary joints and prismatic joints. It can compare equations (4.25) to (4.27) related to the calculation $\frac{\partial T_p}{\partial q_j}, \frac{\partial^2 T_p}{\partial q_j \partial q_k}$ of the fourth-order square matrix $\boldsymbol{J}_j^k$ (it can take any k-th square matrix, the vector represented by the two coordinate systems is transformed into the same vector represented by the j-th coordinate system), which is simplified in advance [3]. Since many elements in the fourth-order square matrix are zero, the obtained expressions of $_i D_i$, D_{ij}, and D_{ijk} are not as complicated as the original ones. The hybrid method proposed by Luh and Lin first compares all the terms of the Newton-Euler formula in the dynamic equation with a computer, and then deletes some of them according to various criteria [30]. Finally, put the remaining items back into the Lagrangian equation. The result of this method is a computer output of a simplified equation expressed in symbolic form. Paul used differential transformation to simplify terms such as D_i, D_{ij} and D_{ii}; this method of simplification of differential transformation has been discussed in the second section of Chapter 4 of this book [34].

5.3 Force and Position Hybrid Control of the Manipulator

5.3.1 *Force and position hybrid control scheme* [7–9, 14, 40]

There are many ways to control the force of the manipulator. Here are a few typical solutions.

1. Active rigid control [43]

Figure 5.11 shows a block diagram of an active stiffness control system. In the figure, J is the Jacobian matrix of the end effector of the manipulator; K_p is the rigid diagonal matrix defined in the end Cartesian coordinate system, and its elements are determined by

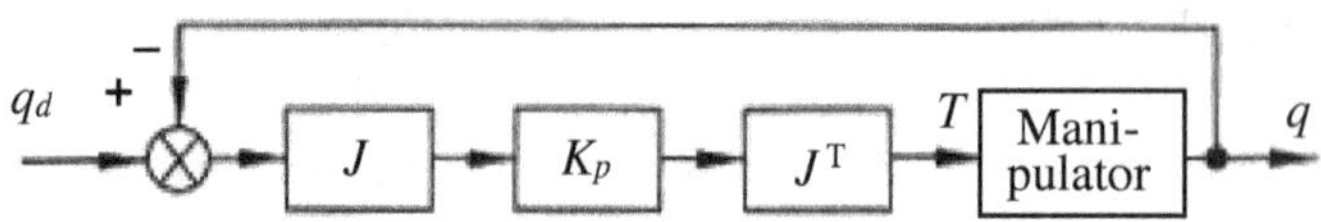

Figure 5.11.   Active rigid control block diagram.

humans. If you want to encounter actual constraints in a certain direction, then the rigidity in this direction should be reduced to ensure lower structural stress; on the contrary, in some directions that do not want to encounter actual constraints, the rigidity should be increased. This allows the manipulator to closely follow the desired trajectory. Therefore, it is possible to adapt to changing work requirements by changing the rigidity.

2. Raibert-Craig position/force hybrid controller [15, 36]

Raibert and Craig conducted important experiments on the position and force hybrid control of the robotic manipulator in 1981 and achieved good results. Later, this kind of controller was called R-C controller.

Figure 5.12 shows the structure of the R-C controller. In the figure, S and $\bar{S}$ are the adaptive selection matrix; x_d and F_d are the desired position and force trajectory defined in the Cartesian coordinate system; $P(q)$ is the kinematics equation of the manipulator; ^cT is the force transformation matrix.

This kind of R-C controller does not consider the influence of the dynamic coupling of the manipulator, which will cause the manipulator to be unstable in some non-singular positions in the working space. After in-depth analysis of the problems in the R-C system, the following improvements can be made:

(1) Consider the dynamic influence of the manipulator in the hybrid controller, and compensate the gravity, Coriolis force and centripetal force of the manipulator.
(2) Considering the under-damping characteristics of the force control system, damping feedback is added to the force control loop to weaken the oscillation factor.
(3) Introduce acceleration feedforward to meet the requirements of the operation task for acceleration, and it can also make the speed transition smoothly.

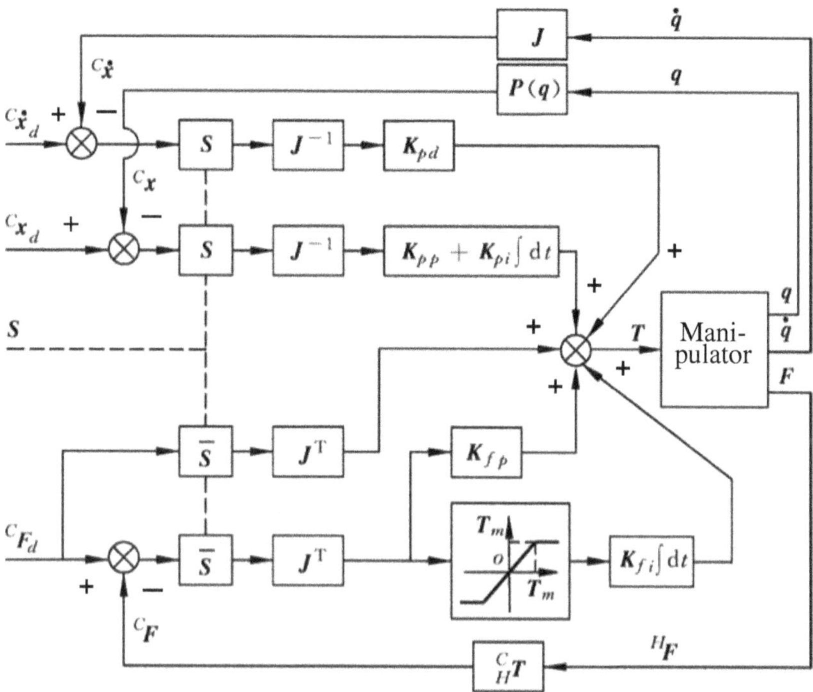

Figure 5.12.   Structure of R-C controller.

The structure diagram of the improved R-C force/position hybrid control system is shown in Figure 5.13. In the figure, $\hat{M}(q)$ is the inertia matrix model of the manipulator.

3. Operating space force and position hybrid control system [11, 12, 22]

Since the robotic manipulator is operated by tools, the dynamic performance of the end tool will directly affect the quality of operation. And because the motion of the end is a complex function of all joint motions, even if the dynamic performance of each joint is feasible, the dynamic performance of the end may not meet the requirements. When the dynamic friction and the flexibility of the link are particularly significant, the use of traditional servo control technology will not be able to guarantee the operating requirements. Therefore, it is necessary to directly establish a control algorithm in the $\{C\}$ coordinate system to meet operational performance requirements. Figure 5.14 is the structure diagram of the operating space force and

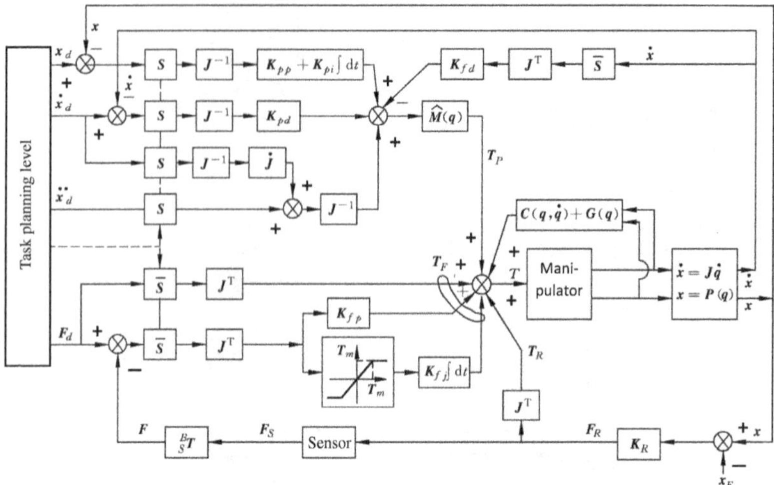

Figure 5.13.　Improved R-C hybrid control system structure.

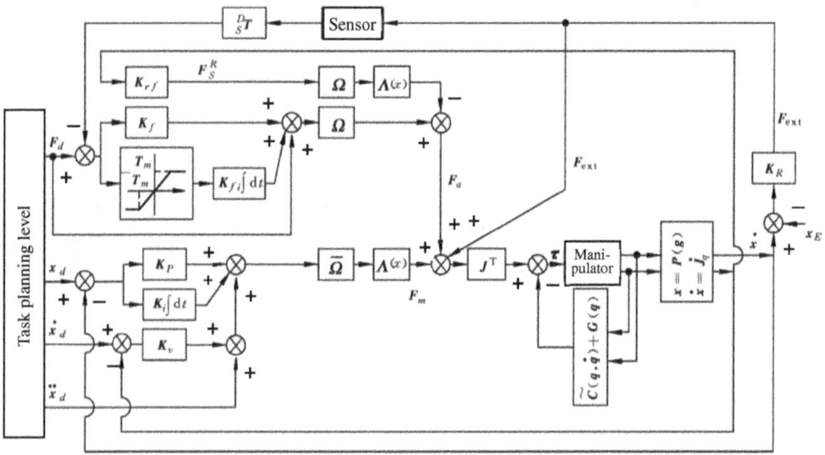

Figure 5.14.　Block diagram of operating space force/position hybrid control system.

position hybrid control system designed by O. Khatib [22]. In the figure, $\Lambda(x) = J^{-T}M(q)J^{-1}$ is the kinetic energy matrix at the end of the manipulator; $\tilde{C}(q,\dot{q}) = C(q,\dot{q}) - J^T\Lambda(x)\dot{J}\dot{q}$; K_p, K_v, K_i and K_i, K_{vf} and K_{ji} are the PID constant gain diagonal matrix.

In addition, there are resistance control and speed/force hybrid control.

5.3.2 *Synthesis of control law of force and position hybrid control system* [36, 47]

Take R-C controller as an example to discuss the control law of force/position hybrid control system.

1. Position control law

Disconnect all the force feedforward and force feedback channels in Figure 5.14 and set $\bar{S}$ the zero matrix, S the identity matrix, and the constraint reaction force is zero, then the system becomes a Coriolis force, gravity and centripetal force compensation and acceleration feedforward standard PID position control system. The integral component in the figure is used to improve the steady-state accuracy of the system. When the role of the integral component is not considered, the system controller equation is:

$$T = \hat{M}(q)[J^{-1}(\ddot{x}_d - \dot{J}J^{-1}\dot{x}_d) + K_{pd}J^{-1}(\dot{x}_d - \dot{x}) + K_{pp}J^{-1}(x_d - x)]$$
$$+ C(q, \dot{q}) + G(q)$$

Or

$$T = \hat{M}(q)[\ddot{q}_d + K_{pd}(\dot{q}_d - \dot{q}) + K_{pp}(q_d - q)] + C(q, \dot{q}) + G(q) \quad (5.41)$$

Make

$$\Delta q = J^{-1}(x - x_d) = J^{-1}\Delta x = q - q_d \quad (5.42)$$

Substitute formula (5.41) into the following dynamic equation of the manipulator

$$T = M(q)\ddot{q} + C(q, \dot{q}) + G(q) - J^T F_{ext} \quad (5.43)$$

In the formula, take $\hat{M}(q) = M(q)$, and F_{ext} is the external restraint reaction moment imposed on the end. Then the dynamic equation of the closed-loop system can be obtained:

$$\Delta \ddot{q} = K_{pd}\Delta \dot{q} + K_{pp}\Delta q = 0 \quad (5.44)$$

Taking K_{pp} and K_{pd} as diagonal matrices, the system becomes a decoupled unit mass second-order system. The choice of gain matrix K_{pp} and K_{pd} is best to make the dynamic response characteristics of each joint of the manipulator approximate to a critically damped state, or a bit over-damped state, because the control of the manipulator is not allowed to overshoot.

Take

$$K_{pd} = 2\xi\omega_n I$$
$$K_{pp} = \omega_n^2 I$$

$$(5.45)$$

In the formula, I is the identity matrix, ω_n is the natural oscillation frequency of the system, and ξ is the damping ratio of the system. Generally $\xi \geq 1$. If $\omega_n = 20$ and $\xi = 1$, then $K_{pd} = 40I$, $K_{pp} = 400I$.

The integral gain K_{pi} should not be selected too large, otherwise, when the initial deviation of the system is large, it will cause instability.

2. Force control law

Suppose the position compliant selection matrix $S = 0$ in Figure 5.13, and the control end receives a reaction force in the direction of the base coordinate system z_0. Assuming that the constrained surface is a rigid body, and the end force is shown in Figure 5.15, then the force control selection matrix for force control of the three-link manipulator is:

$$\bar{S} = \begin{bmatrix} 0 & 0 & 0 \\ 0 & 0 & 0 \\ 0 & 0 & 1 \end{bmatrix}$$

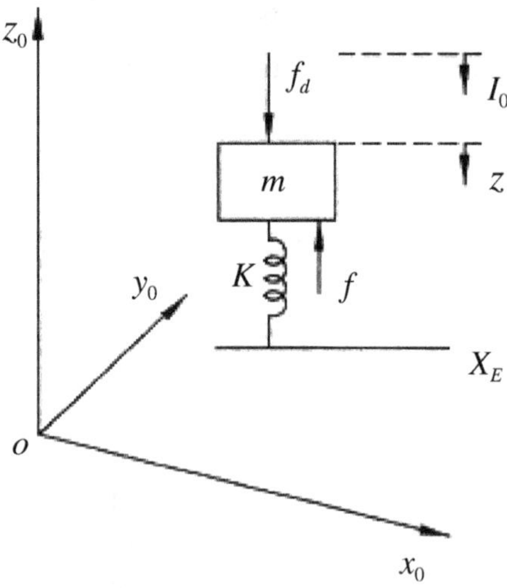

Figure 5.15.   Force diagram at the end of the manipulator.

Expectation is:

$$F_d = \begin{bmatrix} 0 \\ 0 \\ -f_d \end{bmatrix} \tag{5.46}$$

The binding reaction force is:

$$F_R = \begin{bmatrix} 0 \\ 0 \\ f \end{bmatrix} \tag{5.47}$$

In the formula, the value of f is determined by the length of the spring and whether the end is in contact with the restraining surface. Then the controller equation without considering the integral action is:

$$T = J^T \bar{S} F_d + K_{fp} J^T \bar{S}(F_d + F_R) + C(q,\dot{q}) + G(q) - M(q) K_{fd} J^T \bar{S} J \dot{q} \tag{5.48}$$

Substituting formula (5.48) into $\delta x = J \delta q$ [7,9], then get:

$$\begin{bmatrix} \ddot{q}_1 \\ \ddot{q}_2 \\ \ddot{q}_s \end{bmatrix} + K_{fd} J^T \bar{S} J \begin{bmatrix} \dot{q}_1 \\ \dot{q}_2 \\ \dot{q}_3 \end{bmatrix} = M^{-1}(q)(I + K_{fp}) J^T \begin{bmatrix} 0 \\ 0 \\ f - f_d \end{bmatrix} \tag{5.49}$$

Where

$$K_{fd} J^T \bar{S} J = \begin{bmatrix} K_{fd1} & 0 & 0 \\ 0 & K_{fd2} & 0 \\ 0 & 0 & K_{fd3} \end{bmatrix} \begin{bmatrix} J_{11} & J_{21} & J_{31} \\ J_{12} & J_{22} & J_{32} \\ J_{13} & J_{23} & J_{33} \end{bmatrix} \begin{bmatrix} 0 & 0 & 0 \\ 0 & 0 & 0 \\ 0 & 0 & 1 \end{bmatrix}$$

$$\times \begin{bmatrix} J_{11} & J_{12} & J_{13} \\ J_{21} & J_{22} & J_{23} \\ J_{31} & J_{32} & J_{33} \end{bmatrix} = \begin{bmatrix} 0 & 0 & 0 \\ 0 & K_{fd2} J_{32}^2 & K_{fd2} J_{32} J_{33} \\ 0 & K_{fd3} J_{32} J_{33} & K_{fd3} J_{33}^2 \end{bmatrix}$$

Let

$$M^{-1}(q) = \begin{bmatrix} a & 0 & 0 \\ 0 & b & c \\ 0 & c & d \end{bmatrix}$$

For the resistance control, have

$$\dot{V} = \dot{x}^T K_p \tilde{x} - \dot{q}^T J^T (K_q \tilde{x} + K_D \dot{\tilde{x}})$$

Make the left side of formula as

$$\begin{bmatrix} 0 \\ [J_{32}b(1 + K_{fp2}) + J_{33}c(1 + K_{fp3})](f - f_d) \\ [J_{32}c(1 + K_{fp2}) + J_{33}d(1 + K_{fp3})](f - f_d) \end{bmatrix} \triangleq \begin{bmatrix} 0 \\ H_1(f - f_d) \\ H_2(f - f_d) \end{bmatrix}$$

The left side of formula (5.110) equal:

$$\begin{bmatrix} \ddot{q}_1 \\ \ddot{q}_2 + J_{32}^2 K_{fd2}\dot{q}_2 + J_{32}J_{33}K_{fd2}\dot{q}_3 \\ \ddot{q}_3 + J_{33}^2 K_{fd3}\dot{q}_3 + J_{32}J_{33}K_{fd3}\dot{q}_2 \end{bmatrix}$$

Then the dynamic equation of the closed-loop system:

$$\begin{cases} \ddot{q}_1 = 0 \\ \ddot{q}_2 + J_{32}^2 K_{fd2}\dot{q}_2 + J_{32}J_{33}K_{fd2}\dot{q}_3 = H_1(f - f_d) \\ \ddot{q}_3 + J_{33}^2 K_{fd3}\dot{q}_3 + J_{32}J_{33}K_{fd3}\dot{q}_2 = H_2(f - f_d) \end{cases} \qquad (5.50)$$

In the formula, $H_1 > 0, H_2 > 0$.

Equation (5.50) shows that joint 1 has no effect on force control, and joints 2 and 3 have an effect on force control. At the beginning of the dynamic equation or when the contact has just occurred, if the constraint surface is rigid, there are often $f \gg f_d$; If the feedback proportional gain is selected too large, joints 2 and 3 will inevitably accelerate or decelerate, and the end of the manipulator will keep on collision with the contact surface may even cause the system to oscillate. The larger the force feedback damping gain K_{fd} is, the more stable the system is, but the speed becomes worse. The choice of K_{fd} is related to many factors. The integral gain K_{ji} should not be too large, and a non-linear limiter should be connected in series in front of it. Because when the end collides with the constrained surface, the force deviation signal is very large.

3. Hybrid control law of force and position

Set the constraint coordinate system to coincide with the base coordinate system. If the job requires force control in the z_0 direction of the base coordinate system and position control on a constrained surface parallel to the x_0y_0 plane, the compliant selection matrix is

$$\text{Position: } S = \begin{bmatrix} 1 & 0 & 0 \\ 0 & 1 & 0 \\ 0 & 0 & 0 \end{bmatrix}, \text{ and Force: } \bar{S} = \begin{bmatrix} 0 & 0 & 0 \\ 0 & 0 & 0 \\ 0 & 0 & 1 \end{bmatrix}$$

Expected end trajectory:

$$x_d(t) = \begin{bmatrix} x_d & y_d & z_d \end{bmatrix}^T$$
$$F_d(t) = \begin{bmatrix} 0 & 0 & -f_d \end{bmatrix}^T \tag{5.51}$$

Actual end trajectory:

$$x(t) = \begin{bmatrix} x & y & z \end{bmatrix}^T$$
$$F_{ext}(t) = \begin{bmatrix} 0 & 0 & f \end{bmatrix}^T$$

Therefore, for the system in Figure 5.12:

$$T_p = M(q)[J^{-1}(S\ddot{x}_d - \dot{J}J^{-1}S\dot{x}_d] + K_{pd}J^{-1}S(\dot{x}_d - \dot{x})$$
$$+ K_{pp}J^{-1}S(x_d - x) - M(q)K_{pd}J^T\bar{S}\dot{x}$$
$$T_F = J^T\bar{S}F_d + K_{FP}J^{-1}\bar{S}(F_d + F) \tag{5.52}$$

Control input to the manipulator:

$$T = T_p + T_F + C(q,\dot{q}) + G(q) \tag{5.53}$$

Substituting formula (5.53) into formula $\delta x = J\delta q$ [7,9], we can get:

$$M(q)\ddot{q} = T_p + T_F + J^T F_{ext} \tag{5.54}$$

Or

$$M(q)J^{-1}(\ddot{x} - \dot{J}J^{-1}\dot{x}) = T_p + T_F + J^T F_{ext} \tag{5.55}$$

Substituting formula (5.52) into formula (5.55), we get:

$$J^{-1}\begin{bmatrix} \Delta\ddot{x} \\ \Delta\ddot{y} \\ 0 \end{bmatrix} + (K_{pd} - J^{-1}\dot{J})J^{-1}\begin{bmatrix} \Delta\dot{x} \\ \Delta\dot{y} \\ 0 \end{bmatrix} + K_{pp}J^{-1}\begin{bmatrix} \Delta x \\ \Delta y \\ 0 \end{bmatrix} + J^{-1}\begin{bmatrix} 0 \\ 0 \\ \ddot{z} \end{bmatrix}$$

$$+ (K_{fd} - J^{-1}\dot{J})J^{-1}\begin{bmatrix} 0 \\ 0 \\ \dot{z} \end{bmatrix} = \begin{bmatrix} 0 \\ H_1 \cdot \Delta f \\ H_2 \cdot \Delta f \end{bmatrix} \tag{5.56}$$

In the formula, $\Delta x = x - x_d, \Delta y = y - y_d, \Delta f = f - f_d, H_1$ and H_2 see formula (5.50).

If take

$$\begin{cases} K_{pd} = J^{-1}K'_{pd}J + J^{-1}\dot{J} \\ K_{pp} = J^{-1}K'_{pp}J \\ K_{fd} = J^{-1}K'_{fd}J + J^{-1}\dot{J} \end{cases} \qquad (5.57)$$

In the formula, K'_{pd}, K'_{pp} and K'_{fd} are all positive definite diagonal matrices. Substituting formula (5.57) into formula

$$H\ddot{q} + C\dot{q} + q(q) = \tau + J^T F_E = \tau - J^T K_E \tilde{x}_E$$

can get:

$$\begin{bmatrix} \Delta\ddot{x} + K'_{pd1}\Delta\dot{x} + K'_{pp1}\Delta x \\ \Delta\ddot{y} + K'_{pd2}\Delta\dot{y} + K'_{pp2}\Delta y \\ \ddot{z} + K'_{fd3}\dot{z} \end{bmatrix} = J \begin{bmatrix} 0 \\ H_1 \cdot \Delta f \\ H_2 \cdot \Delta f \end{bmatrix} = \begin{bmatrix} J_{12}H_1 + J_{13}H_2 \\ J_{22}H_1 + J_{23}H_2 \\ J_{32}H_1 + J_{33}H_2 \end{bmatrix} \Delta f \qquad (5.58)$$

Because of $J_{32}H_1 + J_{33}H_2 > 0$, take $K'_{fd3} > 0$, then the third equation in equation (5.48) is stable.

Equation (5.58) shows that only when $\Delta f = 0$, the force control and position control in the force and position hybrid control system do not affect each other. Simulation experiments confirmed this conclusion. Therefore, the performance of the force control subsystem in the hybrid control system plays an important role in the entire system.

5.4   Resolved Motion Control of Manipulator

In actual operation, each joint of the robotic manipulator does not move independently, but moves in a coordinated manner, that is, each joint is controlled at a coordinated position and speed. It is necessary to study the resolved motion control of manipulator, including resolved motion speed control, resolved motion acceleration control, and resolved motion force (or torque) control. This section will discuss these issues one by one [2, 17, 20, 33].

5.4.1 *Principle of resolved motion control*

Resolved motion means the joint motion of each joint motor and decomposed into independent controllable motion along each Cartesian coordinate axis. This requires that the drive motors of several joints must run at different time-varying speeds at the same time in order to achieve the required coordinated movement along each coordinate axis. This kind of motion allows the manipulator user to specify the direction and speed along the path in any direction followed by the manipulator. Since users are usually more suitable for using Cartesian coordinates instead of manipulator's joint angle coordinates, resolved motion control can greatly simplify the technical requirements for the motion sequence to complete a certain task.

Generally, the expected motion of the manipulator is specified in accordance with the trajectory of the gripper expressed in Cartesian coordinates, but the servo control system requires the reference input to be specified by the joint coordinates. To design an effective control in the Cartesian coordinate space, it is very important to handle the mathematical relationship between the two coordinate systems. We will recall and further describe the basic kinematics theory between the two coordinate systems of a six-axis manipulator to help us understand various important resolved motion control methods.

The position of the gripper of the manipulator relative to the fixed reference coordinate system can be realized by establishing an orthogonal coordinate system for the gripper. Figure 3.1, Figure 3.3, formula (3.2), formula (3.62), formula (3.102), and formula (3.103) have shown these elements, which are represented by the 4×4 homogeneous transformation matrix as:

$$
{}^{0}\boldsymbol{T}_6(t) = \begin{bmatrix} n_x(t) & o_x(t) & a_x(t) & p_x(t) \\ n_y(t) & o_y(t) & a_y(t) & p_y(t) \\ n_z(t) & o_z(t) & a_z(t) & p_z(t) \\ 0 & 0 & 0 & 1 \end{bmatrix} = \begin{bmatrix} \boldsymbol{n}(t) & \boldsymbol{o}(t) & \boldsymbol{a}(t) & \boldsymbol{p}(t) \\ 0 & 0 & 0 & 1 \end{bmatrix}
$$

$$
= \begin{bmatrix} {}^{0}\boldsymbol{R}_6^{(t)} & \boldsymbol{p}(t) \\ 0 & 1 \end{bmatrix} \tag{5.59}
$$

$$
{}^{0}\boldsymbol{R}_6(t) = \begin{bmatrix} n_x(t) & o_x(t) & a_x(t) \\ n_y(t) & o_y(t) & a_y(t) \\ n_z(t) & o_z(t) & a_z(t) \end{bmatrix} \tag{5.60}
$$

In the formula, the meaning of each vector has been explained in Chapter 3. The rotation matrix $^0\boldsymbol{R}_6(t)$ describes the azimuth of the gripper, and can be expressed by roll angle $\phi(t)$, pitch angle $\theta(t)$, and yaw angle $\psi(t)$, refer to equation (3.5), which is expressed by three Euler transformation angles:

$$
\begin{aligned}
^0\boldsymbol{R}_6(t) &= \begin{bmatrix} n_x(t) & s_x(t) & a_x(t) \\ n_y(t) & s_y(t) & a_y(t) \\ n_z(t) & s_z(t) & a_z(t) \end{bmatrix} \\[2mm]
&= \begin{bmatrix} c\varphi & -s\varphi & 0 \\ s\varphi & c\varphi & 0 \\ 0 & 0 & 1 \end{bmatrix} \begin{bmatrix} c\theta & 0 & s\theta \\ 0 & 1 & 0 \\ -s\theta & 0 & c\theta \end{bmatrix} \begin{bmatrix} 1 & 0 & 0 \\ 0 & c\psi & -s\psi \\ 0 & s\psi & c\psi \end{bmatrix} \\[2mm]
&= \begin{bmatrix} c\varphi c\theta & -s\varphi c\psi + c\varphi s\theta s\psi & s\varphi s\psi + c\varphi s\theta c\psi \\ s\varphi c\theta & c\varphi c\psi + s\varphi s\theta s\psi & -c\varphi s\psi + s\varphi s\theta c\psi \\ -s\theta & c\theta s\psi & c\theta c\psi \end{bmatrix}
\end{aligned} \tag{5.61}
$$

The relative reference coordinate system defines the position vector $\boldsymbol{p}(t)$, Euler angle vector $\boldsymbol{\Gamma}(t)$, linear velocity vector $\boldsymbol{v}(t)$ and angular velocity vector $\boldsymbol{\Omega}(t)$ respectively as follows:

$$
\begin{aligned}
\boldsymbol{p}(t) &\underline{\underline{\Delta}} [p_x(t), p_y(t), p_z(t)]^T, \quad \boldsymbol{\Gamma}(t)\underline{\underline{\Delta}}[\phi(t), \theta(t), \psi(t)]^T \\
\boldsymbol{v}(t) &\underline{\underline{\Delta}} [v_x(t), v_y(t), v_z(t)]^T, \quad \boldsymbol{\Omega}(t)\underline{\underline{\Delta}}[\omega_x(t), \omega_y(t), \omega_z(t)]^T
\end{aligned} \tag{5.62}
$$

The linear velocity of the gripper relative to the reference coordinate system is equal to the derivative of the gripper position with respect to time, that is

$$
\boldsymbol{v}(t) = \frac{d\boldsymbol{p}(t)}{dt} = \dot{\boldsymbol{p}}(t) \tag{5.63}
$$

Since the inverse of the direction cosine matrix is equivalent to its transposed matrix, the instantaneous angular velocity of the gripper coordinate system to the main axis of the reference coordinate system can be obtained from equation (5.61):

$$
\boldsymbol{R}\frac{d\boldsymbol{R}^T}{dt} = -\frac{d\boldsymbol{R}}{dt}\boldsymbol{R}^T = -\begin{bmatrix} 0 & -\omega_z & \omega_y \\ \omega_z & 0 & -\omega_x \\ -\omega_y & \omega_x & 0 \end{bmatrix}
$$

$$
= \begin{bmatrix} 0 & -s\theta\dot{\psi} + \dot{\phi} & -s\phi c\theta\dot{\psi} - c\phi\dot{\theta} \\ s\theta\dot{\psi} & 0 & c\phi c\theta\dot{\psi} - s\phi\dot{\theta} \\ s\phi c\theta\dot{\psi} + c\phi\dot{\theta} & -c\phi c\theta\dot{\psi} + s\phi\dot{\theta} & 0 \end{bmatrix} \quad (5.64)
$$

In the above formula, let the corresponding non-zero elements of the matrix be equal, and the relationship between $[\omega_x(t), \omega_y(t), \omega_z(t)]^T$ and $[\dot{\psi}(t), \dot{\theta}(t), \dot{\varphi}(t)]^T$ can be obtained as follows:

$$
\begin{bmatrix} \omega_x(t) \\ \omega_y(t) \\ \omega_z(t) \end{bmatrix} = \begin{bmatrix} c\phi c\theta & -s\phi & 0 \\ s\phi c\theta & c\phi & 0 \\ -s\theta & 0 & 1 \end{bmatrix} \begin{bmatrix} \dot{\psi}(t) \\ \dot{\theta}(t) \\ \dot{\varphi}(t) \end{bmatrix} \quad (5.65)
$$

It is not difficult to find the inverse relationship as follows:

$$
\begin{bmatrix} \dot{\psi}(t) \\ \dot{\theta}(t) \\ \dot{\phi}(t) \end{bmatrix} = \sec\theta \begin{bmatrix} c\varphi c & -s\varphi & 0 \\ s\varphi c c\theta & c\varphi c\theta & 0 \\ -c\varphi s\theta & s\varphi s\theta & c\theta \end{bmatrix} \begin{bmatrix} \omega_x(t) \\ \omega_y(t) \\ \omega_z(t) \end{bmatrix} \quad (5.66)
$$

The above formula can also be expressed as a matrix vector:

$$
\dot{\Gamma}(t) \underline{\underline{\Delta}} [S(\Gamma)]\Omega(t) \quad (5.67)
$$

With the help of the concept of moving coordinate system, the linear velocity and angular velocity of the gripper can be obtained from the velocity of each joint in front:

$$
\begin{bmatrix} v(t) \\ \Omega(t) \end{bmatrix} = [J(q)]\,\dot{q}(t) = [J_1(q), J_2(q), \ldots, J_6(q)]\,\dot{q}(t) \quad (5.68)
$$

Where $\dot{q}(t) = (\dot{q}_1, \ldots, \dot{q}_6)^T$ is the joint velocity vector of the manipulator: $J(q)$ is a 6×6 Jacobian matrix, and the i-th column vector $J_i(q)$ can be obtained from the following formula:

$$
J_i(q) = \begin{cases} \begin{bmatrix} z_{i-1} \times (p - p_{i-1}) \\ z_{i-1} \end{bmatrix} & \text{For rotation joint } i \\ \begin{bmatrix} z_{i-1} \\ 0 \end{bmatrix} & \text{For translation joint } i \end{cases} \quad (5.69)
$$

Among them, p_{i-1} is the origin of the $(i-1)$th coordinate system relative to the reference coordinate system, z_{i-1} is the unit vector along the motion axis of the joint i, and p is the position of the gripper relative to the reference coordinate system.

If the inverse Jacobian matrix pair $q(t)$ exists, then the joint speed $\dot{q}(t)$ of the manipulator can be calculated from the gripping speed according to equation (5.67):

$$\dot{q}(t) = J^{-1}(q) \begin{bmatrix} v(t) \\ \Omega(t) \end{bmatrix} \tag{5.70}$$

According to the stated requirements of the gripper's linear velocity and angular velocity, the above formula can calculate the speed of each joint and specify the speed that the joint motor must maintain to ensure that the gripper achieves stable movement along the desired Cartesian direction.

Derivation of the velocity vector shown in formula (5.68), the acceleration of the gripper can be obtained:

$$\begin{bmatrix} \dot{v}(t) \\ \dot{\Omega}(t) \end{bmatrix} = \dot{J}(q, \dot{q})\dot{q}(t) + J(q)\ddot{q}(t) \tag{5.71}$$

Where $\ddot{q}(t) = [\ddot{q}_1(t), \ldots, \ddot{q}_6(t)]^T$ is the joint acceleration vector of the manipulator. Substituting $\dot{q}(t)$ in formula (5.70) into formula (5.71), we can get:

$$\begin{bmatrix} \dot{v}(t) \\ \dot{\Omega}(t) \end{bmatrix} = \dot{J}(q, \dot{q})J^{-1}(q) \begin{bmatrix} v(t) \\ \Omega(t) \end{bmatrix} + J(q)\ddot{q}(t) \tag{5.72}$$

According to the above formula, the acceleration of the joint can be calculated from the speed and acceleration of the gripper as follows:

$$\ddot{q}(t) = J^{-1}(q) \begin{bmatrix} \dot{v}(t) \\ \dot{\Omega}(t) \end{bmatrix} - J^{-1}(q)\dot{J}(q, \dot{q})J^{-1}(q) \begin{bmatrix} v(t) \\ \Omega(t) \end{bmatrix} \tag{5.73}$$

The kinematic relationship between the joint coordinate system and the Cartesian coordinate system obtained by the above analysis and calculation will be used to study the subsequent part of this section, that is, to provide control methods for various resolved motions, and to find out that the manipulator gripper in the Cartesian coordinate system.

5.4.2 *Decomposition motion speed control*

Resolved motion rate control (RMRC) means that the motions of the joint motors are combined and run at different speeds at the same time to ensure stable movement of the gripper along the Cartesian coordinate axis. Resolved motion speed control first decomposes the desired gripper (or other end tool) movement into the desired speed of each joint, and then performs speed servo control on each joint. The mathematical relationship between the world coordinates (such as $p_x, p_y, p_z, \phi, \theta, \phi$) and the joint angle coordinates of a 6-link manipulator is inherently non-linear, and can be expressed by a certain non-linear vector value function as follows:

$$x(t) = f[q(t)] \tag{5.74}$$

In the formula, $f(q)$ is a 6×1 vector-valued function, $x(t) = [p_x, p_y, p_z, \psi, \theta, \phi]^T$ is the world coordinate and $q(t) = [q_1, q_2, \ldots, q_n]^T$ is the generalized coordinate. For a 6-link manipulator, the relationship between its linear velocity and angular velocity and joint velocity is given by equation (5.74). For a more general discussion, if it is assumed that the manipulator has m degrees of freedom and the world coordinates have n dimensions, then the relationship between the joint angles and the world coordinates is represented by the non-linear function of equation (5.74). If we take the derivative of equation (5.74), we have:

$$\frac{dx(t)}{dt} = \dot{x}(t) = J(q)\dot{q}(t) \tag{5.75}$$

In the formula, $J(q)$ is the Jacobian matrix of $q(t)$, namely:

$$J_{ij} = \frac{\partial f_i}{\partial q_i}, \quad 1 \leq i \leq n, \quad 1 \leq j \leq m$$

It can be seen from equation (5.75) that there is a linear relationship when speed control is performed. If $x(t)$ and $q(t)$ have the same dimension, that is, when $m = n$, then the manipulator is non-redundant, and its Jacobian matrix can be in a special non-singular position $q(t)$ is the inverse, namely

$$\dot{q}(t) = J^{-1}(q)\dot{x}(t) \tag{5.76}$$

Given the desired speed along the world coordinates, the combination of the speeds of the joint motors that realize the desired

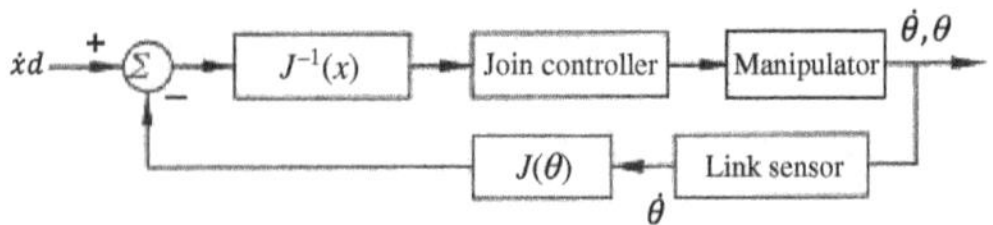

Figure 5.16.   Exploded motion speed control block diagram.

tool movement can be easily obtained according to equation (5.75). Different methods of calculating the inverse Jacobian matrix can be used.

Figure 5.16 shows a block diagram that resolves the motion speed control.

If $m > n$, then the manipulator is redundant, and its inverse Jacobian matrix does not exist. This simplifies the problem of finding the general inverse Jacobian matrix. In this case, if the rank of $J(q)$ is n, you can add a Lagrangian multiplier λ to a cost criterion to formula (5.74) to form an error criterion, and this error criterion minimization, $\dot{q}(t)$ can be obtained, that is:

$$C = \frac{1}{2}\dot{q}^T A\dot{q} + \lambda^T[\dot{x} - J(q)\dot{q}] \tag{5.77}$$

In the formula, A is an $m \times m$ symmetric positive definite matrix, and C is the cost criterion.

Respectively minimizing C with respect to $\dot{q}(t)$ and λ, we can get:

$$\dot{q}(t) = A^{-1}J^T(q)\lambda \tag{5.78}$$

$$\dot{q}(t) = J(q)\dot{q}(t) \tag{5.79}$$

Substitute equation (5.78) into equation (5.79), and solve for λ, and get:

$$\lambda = [J(q)A^{-1}J^T(q)]^{-1}\dot{x}(t) \tag{5.80}$$

Substituting the above formula into the formula (5.78), we get:

$$\dot{q}(t) = A^{-1}J^T(q)[J(q)A^{-1}J^T(q)]^{-1}\dot{x}(t) \tag{5.81}$$

If A is an identity matrix, then the above equation is simplified to equation (5.76).

It is often desirable to command the tool to move along the gripper coordinate system instead of the world coordinate system. At this time, the expected motion speed $\dot{\boldsymbol{h}}(t)$ of the gripper (tool) along the gripper coordinate system has the following relationship with the motion speed $\dot{\boldsymbol{x}}(t)$ of the world coordinate system:

$$\dot{\boldsymbol{x}}(t) = {}^0\boldsymbol{R}_h\dot{\boldsymbol{h}}(t) \tag{5.82}$$

In the formula, ${}^0\boldsymbol{R}_h$ is an $n \times 6$ matrix, which represents the directional relationship between the gripping coordinate system and the world coordinate system. For the gripper coordinate system, given the desired movement speed $\dot{\boldsymbol{h}}(t)$ of the gripper, according to equations (5.80) and (5.82), the joint speed can be obtained

$$\dot{\boldsymbol{q}}(t) = \boldsymbol{A}^{-1}\boldsymbol{J}^T(\boldsymbol{q})[\boldsymbol{J}(\boldsymbol{q})\boldsymbol{A}^{-1}\boldsymbol{J}^T(\boldsymbol{q})]^{-1}\,{}^0\boldsymbol{R}_h\dot{\boldsymbol{h}}(t) \tag{5.83}$$

Because in equations (5.81) and (5.83), the angle position $\boldsymbol{q}(t)$ is related to time, so to calculate $\dot{\boldsymbol{q}}(\boldsymbol{t})$, it is necessary to estimate the value of $\boldsymbol{J}^{-1}(\boldsymbol{q})$ at each sampling time t.

5.4.3 *Resolved motion acceleration control*

Resolved motion acceleration control (RMAC) extends the concept of resolved motion speed control to include acceleration control. For position control problems that directly involve the position and direction of the manipulator, this is an alternative. Resolved motion acceleration control first calculates the control acceleration of the tool, and then decomposes it into the corresponding acceleration of each joint, and then calculates the control torque according to the dynamic equation.

The actual pose $\boldsymbol{H}(t)$ and the expected pose $\boldsymbol{H}^d(t)$ of the gripper (tool) of the manipulator can be represented by a 4×4 homogeneous transformation matrix as follows (refer to Equation 5.59):

$$\boldsymbol{H}(t) = \begin{bmatrix} \boldsymbol{n}(t) & \boldsymbol{o}(t) & \boldsymbol{a}(t) & \boldsymbol{p}(t) \\ 0 & 0 & 0 & 1 \end{bmatrix}$$

$$\boldsymbol{H}^d(t) = \begin{bmatrix} \boldsymbol{n}^d(t) & \boldsymbol{o}^d(t) & \boldsymbol{a}^d(t) & \boldsymbol{p}^d(t) \\ 0 & 0 & 0 & 1 \end{bmatrix}$$

The position error of the gripper is defined as the difference between the desired position and the actual position of the gripper, and can be expressed as:

$$e_p(t) = p^d(t) - p(t) = \begin{bmatrix} p_x^d(t) - p_x(t) \\ p_y^d(t) - p_y(t) \\ p_z^d(t) - p_z(t) \end{bmatrix} \tag{5.84}$$

Similarly, the error of the direction (posture) of the gripper is defined as the deviation between the desired direction and the actual direction of the gripper, and can be expressed as:

$$e_0(t) = 1/2[n(t) \times n^d + o(t) \times o^d + a(t) \times a^d] \tag{5.85}$$

Therefore, the control of the manipulator can be achieved by reducing these errors to zero.

According to the formula (3.83) in Chapter 3, a 6-dimensional vector can be used to jointly express the linear velocity $v(t)$ and angular velocity $\omega(t)$ of a 6-joint manipulator, that is, there is a generalized velocity:

$$\dot{x}(t) = \begin{bmatrix} v(t) \\ a(t) \end{bmatrix} = J(q)\dot{q}(t) \tag{5.86}$$

Among them, $J(q)$ is a 6×6 matrix and is given by equation (5.69). Equation (5.86) is the basis of resolved motion speed control, and its joint speed is solved by the speed of the gripper. If extend this idea further, we can use it to obtain the acceleration of each joint from the acceleration $\ddot{x}(t)$ of the gripper. Then, we can obtain the acceleration of the gripper by taking the derivative of $\dot{x}(t)$:

$$\ddot{x}(t) = J(q)\ddot{q}(t) + \dot{J}(q,\dot{q})\dot{q}(t) \tag{5.87}$$

The guiding ideology of closed-loop resolved motion acceleration control is to reduce the position error and direction error of the manipulator to zero. If the Cartesian coordinate path of a manipulator is pre-planned, then the desired position $p^d(t)$, desired velocity $v^d(t)$ and desired acceleration $\dot{v}^d(t)$ of the gripper are known to the base coordinate system. In order to reduce the position error, the joint torque and force can be applied to the driver of each joint of the manipulator, so that the actual linear acceleration of the gripper meets the following equation:

$$\dot{v}(t) = \dot{v}^d(t) + k_1[v^d(t) - v(t)] + k_2[p^d(t) - p(t)] \tag{5.88}$$

In the formula, k_1 and k_2 are proportional coefficients. The above formula can be rewritten as:

$$\ddot{e}_p(t) + k_1\dot{e}_p(t) + k_2 e_p(t) = 0 \tag{5.89}$$

Among them, $e_p(t) = p^d(t) - p(t)$. The selection of input torque and force must ensure that the position error of the manipulator is asymptotically convergent. This requires choosing the coefficients k_1 and k_2 so that the real part of the characteristic root of equation (5.89) is negative. In the same way, in order to reduce the direction error of the manipulator, the acting torque and force of the manipulator must be selected so that its angular acceleration satisfies the following formula:

$$\dot{\omega}(t) = \dot{\omega}^d(t) + k_1[\omega^d(t) - \omega(t)] + k_2 e_0 \tag{5.90}$$

Let v^d and ω^d form a six-dimensional velocity vector, and position error and direction error form an error vector:

$$\dot{x}^d(t) = \begin{bmatrix} v^d(t) \\ \omega^d(t) \end{bmatrix}, \quad e(t) = \begin{bmatrix} e_p(t) \\ e_0(t) \end{bmatrix} \tag{5.91}$$

Combined formula (5.88) and formula (5.90) can be obtained:

$$\ddot{x}(t) = \ddot{x}^d(t) + k_1[\dot{x}^d(t) - \dot{x}(t)] + k_2 e(t) \tag{5.92}$$

Substitute equation (5.86) and equation (5.87) into equation (5.92), and solve for $\ddot{q}(t)$

$$\begin{aligned} \ddot{q}(t) &= J^{-1}(q)[\ddot{x}^d(t) + k_1(\dot{x}^d(t) - \dot{x}(t)) + k_2 e(t) - \dot{J}(q, \dot{q})\dot{q}(t)] \\ &= -k_1\dot{q}(t) + J^{-1}(q)[\ddot{x}^d(t) + k_1\dot{x}^d(t) + k_2 e(t) - \dot{J}(q, \dot{q})\dot{q}(t)] \end{aligned} \tag{5.93}$$

The above formula is the basis of the closed-loop resolved motion acceleration control of the manipulator. In order to calculate the torque and force applied to each joint driver of the manipulator, it is necessary to apply the recursive Newton-Euler equation of motion.

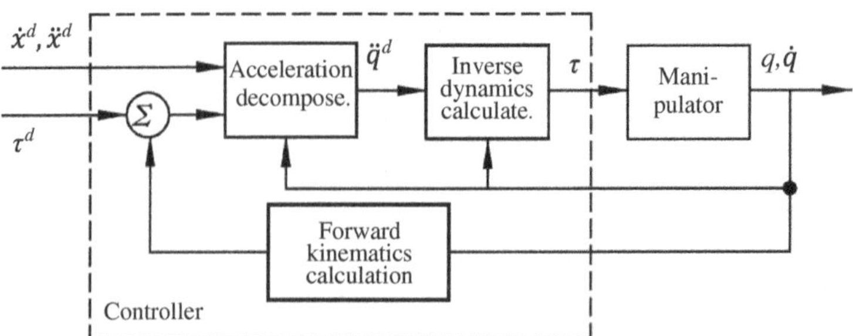

Figure 5.17.   A block diagram of a resolved motion acceleration control scheme.

The joint position $q(t)$ and joint speed $\dot{q}(t)$ are measured by poten-tiometers or optical encoders. The values of $V, \omega, J, J^{-1}, \dot{J}, H(t)$ can be calculated according to the above equations. Substituting these values and the expected position, expected speed and expected acceleration of the gripper obtained from the planned trajectory into equation (5.93), the acceleration of the joint can be calculated. The added joint torque and force can be obtained by recursively solving the Newton-Euler equation of motion. Just like resolved motion speed control, the resolved motion acceleration control method has a wide range of calculation requirements and singularities related to the Jacobian matrix. This requires the application of acceleration information to plan the trajectory of the manipulator gripper.

Figure 5.17 depicts a block diagram of a resolved motion acceler-ation control scheme. According to the above related equations, the expected acceleration, joint acceleration and control torque of the gripper can be calculated.

5.4.4   *Resolved motion force control*

The concept of resolved motion force control (RMFC) is to deter-mine the control torque applied to each joint driver of the manip-ulator, so that the gripper or tool of the manipulator can perform the desired Cartesian position control [37, 40]. The advantage of this control method is that it is not based on the complex dynamics equa-tions of the manipulator, but still has the ability to compensate for the change of the arm structure, the gravity of the link and the inter-nal friction.

The resolved motion force control is based on the relationship between the resolved force vector F (obtained by the wrist force sensor) and the joint torque τ of the joint driver. The control technology consists of Cartesian position control and force convergence control. The position control calculates the desired force and torque added to the end effector (tool) in order to track a desired Cartesian trajectory. The force convergence control determines the joint torque required by each driver, so that the end effector (tool) can maintain the desired force and torque obtained by the position control. Figure 5.18 shows a block diagram of the RMFC control system.

In Figure 5.18, the resolved force vector $\boldsymbol{F} = (F_x, F_y, F_z, M_x, M_y, M_z)^T$, the joint torque $\boldsymbol{\tau} = (\tau_1, \tau_2, \ldots, \tau_n)^T$. They act on the driver of each joint to offset the load force on the tool. Among them, $(F_x, F_y, F_z)^T$ and $(M_x, M_y, M_z)^T$ are the Cartesian forces and moments in the tool (gripper) coordinate system, respectively. The relationship between $\boldsymbol{F}$ and τ can be expressed by the following formula:

$$\boldsymbol{\tau}(t) = \boldsymbol{J}^T(\boldsymbol{q})\boldsymbol{F}(t) \tag{5.94}$$

Because the goal of resolved motion acceleration control is to track the Cartesian position of the end effector (tool), a suitable position-time trajectory must be specified as a function of the arm transformation matrix $^0\boldsymbol{A}_6(t)$ and the speed $(v_x, v_y, v_z)^T$ and angular velocity $(\omega_x, \omega_y, \omega_z)^T$ of the arm coordinate system. In other words, the expected time-varying arm transformation matrix $^0\boldsymbol{A}_6(t + \Delta t)$ can be expressed as:

$$^0\boldsymbol{A}_6(t+\Delta t) = {}^0\boldsymbol{A}_6(t) \begin{bmatrix} 1 & -\omega_z(t) & \omega_y(t) & v_x(t) \\ \omega_z(t) & 1 & -\omega_x(t) & v_y(t) \\ -\omega_y(t) & \omega_x(t) & 1 & v_z(t) \\ 0 & 0 & 0 & 1 \end{bmatrix} \Delta t \tag{5.95}$$

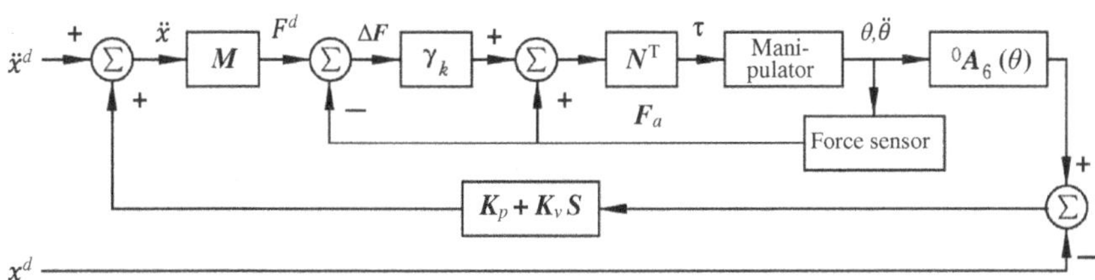

Figure 5.18. Block diagram of resolved motion force control system.

Then, the desired Cartesian velocity $\dot{\boldsymbol{x}}^d(t) = (v_x, v_y, v_z, \omega_x, \omega_y, \omega_z)^T$ can be obtained from the elements of the following equation:

$$\begin{bmatrix} 1 & -\omega_z(t) & \omega_y(t) & v_x(t) \\ \omega_z(t) & 1 & -\omega_x(t) & v_y(t) \\ -\omega_y(t) & \omega_x(t) & 1 & v_z(t) \\ 0 & 0 & 0 & 1 \end{bmatrix} = \frac{1}{\Delta t}\left[(^0\boldsymbol{A}_6)^{-1}(t)\,^0\boldsymbol{A}_6(t+\Delta t) \right] \tag{5.96}$$

According to the above equation, the Cartesian velocity error $\dot{\boldsymbol{x}}^d - \dot{\boldsymbol{x}}$ can be obtained, which is different from the velocity error in equation (5.88); because equation (5.96) adopts the homogeneous transformation matrix method, and the velocity error in equation (5.88) only needs to be derived from $\boldsymbol{p}^d(t) - \boldsymbol{p}(t)$ you can get it.

Similarly, the expected Cartesian acceleration $\ddot{\boldsymbol{x}}^d(t)$ can be obtained as:

$$\ddot{\boldsymbol{x}}^d(t) = \frac{\dot{\boldsymbol{x}}^d(t+\Delta t) - \dot{\boldsymbol{x}}^d(t)}{\Delta t} \tag{5.97}$$

Based on the proportional-derivative control method, if there is no error of arm position and velocity, then the actual Cartesian acceleration can be made to track the desired Cartesian acceleration as closely as possible. This can be achieved by setting the following actual Cartesian acceleration:

$$\ddot{\boldsymbol{x}}(t) = \ddot{\boldsymbol{x}}^d(t) + K_v[\dot{\boldsymbol{x}}^d(t) - \dot{\boldsymbol{x}}(t)] + K_p[\boldsymbol{x}^d(t) - \boldsymbol{x}(t)] \tag{5.98}$$

Or

$$\ddot{\boldsymbol{x}}_e(t) + K_v\dot{\boldsymbol{x}}_e(t) + K_p\boldsymbol{x}_e(t) = 0 \tag{5.99}$$

Choose K_v and K_p values so that the real part of the characteristic root of equation (5.99) is negative, then $\boldsymbol{x}(t)$ will asymptotically converge to $\boldsymbol{x}^d(t)$.

Based on the above control technology, apply Newton's second law

$$\boldsymbol{F}^d(t) = \boldsymbol{M}\ddot{\boldsymbol{x}}(t) \tag{5.100}$$

The desired Cartesian force and moment required to correct the position error can be obtained. In the above formula, M is the mass matrix, and its diagonal elements are the total mass of the load m and the moment of inertia I_{xx}, I_{yy}, I_{zz} of the load spindle. Therefore, according to equation (5.94), the expected Cartesian force F^d can be decomposed into joint torque:

$$\tau(t) = \boldsymbol{J}^T(\boldsymbol{q})\boldsymbol{F}^d = \boldsymbol{J}^T(\boldsymbol{q})\boldsymbol{M}\dot{\boldsymbol{x}}(t) \tag{5.101}$$

In general, compared with the mass of the manipulator, the load mass is negligible. At this time, the revolved motion force control works well. However, if the load mass is close to the manipulator mass, the arm position usually does not converge to the desired position. This is because some joint moments are used for link acceleration. In order to compensate for the effects of these loads and accelerations, the force convergence control is introduced into the resolved motion force control as its second part.

The force convergence control method is based on the Robbins-Monro's random approximation method, which determines the actual Cartesian force $\boldsymbol{F}_a$ to make the observed arm Cartesian force $\boldsymbol{F}_o$ (measured by the wrist force sensor) converge the desired Cartesian force $\boldsymbol{F}^d$ (obtained from the position control technique described above). If the error between the measured force vector $\boldsymbol{F}_o$ and the expected Cartesian force is greater than the user-designed threshold $\Delta \boldsymbol{F}(k) = \boldsymbol{F}^d(k) - \boldsymbol{F}_o(k)$, then the actual Cartesian force is corrected by the following formula:

$$\boldsymbol{F}_a(k+1) = \boldsymbol{F}_a(k) + \gamma_k \Delta \boldsymbol{F}(k) \tag{5.102}$$

In the formula, $\gamma_k = 1/(k+1), k = 1, \ldots, N$. Theoretically, the value of N must be large; however, in practice, the selection of the value of N is based on force convergence. According to computer simulation studies, N taking 1 and 2 can provide quite good force vector convergence.

In short, the resolved motion force control with force convergence control has the following advantages: this control method can be extended to manipulators with various load conditions and any number of degrees of freedom without increasing the complexity of calculations.

5.5   Adaptive Control of Manipulator

Adaptive control has been widely used in manipulators. This section will review the progress of adaptive control of manipulators and discuss the various design results achieved.

According to different design technologies, the adaptive control of the robotic manipulator can be divided into three categories, namely model reference adaptive control, self-tuning adaptive control, and linear perturbation adaptive control [10, 40, 46].

Adaptive manipulators are controlled by adaptive controllers. The adaptive controller has a sensory device that can maintain automatic adaptation to the environment in an incompletely determined and locally changing environment, and perform different cyclic operations in various search and automatic guidance methods.

There are nonlinear and uncertain factors in the dynamic model of the manipulator. These factors include unknown system parameters (such as friction), nonlinear dynamic characteristics (such as gear backlash and gain nonlinearity), and environmental factors (such as load changes and other disturbances). Using adaptive control to automatically compensate for the above factors can significantly improve the performance of the manipulator.

5.5.1   *State model and structure of adaptive controller*

The adaptive control of the robotic manipulator is closely related to the dynamics of the manipulator. The dynamic equation of a rigid manipulator with n degrees of freedom and n joints individually driven can be expressed by the following formula, refer to formula (4.24):

$$F_i = \sum_{j=1}^{n} D_{ij}(q)\ddot{q}_j + \sum_{j=1}^{n}\sum_{k=1}^{n} C_{ijk}(q)\dot{q}_j\dot{q}_k + G_i(q) \quad i = 1,\ldots,n$$

$$(5.103)$$

The vector form of this dynamic equation is:

$$\boldsymbol{F} = \boldsymbol{D}(\boldsymbol{q})\ddot{\boldsymbol{q}} + \boldsymbol{C}(\boldsymbol{q},\dot{\boldsymbol{q}}) + \boldsymbol{G}(\boldsymbol{q}) \qquad (5.104)$$

Redefine as:

$$C(q,\dot{q})\underline{\underline{\Delta}}C^1(q,\dot{q})\dot{q}$$

$$G(q)\underline{\underline{\Delta}}G^1(q)q \tag{5.105}$$

Substituting into the formula (5.104) can be obtained:

$$F = D(q)\ddot{q} + C^1(q,\dot{q})\dot{q} + G^1(q)q \tag{5.106}$$

This is a quasi-linear system expression.

Define

$$x = [q,\dot{q}]^T \tag{5.107}$$

For the $2n \times 1$ state vector, the equation (5.106) can be expressed as the following state equation:

$$\dot{x} = A_p(x,t)x + B_p(x,t)F \tag{5.108}$$

Where, $A_p(x,t) = \begin{bmatrix} 0 & I \\ -D^{-1}G^1 & -D^{-1}C^1 \end{bmatrix}_{2n \times 2n}$, $B_p(x,t) = \begin{bmatrix} 0 \\ D^{-1} \end{bmatrix}_{2n \times n}$ is a very complex nonlinear function of the state vector x.

The above-mentioned manipulator dynamics model is the adjustment object of the manipulator adaptive controller.

In fact, the dynamics of the driving device must be included in the control system model. For a manipulator with n drive joints, the dynamic action of its riving device can be expressed as:

$$M_a u - \tau = J_a\ddot{q} + B_a\dot{q} \tag{5.109}$$

Where u, q and τ are the $n \times 1$ vectors of the input voltage, displacement and disturbance torque of the driving device respectively; and M_a, J_a, B_a are the $n \times n$ diagonal matrixes, which are determined by the parameters of the driving device. It consists of two parts:

$$\tau = F(q,\dot{q},\ddot{q}) + \tau_d \tag{5.110}$$

Among them, F is determined by equation (5.106), which represents the torque related to the motion of the link; it includes the nonlinearity and friction torque of the motor.

Solve formulas (5.106), (5.109) and (5.110) simultaneously, and define:

$$\begin{aligned}
\boldsymbol{J}(\boldsymbol{q}) &= \boldsymbol{D}(\boldsymbol{q}) + J_a \\
\boldsymbol{E}(\boldsymbol{q}) &= \boldsymbol{C}^1(\boldsymbol{q}) + B_a \\
\boldsymbol{H}(\boldsymbol{q})\boldsymbol{q} &= \boldsymbol{G}^1(\boldsymbol{q})\boldsymbol{q} + \boldsymbol{\tau}_d
\end{aligned} \tag{5.111}$$

The time-varying nonlinear state model of the manipulator driving system can be obtained as follows:

$$\dot{\boldsymbol{x}} = A_p(\boldsymbol{x}, t)\boldsymbol{x} + B_p(\boldsymbol{x}, t)\boldsymbol{u} \tag{5.112}$$

Where

$$A_p(\boldsymbol{x}, t) = \begin{bmatrix} 0 & I \\ -J^{-1}H & -J^{-1}E \end{bmatrix}_{2n \times 2n} \tag{5.113}$$

$$B_p(\boldsymbol{x}, t) = \begin{bmatrix} 0 \\ J^{-1}M_a \end{bmatrix}_{2n \times n}$$

The state models (5.108) and (5.112) have the same form, and both can be used for the design of adaptive controllers.

There are two main structures of the adaptive controller, namely model reference adaptive controller (MRAC) and self-tuning adaptive controller (STAC), as shown in Figure 5.19(a) and (b) respectively. The existing robotic manipulator adaptive control system is basically established by applying these design methods.

Based on the above two basic structures, many design methods of adaptive controllers for manipulator have been proposed, and corresponding progress has been made.

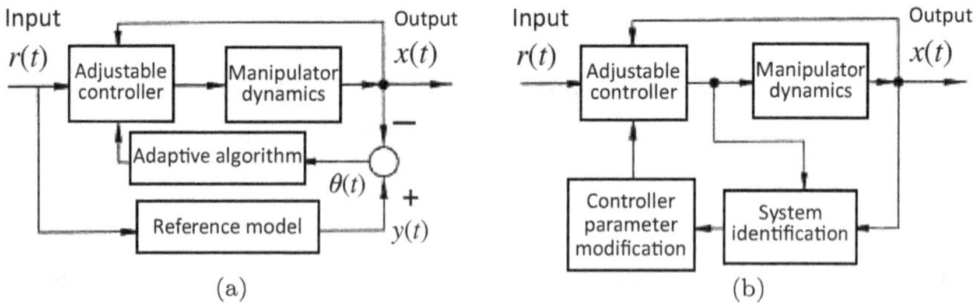

Figure 5.19. Structure of the manipulator adaptive controller. (a) Model reference adaptive controller (b) Self-tuning adaptive controller.

5.5.2 *Manipulator model reference adaptive controller*

MRAC is the earliest adaptive control technology for manipulator control. Its basic design idea is to synthesize a control signal $\boldsymbol{u}$ for the state equation (5.112) of the manipulator, or synthesize an input $\boldsymbol{F}$ for the state equation (5.108). This control signal will force the system to have the required characteristics in a certain desired manner specified by the reference model. Based on the stated goal and the structure represented by equation (5.113), the selected reference model can be a stable linear time-invariant system:

$$\dot{\boldsymbol{y}} = A_M \boldsymbol{y} + B_M \boldsymbol{r} \tag{5.114}$$

In the formula, $\boldsymbol{y}$ is the state vector of the $n \times 1$ reference model, $\boldsymbol{r}$ is the input vector of the $n \times 1$ reference model, and

$$A_M = \begin{bmatrix} 0 & I \\ -\wedge_1 & -\wedge_2 \end{bmatrix}, \quad B_M = \begin{bmatrix} 0 \\ \wedge_1 \end{bmatrix} \tag{5.115}$$

Among them, $\wedge_1$ is the $n \times n$ diagonal matrix containing the term ω_i, and $\wedge_2$ is the $n \times n$ diagonal matrix containing the term $2\xi_i\omega_i$.

Equation (5.114) represents n decoupled second-order differential equations with specified parameters ξ_i and ω_i:

$$\ddot{y}_i + 2\xi_i\omega_i\dot{y}_i + \omega_i^2 y_i = \omega_i^2 r \tag{5.116}$$

In the formula, the input variable r represents the ideal manipulator motion trajectory prescribed in advance by the designer. When the input terminal introduces appropriate state feedback, the state equation of the manipulator becomes adjustable by adjusting the feedback gain. The state variable x of this system is compared with the state y of the reference model, and the obtained state error e is used to drive the adaptive algorithm, as shown in Figure 5.19(a), to maintain the state error close to zero.

The adaptive algorithm is designed according to the asymptotic stability requirements of MRAC. Commonly used stability criteria include Lyapunov stability criterion and Popov superstability criterion.

1. Design of Lyapunov MRAC

Let the control input $\boldsymbol{u}$ be

$$\boldsymbol{u} = K_x \boldsymbol{x} + K_u \boldsymbol{r} \tag{5.117}$$

In the formula, K_x and K_u are respectively $n \times n$ time-varying adjustable feedback matrix and feedforward matrix.

According to formula (5.117), the closed-loop system state model of formula (5.112) can be obtained:

$$\dot{\boldsymbol{x}} = A_s(\boldsymbol{x}, t)\boldsymbol{x} + B_s(\boldsymbol{x}, t)\boldsymbol{r} \tag{5.118}$$

Among them

$$A_s = \begin{bmatrix} 0 & I \\ -J^{-1}(H + M_a K_{x1}) & -J^{-1}(E + M_a K_{x2}) \end{bmatrix},$$

$$B_s = \begin{bmatrix} 0 \\ J^{-1} M_a K_u \end{bmatrix}$$

Appropriate design of K_{xi} and K_u can make the system shown in equation (5.118) completely match the reference model represented by equation (5.114).

Define the $2n \times 1$ state error vector $\boldsymbol{e}$ as

$$\boldsymbol{e} = \boldsymbol{y} - \boldsymbol{x} \tag{5.119}$$

You can get:

$$\dot{\boldsymbol{e}} = A_M \boldsymbol{e} + (A_M - A_s)\boldsymbol{x} + (B_M - B_S)\boldsymbol{r} \tag{5.120}$$

The control objective here is to find an adjustment algorithm for K_x and K_u so that

$$\lim_{t \to \infty} \boldsymbol{e}(t) = 0$$

Define the positive definite Lyapunov function $\boldsymbol{V}$ as:

$$\boldsymbol{V} = \boldsymbol{e}^T P \boldsymbol{e} + tr\left[(A_M - A_S)^T F_A^{-1}(A_M - A_S)\right]$$

$$+ tr\left[(B_M - B_S)^T (P \boldsymbol{e} \boldsymbol{r}^T - F_B^{-1} \dot{B}_S)\right] \tag{5.121}$$

Therefore, from equation (5.120) and equation (5.121), we can get:

$$\dot{V} = e^T(A_M P + P A_M)e + tr\left[(A_M - A_S)^T(P e \boldsymbol{x}^T - F_A^{-1}\dot{A}_S)\right]$$

$$+ tr\left[(B_M - B_S)^T(P e \boldsymbol{r}^T - F_B^{-1}\dot{B}_S)\right] \tag{5.122}$$

According to Lyapunov's stability theory, the necessary and sufficient condition to ensure that Eq. (5.114) is satisfied is negative definite. It can be obtained from this:

$$A_M^T P + P A_M = -Q \tag{5.123}$$

$$\dot{A}_S = F_A P e \boldsymbol{x}^T \approx B_P \dot{K}_x, \quad \dot{B}_S = F_B P e \boldsymbol{r}^T \approx B_p \dot{K}_u \tag{5.124}$$

As well as

$$\dot{K}_u = K_u B_m^+ F_B P e \boldsymbol{r}^T, \quad \dot{K}_x = K_u B_m^+ F_A P e \boldsymbol{x}^T \tag{5.125}$$

Among them, P and Q are positive definite matrices, and P satisfies formula (5.105), B_m^+ is the Penrose pseudo-inverse matrix of B_M, and F_A and F_B are positive definite adaptive gain matrices.

Although the Lyapunov design method can guarantee asymptotic stability, large state errors and oscillations may occur during the transition process. When appropriate additional control inputs are introduced, the convergence rate of the system can be improved.

2. Design of Popov's super stable MRAC

This is another way to design an adaptive control law with stability conditions. When applying Popov's superstability theory, the Manipulator equation (5.122) is regarded as a nonlinear time-varying system equation.

Let the control input be

$$\boldsymbol{u} = \phi(\boldsymbol{v}, \boldsymbol{x}, t)\boldsymbol{x} - K_x \boldsymbol{x} + \varphi(\boldsymbol{v}, \boldsymbol{r}, t)r + K_u \boldsymbol{r} \tag{5.126}$$

Where ϕ and φ are the $n \times 2n$ and $n \times n$ matrices generated by the adaptive algorithm; K_x and K_u are the $n \times 2n$ and $n \times n$ constant coefficient feedback gain matrix and feedforward gain matrix; $\boldsymbol{v}$ is the $n \times 1$ generalized state error vector

$$\boldsymbol{v} = De = D(\boldsymbol{y} - \boldsymbol{x}) \tag{5.127}$$

Among them, D is the transfer matrix of the $n \times 2n$ linear compensator.

According to formula (5.112), formula (5.114) and formula (5.126), the MRAC equation can be expressed by the state error:

$$\dot{e} = A_M e + \begin{bmatrix} 0 \\ I \end{bmatrix} w_1$$

$$v = De \tag{5.128}$$

$$w_1 = -w = \bar{B}_p \left[B_p^+(A_M - A_p) + K_x - \phi \right] x$$

$$+ \bar{B}_p [B_p^+ B_M - K_u - \psi] r$$

Where $\bar{B}_p$ is defined as $B_p = \begin{bmatrix} 0 \\ \bar{B}_p \end{bmatrix}, B_p^+ = (B_p^T B_p)^{-1} B_p^T$

According to Popov's superstability theory, the necessary and sufficient conditions for the system shown in equation (5.110) to satisfy the stability of the control target $\lim\limits_{t \to \infty} e(t) = 0$ are:

(1) $n \times n$ transfer matrix is defined as

$$G(s) = D(SI - A_M)^{-1} \begin{bmatrix} 0 \\ I \end{bmatrix} \tag{5.129}$$

is a strictly positive real matrix;

(2) Integral inequality

$$\int_0^t v^T w d\tau > -r_0^2 \tag{5.130}$$

It is true for all $t > 0$.

It can be obtained from the above two conditions:

$$\phi(v, x, t) = q \frac{v}{\|v\|} (\text{sgn} x)^T, \quad \psi(v, r, t) = p \frac{v}{\|v\|} (\text{sgn} r)^T \tag{5.131}$$

Where

$$q \geq \frac{[\lambda_{\max}(RR^T)]^{\frac{1}{2}}}{\lambda_{\min}(\bar{B}_p)}, \quad p \geq \frac{[\lambda_{\max}(SS^T)]^{\frac{1}{2}}}{\lambda_{\min}(\bar{B}_p)} \tag{5.132}$$

As well as

$$R = \bar{B}_P B_p^+(A_M - A_P) + \bar{B}_P K_x, \quad S = \bar{B}_p B_p^+ B_M - \bar{B}_p K_d \tag{5.133}$$

Due to the intermittent nature of ϕ and φ, MRAC uses two models (distinguished by $v \neq 0$ and $v = 0$) to reach the asymptotic stability. The subset of $v = 0$ defines a "siding set." The control system

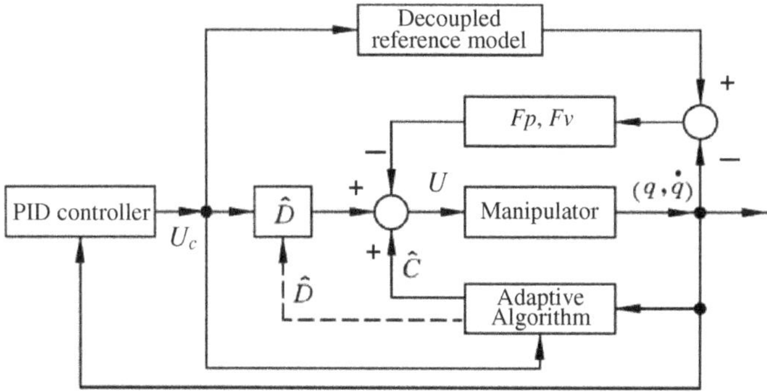

Figure 5.20. Structure diagram of non-linear compensation and decoupling MRAC.

transforms this subset at a high operating frequency, resulting in the trajectory is the same as that of the variable structure system. The decaying speed of the error e along the sliding set is only affected by Q in equation (5.123); therefore, the convergence speed can be controlled by the designer.

In addition to the above two MRAC design methods, there are other design techniques that can also be used for MRAC design. These technologies include Dubowsky's simplified MRAC approximate design method, and Horwitz's proposed non-linear compensation and decoupling MRAC system. Figure 5.20 shows the block diagram of this nonlinear compensation and decoupling MRAC system.

5.6 Intelligent Control of Manipulator

This section takes the neural control of the multi-finger dexterous hand as an example to introduce the intelligent control of the manipulator [16, 22].

5.6.1 *Overview of manipulator control based on deep learning*

With the in-depth development of machine learning research, more and more machine learning algorithms, especially deep learning and deep reinforcement learning algorithms, have been widely used in the field of manipulator control. In the past three years alone, there have been dozens of research and application results in this area in China. The following is an overview of the research and application examples of manipulator path and position control, manipulator trajectory control, manipulator target tracking control, and manipulator motion control.

1. Manipulator path and position control

Based on deep reinforcement learning strategy, Yang S.Z. *et al.* studied the problem of manipulator arm control. Combining deep learning and deterministic ploy gradient reinforcement learning (DDPG), design the deep learning step of the deterministic ploy gradient, so that the manipulator arm has high environmental adaptability after training and learning, and can quickly and accurately find the moving target point in the environment [48]. Qian L.T. invented a manipulator control system based on deep learning to deploy and control the entire system to achieve all-round control of the manipulator's speed, angle and strength, while ensuring the smooth and safe operation of the system [35]. Liu H. *et al.* proposed a deep learning control planning method for manipulator motion path in an intelligent environment. By separately establishing a global static path planning model and a local dynamic obstacle avoidance planning model, using the nonlinear fitting characteristics of deep learning, the global optimal path can be quickly found. It avoids the problem of falling into local optimality in common path planning [24].

2. Manipulator trajectory control

Zhang Haojie *et al.* fused reinforcement learning and deep learning methods and proposed an end-to-end control method for manipulators based on deep Q network learning, which improved the accuracy of collision-free movement of manipulators without obstacle maps or sparse lidar data. The model generated by this method training effectively establishes the mapping relationship between the lidar data and the manipulator motion speed, so that the manipulator selects the action with the largest Q value to execute in each control cycle, and can smoothly move to avoid obstacles [49]. Tang Chaoyang and others invented a robot obstacle avoidance control method and device based on deep learning, which can accurately predict the position information of moving obstacles, quickly generate control instructions for controlling the robot to avoid moving obstacles, and quickly control the robot to complete obstacles. Avoid, improve the accuracy of obstacle avoidance [42].

3. Manipulator motion control

Aiming at the problem that the traditional mechanical control method is difficult to effectively control the behavior of the yellow peach core digging manipulator, Ge Hongwei *et al.* proposed a method based on deep reinforcement learning to control the behavior of the yellow peach core digging robot with visual function. The invention exerts the deep learning perception ability and the reinforcement learning decision-making ability, so that the robot can use deep learning to recognize the state of the peach core, and guide the single-chip microcomputer to control the motor to dig out the peach core through the reinforcement learning method to complete the core digging task [18]. Zhang Songlin believes that the technology represented by convolutional neural networks can be trained according to different control requirements to improve the control effect of the system. It has been widely used in fields such as manipulator control and target recognition. With the complexity of the manipulator application environment, the design is based on a convolutional neural network robot control algorithm to achieve precise object grasping in an unstructured environment, and a complete manipulator automatic grasping planning system is established [50].

4. Foot robot walking control and gait planning

Song Guangming *et al.* proposed a fall self-reset control method for quadruped robots based on deep reinforcement learning, which uses deep reinforcement learning algorithms to enable the robot to automatically reset on flat ground under any fall posture without pre-programming or human intervention. The robot's intelligence, flexibility and environmental adaptability [39]. Bi Sheng *et al.* invented a gait planning method for predictive control of humanoid robots based on deep reinforcement learning, which can effectively solve the walking problem of humanoid robots in complex environments [4]. Liu Huiyi and others invented a gait control method for a humanoid robot based on a deep Q network, including constructing a gait model, learning and training the deep Q network based on training samples, obtaining the state parameters of the humanoid robot in the action environment, and using the built gait model of the humanoid robot controls the gait and updates the deep Q network by generating a reward function. The invention can improve the walking speed of the humanoid robot and realize the fast and stable walking of the humanoid robot [25].

In addition, Li Yingying proposed a voice interaction and control method for intelligent industrial manipulators based on deep learning [23]. Fang designed a cyclic emotion CMAC neural network controller for mobile robots based on vision. Adding a loop structure to the sliding model neural network controller ensures that the previous state of the robot is preserved to improve its dynamic mapping ability. Experiments show that the proposed system is superior to other popular neural network-based control systems and has advantages in approaching highly nonlinear dynamics [16]. Wang *et al.* proposed a T-S-based FNN multi-sensor information fusion method based on fuzzy logic and neural network for robot obstacle avoidance and navigation control. This method is feasible and effective in robot obstacle avoidance and path planning [45]. There are many results of optimal control of manipulators. For example, Bobrow *et al.* proposed path optimization control of manipulators [6], and Luo *et al.* designed the optimal/PID formula of manipulators [31].

5.6.2 *Neural control of multi-fingered dexterous manipulators* [5, 26, 41, 51]

Multi-fingered dexterous manipulator, also known as multi-fingered multi-joint hand, is a type of manipulator in the form of association and series. It is generally composed of a palm and three to five fingers, each with three to four joints. Because it has multiple joints (≥ 9), it can grasp and operate almost any object. If a fingertip force sensor and a tactile sensor are installed to control the grasping force, it can grasp and operate fragile objects (such as eggs, etc.). The mechanical body of the multi-finger dexterous hand is generally small and has more degrees of freedom. Therefore, the servo motor is often used for long-distance drive through a steel wire or nylon rope with a sleeve, and the servo motor is controlled to rotate in an orderly manner, which can make the multi-finger dexterous hand complete various grasping and operations. Due to the deformation of the rope and the friction between the rope and the casing, the coupling between the joints makes the multi-finger dexterous hand more nonlinear than ordinary manipulators. At present, the research on intelligent grasping of multi-fingered dexterous hands and the research on coordinated control of position/force are one of the hot topics in robotics research.

The following introduces the use of a trained multilayer feedforward network as a controller to control the joints of a multi-finger dexterous hand to track a given trajectory, including the network structure, learning algorithm, control system software and hardware components, and experimental results.

1. Network structure and learning algorithm

This system uses a $3 \times 20 \times 1$ three-layer feedforward network to learn the input-output relationship of the original controller. The neuron uses a sigmoid function, namely $y = 1/(1 + e^{-x})$. After learning, use the feedforward network as the controller. The controller, which is a model for network learning, is a successful controller that has been proven in practice. The input and output data pairs generated by this controller are used for the network to learn. The trained network can well approximate the input and output mapping relationship of the original controller.

Learning adopts a hybrid learning algorithm that combines BP algorithm and chemotaxis algorithm, that is, first train the network with BP algorithm, and then train with chemotaxis algorithm. Practice has proved that this hybrid learning algorithm can avoid local minima and has a faster convergence rate than using either algorithm alone. BP algorithm is the most common learning algorithm. The chemotaxis algorithm was proposed by Bremermann and Anderson, and it is especially suitable for dealing with the training problem of dynamic networks. The chemotaxis algorithm used here is as follows:

1. Set the weight W to the random initial value on $[-0.1, 0.1]$, namely W_0;
2. Input the sample into the network and calculate the network output;
3. Find the value of the objective function J, and set $B_1 = J$;
4. Generate a normal distributed random vector W' on $[-1, +1]$ with the same dimension as the weight W and zero mean value;
5. Let $W = W_0 + a \times W', a < 1$, is a real coefficient;
6. Find the value of the objective function J, set $B_2 = J$;
7. If $E_2 < E_1$, then let $W_0 = W$, turn to (4); if $E_2 \geq E_1$, turn to (4).

The weight matrix learned by this algorithm is W_0, and the objective function is defined as $J = \sum_{i=1}^{N} (E_{rr}(i))^2$, $E_{rr}[i]$ is the learning error of the i-th sample, and N is the number of samples.

2. Controller design based on neural network [16]

1. Control system hardware

This system uses the three-finger dexterous hand of the Robotics Institute of Beijing University of Aeronautics and Astronautics as the test bed, and its controller adopts a hierarchical structure. The upper host is PC-386, which is responsible for man-machine information exchange, mission planning and path planning. The lower layer is the servo controller,, there is a PC bus-based 8031 single-chip position servo controller for each motor. Figure 5.21 shows the hardware diagram of the controller. The finger joints in the picture are equipped with a potentiometer, which is used as an angle sensor, and its output signal is used as the feedback signal of the servo controller.

2. Control system software design

The control software is divided into two parts, the upper computer software is written in C language, and the servo controller software is written in MCS-51 single-chip assembly language. Figure 5.22 is the structure diagram of the controller. The upper computer software is responsible for calculating the network output and generating the corresponding control signal according to the error signal. The servo controller obtains the control instruction from the host, and after appropriate processing, generates the corresponding PWM motor control signal to control the motor rotation. The calculation of the neural network is all done by the host computer. This is because the calculation of the neural network includes a large number of non-linear functions, which is very difficult and slow to implement in assembly language. Figure 5.23 is a flowchart of the host software. The function of the timer is to ensure that an interpolation is

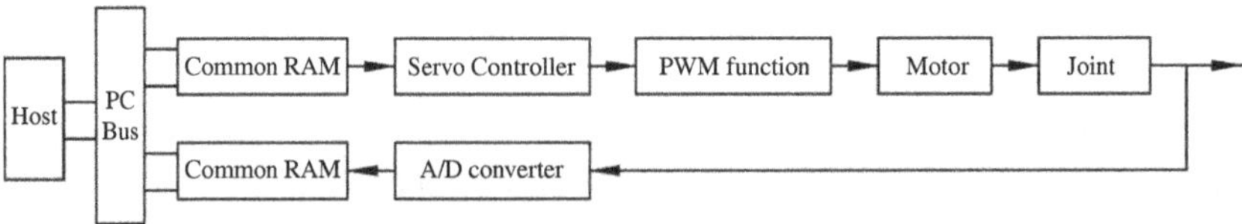

Figure 5.21.　Hardware diagram of the control system.

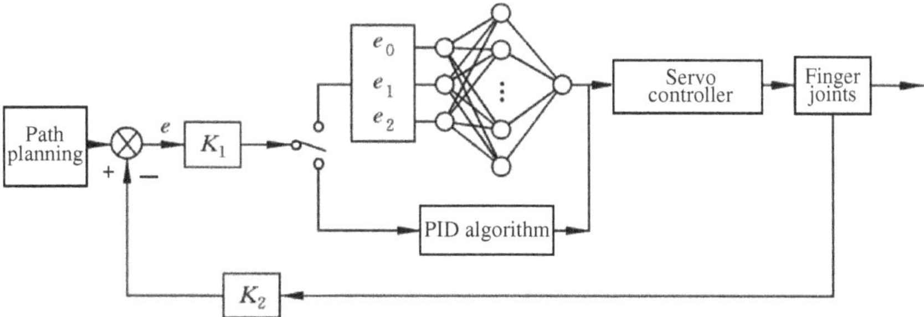

Figure 5.22.   Controller structure diagram.

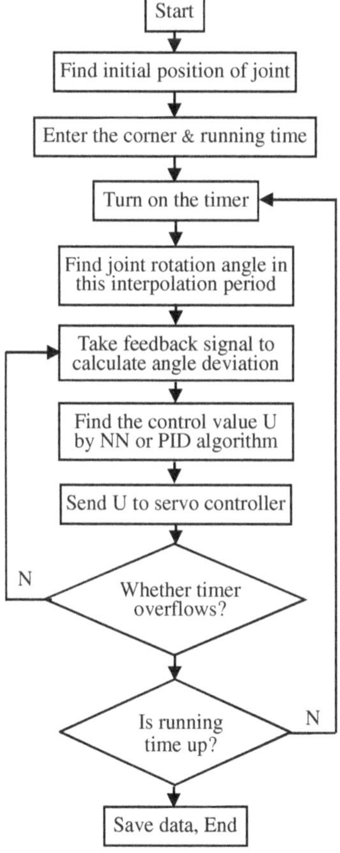

Figure 5.23.   Software flow chart of Host.

performed at 40 ms. It is realized by the CMOS timing of the host computer, which can be accurate to microseconds.

3. Compound control method

Through experimentation, it is found that simply using neural network controller for control, the response of the system can track the given trajectory well in the tracking stage, but the steady-state effect is not good, and there is a large steady-state error. This is because the neural network can learn the input and output mapping relationship of the original controller, but it cannot fully reproduce this relationship. There will always be a certain error, and after the error is small to a certain range, it will become very difficult to reduce it further. Due to time constraints, network learning can only obtain an approximate optimal solution, and it is impossible to obtain a true optimal solution. In order to make the system have a good steady-state response, a PID controller is used to control the system in the steady state, and uses its integral action to eliminate the steady-state error. The experimental results show that this composite controller can ensure that the system has a good stability state response.

5.7　Chapter Summary

This chapter begins to study the problem of manipulator control. Section 5.1 summarizes the basic principles of manipulator control and transmission. After briefly describing the classification of manipulator controllers, it focuses on analyzing the relationship between the control variables.

Since manipulator control is one of the key research issues in this book, this issue is discussed in two chapters. Section 5.2 discusses the position control, position and force control, and decomposition motion control of the manipulator.

Position control is the most basic control of the manipulator. Two structures of manipulator position control are mainly discussed, joint space control structure and Cartesian coordinate space control structure. Taking PUMA manipulator as an example, the servo control structure is introduced. On this basis, the single-joint position controller and the multi-joint position controller are discussed separately, involving the structure, mathematical model, coupling and compensation of these controllers.

Section 5.3 discusses the position and force control of the manipulator and studies the comprehensive problem of the control law of the force and position hybrid control system, involving the position control law, the force control law and the synthesis of the two hybrid control laws.

Section 5.4 analyzes the resolved motion control of the manipulator, and discusses the resolved motion speed control, resolved motion acceleration control and resolved motion force control of the manipulator based on the principle of decomposition resolved motion control. The resolved movement decomposes the movement of the manipulator into independent movements along each Cartesian coordinate axis, and each movement is coordinated at different speeds. For the decomposition of motion speed control, acceleration control and force control, the system control block diagram and dynamic relationship are studied one by one.

Adaptive control is one of the important methods of modern control. It can maintain automatic adaptation to the environment in an incompletely determined and locally changing environment, and can combine sensors or sensing systems to search and guide ways to perform different cyclic operation. Section 5.5 first discusses the state model and structure of the adaptive controller, and then studies the structure of the model reference adaptive controller and the self-tuning adaptive controller, and emphatically discusses the model reference adaptive controller and the self-tuning adaptive controller design.

Intelligent control is a completely new type of control method, and its use in the manipulator system needs to be further studied and improved. Manipulators have widely used fuzzy control and neural control. Section 5.6 first introduces the manipulator adaptive fuzzy control situation and discusses the design and system structure of the manipulator adaptive fuzzy control system with examples. Then it outlines neural control, especially the various controls of manipulators based on deep learning, as an application example of manipulator neural control, introduces the network structure, learning algorithm and neural controller design of the neural control system of the multi-finger dexterous hand.

References

[1] Adasa, H. and Slotine, J.J.E. (1986). *Robot Analysis and Control.* John Wiley and Sons Inc.

[2] Bajd, T., Mihelj, M., Lenarcic, J., *et al.* (2010). *Robotics (Intelligent Systems, Control and Automation: Science and Engineering).* Springer.

[3] Bejczy, A.K. and Paul, R.P. (1981). Simplified robot arm dynamics for control, Proceedings of 20th IEEE Conference on Decision and Control, pp. 261–262.

[4] Bi, S., Liu, Y.D., Dong, M., *et al.* (2018). *Gait planning method of preview control humanoid robot based on deep reinforcement learning (invention patent).* Publication/Announcement Date: 2018-09-18 (in Chinese).

[5] Bi, S.S. and Zong, G.H. (1999). Research and development of micro-manipulation robot system. *China Mechanical Engineering,* 10(9):1024–1027.

[6] Bobrow, J.E., Dubowsky, J.E. and Gibson, J.S. (1985). Time-optimal control of robotic manipulators along specified path. *International Journal of Robotics Research,* 4(3):3–17.

[7] Cai, Z.X. (1988). *Robotics: Principles and Applications, Chapter 3 and Chapter 5.* Changsha: Central South University of Technology Press (in Chinese).

[8] Cai, Z.X. (2021). *Fundamentals of Robotics,* 3rd Edition: Beijing: Mechanical Industry Press (in Chinese).

[9] Cai, Z.X. (2021). *Robotics,* 4th Edition, Chapter 3 and Chapter 5. Beijing: Tsinghua University Press (in Chinese).

[10] Cai, Z.X. and Zhang, Z.J. (1987). *Design of robot adaptive controller,* Proceedings of the First National Robotics Symposium, China, pp. 239–246 (in Chinese).

[11] Cai, Z.X., Xie, G.H., Wu, Z.H., *et al.* (1998). Control the robot directly in the position to realize the force/position adaptive fuzzy control. *Robot,* 20(4):297–302 (in Chinese).

[12] Chen, G.D., Chen, G.D., Chang, W.S., Zhang, P., *et al.* (1996). Force/position hybrid control algorithm for symmetrical coordination of both hands. *Journal of Automation,* 22(4):418–427 (in Chinese).

[13] Coiffet, P. and Chirouze. (1983). *An Introduction to Robot Technology.* Hermes Publishing.

[14] Craig, J.J. (1986). *Introduction to Robotics Mechanics and Control.* Addison-Wesley Publishing Company.

[15] Craig, J.J. (2004). *Introduction to Robotics: Mechanics and Control,* 3rd Edition. Prentice-Hall.

[16] Fang, W.B., Chao, F., Yang, L.Z., *et al.* (2019). A recurrent emotional CMAC neural network controller for vision-based mobile robots. *Neurocomputing*, 334:227–238.

[17] Fu, K.S., Gonzalez R.C. and Lee, C.S.G. (1987). *Robotics: Control, Sensing, Vision, and Intelligence*. Wiley.

[18] Ge, H.W., Lin, J.J., Sun, L. and Zhao, M.D. (2018). A behavior control method for a yellow peach core-digging robot based on deep reinforcement learning (invention patent). Public/Announcement Date: 2018-04-20 (in Chinese).

[19] Hunt, V.D. (1983). *Industrial Robotics Handbook*. Industrial Press Inc.

[20] Huo, W. (2005). *Robot Dynamics and Control*, Beijing: Higher Education Press (in Chinese).

[21] Jiang, X.S. (1994). *Introduction to Robotics*. Shenyang: Liaoning Science Press (in Chinese).

[22] Katic, D. and Vukobratovic, M. (2010). *Intelligent Control of Robotic Systems*. Springer.

[23] Li, Y.Y. and Xiao, N.F. (2017). A voice interaction and control method for intelligent industrial robots based on deep learning (invention patent). Public/Announcement Date: 2017-06-27 (in Chinese).

[24] Liu, H., Li, Y.F., Huang, J.H., *et al.* (2017). A deep learning control planning method for robot motion path in an intelligent environment (invention patent). Public/Announcement Date: 2017-11-21 (in Chinese).

[25] Liu, X.Y., Yuan, W., Tao, Y. and Liu, X.Y. (2020). Humanoid robot gait optimization control method based on deep Q network (invention patent). Public/Announcement Date: 2020-02-07 (in Chinese)

[26] Lu, J.Z., Xiong, Y.L. and Yang, S.Z. (1995). Control and sensor technology of robotic multi-finger hands. *Robot*, 17(3):184–192 (in Chinese).

[27] Luh, J.Y.S. (1983). An anatomy of industrial robots and their controls. *IEEE Transactions on Automatic Control*, AC-28(2):133–153.

[28] Luh, J.Y.S. (1983). Conventional controller design for industrial robots — A Tutorial. *IEEE Transactions on Systems, Man, and Cybernetics*, SMC-13:298–316.

[29] Luh, J.Y.S. and Lin, C.S. (1981). *Automatic generation of dynamic equations for mechanical manipulators*, Proceedings of Joint Automatic Control Conference, TA-2D.

[30] Luh, J.Y.S., Walker, M.W. and Paul, R.P. (1980). On-line computational scheme for mechanical manipulators. *Transactions ASME Journal of Dynamics Systems, Measurement, Control*, 102:69–76.

[31] Luo, G.L. and Saridis, G.N. (1985). Optimal/PID formulation for control of robotic manipulator. *Proceedings IEEE 1985 International Conference on Robotics and Automation*, 621.

[32] Lynch, K.M. and Park, F.C. (2017). *Modern Robotics: Mechanics, Planning, and Control*. Cambridge University Press.

[33] Niku, S.B. (2010). *Introduction to Robotics: Analysis, Control, Applications*. Wiley.

[34] Paul, R.P. (1981). *Robot Manipulators: Mathematics, Programming and Control*. MIT Press.

[35] Qian, L.D. (2019). A robot control system based on deep learning (invention patent). Public/Announcement Date: 2019-05-17 (in Chinese).

[36] Raibert, M.H. and Craig, J.J. (1981). Hybrid position/force control of manipulators. *ASME Journal of Dynamic Systems, Measurement, and Control*, 102:126–133.

[37] Reza, N.J. (2010). *Theory of Applied Robotics: Kinematics, Dynamics, and Control*, 2nd Edition. Springer.

[38] Siciliano, B., Sciavicco, L., Villani, L., *et al.* (2011). *Robotics: Modeling, Planning and Control (Advanced Textbooks in Control and Signal Processing)*. Springer.

[39] Song, G.M., He, M., Wei, Z. and Song, A.G. (2020). A self-reset control method for a quadruped robot based on deep reinforcement learning (invention patent). Public/Announcement Date: 2020-03-06 (in Chinese).

[40] Sun, F.C., Lu, W.J. and Zhu, Y.Y. (1999). Neural adaptive control of manipulator based on ideal trajectory learning. *Journal of Tsinghua University (Natural Science Edition)*, 39(11):123–126 (in Chinese).

[41] Sun, L.N., Tang, T. and Xu, Q. (1998). Research on the tiny manipulator with force perception function. *Robot*, 20(S1):570–574 (in Chinese).

[42] Tang, Z.H., Chen, Y., Duan, X., *et al.* (2020). A method and device for robot obstacle avoidance control based on deep learning (invention patent). Public/Announcement Date: 2020-04-17 (in Chinese).

[43] Villani, L. and De Schutter, J. (2016). Force control. *Handbook of Robotics*, 2nd Edition, Siciliano, B. and Khatib, O. (Eds.) (pp. 195–219). Springer.

[44] Vukobratovic, M. and Stokic, D. (1982). *Scientific Fundamentals of Robotics 2, Control of Manipulation Robots, Theory and Application*. Springer-Verlag.

[45] Wang, Z.Y. (2018). Robot obstacle avoidance and navigation control algorithm research based on multi-sensor information fusion, 11th International Conference on Intelligent Computation Technology and Automation (ICICTA), Changsha, Peoples R China, Sep. 22–23, 2018.

[46] Xia, T.C. (1986). *Adaptive control of robot manipulators: A review. Report No.RRL-86-1*. University of California-Davis.

[47] Xiong, Y.L., Li. W.L., Chen, W.B., *et al.* (2017). *Robotics Modeling, Control and Vision*. Huazhong University of Science and Technology Press (in Chinese).

[48] Yang, S.Z., Han, J.Y., Liang, P., *et al.* (2019). Robot arm control based on deep reinforcement learning. *Fujian Computer*, 35(1):28–29 (in Chinese).

[49] Zhang, H.J., Su, Z.B. and Su, B. (2018). Robot end-to-end control method based on deep Q network learning. *Chinese Journal of Scientific Instrument*, 39(10):6–43 (in Chinese).

[50] Zhang, S.L. (2019). Robot system control based on convolutional neural network algorithm. *Journal of Changchun University (Natural Science Edition)*, 29(2):14–17 (in Chinese).

[51] Zhang, Y.D. (1999). Research on the structure optimization design and grasping mechanism of the robot multi-fingered dexterous hand. *Harbin Institute of Technology PhD Thesis* (in Chinese).

[52] Zhu, S,Q. and Wang, X.Y. (2019). *Robot Technology and Its Application*, 2nd Edition. Zhejiang University Press (in Chinese).

Chapter 6

Manipulator Planning

Manipulator planning is divided into high-level planning and low-level planning, namely intelligent planning and trajectory planning. This chapter only discusses the high-level planning of the manipulator, because the trajectory planning of the manipulator belongs to the low-level planning and will not be introduced in this book.

6.1 Overview of Manipulator Planning

Automatic planning starts from a specific problem state, seeks a series of actions, and establishes an operation sequence until the target state is obtained. Compared with general problem solving, automatic planning pays more attention to the problem-solving process rather than the solution result. In addition, the problems to be solved by planning, such as the manipulator world problem, are often real-world problems rather than more abstract mathematical model problems. The automatic planning system is an advanced solution system and technology. When studying automatic planning, manipulator planning and problem solving are generally discussed as typical examples. This is not only because manipulator planning is the most important research object of automatic planning, but also because manipulator planning can be visually and intuitively tested. In view of this, automatic planning is often referred to as manipulator planning or robot planning. Manipulator planning is an important problem-solving technology, and its principles, methods and techniques can be popularized and applied to other planning objects or systems [39, 48, 56].

1. The concept and function of planning [30, 38, 59]

Robot planning is an important research field of robotics, and it is also an interesting combination of artificial intelligence and robotics. Some high-level planning systems for robots have been developed [30, 36, 37, 45, 60, 62, 70]. Among them, some focus on the proof-of-principle machine, and they apply general search heuristic technology to express the desired goal with logical calculus. STRIPS and ABSTRIPS fall into this category. This system represents the world model as an arbitrary set of first-order predicate calculus formulas, uses resolution-refutation to solve specific model problems, and uses means-ends analysis strategies to guide the solution system to seek the goal. Another type of planning system uses supervised learning to speed up the planning process and improve problem-solving capabilities. Since the 1980s, other planning systems have been developed, including non-linear planning [61], induction based planning systems [60] and hierarchical planning systems [36, 63, 68]. In the past 20 years, expert systems have been applied to many different levels of robot planning. In recent years, robot planning based on machine learning, especially deep learning, has been widely used. In addition, many results have been achieved in the research on path planning of robots, especially mobile robots and trajectory planning of robot manipulators [43].

Automatic planning is called high level planning in robot planning, and it has different planning goals, tasks, and methods from low-level planning. In this section, first introduce the concept of planning, and then discuss the tasks of the automatic planning system.

In human daily life, the planning means that deciding the course of action before taking action. In other words, the term planning refers to the process of calculating the steps of a problem solving procedure before executing any step in the procedure. A plan is a description of the course of action. It can be an unordered list of goals like a department store list; but generally speaking, a plan has an implicit order of a certain planning goal. For example, for most people, wash your face and brush your teeth or rinse your mouth before eating breakfast. For another example, if a manipulator wants to move a certain work piece, it must first move to the vicinity of the work piece, then grasp the work piece, and then move the work piece with it.

Lack of planning may lead to suboptimal problem solving; for example, if because of lack of planning, someone ran to the library twice in order to borrow a book and return a book. In addition, if the goal is not independent, then the lack of planning before the action may actually rule out a certain solution to the problem. For example, the planning of building a power substation includes sub-plans such as laying walls, installing transformers, and laying cables. These sub-plans are not independent of each other. First, the cables must be laid, then the walls must be laid, and finally the transformer must be installed. If there is no planning and the order is reversed, the power substation cannot be built.

Planning can be used to monitor the problem-solving process and find errors before they cause greater harm. If the problem-solving system is not the only actor in the problem-solving environment, and if the environment may change in unexpected ways, then this monitoring is particularly important. For example, consider an aircraft operating on a distant planet. It must be able to plan a route, and then re-plan it when it finds that the state of the environment does not match expectations. The feedback on the state of the environment is compared with the expected state of the plan, and when there is a difference between the two, the plan is revised. The benefits of planning can be summarized as simplifying search, solving target contradictions, and providing a basis for error compensation.

2. Tasks and methods of the robot planning system

In the robot planning system, there must be methods to perform the following tasks:

(1) According to the most effective heuristic information, choose the best rule to apply to the next step.
(2) Apply the selected rule to calculate the new state generated by applying the rule.
(3) Check the solution obtained.
(4) Check the empty ends (end points that cannot reach the target) in order to discard them and make the system's solution work proceed in a more effective direction.
(5) Check the almost correct answer and apply specific techniques to make it completely correct.

6.2 Robot Planning in the Block World [38, 45, 60]

Robotic problem solving is to seek the action sequence of a certain robot, this sequence can make the robot achieve the expected work goal and complete the specified work task. Robot planning is divided into high-level planning and low-level planning. The planning discussed here belongs to the category of high-level planning.

6.2.1 *The robot problem in the block world*

Many problem solving system concepts can be experimentally studied and applied in robot problem solving. The robot problem is relatively simple and intuitive. In a typical representation of a robot problem, the robot is able to perform a set of actions. For example, imagine a world of building blocks and a robot. The world is a few marked cube blocks (assumed to be the same size here), which are either stacked on top of each other or placed on the table; the robot has a movable manipulator, which can pick up the blocks and move the blocks from one place to another. Examples of actions that the robot can perform in this example are as follows:

unstack(a, b): Pick up (remove) block a stacked on block b. Before performing this action, the hand of the robot is required to be empty, and the top of the building block should be empty.

stack(a, b): Stack block a on block b. Before the action, the manipulator must have grasped block a, and the top of block b must be empty.

pickup(a): Pick up block a from the table and hold it. Before the action, the manipulator is required to be empty-handed, and there is nothing on top of block a.

putdown(a): Put block a on the desktop. The manipulator has grasped block a before the action is required.

Robot planning includes many functions, such as recognizing the world around the robot, expressing action plans, and monitoring the execution of these plans. The main thing to be studied is the action sequence of the integrated robot, that is, in a given initial situation, after a certain sequence of actions to achieve the specified goal.

The production system using state description as the database is the simplest problem solving system. Both the state description and goal description of the robot problem can be formed by predicate logic formulas. In order to specify the actions performed by the robot and the results of the performed actions, the following predicates need to be applied:

ON(a, b): Block a is above block b.
ONTABLE(a): Block a is on the desktop.
CLEAR(a): There is nothing on top of block a.
HOLDING(a): The manipulator is holding block a.
HANDEMPTY: The manipulator is empty-handed.

Figure 6.1(a) shows the robot problem of the initial layout. This layout can be represented by the conjunction of the following predicate formulas:

CLEAR(B): The top of block B is empty
CLEAR(C): The top of block C is empty
ON(C, A): Building block C is stacked on building block A
ONTABLE(A): Block A is placed on the desktop
ONTABLE(B): Block B is placed on the desktop
HANDEMPTY: The manipulator is empty-handed

The goal is to build a pile of blocks, in which block B is piled on top of block C, and block A is piled on top of block B, as shown in Figure 6.1(b). You can also use predicate logic to describe the goal as:

$$ON(B, C) \land ON(A, B)$$

6.2.2 *The solution of robot planning in the block world*

A rule called the STRIPS planning system will be used to represent the robot's actions, which is composed of three parts. The first

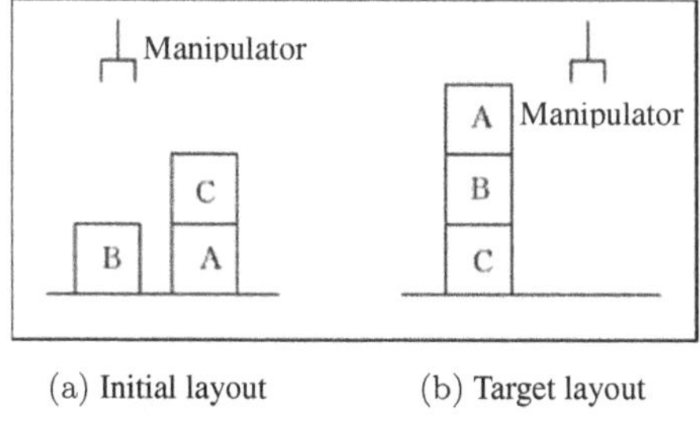

(a) Initial layout   (b) Target layout

Figure 6.1.   Robot question in the block world.

part is the prerequisites. In order to enable the F rule to be applied to the state description, this precondition formula must be a predicate calculus expression that logically follows the facts in the state description. Before applying the F rule, you must make sure that the prerequisites are true. The second part of the F rule is a predicate called delete table. When a rule is applied to a state description or database, the contents of the deleted table are deleted from the database. The third part of the F rule is called adding a table. When a rule is applied to a database, the content of the added table is added to the database. For the example of piled wood, the move action can be expressed as follows:

move(x, y, z): Move object x from object y to z.
Prerequisites: CLEAR(x), CLEAR(z), ON(x, y)
Delete table: ON(x, z), CLEAR(z)
Add table: ON(x, z), CLEAR(y)

If move is the only operator or applicable action for this robot, then a search graph or search tree as shown in Figure 6.2 can be generated.

Consider the example shown in Figure 6.1 in more detail below. The 4 actions (or operators) of the robot can be expressed in the form of STRIPS as follows:

(1) stack(x, y)
Prerequisites and delete table: HOLDING(x)∧CLEAR(y)
Add table: HANDEMPTY, ON(x, y)
(2) unstack(x, y)
Prerequisite: HANDEMPTY∧ON(x,y)∧CLERA(x)
Delete table: ON(x, y), HANDEMPTY
Add table: HOLDING(x), CLEAR(y)

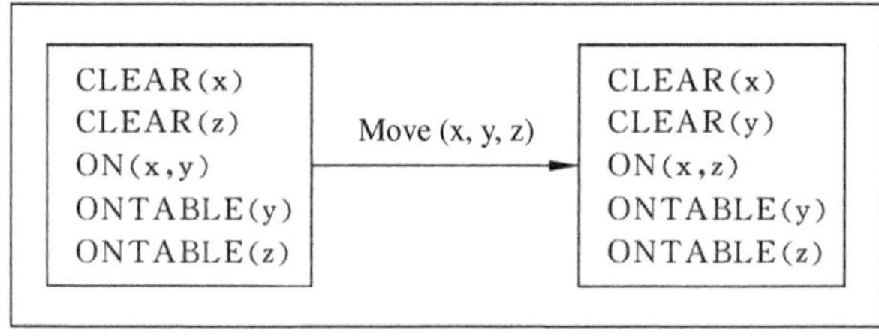

Figure 6.2.   Search tree representing the move action.

(3) pickup(x)

Prerequisite: ONTABLE(x)∧CLEAR(x)∧HANDEMPTY

Delete table: ONTABLE(x)∧HANDEMPTY

Add table: HOLDING(x)

(4) putdown(x)

Prerequisites and delete table: HOLDING(x)

Add table: ONTABLE(x), HANDEMPTY

Assume that the target is in the state shown in Figure 6.1(b), namely ON(B,C)∧ON(A, B). Starting from the initial state description shown in Figure 6.1(a), the F rule can only be applied to the two actions of unstack(C, A) and pickup(B). Figure 6.3 shows the entire state space of this problem, and uses a thick line to indicate the solution path from the initial state (marked by S0) to the target state (marked by G). Different from the habitual state space drawing method, this state space shows the symmetry of the problem, instead

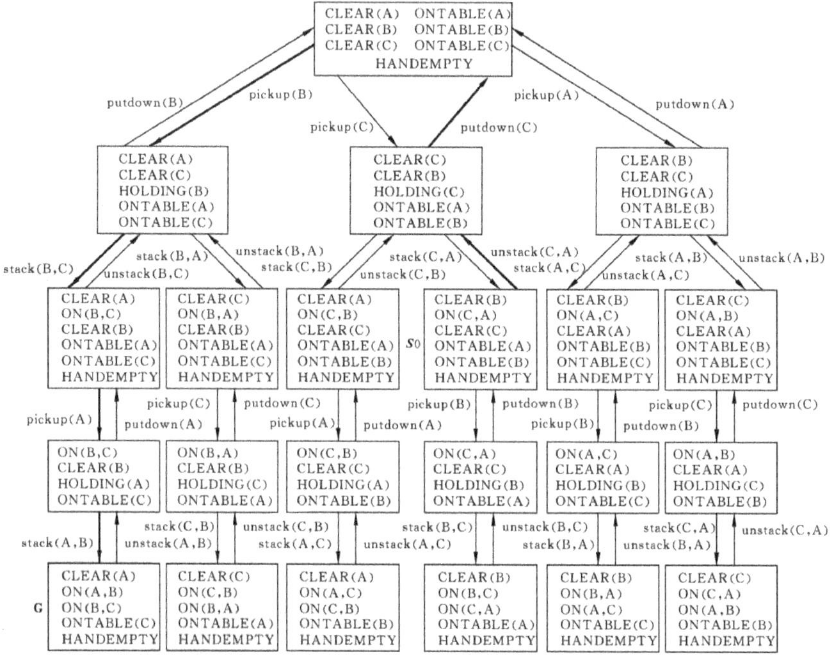

Figure 6.3.   State space of the robot problem in the block world.

of placing the initial node S0 on the vertex of the graph. Also, note that each rule in this example has an inverse rule.

Along the branch shown by the thick line, starting from the initial state, and reading the F rule on the connecting arc in a positive direction, an action sequence that can reach the target state is as follows:

$$\{\text{unstack(C, A), putdown(C), pickup(B), stack(B, C),}$$
$$\text{pickup(A), stack(A, B)}\}$$

Call this action sequence the plan to achieve the goal of the robot problem in the block world.

6.3　Robot Planning System Based on Resolution Principle [36–38]

A robot planning system based on the principle of resolution, STRIPS (Stanford Research Institute Problem Solver), that is, the Stanford Research Institute problem solving system, is designed to draw general conclusions from the problem to be solved [10, 36, 37].

6.3.1　*Composition of STRIPS system*

Fikes, Hart and Nilsson successfully researched STRIPS in 1971 and 1972, and it is an integral part of Shakey's robot program control system. This robot is a self-carrying cart designed to move around a simple environment, and it can move in accordance with simple English commands. A mobile car (robot) that is a component of STRIPS. Shakey consists of the following four main parts:

(1) Wheels and their propulsion system;
(2) The sensor system is composed of a TV camera and a contact rod;
(3) A computer that is not on the car body to execute programming. It can analyze the feedback information and input instructions obtained by the sensors on the car, and send a trigger signal to the wheels to make the propulsion system;
(4) The radio communication system is used for data transmission between the computer and the wheels.

STRIPS is the programming that decides which instruction to send to the robot. The robot world includes some rooms, doors between rooms, and movable boxes; in more complicated situations, there are lights and windows. For STRIPS, the actual world that exists at any time is described by a set of predicate calculus clauses. For example, the clause

INROOM(ROBOT, R2)

There is an assertion in the database, indicating that the robot is in room 2 at that moment. As the actual situation changes, the database must be revised in time. In general, the database describing the world at any moment is called the world model.

The control program contains many subroutines. When these subroutines are executed, the robot will move through a door, push a box through a door, turn off a light or perform other actual actions. These procedures themselves are very complicated, but do not directly involve problem solving. For robot problem solving, these programs are a bit like the relationship between walking and picking up objects in human problem solving.

Section 6.2 introduced the composition of the F rules (ie operators) of the STRIPS system. The composition of the entire STRIPS system is as follows:

(1) World model. Is a first-order predicate calculus formula;
(2) Operator (F rule). Including prerequisites, delete tables and add tables;
(3) Operation method. Apply state space representation and means-ends analysis. E.g.:
State: (M, G), including initial state, intermediate state and target state.
Initial state: (M0, (G0))
Goal status: get a world model, which leaves no unsatisfied goals.

6.3.2 *Planning process of STRIPS system*

Each solution to the STRIIPS problem is a sequence of operators to achieve a goal, that is, a rule to achieve the goal. The following examples illustrate the process of solving the STRIPS system rules.

Example 6.1. Consider a relatively simple case of the STRIPS system, which requires the robot to go to the adjacent room to retrieve a box. The world model of the robot's initial state and goal state is shown in Figure 6.4. There are two operators, namely gothru and pushthru ("walking through" and "pushing through"), which are described below:

(1) OP1: gothru(d, r1, r2);

The robot passes through the door d between the room r1 and the room r2, that is, the robot walks from the room r1 through the door d and enters the room r2.

Prerequisite: INROOM(ROBOT, r1) $\wedge$ CONNECTS(d, r1, r2), that is, the robot is in room r1, and the door d connects the two rooms r1 and r2.

Delete table: INROOM(ROBOT, S), for any S value.

Add table: INROOM(ROBOT, r2).

(2) OP2: pushthru(b, d, r1, r2)

The robot pushes the object b from the room r1 through the door d to the room r2.

Prerequisite: INROOM(b, r1)$\wedge$INROOM(ROBOT, r1)$\wedge$ CONNECTS(d, r1, r2)

Delete table: INROOM(ROBOT, S), INROOM(b, S); for any S.

Add table: INROOM(ROBOT, r2), INROOM(b, r2).

The difference table for this problem is shown in Table 6.1.

Assume that the initial state M0 and goal G0 of this problem are as follows:

$$
M0 : \begin{cases} \text{INROOM(ROBOT, R1)} \\ \text{INROOM(BOX1, R2)} \\ \text{CONNECTS(D1, R1, R2)} \end{cases}
$$

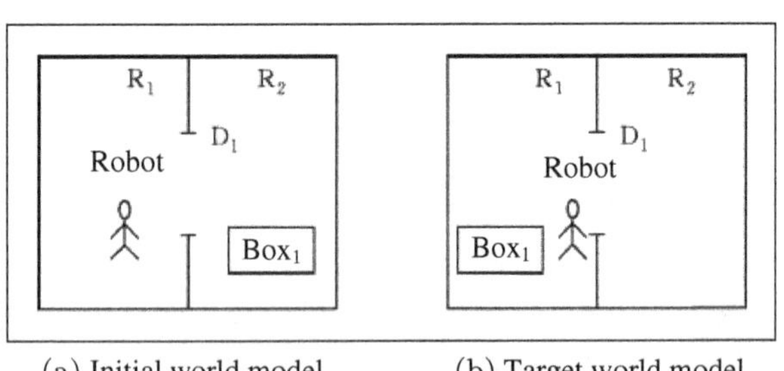

(a) Initial world model   (b) Target world model

Figure 6.4.   A simplified model of STRIPS.

Table 6.1. Difference table.

Difference	Operator	
	gothru	pushthru
The robot and the object are not in the same room	×	
The object is not in the target room	×	
The robot is not in the target room	×	
The robot and the object are not in the same room, but not the target room		×

G0: INROOM(ROBOT, R1)∧INROOM(BOX1, R1)∧CONN-ECTS(D1, R1, R2)

Next, use the means-ends analysis method to solve this robot plan step by step.

(1) The main loop iteration of do GPS, until M0 matches G0.

(2) begin.

(3) G0 cannot satisfy M0, find the difference between M0 and G0. Although this problem cannot be solved immediately, if the initial database contains the phrase INROOM(BOX1, R1), then the process of solving this problem can be continued. GPS finds their difference d1 as INROOM(BOX1, R1), that is, put the box (object) in the target room R1.

(4) Selection operator: an operator related to reducing the difference d1. According to the difference table, the STRIPS selection operator is OP2: pushthru(BOX1, d, r1, R1)

(5) Eliminate the difference d1 and set the prerequisite G1 for OP2 as:

G1: INROOM(BOX1, r1)∧INROOM(ROBOT, r1)∧CONNECTS (d, r1, R1)

This prerequisite is set as a sub-goal, and STRIPS tries to reach G1 from M0. Although G1 still cannot be satisfied, it is impossible to find a direct answer to this question immediately. However, STRIP found:

if

$$r1 = R2$$
$$d = D1$$

The current database contains INROOM(ROBOT, R1)

Then the process can continue. Now the new sub-goal G1 is:

G1: INROOM(BOX1, R2)∧INROOM(ROBOT, R2)∧CONNECTS(D1, R2, R1)

(6) GPS(p): Repeat steps 3 to 5, and iteratively call to solve this problem.

Step 3: The difference d2 between G1 and M0 is

INROOM(ROBOT, R2)

That is, the robot is required to move to room R2.

Step 4: According to the difference table, the relevant operator corresponding to d2 is

OP1: gothru(d, r1, R2)

Step 5: The prerequisites for OP1 are:

G2: INROOM(ROBOT, R1) ∧ CONNECTS(d, r1, R2)

Step 6: Apply the permutation formula r1 = R1 and d = D1, the STRIPS system can reach G2.

(7) Apply the operator gothru(D1, R1, R2) to M0 to find the intermediate state M1:

Delete table: INROOM(ROBOT, R1)

Add table: INROOM(ROBOT, R2)

$$M1 : \begin{cases} \text{INROOM(ROBOT, R2)} \\ \text{INROOM(BOX1, R2)} \\ \text{CONNECTS(D1, R1, R2)} \end{cases}$$

Apply the operator pushthru to the intermediate state M1,

Delete table: INROOM(ROBOT, R2), INROOM(BOX1, R2)

Add table: INROOM(ROBOT, R1), INROOM(BOX1, R1)

Obtain another intermediate state M2 as:

$$M2 : \begin{cases} \text{INROOM(ROBOT, R1)} \\ \text{INROOM(BOX1, R2)} \\ \text{CONNECTS(D1, R1, R2)} \end{cases}$$

M2 = G0

(8) end.

Since M2 matches G0, this robot planning problem is solved through middle-outcome analysis. In the solution process, the STRIPS rules used are operators OP1 and OP2, namely

gothru(D1, R1, R2), pushthru(BOX1, D1, R2, R1)

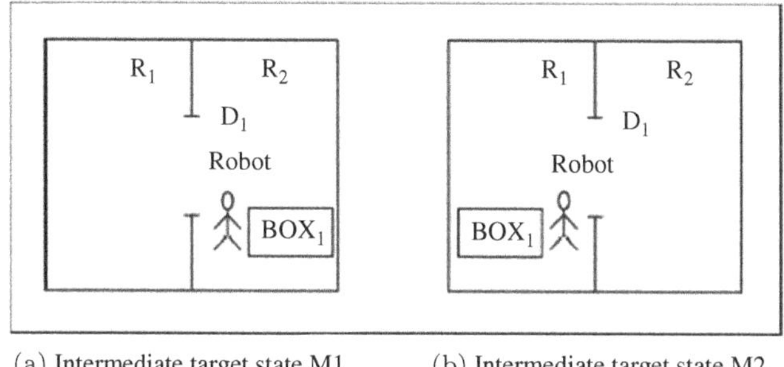

(a) Intermediate target state M1   (b) Intermediate target state M2

Figure 6.5.   The world model of the intermediate target state.

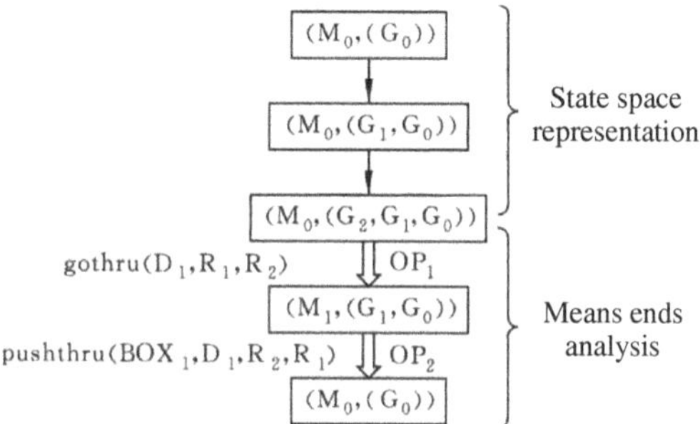

Figure 6.6.   Search diagram of a robot planning example.

The intermediate state models M1 and M2, namely the sub-goals G1 and G2, are shown in Figure 6.5.

It can be seen from Figure 6.5 that M2 is the same as the target world model G0 in Figure 6.4.

Therefore, the final plan obtained is {OP1, OP2}, that is

$$\{\text{gothru}(D1,\ R1,\ R2),\ \text{pushthru}(BOX1,\ D1,\ R2,\ R1)\}$$

The search diagram for this robot planning problem is shown in Figure 6.6, and the AND-OR tree is shown in Figure 6.7.

6.4 Robot Planning Based on Expert System

Since the 1980s, the application of expert system technology for robot planning and programming at different levels has been studied [5, 6, 9, 41, 72, 74, 78]. This section will combine the author's research on

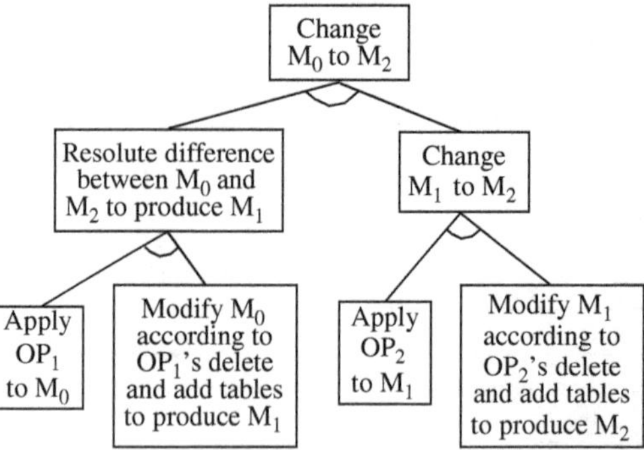

Figure 6.7.　AND diagram of a robot planning example.

the robot planning expert system, and introduce the robot planning based on the expert system [4–22].

6.4.1　*Structure and mechanism of the planning system* [3–9, 16, 68]

The robot planning expert system is a robot planning system built with the structure and technology of the expert system. At present, most expert systems are based on the structure of the planning system (rule-based system) to imitate the comprehensive mechanism of human beings. Here, a rule-based expert system is also used to establish a robot planning system.

1. System structure and planning mechanism

The rule-based robot planning expert system consists of five parts, as shown in Figure 6.8.

(1) Knowledge base. It is used to store expert knowledge and experience in specific areas, including world models of robotic work environments, initial states, object descriptions, and the possible actions or rules. To simplify the structure diagram, we consider the total database or global database that characterizes the current state of the system as part of the knowledge base.

(2) Control strategy. It contains a comprehensive mechanism to determine what rules the system should apply and how to find it. When using the PROLOG language, its control strategy is search, matching and backtracking.

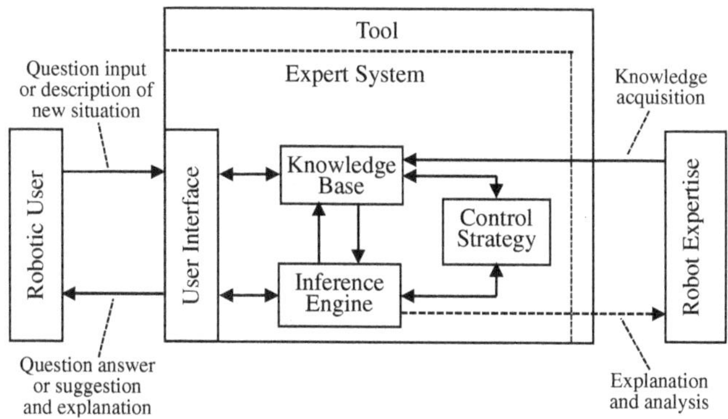

Figure 6.8.   Robot planning based on expert system.

(3) Inference engine. It is used to remember the rules and control strategies and reasoning strategies used. Based on the information in the knowledge base, the inference engine enables the entire robot planning system to work logically and coordinately, make decisions, and find the ideal sequence of robot operations. Sometimes this part is called a planner.

(4) Knowledge acquisition. First get expert knowledge of a particular domain. This knowledge is then transformed into a computer program using programming languages such as PROLOG [28, 29, 34, 64] and LISP. Finally, save them in the knowledge base for use.

(5) Explanation and description. Through the user interface, interactions between the expert system and the user enable the user to input data, ask questions, know the reasoning results, and understand the reasoning process.

In addition, to build an expert system, certain tools are needed, including computer systems or networks, operating systems and programming languages, and other supporting software and hardware. For the robot planning system studied in this section, we use DUALVAX11/780 computer, VM/UNIX operating system and C-PROLOG programming language.

The total database changes as each rule is adopted or an operation is performed. The goal of a rule-based expert system is to gradually change the state of the total database by executing the rules one by one and its related operations until an acceptable database (called the goal database) is obtained. Combining these related operations in turn forms an operational sequence that gives the operations that the robot must follow and the sequence of its operations. For instance, the robot transfer operation system, the planning sequence gives the process actions required by the handling robot to carry one or more specific parts or artifact from the initial position to the target position.

2. Three key elements of task-level robot planning [31]

Task-level robot planning is to find a way to simplify the programming of the robot. The task-level programming language is used to make the robot easy to program in order to open up the universality and adaptability of the robot.

Task-level robot planning is to find a way to simplify robot programming, using a programming language of task-level to make the robot easy to program, in order to develop the versatility and adaptability of the robot.

Task planning is one of the most important aspects of robot high-level planning. It consists of the following three elements:

(1) Establish model. The world model for establishing a robot working environment involves a large number of knowledge representations, including: geometric descriptions of all objects and robots in the task environment (such as the shape size of the object and the mechanical structure of the robot), and descriptions of the robot's motion characteristics (e.g., joint limits, speed and acceleration limits, and sensor characteristics, etc.) as well as object intrinsic characteristics and robot link descriptions (such as object mass, inertia, and linkage parameters).

In addition, the geometry, kinematic, and physical models of other objects must be provided for each new task.

(2) Task description. The state of the model is defined by the relative positions of the objects within the robot's working environment, and the tasks are specified by the order of transformation of the states. These states include an initial state, each intermediate state, and a goal state. In order to illustrate the task, the CAD system can be used to determine the position of the object within the model in a desired posture; the relative position of the robot and the characteristics of the object can also be specified by the robot itself. However, this practice is difficult to explain and correct. A better approach is to use a set of symbolic spatial relationships needed to maintain the

relative position between objects. In this way, a sequence of symbolic operations can be used to illustrate and define tasks, simplifying the problem.

(3) Program synthesis. The final step in task-level robot planning is to integrate the robot's procedures. For example, for a grab planning, a program to grasp the point is designed, which is related to the posture of the robot and the description characteristics of the object being grasped. This grab point must be stable. For example, for motion planning, if it belongs to free movement, it is necessary to integrate the procedure to avoid obstacles; if it is guidance and compliance movement, then it is necessary to consider the motion of the sensor to carry out program integration.

6.4.2 *ROPES robot planning system* [4, 7–10, 17, 20]

An example of a robot planning system using an expert system is now given. This is a less complicated example. We built this system using a rule-based system and the C-PROLOG programming language, called the ROPES system, the RObot Planning Expert Systems.

1. System simplified block diagram

A simplified block diagram of the ROPES system is shown in Figure 6.9.

To build an expert system, you must first acquire expert knowledge carefully and accurately. Expert knowledge of the system includes knowledge from experts and personal experiences, textbooks, manuals, papers and other references. The acquired expert

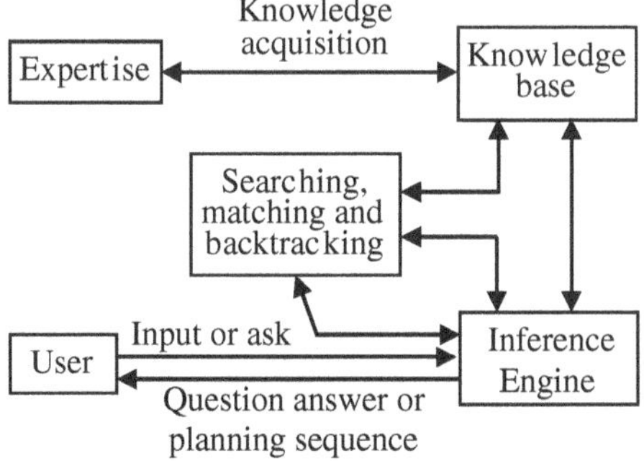

Figure 6.9. Simplified block diagram of the ROPES system.

knowledge is represented by computer programs and statements and stored in the knowledge base. Inference rules are also placed in the knowledge base. These procedures and rules are compiled in the C-PROLOG language. The main control strategies of this system are searching, matching and backtracking.

The program operator (user) at the system terminal, input initial data, ask questions, and talk to the inference engine; then, the answer and the inference result, that is, the planning sequence, are obtained at the terminal from the inference machine.

2. World models and assumptions

The ROPES system contains several subsystems for task planning, path planning, transfer operation planning, and finding a collision-free path for the robot [20]. Here, the transfer operation planning system is taken as an example to illustrate some specific problems of the system.

Figure 6.10 shows the world model of the robot assembly line. As can be seen from the figure, the assembly line passes through six working sections (section 1 to section 6). There are six doorways to connect with each relevant section. Ten assembly robots (robot 1 to robot 10) and ten work stations (station 1 to station 10) are installed beside the assembly line. On the racks on both sides of the shop where the assembly line is located, there are ten parts to be assembled, which have different shapes, sizes and weights. In addition, there is a mobile handling robot and a transport trolley. This robot can send the required parts from the rack to the designated workbench for assembly robots to assemble. When the size of the parts being transported is large or heavy, the handling robot needs to transport them with a small truck. We call this part "heavy."

In addition to the assembly line model presented in Figure 6.10, we can also use a mobile handling robot and a handling trolley. This robot can send the required parts from the rack to the designated workbench for the assembly robot to use for assembly.

In order to express knowledge, describe rules and understand planning results, some definitions of this system are given as follows:

go(A, B): The handling robot moves from position A to position B, among them

A = (area A, Xa, Ya): the position (Xa, Ya) in the section A,
B = (area B, Xb, Yb): the position (Xb, Yb) in the section B,

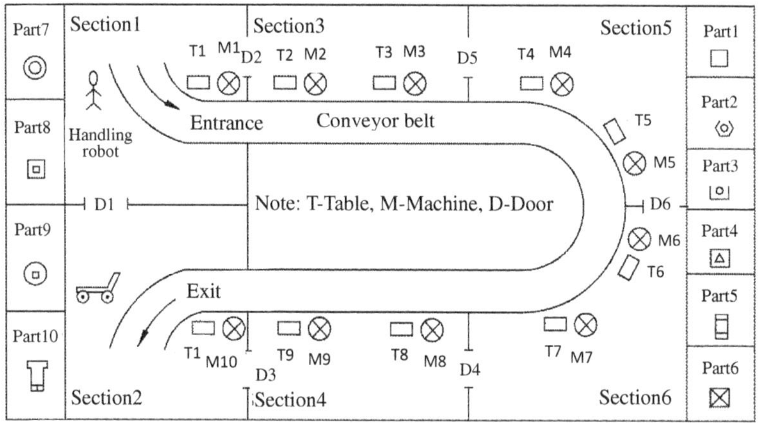

Figure 6.10.   Environmental model of robot assembly line.

Xa, *Ya*: the horizontal and vertical coordinate meters of the Cartesian coordinates in the section A;

Xb, *Yb*: The number of coordinate meters in the section B.

gothru(A, B): The handling robot moves from position A through a door to position B.

carry(A, B): The handling robot grabs the object from position A to position B.

carrythru(A, B): The handling robot grabs the object from position A through a certain door and arrives at position B.

move(A, B): The transport robot moves the cart from position A to position B.

movethru(A, B): The handling robot moves the car from position A through a certain door to position B.

push(A, B): The handling robot pushes the heavy parts from position A to position B with the trolley.

pushthru(A, B): The handling robot uses a trolley to push heavy parts from position A through a door to position B.

loadon(M, N): The handling robot loads a heavy part M onto the trolley N.

unload(M, N): The handling robot removes a heavy part M from the cart N.

transfer(M, cartl, G): The handling robot unloads the heavy part M from the car (cartl) to the target position G.

3. Planning and implementation results

As mentioned above, the planning system uses a rule-based expert system and the C-PROLOG language to generate a planning sequence. The planning system uses a total of 15 rules, each of which contains two sub-rules, so in fact 30 rules are used in total. Store these rules in the system's knowledge base. These rules are used in conjunction with C-PROLOG's evaluated predicates to quickly derive inference results. The planning performance of several systems is compared below.

The ROPES system is implemented on the DUAL-VAX11/780 computer and VM/UNIX (4.2BSD) operating system on the Purdue University Computer Network (PECN) at Purdue University using the C-PROLOG language. The PULP-I system was implemented on a CDC-6500 computer explaining the LISP on Purdue University's Purdue Computer Network (PCN). The STRIPS and ABSTRIPS systems were solved on a PDP-10 computer using partially compiled LISP (excluding garbage collection). It is estimated that the actual average operating speed of the CDC-6500 computer is eight times faster than the PDP-10. However, due to the PDP-10's ability to partially compile and clean up the junk, its data processing speed is actually slightly slower than the CDC-6500. The DUAL-VAX11/780 and VM/UNIX systems are also many times slower than the CDC-6500. However, for comparison purposes, we use the same computation time unit to process the four systems and compare them directly.

Table 6.2 compares the complexity of these four systems, where the PULP-24 system is used to represent the ROPES system. It is clear from Table 6.2 that the ROPES (PULP-24) system is the most complex, the PULP-I system is the second, and the STRIPS and ABSTRIPS systems are the simplest.

The planning speeds of these four systems are graphically represented on the logarithmic coordinates of Figure 6.11. From the curve, PULP-I is planned much faster than STRIPS and ABSTRIPS.

Table 6.3 carefully compares the planning speeds of the two systems, PULP-I and ROPES. As can be seen from Table 6.3, the planning speed of the ROPES (PULP-24) system is much faster than the PULP-I system.

Table 6.2. Comparison of world models of planning systems.

| | System name | | | | |
System name	Room	Door	Box	Others	Total
STRIPS	5	4	3	1	13
ABSTRIPS	7	8	3	0	18
PULP-I	6	6	5	12	27
PULP-24	6	7	5	15	33

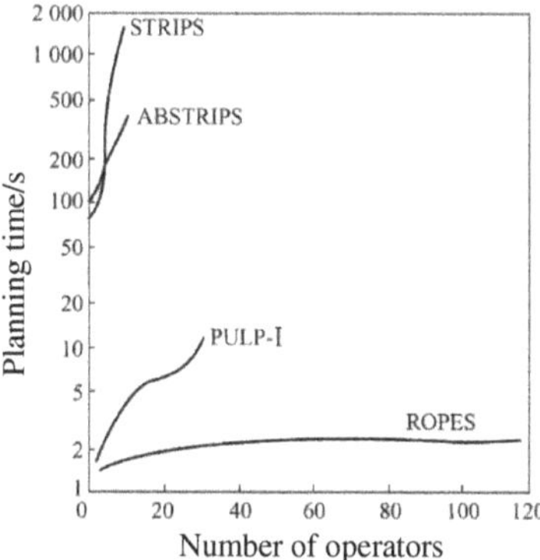

Figure 6.11. Comparison of planning speed.

Table 6.3. Comparison of planning time.

| Operator Number | CPU planning time (sec.) | | Operator Number | CPU planning time (sec.) | |
	PULP-I	PULP-24		PULP-I	PULP-24
2	1.582	1.571	49	...	2.767
6	2.615	1.1717	53	...	2.950
10	4.093	1.850	62	...	3.217
19	6.511	1.967	75	...	3.233
26	6.266	2.150	96	...	3.483
34	12.225	...	117	...	3.517

6.5   Conclusion and Discussion

(1) This planning system is a subsystem of the ROPES system. It is based on C-PROLOG and was implemented in the DUAL-VAX11/780 computer of Purdue University. It has been well planned result. Compared with STRIPS, ABSTRIPS and PULP-I, the system has better planning performance and faster planning speed.

(2) The system is capable of outputting all possible sequence of answers for a given task, whereas previous systems could only give an arbitrary solution. When the "cut" predicate is introduced, the system also outputs only a single solution; it is not an "optimal" solution, but a "satisfactory" solution.

(3) Planning will become complicated when it comes to certain uncertain tasks. At this time, probability, credibility, and/or fuzzy theory can be used to represent knowledge and tasks and solve such problems.

(4) The C-PROLOG language is well suited and effective for many planning and decision making systems, and is more efficient and simpler than LISP. Establishing an efficient planning system on a microcomputer should be a direction of research.

(5) When the number of operators in the planning system increases, the planning time of the system increases little, while the planning time of the PULP-I system increases almost linearly. Therefore, the ROPES system is particularly suitable for large-scale planning systems, while PULP-I can only be used for systems with a small number of operators.

6.6   Path Planning

When moving various mobile bodies in complex environments, navigation and control are required, and effective navigation and control require optimized decision making and planning [33]. A mobile intelligent robot is a typical mobile body. The mobile intelligent robot is a kind of robot system that can realize the certain operation function by realizing the autonomous movement of the target in the obstacle environment by sensing the environment and its own state through the sensor. This section uses robots as an example to discuss the planning of moving objects. Sometime this kind of planning is called robot planning.

Navigation technology is the core of mobile robot technology, and path planning is an important link and topic of navigation research. The so-called path planning means that the mobile robot searches for an optimal or sub-optimal path from the initial state to the goal state according to a certain performance index (such as distance, time, energy, etc.). The main problems involved in path planning include: using the obtained environment information of the mobile robot to establish a more reasonable model, and then using an algorithm to find an optimal or near-optimal collision-free path from the initial state to the goal state; dealing with the uncertainties in the model and errors in path tracking to minimize the impact of external objects on the robot; how to use all known information to guide the robot's actions, resulting in relatively better behavioral decisions. How to quickly and effectively complete the navigation task of mobile robots in complex environments will still be one of the main directions for future research. How to combine the advantages of various methods to achieve better results is also a problem to be explored. This section presents some of our latest research on path planning.

6.6.1 *The main methods and development trends of robot path planning*

1. The main methods of robot path planning

There are three main types of robot path planning methods.

(1) Case-based learning planning method

Case-based learning planning methods rely on past experience for learning and problem solving. A new case can be obtained by modifying the old case in the case base that is similar to the current situation. Applying case-based learning to the path planning of a mobile robot can be described as follows: first, an instance library is built using the information used or generated by the path planning, and any instance in the library contains environmental information and path information for each planning. These examples can be obtained through a specific index. Then, the examples generated by the current planning task and environment information are matched with the cases in the case base to find an optimal matching case, and then the case is corrected and used as the final result. Mobile robot navigation requires good adaptability and stability, and a case-based approach can meet this demand [26, 65].

(2) Planning method based on environment model [24, 46, 71]
The planning method based on the environmental model first needs to establish an environmental model about the robot's motion environment. In many cases, due to the uncertainty of the working environment of the mobile robot (including non-structural, dynamic, etc.), the mobile robot cannot establish a global environment model, but can only establish a local environment model based on sensor information in real time, so the real-time and reliability of the local model become the key to affect whether the mobile robot can move safely, continuously and smoothly. The methods of environmental modeling can be basically divided into two categories, namely, network/graph modeling methods and grid-based modeling methods. The former mainly includes free space method, vertex image method, generalized cone method, etc., which can obtain relatively accurate solutions, but the computational cost is quite large, which is not suitable for practical applications. The latter is much simpler to implement, so it is widely used, and its typical representative is the quaternary tree modeling method and its extension algorithm [24, 46, 71].

The planning method based on the environmental model can be subdivided into a global path planning with completely known environmental information and a partial path planning with completely unknown or partially unknown environmental information according to the completeness of the environmental information. Since the environmental model is known, the design criteria for global path planning would be to maximize the effectiveness of the planning as possible. There are many mature methods in this field, including viewable method, tangential method, Voronoi diagram method, topological method, penalty function method, grid method and so on.

As a hot issue in current planning research, local path planning has been studied intensively. In the case where the environmental information is completely unknown, the robot does not have any prior information, so the plan is to improve the obstacle avoidance ability of the robot, and the effect is second. The methods that have been proposed and applied are incremental D*Lite algorithm and rolling window-based planning method [77]. When the environment is partly unknown, the planning methods mainly include artificial potential field method [58], fuzzy logic algorithm [58, 82], evolutionary computing and genetic algorithm [32, 50, 88], artificial neural network algorithm [83], simulated annealing algorithm, ant colony optimization algorithm [54, 73, 79, 84, 85], particle swarm algorithm [52, 73], immune evolutionary algorithm [51] and heuristic search methods, etc. Heuristic methods include Algorithm A*, incremental graph search algorithm (also known as Dynamic Algorithm A*), Algorithms D* and focused D*. The United States launched the "Mars Pathfinder" detector in December 1996. The path planning method adopted by the "Sojna" Mars rover is the Algorithm D*, which can independently determine the obstacles on the road ahead and pass the real-time re-planning to make decisions about the actions that follow.

(3) Behavior-based path planning method
The behavior-based approach was developed by Brooks in his well-known inclusive structure, an autonomous robot design technique inspired by biological systems that uses a bottom-up principle of animal evolution, trying to work from simple agents to build a complex system. Using this method to solve the problem of mobile robot path planning is a new development trend. It decomposes the navigation problem into many relatively independent behavior units, such as tracking, collision avoidance, and target guidance. These behavioral units are complete motion control units consisting of sensors and actuators with corresponding navigation functions. Each behavior unit behaves differently, these units work together to complete the navigation task [57].

Behavior-based methods can be roughly divided into three types: reflective behavior, reactive behavior, and deliberate behavior. The reflective behavior is similar to the frog's knee reflex. It is an instantaneous stress instinct response. It can respond quickly to sudden situations, such as the emergency stop of a mobile robot during exercise, but the method is not intelligent. It is generally used in combination with other methods. The deliberate behavior uses the known global environment model to provide the optimal sequence of actions for the agent system to reach a specific target. It is suitable for planning in complex static environments. The real-time re-planning of mobile robots in motion is a deliberate behavior. There is a reversing action to get out of the danger zone, but since careful planning requires a certain amount of time to execute, it reacts slowly to unpredictable changes in the environment. Reactive behavior and deliberate behavior can be distinguished by sensor data, global knowledge, reaction speed, theoretical power and computational complexity. Recently, in the development of deliberate behavior, a declarative cognitive behavior similar to human brain memory has emerged. The application of such planning depends not only on sensors and existing prior information, but also on the target to be reached. For a target that is far away and temporarily invisible, there may be a behavioral bifurcation point, that is, there are several behaviors to be adopted, and the robot has to choose the optimal choice. This decision-making behavior is a declarative cognitive behavior. It is used for path planning can make mobile robots have higher intelligence, but due to the complexity of decision-making, this method is difficult to use in practice, and this work needs further study.

2. Development Trends of Path Planning

With the expansion of the application range of mobile robots, the requirements of mobile robot path planning are becoming more and more highly. Individual planning methods sometimes cannot solve some planning problems well, so the new development tends to combine multiple methods.

(1) Combination of reactive behavior based planning and deliberate behavior-based planning

The planning method based on reactive behavior can achieve good planning results under the premise of establishing a static environment model, but it is not suitable for the situation where there are some non-model obstacles (such as tables, people, etc.) in the environment. To this end, some scholars have proposed a hybrid control structure, which combines deliberate behavior with reactive behavior, which can better solve this type of problem.

(2) Combination of global path planning and local path planning

Global planning is generally based on known environmental information, and the scope of adaptation is relatively limited; local planning can be applied to situations where the environment is unknown, but sometimes the response speed is not fast, and the quality of the planning system is high, so if the two are combined, then better planning results can be achieved.

(3) Combination of traditional planning methods and new intelligent methods

Some new intelligent technologies have been introduced into path planning in recent years, and have also promoted the integration of various methods, such as the combination of artificial potential field, evolutionary computing, neural network [23,32], and fuzzy control. In addition, trajectory planning is also combined with machine learning to improve the trajectory planning performance.

6.6.2 *Robot path planning based on approximate VORONOI diagram*

When a mobile robot operates in an unknown environment, it first needs to perceive the environment. Based on the hierarchical representation idea of spatial knowledge, the sensory information fusion is carried out in the metric representation method of occupying the grid in the perception layer, and it provides support for dynamic and deliberate planning in real-time navigation, while the environmental topology modeling focuses on regional information topological features are extracted from it.

The following gives an approximate Voronoi boundary network (AVBN) modeling method in raster space, which realizes the automatic extraction from sensor perception information to the environment topology network model. Aiming at the problem of network path planning, a GAs algorithm based on the Elitist competition mechanism is proposed. The algorithm adopts the optimal selection method in the biological community in the natural niche. Only the best performer has the power to cross, which simplifies the population management speeds up the search process of feasible solutions [27].

1. Space representation method of mobile robot operating environment

Mobile robots in unknown environments only have less prior knowledge, so the feature extraction of the environment is an important intelligent behavior to realize autonomous navigation and control such as environment modeling, positioning, and planning. Cognitive psychologists regard a large number of cognitive processes as "perceptual," which is the process of accepting sensory input and converting it into more abstract codes.

The problem of environmental modeling is essentially the organization and utilization of perceptual knowledge. A reasonable environment model can effectively help mobile robots realize navigation control. Environmental modeling belongs to the category of environmental feature extraction and knowledge representation methods. It determines how the system stores, utilizes and acquires knowledge. It is a key problem that needs to be solved for navigation in an unknown environment.

(1) Spatial cognition between humans and animals

With the evolution of the biological and natural worlds, on the one hand, the perception of advanced organisms has improved, and at the same time, the multiplication and interaction of animals and plants has made the natural environment increasingly complex. Animals living in the environment need corresponding spatial knowledge representation as the cognitive basis of navigation.

The unknown degree of the environment is relative. The mobile robot runs in the unknown environment, and can gradually transform from blind behavior to purposeful behavior by continuously perceiving the environment and extracting environmental knowledge. Environmental modeling is to extract cognitive maps reflecting environmental characteristics from the accumulation of environmental information, reducing blind wandering in the navigation process. If an intelligent system has autonomous environmental information processing capabilities, can independently obtain environmental characteristics through sensor information, and establish an environmental model, then it can achieve the requirements for purposeful movement in an unknown environment.

Cognitive map is a knowledge representation of large-scale environmental information, based on a long-term observation and accumulation of environmental information, and is used to find a path and determine the relative position of oneself and a target.

(2) Representation of spatial knowledge

After long-term research, for the representation of mobile robot environment knowledge, researchers have proposed a metric representation method of spatial knowledge representation, a topological structure representation method, and a hybrid representation method combining the characteristics of the two.

Metric representation uses coordinate information in the world coordinate system to describe the characteristics of the environment, which includes spatial decomposition methods and geometric representation methods.

Topological representation uses nodes to represent a specific location, and edges to represent the connection of these locations. It can be expressed as $G = (V, E)$ to describe the characteristics of free space, where V represents a set of vertices, E represents the edge set that connects the vertices. The Voronoi diagram is usually used to represent the skeleton of the feasible region of the environment. The basic idea of this method is to generate the Voronoi boundary from the environment model, which has the same shortest distance to each obstacle, and forms the vertex of the graph at the junction of the edges. The topological representation method is highly efficient in path planning (the path may also be sub-optimal), and requires small storage space, which is suitable for applications in large-scale environments.

Hybrid representation is a method of extracting topological features from a global metric map. In the robot navigation process, the distance information of the sonar sensor is used to extract the features. When the distance information of the sonar is obviously different from the known node, a new node is generated. If the feature information is similar to the feature of the known node, the partial observable Markov decision method is used for probabilistic positioning, and the positioning and the generation of the environment topology map are performed at the same time. Another method is to extract the global topological structure map from the local metric map. The local map is represented by a grid, and an edge set with a similar topology is used to connect multiple existing local maps. Since the local metric map is maintained in a small range, the error can be ignored. Two levels of planning, local and global, are used in path planning, namely, regional planning based on the occupied grid and global planning based on topological connections. The hybrid representation method includes the characteristics of metric representation and topology representation, which can combine the efficiency of topological structure path planning with the higher navigation accuracy of the metric space to realize multi-level planning.

(3) Hierarchical spatial knowledge representation of mobile robots

Mobile robots running in the environment cannot do without the accumulation of environmental knowledge. In the unknown environment, the requirements for spatial knowledge representation methods are embodied in autonomy, scalability, maintainability, economy, and high efficiency.

Integrating the characteristics of environmental knowledge representation, the knowledge representation of the environment is divided into two levels: perception level and feature level, and different environmental representation methods are used respectively. At the perception level, the grid space measurement method is used. At the feature level, an environment model is established by constructing an environment topology map. Since the environmental model is based on the accumulation of regional information, it has a certain lag compared with the actual navigation action. The environment model serves the "quasi-real-time" global planning and decision-making in navigation; correspondingly, the perception-level information fusion and dynamic deliberate planning provide real-time navigation support.

Deliberate planning is a regional planning behavior between reactive behavior and global planning behavior. It still relies on intuitive perceptual hierarchical information, and there is no process of establishing an environmental model. Deliberate planning behavior (such as: D* algorithm, etc.) is difficult to conduct a large-scale planning behavior in a high-resolution raster map environment. Although the planning algorithm can theoretically calculate the planning task of an infinite sequence of steps, the complexity of calculating the path in a large-scale grid environment is often greater than the complexity of modeling.

Environmental modeling in an unknown environment requires robots to autonomously extract environmental models based on the accumulated information of the environment. The topology representation method can effectively reduce the dimensionality of the metric space. The Voronoi diagram can convert the two-dimensional metric perception information into a network model of the feasible area, making the planning a one-dimensional optimal node sequence problem, which is an effective topology representation method and has the characteristics of low storage space and low planning and calculation cost. The Voronoi diagram represents the segmentation state of the neighboring area centered on the obstacle in the environment. Therefore, the incremental topology environment modeling can be achieved through the modeling of sub-regions.

2. Approximate **VORONOI** boundary network **(AVBN)** modeling method

The Voronoi diagram proposed by the Russian mathematician G. Voronoi in 1908 was originally applied to the study of the proximity problem of plane points. The so-called proximity problem refers to n points in a given plane. The plane is divided into areas centered on these points. The distance from any position in the area to the center point is closer than the distance to other center points, as shown in Figure 6.12. The generalized Voronoi diagram is extended to physical objects in multi-dimensional space, and has been widely used in geographic information systems, pattern recognition, machining path planning, mobile robot planning and other fields.

In some documents, the generalized Voronoi graph (GVG) on the two-dimensional plane is also called the generalized Voronoi diagram (GVD). Because in the planning of mobile robots, only the situation on the plane is usually considered, so in this section GVG and GVD are both the same concept.

In the planning of mobile robots, GVG with obstacles as physical objects is used to describe the networked structure of feasible areas. Voronoi diagram is not only a general tool for solving some basic proximity problems, but also a research field in computational

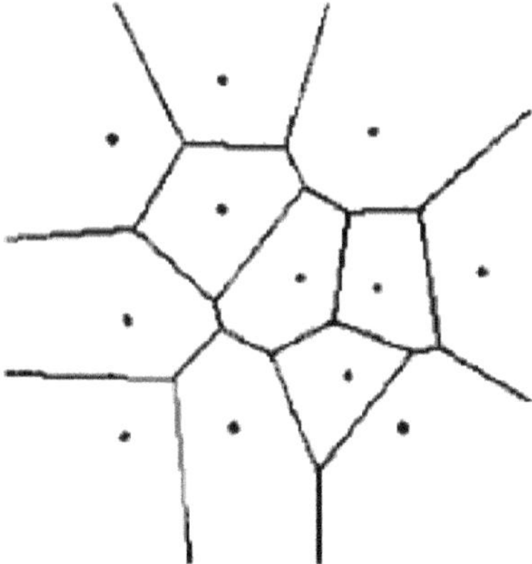

Figure 6.12.   Voronoi diagram of point set.

geometry. Among the many algorithms for constructing Voronoi diagrams, the divide-and-conquer algorithm is widely used. The divide and conquer algorithm first constructs the influence area of a single object, adds other objects based on the influence area of a certain object, and stitches the influence areas of the two objects to obtain a common Voronoi diagram. Then use this picture as the base area, add new objects and stitch them until all objects are included in the Voronoi diagram.

For the GVG structure of physical objects in the obstacle environment, brushfire decomposition in discrete space is often used, as shown in Figure 6.13. The difficulty of using grass fire decomposition method lies in the decomposition of non-convex set obstacles. Because the skeleton graph generated by non-convex set obstacles is difficult to ensure the connectivity of the network, it is necessary to obtain obstacle vertex information to decompose non-convex set obstacles into a combination of convex sets.

Howie Choset proposed the use of sensor ranging information to realize the positioning of the robot and the construction of the GVG model according to the characteristics of the adjacent edges of Voronoi. However, this method is susceptible to the interference of environmental dynamic obstacles and sensor noise, and the number of network nodes that generate environmental models is large. At the same time, real-time construction of GVG requires mobile robots to have the ability to measure obstacles in 360 degrees. In some applications in unstructured environments, it is difficult to obtain comprehensive obstacle distance information.

Figure 6.13.   Using grass fire method to generate GVG diagram.

The specific method of approximate Voronoi Boundary Network (AVBN) modeling modeling is limited to the length of this book and cannot be introduced here. Interested readers can refer to the related literature [86–88].

6.6.3 *Robot path planning based on immune evolution and example learning*

The following introduces a fast and effective path planning algorithm for evolutionary path planning based on example learning and immunity [52].

The convergence speed of evolutionary computation is slow, and it often consumes a lot of machine time, which cannot meet the requirements of online planning and real-time navigation. If only standard evolution calculations of selection, crossover, and mutation are used for path planning, theoretically, the optimal path can be evolved with a probability of 1 when the optimal preservation strategy is used, but the evolutionary algebra will be a huge number. Usually evolution-based path planning and navigation take into account the characteristics of robot navigation and design new evolution operators. In view of the similarities before and after this environment, the experience in the past evolution process (individuals with good performance) is expressed through examples and stored in the example library, and then some examples are selected to be added to the population during the new evolution process, and life science is added at the same time. The immune principle in science is combined with evolutionary algorithms to construct a class of evolutionary algorithms to meet the real-time requirements of online planning. The immune operator in the algorithm is completed by two steps of vaccination and immune selection, and uses the immune selection operator based on the principle of simulated annealing.

1. Individual coding method

A path is a polyline composed of several line segments from the start point to the end point. The end points of the line segments are called nodes (represented by plane coordinates (x, y)), and the path that bypasses obstacles is a feasible path. A path corresponds to an individual in the evolutionary population. A gene is represented by a table composed of its node coordinates (x, y) and state quantities

b which characterizes whether the node is within the obstacle and whether the line segment formed by the current node and the next node intersects the obstacle, and record the immune operation status of bypassing obstacles (detailed later). Individual X can be expressed as follows:

$$X = \{(x_1, y_1, b_1), (x_2, y_2, b_2), \ldots, (x_n, y_n, b_n)\}$$

Among them, (x_1, y_1), (x_n, y_n) are fixed and respectively indicate the start and end nodes.

The size of the population is a predetermined constant N, and n-2 coordinate points $(x_2, y_2), \ldots (x_{n-1}, y_{n-1})$ are generated in a random manner.

2. Fitness function

The problem in question is to find the shortest path, which requires that the path does not intersect with obstacles, and to ensure that the robot can travel safely. Accordingly, the fitness function can be taken as:

$$Fit(X) = dist(X) + r\varphi(X) + c\phi(X) \tag{6.1}$$

Where r and c are normal numbers, and the definitions of $dist(X)$, $\varphi(X)$ and $\phi(X)$ are as follows.

$dist(X) = \sum_{i=1}^{n-1} d(m_i, m_{i+1})$ is the total length of the path, $d(m_i, m_{i+1})$ is the distance between the two suburban nodes m_i and m_{i+1}, $\phi(X)$ is the number of line segments that the path intersects with obstacles, and $\varphi(X) = \max_{i=2}^{n-1} C_i$ is the safety of the node, where

$$C_i = \begin{cases} g_i - \tau, & \text{if } g_i \geq \tau \\ e^{\tau - g_i} - 1, & \text{otherwise} \end{cases}$$

where g_i is the distance from the line segment $\overline{m_i m_{i+1}}$ to all detected obstacles, and τ is the pre-defined safety distance parameter.

3. Immunity and evolution operators

Crossover operator: Two individuals are selected by the selection method, the node number of the shorter of the two is the upper limit, and the lower limit is 1 to generate a random number that obeys a uniform distribution. This number is the crossover point. The two individuals perform crossover operations. Let the probability of crossover operation be p_c.

Type I mutation operator: randomly select a node (non-start and end) on the path and replace the coordinates x and coordinates y of this node with randomly generated values in the entire problem space.

Type II mutation operator: randomly select a node(x, y) (non-start and end) on the path and replace the coordinates x and coordinates y of this node with a random value near the original coordinates.

The immune operator is the key evolutionary operator. How to design the immune operator? First, analyze the problem. The key goal of path planning is to avoid obstacles. Therefore, the information needed to bypass obstacles is important feature information. Design an immune operator (or immune operation) that bypasses obstacles, as shown in Figure 6.14, trying to bypass obstacles that block the road.

From the perspective of robot movement, straight-line operation is the most ideal. As the environment becomes more complex, the route of operation becomes more complicated, especially at points with large turning angles, which make it more difficult to control motion and reduce forward speed. In order to improve the smoothness of the path, the point with a large turning angle (measured by curvature) should be trimmed. The nodes on the path generated by the immune operation to bypass obstacles are sometimes out of order, and the order of some nodes needs to be exchanged; sometimes there are

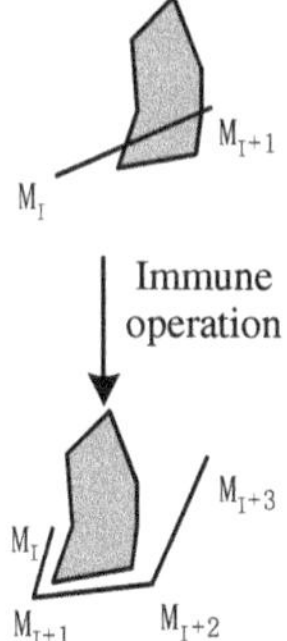

Figure 6.14.   Immune operator.

redundant nodes that need to be deleted. For this purpose, trimming operators, commutating operators, and deleting operators are used.

4. Algorithm description and immune and evolution operator analysis

The constructed algorithm is as follows:

Start
{
Initialize the group;
Evaluate the fitness of the group;
If the shutdown conditions are not met, it will execute in a
   loop:
{
Take out several individuals from the sample library to replace
   the worst individual;
Cross operation
Type I mutation operation;
Type II mutation operation;
Delete operation
Exchange operation
Trimming
Vaccination by immunization operation;
Immune selection
Evaluate the fitness of the group;
Eliminate some individuals and maintain the population size;
}
}

Immunization selection after vaccination in the immunization operation is to compare the individual X' produced by the immunization operation with its paternal parent X. If the fitness value is improved, then the paternal parent will be replaced, otherwise, the paternal parent will be replaced by probability $p(X) = \exp((fit(X) - fit(X'))/T_k)$.

If there is no sample library of similar environment, that is to say, it is a brand new environment, and offline immune evolution planning is required. At this time, "replace several individuals from the sample library to replace the worst individual" in the algorithm. When the shutdown conditions are met, the individuals in the population are added to the sample inventory. When the environment changes, proceed according to the algorithm described in the figure, and constantly take out examples from the example library and add them to the current evolutionary population, and use the experience gained in the past evolution process to speed up the evolution.

If the environment changes many times, the number of examples in the sample library will increase each time it occurs. For this, you can consider storing the sample inventory in a "first in, first out" manner, and delete some of the earliest stored examples, because the environment after many changes, the earliest experience may be out of date.

How to adapt to new changes in the environment while learning experience is the task of evolutionary operators, especially immune operators. The following analyzes the role of immune operators, etc. From the perspective of the operator structure of the entire immune evolution algorithm, the main role of the immune operator is local, and the evolutionary algorithm plays a global role. Therefore, the constructed algorithm has better global convergence performance. The combination of evolutionary algorithm and immune operator with strong local optimization ability; from the analysis of antibody fitness improvement ability, combined with formula (6.1), the immune operator that bypasses obstacles can transform an infeasible path into a feasible path. Operators can make the motion path smoother, but the smoothing immune operation on the infeasible path is less meaningful than the smoothing of the feasible path, which is reflected in the determination of the coefficient of formula (6.1). In this case, the probability of the immune operation for smoothing the infeasible path should be less than the probability of the immune operation for bypassing the obstacle. If all paths are feasible, the probability of smoothing the immune operation will increase appropriately.

The state quantity b saves the record of immune operation bypassing obstacles, indicating the frequency of use of smoothing and deleting node operations, and several nodes generated by immune operation bypassing obstacles. In the subsequent generations of evolution, it should use greater probability for deletion and smoothing operations.

State b is composed of the following parts: whether the line segment from this node to the next node intersects with obstacles; records of immune operations that bypass obstacles, if this node uses immune or type I mutation operations that bypass obstacles in the contemporary era, Set this to an integer k ($k = 6$ in the following simulation experiment). If it is a type II mutation operation, set this to $k/2$. If smoothing and deleting nodes are used, this value is reduced 1; When this value is 0, the probability of smoothing and deleting nodes is p_{d0} (the simulation experiment takes 0.2), otherwise it is p_{d1} (the simulation experiment takes 0.8).

6.7 Robot Planning Based on Machine Learning

In recent years, machine learning has been increasingly used in automatic planning. This section summarizes the research progress and application overview of intelligent planning based on machine learning.

6.7.1 *Overview of intelligent planning applications based on machine learning*

Machine learning has been successfully applied in the fields of pattern recognition, speech recognition, expert systems and automatic planning. Deep reinforcement learning (DRL) can effectively solve the path planning problem in continuous state space and action space. It can directly use raw data as input and output results as execution functions, realizing an end-to-end learning model, which greatly improves the efficiency and convergence of the algorithm. In recent years, DRL has been widely used in robot control, intelligent driving and traffic control planning, control and navigation. First of all, machine learning has been widely used in various robots and intelligent mobile planning; for example, inspection robot path planning based on HPSO and reinforcement learning [69], robot path planning in unknown environments based on deep reinforcement learning [3], based on The fog robot method of deep learning is used for object recognition and grasping planning of robot surface deburring [1], deep reinforcement learning from virtual to reality is used for continuous control of mobile robots without map navigation [49], novel based on learning planetary rover global path planning [80], autonomous path planning for unmanned ships based on deep reinforcement learning [42], intelligent collision avoidance navigation for unmanned ships considering navigation experience rules [66,67], intelligence based on deep reinforcement learning body obstacle avoidance and path planning [35], etc.

Secondly, intensity deep learning is also more applied to the bottom-level planning and control of moving objects (robots); for example, the model migration trajectory planning of deep reinforcement learning for intelligent vehicles [76], and the bottom-level control of quadrotor aircraft based on deep reinforcement learning [47] etc. In addition, machine learning is also applied to non-robot planning; for example, social awareness motor planning with deep reinforcement learning [25], artificial intelligence-assisted planning based on machine learning [44], and goal-oriented integrated planning based on intention networks and deep learning autonomous navigation [40] and so on.

Reinforcement learning has attracted widespread attention in recent years. It can realize the mapping of learning from the environment to the behavior, and seek the most accurate or best action decision through the following methods of "maximizing value function" and "continuous action space."

Maximizing value function Mnih *et al.* proposed a deep Q network (DQN) algorithm, which opened the wide application of DRL. The DQN algorithm uses the powerful fitting ability of the deep neural network, avoids the huge storage space of the Q table, and uses experience replay memory and the target network to enhance the stability of the training process. At the same time, DQN implements an end-to-end learning method, using only raw data as input and output results as the Q value of each action. The DQN algorithm has achieved great success in discrete actions, but it is difficult to achieve high-dimensional continuous actions. If the continuously changing actions are split infinitely, the number of actions will increase exponentially with the increase of the degrees of freedom, which will cause catastrophic problems in latitude and may lead to great training difficulty. In addition, only the discretization of the action can delete important information about the structure of the action domain. The Actor-Critic (AC) algorithm has the ability to deal with continuous action problems and is widely used in continuous action spaces [2]. The network structure of AC algorithm includes Actor network and Critic network. The Actor network is responsible for outputting the probability value of the action, and the Critic network evaluates the output action. In this way, the network parameters can be continuously optimized and the optimal action strategy can be obtained; however, the random strategy of the AC algorithm makes it difficult for the network to converge. Lillicrap *et al.* provide a deep deterministic policy gradient (DDPG) algorithm to solve the problem of deep reinforcement learning (DRL) in a continuous state [53].

Continuous action space DDPG algorithm is a model-free algorithm that combines the advantages of DQN algorithm with experience replay memory and target network. At the same time, the AC algorithm based on deterministic policy gradient (DPG) is used to make the network output result have a certain action value, so as to ensure that DDPG is applied to the continuous action space field. DDPG can be easily applied to complex problems and larger network structures. Zhu Min *et al.* proposed a framework for a human-like autonomous car following plan based on DDPG [83]. Under this framework, the driverless car learns from the environment through trial and error to obtain the path planning model of the driverless car, which has good experimental results. This research shows that DDPG can gain insight into driver behavior and help develop human-like autonomous driving algorithms and traffic flow models.

6.7.2 *Research progress of autonomous route planning for unmanned ships based on deep learning*

Improving the autonomous driving level of ships has become an important guarantee for enhancing the safety and adaptability of ships' navigation. Unmanned ships can be more adaptable to the complex and changeable harsh environment at sea. This requires unmanned ships to have autonomous path planning and obstacle avoidance capabilities, so as to effectively complete tasks and enhance the overall capabilities of ships.

The research direction of unmanned ships involves autonomous path planning, navigation control, autonomous collision avoidance and semi-autonomous mission execution. As the basis and premise of autonomous navigation, autonomous path planning plays a key role in ship automation and intelligence [55]. In the actual navigation process, ships often encounter other ships, so that it is required reasonable methods to guide the ship to avoid other ships and navigate according to the target. The unmanned ship path planning method can guide the ship to take the best course of action and avoid collisions with other ships and obstacles. Traditional path planning methods usually require relatively complete environmental information as prior knowledge, and it is very difficult to obtain surrounding environmental information in an unknown and dangerous marine environment. In addition, traditional algorithms have a large amount of computation, which makes it difficult to implement real-time behavioral decision-making and accurate path planning of ships.

At present, research on autonomous path planning of unmanned ships has been carried out at home and abroad. These methods include traditional algorithms, such as APF, speed obstacle method, A* algorithm, and some intelligent algorithms, such as ant colony optimization algorithm, genetic algorithm, neural network algorithm and other DRL related algorithms.

In the field of smart ships, the application of DRL in unmanned ship control has gradually become a new research field. For example, the path planning and maneuvering method of unmanned cargo ships based on Q learning [23], the autonomous navigation control of unmanned ships based on the relative value iteration gradient (RVIG) algorithm [75], and the automatic avoidance of multiple ships based on the Dueling DQN algorithm Collision proposed a behavior-based USV local path planning and obstacle avoidance method [81] and so on. DRL overcomes the shortcomings of the usual intelligent algorithms, which require a certain number of samples, and have fewer errors and response time.

Many key autonomous path planning methods have been proposed in the field of unmanned ships. However, these methods are mainly focused on the research of small and medium-sized USVs, while the research on unmanned ships is relatively small. In this paper, DDPG is selected for unmanned vessel planning because of its powerful deep neural network function fitting ability and good generalized learning ability.

This autonomous path planning proposes a DRL-based model to realize intelligent path planning for unmanned ships in unknown environments. The model uses the DDPG algorithm to learn the best action strategy in a simulated environment through continuous interaction with the environment and the use of historical experience data. Navigation rules and situations encountered by ships are converted into navigation constraint areas to achieve the safety of the planned route and ensure the validity and accuracy of the model. The data provided by the ship's automatic identification system (AIS) is used to train this path planning model. Then, an improved DRL is obtained by combining DDPG with an artificial potential field. Finally, the path planning model is integrated into the electronic chart platform for experiments. The results of comparative experiments show that the improved model can realize autonomous path planning with fast convergence and good stability [23].

6.8 Chapter Summary

After explaining the role and tasks of robot planning, this chapter starts with the robot planning in the building block world, and gradually and in-depth discusses the robot planning. The high-level planning of robots that we have discussed includes the following methods:

(1) Rule deduction method, using F rule to solve the planning sequence.
(2) Logic calculation and general search method. STRIPS and ABSTRIPS systems belong to this method.
(3) Rules based on the expert system. Such as ROPES planning system, it has faster planning speed, stronger planning ability and greater adaptability.
(4) Path planning based on approximate Voronoi diagram.
(5) Local path planning based on simulated annealing.
(6) Mobile robot path planning based on immune evolution and example learning.
(7) Robot planning based on machine learning, especially robot planning based on deep learning and deep reinforcement learning.

There are other robot planning systems, such as the triangulation table programming method (with the most preliminary learning ability), nonlinear programming using target sets, and nonlinear programming using minimum constraint strategies. Due to space limitations, we will not introduce them one by one.

Finally, it is worth pointing out: First, robot planning has been developed into a comprehensive application of multiple methods of planning. Second, robot planning methods and technologies have been applied to various fields of image processing, computer vision, production process planning and monitoring, and robotics. Third, robot planning still has some further research issues, such as multi-robot coordinated planning and real-time planning. In the future, more advanced robot planning systems and technologies are bound to emerge.

References

[1] Ajay, K.T., Nitesh, M. and John, K. (2019). A Fog Robotics Approach to Deep Robot Learning: Application to Object Recognition and Grasp Planning in Surface Decluttering. arXiv:1903.09589v1 [cs.RO] 22 Mar 2019, pp. 1–8.

[2] Bahdanau, D., Brakel, P., Xu, K., *et al.* (2016). Actor-critic algorithm for sequence prediction. arXiv 2016, arXiv:1607.07086.

[3] Bu, X.J. (2018). Research on Robot Path Planning in Unknown Environment Based on Deep Reinforcement Learning. Harbin Institute of Technology Master's Degree Thesis (in Chinese).

[4] Cai, Z.X. and Fu, K.S. (1988). Expert system based robot planning. *Control Theory and Applications*, 5(2):30–37.

[5] Cai, Z.X. (1985). Task planning with collision-free using expert system, *1984–1985 Annual Research Summary*, School of EE, Purdue University.

[6] Cai, Z.X. (1986). Some research works on expert system in AI course at Purdue. In *Proc. IEEE Int'l Conf. on Robotics and Automation*, Vol. 3, pp. 1980–1985, San Francisco: IEEE Computer Society Press.

[7] Cai, Z.X. (1987). Design of knowledge base of robot planning system, *National Machine Learning Symposium*, China (in Chinese).

[8] Cai, Z.X. (1988). An expert system for robot transfer planning. *Journal of Computer Science and Technology*, 3(2):153–160.

[9] Cai, Z.X. (1989). Robot path-finding with collision-avoidance. *Journal of Computer Science and Technology*, 4(3):229–235.

[10] Cai, Z.X. (1992). A Knowledge-based flexible assembly planner. *IFIP Transactions, B-1: Applications in Technology*, pp. 365–372, North-Holland.

[11] Cai, Z.X. (2021). *Robotics*, 4th Edition, Chapter 8. Beijing: Tsinghua University Press (in Chinese).

[12] Cai, Z.X. (1988). An expert system for robot transporting planning. *Chinese Journal of Computers*, 11(4):242–250 (in Chinese).

[13] Cai, Z.X. and Fu, K.S. (1986). Robot planning expert system. In *Proc. IEEE Int'l Conf. on Robotics and Automation*, Vol. 3, pp. 1973–1978. San Francisco: IEEE Computer Society Press.

[14] Cai, Z.X. and Fu, K.S. (1987). ROPES: A new robot planning system, *Pattern Recognition and Artificial Intelligence*, 1:77–86 (in Chinese).

[15] Cai, Z.X. and Jiang, Z.M. (1991). A Multirobotic pathfinding based on expert system. *Preprints of IFAC/IFIP/IMACS Int'l Symposium on Robot Control.* Pergamon Press, pp. 539–543.

[16] Cai, Z.X. and Peng, Z.H. (2002). Cooperative coevolutionary adaptive genetic algorithm in path planning of cooperative multi-mobile robot system. *J. Intelligent & Robotic Systems: Theory and Applications*, 33(1):61–71.

[17] Cai, Z.X. and Tang, S.X. (1995). A multi-robotic planning based on expert system. *High Technology Letters*, 1(1):76–81.

[18] Cai, Z.X., Liu, L.J., Cai, J.F. and Chen, B.F. (2020). *Artificial Intelligence and Its Applications*, 6th Edition, Chapter 8. Beijing: Tsinghua University Press (in Chinese).

[19] Cai, Z.X., Wen, Z.-Q. and Zou, X.-B., *et al.* (2008). A mobile robot path-planning approach under unknown environments, *Proc. 17th IFAC World Congress*, Seoul, Korea, pp. 5389–5392.

[20] Cai, Z.X., Liu, L.J., Chen, B.F. and Wang, Y. (2021). *Artificial Intelligence: From Beginning to Date*, Chapter 9. Singapore: World Scientific Publishers and Tsinghua University Press.

[21] Cai, Z.X. (1988). *Robotics: Principles and Applications*, Chapter 9. Changsha: Central South University of Technology (in Chinese).

[22] Cai, Z.X. (1991). High level robot planning based on expert system, In Cun-Xi Xie, H. Makino, and T.P. Leung (Eds.), *Proc. of Asian Conf. on Robotics and Application*, pp. 311–318, April 1991, Hong Kong Polytechnic.

[23] Chen, C., Chen, X.Q., Ma, F., *et al.* (2019). A knowledge-free path planning approach for smart ships based on reinforcement learning. *Ocean Eng.*, 189:106299.

[24] Chen, S.M. and Fang, H.J. (2003). Optimal path planning algorithm in dynamic unknown environment. *Journal of Huazhong University of Science and Technology*, 31(12):29–31 (in Chinese).

[25] Chen, Y.F., Michael, E., Miao, L., *et al.* (2018). Socially aware motion planning with deep reinforcement learning. arXiv:1703.08862v2 [cs.RO] 4 May 2018.

[26] Chen, Z.H. (2000). Research on case-based learning algorithm for path planning of lunar probe. *Aeronautical Computing Technology*, 30(2):1–4 (in Chinese).

[27] Choset, H. and Burdick, J. (2000). Sensor based motion planning: Incremental construction of the hierarchical generalized Voronoi graph. *Int. Journal of Robotics Research*, 19(2):126–148.

[28] Clark, K.L. and McCabe, F.G. (1980). *PROLOG: A language for implementing expert systems*, Machine Intelligence.

[29] Clocksin, W.F. and Mcllish, C.S. (1981). *Programming in PROLOG*, Springer-Verlag.

[30] Cohen, P.R. and Feigenbaum, E.A. (1982). *Handbook of Artificial Intelligence*, Vol. 3, William Kaufmann, Inc.

[31] Craig, J.J. (1986). *Introduction to Robotics Mechanics and Control*. Addison-Wesley Publishing Company.

[32] Cui, K., Wu, L. and Chen, S.B. (1998). Application of genetic algorithm in path planning of redundant arc welding robot. *Robot*, 20(5):362–367 (in Chinese).

[33] Dai, B., Xiao, X.M. and Cai, Z.X. (2005). Research status and prospects of mobile robot path planning technology. *Control Engineering*, (3):198–202 (in Chinese).

[34] Davis, R.H. and Comacho, M. (1984). The application of logic programming to the generation of plans for robots, *Robotica*, 2(93–103):137–146.

[35] Deng, W. (2019). Research and Application of Intelligent Body Obstacle Avoidance and Path Planning Based on Deep Reinforcement Learning. Master's Degree Thesis of University of Electronic Science and Technology of China (in Chinese).

[36] Fikes, R.E. and Nilsson, N.J. (1971). STRIPS: A new approach to the application of theorem proving to problem solving, *2nd IJCAI*, 608–620.

[37] Fikes, R.E. *et al.* (1972). Learning and executing generalized robot plans. *Artificial Intelligence*, 3:251–288.

[38] Fu, K.S., Cai, Z.X. and Xu, G.Y. (1987). *Artificial Intelligence and Its Applications, Chapter 7*. Beijing: Tsinghua University Press (in Chinese).

[39] Fu, K.S., Gonzalez, R.C. and Lee, C.S.G. (1987). *Robotics: Control, Sensing, Vision, and Intelligence*. Wiley.

[40] Gao, W., David, H.W. and Sun, L. (2017). Intention-Net: Integrating planning and deep learning for goal-directed autonomous navigation. arXiv:1710.05627v2 [cs.AI] 17 Oct 2017.

[41] Gevarter, W.B. *et al.* (1985). An expert system for the planning and scheduling of astronomical observation, *Proceedings IEEE 1985 International Conference on Robotics and Automation*, pp. 221–226.

[42] Guo, S.Y., Zhang, X.G., Zheng, Y.S. and Du, Y.Q. (2020). An autonomous path planning model for unmanned ships based on deep reinforcement learning. *Sensors*, 20, 426(1–35); doi:10.3390/s20020426.

[43] Honig, W., Preiss, J.A., Kumar, T.K.S., *et al.* (2018). Trajectory planning for quadrotor swarms. *IEEE Transactions on Robotics*, 34(4): 856–869.

[44] Huang, D.X. (2017). Prospects for artificial intelligence-assisted planning based on machine learning. *Urban Development Studies*, 24(5):50–55 (in Chinese).

[45] Hunt, E.B. (1975). *Artificial Intelligence*, Academic Press.

[46] Ji, J., Khajepour, A., Melek, W. and Huang, Y. (2017). Path planning and tracking for vehicle collision avoidance based on model predictive control with multiconstraints. *IEEE Trans. Vehicle Technology*, 66(2):952–964.

[47] Lambert, N.O., Drewe, D.S., Yaconelli, J., *et al.* (2019). Low-level control of a quadrotor with deep model-based reinforcement learning. *IEEE Robotics and Automation Letters*, 4(4):4224–4230.

[48] Latombe, C. (1991). *Robot Motion Planning*. Boston: Kluwer Academic Publishers.

[49] Lei, T., Giuseppe, P. and Ming, L. (2017). Virtual-to-real deep reinforcement learning: continuous control of mobile robots for mapless navigation. arXiv:1703.00420v4 [cs.RO] 21 Jul 2017.

[50] Li, M.Y. and Cai, Z.X. (2000). Improved evolutionary programming and its application in robot path planning. *Robot*, 22(6):490–494 (in Chinese).

[51] Li, M.Y. and Cai, Z.X. (2005). Evolutionary path planning of mobile robots based on swarm behavior and cloning, *Journal of Central South University (Natural Science Edition)*, 36(5):739–744 (in Chinese).

[52] Li, M.Y. and Cai, Z.X. (2005). Research on mobile robot immune evolution planning combined with example learning, *Computer Engineering and Application*, 41(19):18–21 (in Chinese).

[53] Lillicrap, T.P., Hunt, J.J., Pritzel, A., *et al.* (2015). Continuous control with deep reinforcement learning. *Comput. Sci.*, 1(8):A187.

[54] Liu, G., *et al.* (2005). The ant algorithm for solving robot path planning problem. *Third International Conference on Information Technology and Applications*, 2005(ICITA2005), 2(4-7):25–27.

[55] Liu, Z., Zhang, Y., Yu, X. and Yuan, C. (2016). Unmanned surface vehicles: An overview of developments and challenges. *Annu. Rev. Control*, 41:71–93.

[56] Lynch, K.M. and Park, F.C. (2017). *Modern Robotics: Mechanics, Planning, and Control.* Cambridge University Press.

[57] Mac, T.T., Copot, C., Tran, D.T. and Keyser, R. De. (2016). Heuristic approaches in robot path planning: A survey. *Robotics and Autonomous Systems*, 86:13–28.

[58] Mbede, J.B., Huang, X.H. and Wang, M. (2000). Fuzzy motion planning among dynamic obstacles using artificial potential fields for robot manipulators. *Robotics and Autonomous Systems*, 32(1): 61–72.

[59] Nilsson, N.J. (1971). *Problem Solving Methods in Artificial Intelligence.* McGraw-Hill Book Company.

[60] Nilsson, N.J. (1980). *Principle of Artificial Intelligence.* Tioga Publishing Co.

[61] Rich, E. (1983). *Artificial Intelligence.* McGraw-Hill Book Company.

[62] Sacerdoti, E.D. (1973). Planning in a hierarchy of abstraction spaces. *3rd IJCAI*, 412–422.

[63] Sacerdoti, E.D. (1974). Planning in a hierarchy of abstraction spaces. *Artificial Intelligence*, 5:115–135.

[64] Schlobohm, D.A. (1984). Introduction to PROLOG, *Robotics Age*, Nov. 13–19 and Dec. 24–25.

[65] Shang, Y., Xu, Y.R. and Pang, Y.J. (1998). Case-based learning algorithm for global path planning of autonomous underwater vehicles. *Robot*, 2(6):4272432 (in Chinese).

[66] Shen, H.Q., Guo, C., Li, T.S., *et al.* (2018). Intelligent collision avoidance navigation method for unmanned ships considering navigation experience rules. *Journal of Harbin Engineering University*, 39(6):998–1005 (in Chinese).

[67] Shen, H.Q., Hashimoto, H., Matsuda, A., *et al.* (2019). Automatic collision avoidance of multiple ships based on deep Q-learning. *Appl. Ocean Res.*, 86:268–288.

[68] Simons, G.L. (1980). *Robots in Industry*, NCC Applications Ltd.

[69] Song, Y. (2019). Research on Path Planning of Inspection Robot Based on HPSO and Reinforcement Learning. *Master's Degree Thesis of Guangdong University of Technology* (in Chinese).

[70] Tangwongsan, S. and Fu, K.S. (1979). Application of learning to robotic planning. *International Journal of Computer and Information Science*, 8(4):303–333.

[71] Wagner, G. and Choset, H. (2015). Subdimensional expansion for multirobot path planning. *Artificial Intelligence*, 219:1–24.

[72] Weiss, S.M. and Kulikowski, C.A. (1984). *A Practical Guide to Designing Expert Systems*, Rowmand and Allankeld Publishers.

[73] Wen, Z.Q. and Cai, Z.X. (2006). Global path planning approach based on ant colony optimization algorithm. *Journal of Central South University of Technology*, 13(6):707–712.

[74] Wu, M.C. and Liu, C.R. (1985). Automated process planning and expert system, Proceedings *IEEE 1985 International Conference on Robotics and Automation*, pp. 186–191.

[75] Yang, J., Liu, L., Zhang, Q. and Liu, C. (2019). Research on autonomous navigation control of unmanned ship based on unity 3D. In *Proceedings of the 2019 IEEE International Conference on Control, Automation and Robotics* (ICCAR), Beijing, China, 19–22 April, pp. 2251–2446.

[76] Yu, L.L., Shao, X.Y., Long, Z.W., *et al.* (2019). Model migration trajectory planning method for intelligent vehicle deep reinforcement learning. *Control Theory and Applications*, 36(9):1409–1422 (in Chinese).

[77] Zhang, C.G. and Xi, Y.G. (2001). Robot path planning based on rolling window when the global environment is unknown. *Science in China (Edition E)*, 31(1):51–58 (in Chinese).

[78] Zhang, H. (1985). RP — An expert system for robot programming, In *Some Prototype Examples for Expert Systems*, Vol. 2, K.S. Fu (Ed.), TR-EE 85-1, Purdue University, pp. 317–319.

[79] Zhang, H., Yi, S., Luo, X., *et al.* (2004). Robot path planning, control and decision based on ant colony optimization algorithm in complex environment, *Control and Decisions*, 19(2):166–170 (in Chinese).

[80] Zhang, J., Xia, Y.Q. and Shen, G.H. (2019). A novel learning-based global path planning algorithm for planetary rovers. Neurocomputing, Vol. 361, pp. 69–76, DOI:0.1016/j.neucom.2015.0759.0

[81] Zhang, R.B., Tang, P., Su, Y., *et al.* (2014). An adaptive obstacle avoidance algorithm for unmanned surface vehicle in complicated marine environments. *IEEE CAA J. Autom. Sin.*, 1, 385–396.

[82] Zhao, H., Cai, Z.X. and Zou, X.B. (2003). Path planning based on fuzzy ART and Q-learning. *Proceedings of the 10th Annual Conference of Chinese Association of Artificial Intelligence*. Guangzhou, pp. 834–838 (in Chinese).

[83] Zhu, M., Wang, X. and Wang, Y. (2018). Human-like autonomous car-following model with deep reinforcement learning. *Transp. Res. Part C Emerg. Technol.*, 97:348–368.

[84] Zhu, Q.B. (2005). Ant prediction algorithm for robot path planning in dynamic and complex environment. *Chinese Journal of Computers*, 28(11):1898–1906 (in Chinese).

[85] Zhu, Q.B. and Zhang, Y.L. (2005). Ant colony algorithm for robot path planning based on grid method. *Robot*, 27(2):132–136 (in Chinese).

[86] Zou, X.B. and Cai, Z.X. (2002). Environmental non-smooth modeling and path planning based on sensor information. *Progress in Natural Science*, 12(11):1188–1192 (in Chinese).

[87] Zou, X.B., Cai, Z.X., Liu, J., *et al.* (2001). A local path planning method for mobile robots. *Proceedings of the Ninth Annual Conference of Chinese Association of Artificial Intelligence*. Beijing, pp. 947–950 (in Chinese).

[88] Zou, X.B. and Cai, Z.X. (2002). Evolutionary path-planning method for mobile robot based on approximate Voronoi boundary network. *ICCA2002*. Xiamen, pp. 496–500.

Part II
Mobile Robot

Chapter 7

Architecture and Dynamics Model of Mobile Robot

The robot control system is composed of hardware with certain functions and corresponding software. When designing a control system for a mobile robot, the constraints in the specific application environment must be considered, including the limited payload and space of the mobile robot, the limited energy resource that the robot can carry, and the limited computational power, limited communication bandwidth, etc. [11, 12, 14, 62].

Design of an intelligent control system first needs to consider determining an appropriate architecture to achieve reasonable coordination between system modules, and to have openness and scalability in the software and hardware of the system [33, 79]. The system architecture is derived from architecture, and later in computer science, it specifically refers to the physical structure of a computer, including the organizational structure, capacity, the computer's CPU, memory, and the interconnection between input and output devices. The reasonable design of the system architecture will be able to achieve complex behaviors and have further perfect expansion capabilities [68, 108].

7.1 Architecture of Mobile Robots

The architecture of mobile robots is usually divided into three types: hierarchical, reactive, and deliberative/reactive [70].

7.1.1 *Hierarchical architecture*

Early intelligent mobile robots mostly followed the design idea of hierarchical architecture. The action of the robot is regarded as a hierarchical process of repeated execution, that is, sense-model-plan-act (SMPA) [78]. In the modeling phase, sensor information is fused into a world model. The planner searches the state space to find a way to reach the goal state from the current state. The actuator strictly follows the implementation process of the planned results. The hierarchical architecture has two characteristics: (1) The system has only a one-way cycle process, and each module is regarded as a functional component (such as a visual component of perception, a path planning component), so it is also called functional decomposition architecture [13]; (2) Intelligence is mainly reflected in the planner, and in the execution phase, it runs according to the planning results. In the 1980s, this was the main paradigm of mobile robot design [13, 78]. Due to the complex calculations of planning and modeling, there is a large time lag between planning output and environmental information input, which makes it difficult for planning results to adapt to environmental changes during execution. In addition, environmental perception information is input to the modeling link, and the execution process is open-loop control without perceptual feedback, so it is difficult to overcome the various uncertainties caused by the errors of sensors and actuators.

7.1.2 *Reactive architecture*

Behaviorism is rooted in connectionism in the 1980s, and has shown good performance in the design of robots. And it has developed into three major artificial intelligence technologies (schools) alongside neural network technology (connectionism) and symbolism artificial intelligence technology [9]. Behavior control believes that intelligence depends on perception and action, and proposes the "sense-action" mode of intelligent behavior [8]. Behaviorism believes that intelligence does not require knowledge, representation, or reasoning; intelligent behavior can only be manifested in the interaction between the real world and the surrounding environment.

Behaviorism decentralizes complex control into several basic behavior units. Each behavior unit can achieve a rapid response from perception to behavior according to its own local goals. There is no need to establish an obvious world model to represent the environment, and each behavior unit only deals with perceptual information related to it. The behavior-based method is a bottom-up method of constructing a system. It uses behavior to encapsulate the capabilities of perception, exploration, obstacle avoidance, planning, and task execution in robot control. Behavior-based control structure has faster response ability in actual control and appears to be more adaptable than traditional methods, so it once became the focus of mobile robot research. On the basis of behavior-based structure, many new ideas and control strategies are proposed. Typical representatives of behaviorist architecture are the subsumption architecture proposed by Brooks in 1986 [8] and the schema-based architecture proposed by Arkin in 1989 [1,3].

The inclusive architecture is proposed for how to improve the adaptability of robots in the natural environment, breaking through the computational bottleneck of traditional SMPA systems, and greatly improving the real-time performance of complex systems. It designs a variety of basic behaviors for each local goal, forming different levels of ability. Sensing data is provided in a parallel manner according to the needs of each level, and various behaviors act on the driving device after coordination. Since a single behavior is only designed to complete a partial special task, it can respond quickly to input. The various behavior levels of the system work in parallel synchronously or asynchronously, and the behavior output of the high-level can inhibit the behavior output of the low-level. The defect of the containment structure is that there are multiple control loops, so more cost is needed to design a coordination mechanism to resolve the conflict between each loop competing for control of the same driving device [45]. The coordination mechanism of the inclusive architecture adopts a fixed priority. As the complexity of the task increases and the number of behavior levels increases, the interaction between various behaviors increases the difficulty of predicting the overall behavior. So may often produce non-optimized behavior.

Each action schema in the schema-based reactive behavior architecture has a corresponding perception schema to provide necessary perception information. A motor schema is embodied as a vector with speed and direction in the potential energy field, and the final behavior is determined by fusing the output of each motion schema.

Action schema is the basic unit that constitutes a complex behavior. Each schema works synchronously or asynchronously, and generates corresponding behavior intentions based on the sensor information perception schema. Schema is the basic unit of brain or robot behavior analysis. Schema describes some subsets with similar structural features. It is an intelligent behavior that understands the coordination of perceptual actions from the perspective of neuroscience and psychology. The perception schema is the impression produced by the various sensory information received by the sensory organs, and the various related impressions are organized into a characteristic unit with a certain specific connotation. Perception schemas will have an impact on behavior and stimulate response action schemas. Arkin adopted the design idea of this biological robot in the design of intelligent robots, decomposing complex intelligent behaviors into a series of perceptual-action schema combinations [2]. In order to coordinate multiple modes for the same actuator, the output vector of the action pattern is weighted with a gain factor. In ensuring the safety of the robot or other important behaviors, a higher gain value is assigned, and the output composite vector guides the robot's actions.

7.1.3 *Deliberate/reactive composite architecture*

Since there is no internal state model, the behavior-based system can respond quickly to changes in the environment, so it is meaningful for real-time applications. The behaviorist philosophy believes that "the world is its best model." Since there is no need for an internal global model, and local information often cannot reflect the overall state of the environment, it is difficult to judge the overall performance of certain behaviors. It may be trapped in a local minima in a complex environment and difficult to detect.

The SMPA-based structure can solve this problem, but an internal model needs to be established, which consumes too much time, and the real world may have changed during such complex calculations. Therefore, Connell in 1990, Gat and Bonasso in 1991 used the advantages of these two structures to propose a composite architecture design method [6, 19, 41], while the autonomous robot architecture (AuRA) is developed a composite architecture based on a schema-based behavioral architecture by Arkin *et al.* The three-tier system proposed by Gat includes three main levels [41]: The control layer responds quickly to perception information, usually using reactive behavior. The deliberate layer handles computationally complex planning or other large-scale information processing tasks; the deliberate layer intervenes in the control layer in two ways, that is, the plan is generated by deliberate and the planning result is handed over to the control layer for execution, or deliberate is initiated in response to a request from the bottom layer plan and deal with problems that are difficult to solve with reactive behavior. The coordination layer, also known as the sequencer layer, is like a behavior sequencer, which determines what behavior the controller should currently select and provides the required information.

The composite architecture uses a deliberate way to model and plan complex environments, and uses reactive behavior to overcome the uncertainty of dynamic changes during the execution process, so it has strong functions and application flexibility.

The composite system structure has become the main design paradigm for complex intelligent systems because it takes into account the real-time control of the control and the optimization requirements of the overall performance. It is precisely because the composite architecture requires the system to be compatible with real-time control and complex calculations, distributed control has become an ideal way to achieve [101, 104]. Distributed control has good openness and scalability. At present, robot controllers are generally developing towards an open distributed control structure.

7.2 Composition of Mobile Robot Prototype System

This section introduces and analyzes the software and hardware structure of the mobile mechanism, sensor system, and control system of the mobile robot prototype system. According to the complex calculation performance requirements of sensor information processing and navigation control, the design idea of distributed control system is given. Aiming at the software architecture of the intelligent system, the design idea of the deliberate/reaction compound paradigm is analyzed, and a compound architecture design method based on the hierarchical representation of spatial knowledge is given.

7.2.1 *Mobile system and sensor system of mobile robot*

1. The mobile mechanism of the mobile robot

In order to allow mobile robots to operate in complex environments, in addition to wheeled structures, researchers have also proposed a variety of mobile mechanisms, such as an upright walking mechanism, and a crawling mechanism that mimics insect legs. The wheeled mechanism has high efficiency and low energy loss on flat roads, but its ability to overcome obstacles is poor. The legged structure has a good ability to overcome obstacles, but the operation efficiency is relatively low, and the control is complicated. People try to combine the efficiency of wheels with the adaptability of legs to design moving mechanisms. Sojourner robot [60] and Shrimp robot [32] represent a passive hybrid structure. Their movement mechanism can adaptively change the center of gravity of the body when overcoming obstacles, and can maintain balance on bumpy roads. In addition, researchers have also studied reconfigurable mobile robots that can actively adjust their structural form [50], so that they have better obstacle crossing capabilities and smooth running performance.

In order to meet the needs of carrying out experiments in indoor and outdoor environments, the mobile robot adopts a movement mechanism consisting of four driving wheels and a follower wheel. At present, the commonly used motors in robot drive are DC servo motors, AC servo motors and stepper motors. As a control system actuator, stepper motor has a wide range of applications [65]. Although the stepper motor has some shortcomings, such as the starting frequency is too high or the load is too large, it is easy to lose steps or stall, and the overload capacity is small. However, the stepping motor has good controllability of the rotation position, and can realize high-precision open-loop control when the load torque is appropriate. The comprehensive cost of the stepper motor and its drive controller is lower than other types of drive systems, so the stepper motor is the best choice as the drive motor of the car body.

2. Track estimation system for mobile robots

In the navigation process, the track estimation system of the mobile robot is similar to the animal's proprioceptive and distinguishing ability of spatial orientation. The basic trajectory estimation technique was originally designed to meet the needs of sailing in the vast sea [46,49]. In the past, ships used their course and speed to calculate their track, observe the sun or other stars to correct their course, and use various natural or artificial signs such as coasts, islands and reefs for absolute positioning. Similar to maritime navigation, the navigation technology of mobile robots is still based on track estimation and absolute information correction methods. If the track estimation system can maintain high accuracy, it will undoubtedly greatly reduce the amount of calculation for absolute positioning. Designing a cost-effective track estimation system is still a key link in the design of mobile robot systems.

The measurement of the track estimation system mainly includes the following three types:

(1) Basic attitude measurement

Basic state measurement includes the measurement of wheel rotation speed and rocker angle. The odometer information of the mobile robot can be determined by measuring the rotation of the wheels. The encoder adopts a 360-pulse/revolution incremental photoelectric encoder, which has a dual-channel pulse output with a phase difference of 90 degrees. The rotation direction can be judged by the phase change of the dual-channel pulse. The encoder is installed on the shaft of the drive motor and directly measures the rotation of the motor. The motor drives the wheel through a 15-fold reducer, and the wheel diameter is 250 mm. From this, the resolution of the odometer can be calculated as:

$$\delta = \frac{2\pi r}{\eta P} = \frac{2 \times 3.14 \times 125}{15 \times 360} = 0.14537 \,(\text{mm})$$

The rotation range of the rocker is $\pm 15°$, and a precision angle potentiometer is used to convert the rotation amount into a $-5 \sim +5$V electrical signal, and a ten-digit precision A/D conversion card is used to obtain the digital value of the rotation angle.

(2) Heading measurement

In order to solve the problem of heading measurement of mobile robots, sensors such as electronic compasses and gyroscopes can be used. Electronic compasses are susceptible to interference from electromagnetic fields, and their applications are subject to certain restrictions. In high-performance mobile robots, inertial sensors such as gyroscopes have begun to detect changes in the robot's running direction. The inertial navigation system (INS) can be isolated from the environment, so that it can be used in a very harsh environment to avoid the influence of uncertain factors such as wheel and ground sliding on the heading measurement.

A gyroscope is a sensor used to detect the orientation or angular velocity of a moving object. Fiber optic gyroscope is one of the most important components in the track estimation system, and its performance directly affects the positioning accuracy of the mobile robot. E.CoreRD1100 gyroscope can be selected as heading measurement sensor.

(3) Tilt attitude measurement

Mobile robots running under complex terrain need to detect the pitch and roll attitude to prevent the robot from being out of balance. At the same time, when a mobile robot performs track estimation under undulating terrain, it also needs pitch and roll attitude parameters. The CXTILT02E/02EC sensor can measure the tilt angle in two directions, pitch and roll, with an accuracy of 0.1 degrees. The tilt sensor adopts the silicon material micro-mechanical technology manufacturing process, which has high reliability, stability and compact shape. The data communication adopts the RS232 interface, and the horizontal zero point of the inclination is calibrated through the serial port command [36 Tilt]. CXTIL02EC has high measurement accuracy and linearity, and adopts on-board temperature sensor for temperature compensation to avoid drift caused by temperature changes.

In order to solve the difficulty of obstacle detection in unstructured environment and improve the accuracy of environment modeling, a 3-D environment perception and analysis system combining laser ranging instrument and precision rotating pan/tilt was designed. The system uses a 2-D scanning laser rangefinder LMS291, and the 3-D information of the environment can be obtained through the pitch and horizontal rotation of the sensor. The sensor pan/tilt can achieve a horizontal rotation of $\pm150°$ under the drive of a stepper motor, so it can realize a panoramic observation of the environment without the car body rotating, as shown in Figure 7.1.

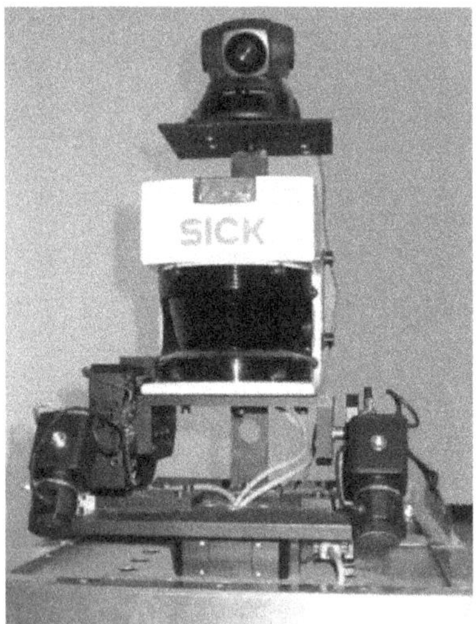

Figure 7.1.　Lidar and vision sensor of mobile robot.

7.2.2　*Software and hardware structure of mobile robot system*

1. The software structure of the mobile robot control system

The mobile robot is a complex computer control system. The realization of the robot's intelligent behavior is not only accomplished by integrating the required hardware and corresponding driving software. The realization of intelligent behavior depends to a large extent on the design of system software functions [10, 93]. The architecture determines how the system is structured and determines the distribution of module functions. The reasonable design of the architecture will be able to realize complex intelligent behaviors and have further perfect expansion capabilities [110, 111].

(1) Composite architecture of intelligent system

The composite architecture uses a deliberate approach to model and plan the complex environment, and uses reactive behavior to overcome the uncertainty of dynamic changes in the execution process [42]. A typical three-layer composite intelligent system is shown in Figure 7.2.

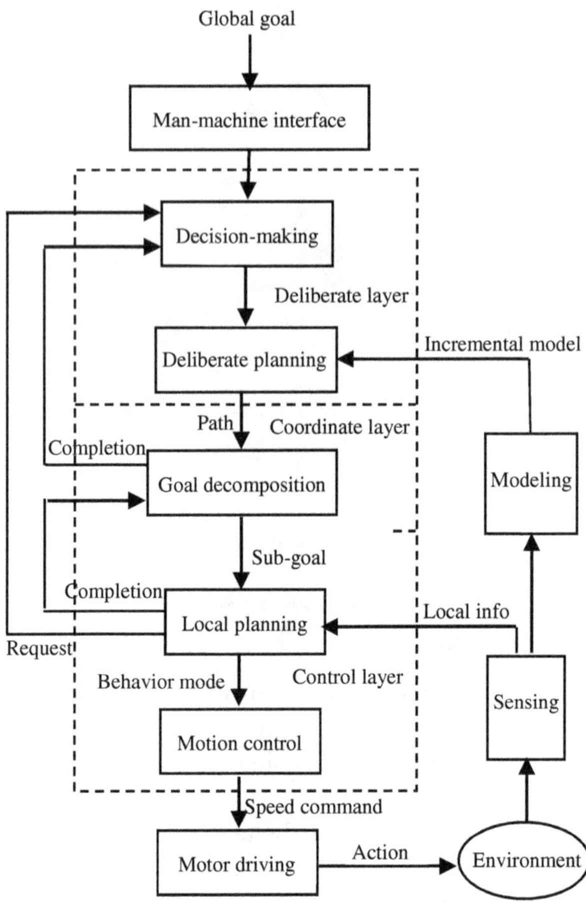

Figure 7.2. Three-layer control model of mobile robot with compound architecture.

The core of the deliberative layer is deliberative planning based on map-based planning behavior. The core of the control layer is the local planning of reactive behavior based on sensor information. Deliberate planning periodically executes the deliberate planning algorithm, and delivers the planning results to the control layer for execution, or initiates planning behavior in response to a request from the control layer. The deliberate layer and the control layer run asynchronously and independently, and a deliberately planned execution cycle may experience multiple reaction behavior control cycles.

The coordination layer connects the control layer and the deliberate layer, feeds back the state of the control layer to the deliberate layer, and decomposes the planning results of the deliberate layer into sub-goal sequences for execution by the control layer. After fusion, the environmental information of the sensors is sent to the deliberate layer and the control layer according to different needs, to meet the needs of modeling and real-time control.

The composite architecture has become the main paradigm of intelligent robot system design, and it is still under continuous development. At present, most intelligent robot systems are layered mainly based on the update rate of the information processing process, and the relationship between the layering of the architecture and the representation of spatial knowledge is fragmented. Algorithms with slow refresh are placed at the deliberate layer, and algorithms with fast refresh are placed at the control layer. The processing of large-information sensor data (such as visual information) is often placed on a deliberate layer because it occupies more computing resources. In addition, the control layer of most systems still relies on high-level planning and needs to initialize a sequence of actions. If the high-level planner is damaged, it may be difficult for the system to continue to operate. In these systems, the spatial knowledge form is usually single, either a metric representation or a topological representation method, and a single knowledge representation form is often difficult to meet the requirements of real-time control and knowledge expansion at the same time.

(2) Architecture based on spatial knowledge representation

The knowledge representation method determines how the system stores, utilizes and acquires knowledge. Traditional artificial intelligence limits the form of knowledge representation to symbolic representation, while behaviorist intelligence breaks through the traditional symbolic knowledge representation restrictions [8]. Cognitive researches have shown that spatial behavior ability is often closely related to the corresponding knowledge representation form. Figure 7.3 shows the cognitive process of people or intelligent systems in the environment.

This process includes three closed-loop feedback structures: (1) The environmental information from the sensor is fused to form a space vector that guides the action, which embodies the reactive

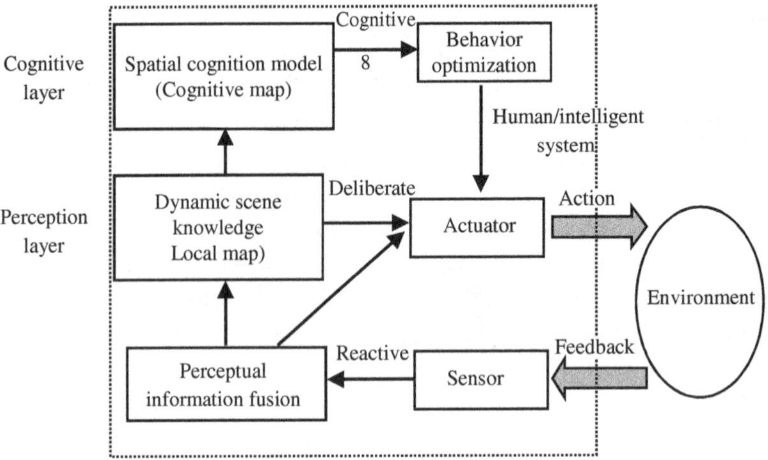

Figure 7.3.   Human/intelligent system's cognitive process of the environment.

behavioral intelligence. (2) Perception information constructs a local environment map through perceptual level modeling. This map intuitively describes the environment in the form of dynamic scene knowledge and can guide complex behavior judgments. This process reflects dynamic deliberate intelligence. (3) After the accumulated information is further processed by feature modeling, a cognitive map of the environment is obtained; Through the cognitive map, the expansion of spatial knowledge and support for global optimization and decision-making are realized, reflecting more complex cognitive behaviors intelligence.

Cognitive intelligence is the most advanced form of navigation, and it often requires long-term information accumulation in the environment to achieve feature-level modeling and obtain a feature impression of the environment. For real-time navigation, reactive behavior and deliberate behavior are generally indispensable, while cognitive behavior is not necessarily involved in the real-time navigation process. Cognitive maps focus on the expansion of spatial knowledge and support global navigation decisions in large-scale environments.

Further research has shown that different levels of environmental knowledge representation determine the spatial behavior capabilities of different levels of complexity. Cognitive scientists divide the spatial behavior competences of animals into four different levels: reflex-like behavior, integration behavior, learning behavior, and cognition behavior [75].

Reflective behaviors are actions that are directly triggered by sensors that do not require knowledge. The most typical example is the knee-jerk reaction. The fusion behavior simply synthesizes the perceptual information and makes action judgments, such as the simple obstacle avoidance behavior to avoid collisions during walking. Fusion behavior embodies reactive behavioral intelligence, requiring a temporary work area memory to realize the processing of perceptual information, and the form of knowledge representation is a simplified vector for guiding actions. Learning behavior is based on long-term information memory, and spatial knowledge is represented as dynamic scene knowledge. Although the memory of information is long-term, the difference between the performance form and the temporary memory in the fusion behavior is reflected in the "quantity" rather than the "qualitative" change. The behavior of learning is usually embodied in comparing the current state of the scene with the state of the historical scene through "memory" to make more feasible behavioral judgments. In cognitive behavior, through the feature extraction of long-term accumulated information, the representation form of spatial knowledge will undergo a "qualitative" change. Intuitive large-capacity scene knowledge is abstracted as a characteristic model of the environment and becomes relatively persistent knowledge.

The above four behavioral abilities often exist in parallel. According to the difference of the spatial knowledge representation form, the environmental spatial knowledge representation can be divided into the perception layer and the characteristic layer.

(1) Perception layer: Sensor information is intuitively expressed as coordinate measurement information corresponding to the actual environment, and the fusion of sensor information is the process of environmental reconstruction at the perception layer. This process requires the ability to fuse sensor information in real time and reproduce the environment as realistically as possible. At the perception layer, the occupancy grid method is often used to realize the representation of scene knowledge.

(2) Feature layer: The modeling of the environment is to realize the organization and utilization of perceptual information at the cognitive layer. It extracts information processing from discrete, one-sided, and incomplete spatial information that can be applied to information processing process of the knowledge form of decision-making and planning. Environmental information is expressed in a relatively small and effective feature form, for example, a topological map is used to express the environment.

The scene knowledge representation at the perception layer realizes the integration of environmental information and directly serves the real-time navigation process, while the knowledge representation at the feature layer realizes the expansion of spatial knowledge and serves global planning and decision-making. Navigating in an unknown environment and realizing autonomous spatial knowledge feature extraction are important intelligent behaviors of mobile robots. Based on the above considerations, a hierarchical control architecture oriented to spatial knowledge representation is designed, as shown in Figure 7.4

In the layering of the architecture, from the bottom to the top are the control layer, the deliberate layer and the feature layer. The control layer uses reflective or reactive behavior to achieve obstacle avoidance-based path planning and path tracking control. The deliberate layer processes large amounts of sensory information, and merges it with track estimation information to obtain a dynamic local environment knowledge base. The information in the local environment knowledge base adopts the occupying grid representation method, which can realize the reproduction of the local environment intuitively and quickly. Dynamic deliberate planning is a planning behavior based on grid representation to provide support for real-time navigation. In the feature layer, the feature representation of the environment is obtained from the local perception information, which solves the limitation that the grid representation method is difficult to expand the spatial knowledge. Topological structure diagram is an effective method to realize large-scale environmental spatial representation, and it has the characteristics of low storage cost, good scalability, and fast planning speed. The knowledge form of the environmental model in the feature layer is no longer an intuitive measurement information. The environmental model provides support

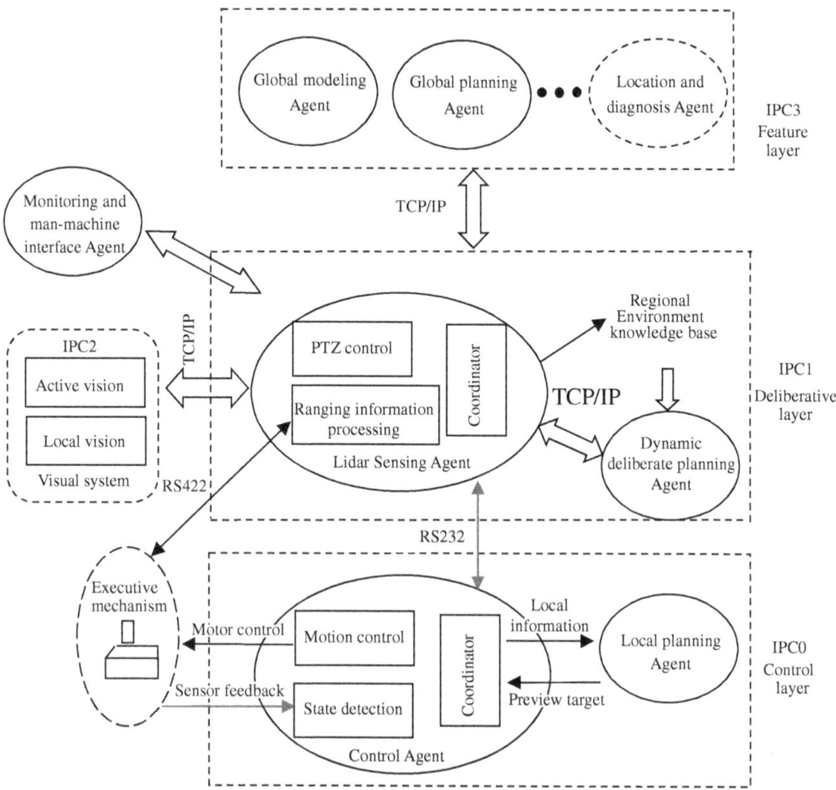

Figure 7.4. Hierarchical control architecture of mobile robot oriented to spatial knowledge representation.

for global planning and decision-making behavior, and does not necessarily intervene in the real-time navigation process. The most typical application is that after the robot completes the task through exploration, it can return home with the help of an autonomously established environment model.

2. The hardware structure of the mobile robot control system

The hardware is the executor of each part of the system. How to effectively organize and coordinate various sensors and computer system resources to realize the flexible combination of basic functional modules is an important factor considered in the design of the control system. Due to different application backgrounds, the structure of the mobile robot control system also has its own characteristics. The control form can be divided into centralized control, hierarchical control and distributed control [10]. Compared with the traditional multi-layer and centralized control system, the distributed control system has the advantages of strong integration ability and good scalability.

(1) On-board distributed control system structure

The mobile robot system adopts a bus backplane that can accommodate 4 industrial computer systems, and can install 4 industrial computer CPU full-length cards. Each industrial computer system can install 2 PCI slot function boards and 2 ISA slot function boards. A low-power industrial computer mainboard with a PIII800 MHz CPU and 256 M memory was selected to be used in motion control, lidar information processing, multi-eye vision information processing, and decision support systems. The development environment of the system is W98/2000 system, using Visual C++ as the software development tool. Organize the on-board computer into a local area network control system through a hub. The hardware structure of the system is shown in Figure 7.5.

The vehicle-mounted distributed control system includes a motion control system, a lidar information processing system, a visual information processing system, and a decision support system.

(1) Motion control system Industrial computer IPC0 is used as a motion control system, and the operating system is based on the W98 platform. Two RS232 serial communication interfaces are integrated on the CPU mainboard, one is connected with an external gyroscope as the robot heading sensor, and the other serial port is used for point-to-point real-time communication with IPC1. The main function of the motion control system is to calculate and feedback the real-time motion posture of the robot car body, while achieving dynamic obstacle avoidance and tracking control. The main control boards are ADT850 stepping motor control card based on PCI bus and A/D acquisition card PCI1710. The ADT850 motion control card is a 32-bit four-channel stepping motor control card based on PCI bus. It controls the four drive motors on the left and right joysticks of the mobile robot to realize the movement of the car body.

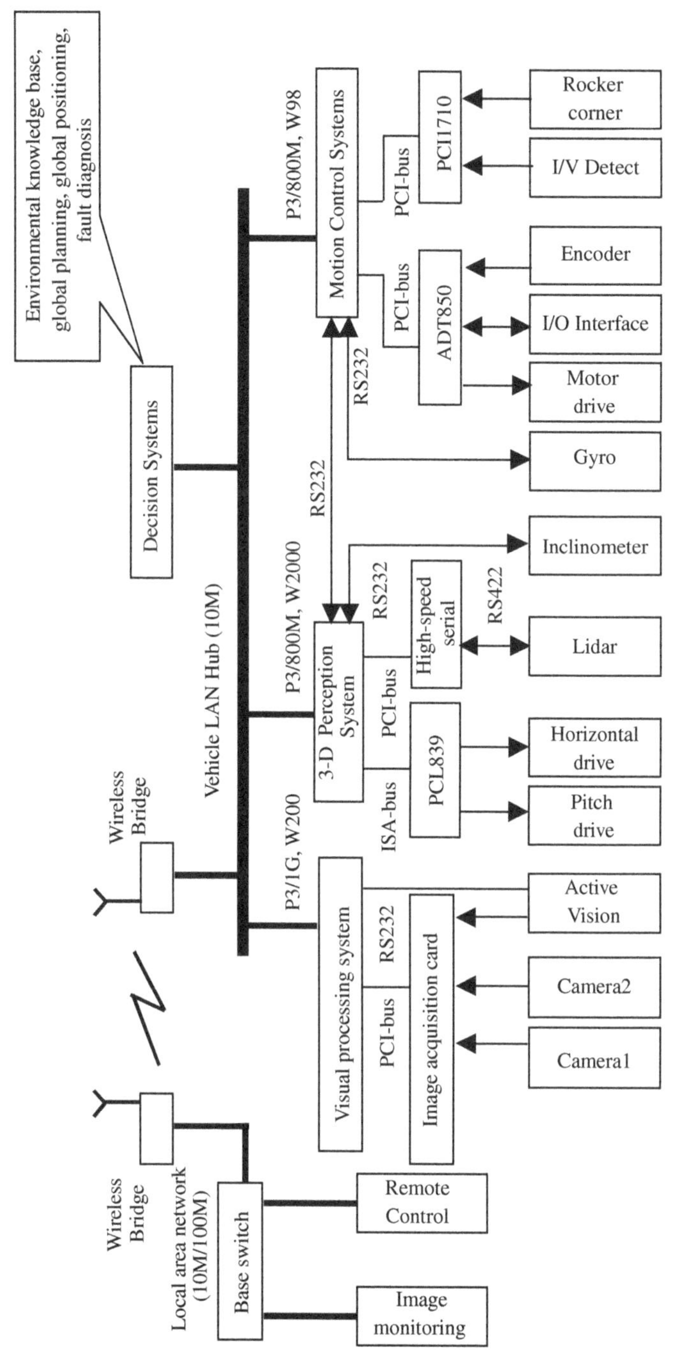

Figure 7.5.   Hardware structure of the mobile robot control system.

(2) Lidar information processing system Industrial computer IPC1, as a 3-D environment-aware information processing system, runs under the Win2000 operating system, and the memory is a 40G hard disk. Install the stepping motor control card PCL839 to control the rotation of the lidar. Install a PCI bus-based high-speed RS-422 interface card, receive lidar information at a communication speed of 500K baud rate and perform local 3-D environmental flatness analysis, distinguish feasible areas and obstacle areas in the environment, and establish a database of local map.

(3) Visual information processing system Industrial computer IPC2 as a visual information processing system, works in the Win2000 operating environment, with a memory of 40G. The CPU adopts PIII, 1 GHz. Install a multi-channel image acquisition card to process global visual information and pre-local visual information, and establish a visual feature database. The visual computer system controls the active vision PTZ through the RS-232 interface.

(4) Decision support system This system mainly performs non-real-time, large-scale intelligent calculations, such as high-level intelligent tasks such as global positioning, fault diagnosis, environmental modeling, and global path planning. The system has no special control board, and the functions of the system are realized by multiple software agents.

(2) Communication and monitoring system

The communication of the control system includes the communication between the mobile robot vehicle-mounted control system and the laboratory monitoring system and the communication between the control computers in the vehicle-mounted system. The vehicle control system is equipped with a hub, which organizes the various computer systems into a distributed control system based on the local area network, and they can communicate through the TCP/IP protocol. In order to ensure the reliable interaction of real-time information between IPC0 and IPC1, RS232 point-to-point communication is adopted between them, and the baud rate is set to 128 K baud.

(3) Network-based open experimental platform

In the navigation control system, IPC1 has become the robot information integration center because it is responsible for the fusion of lidar distance information and track estimation information.

7.3 Architecture of Four-layer Hierarchical Mobile Robot Navigation Control System

7.3.1 *Decomposition of the control task of the navigation system*

The navigation control system should replace the driver to complete the driving task from accepting the destination to controlling the operation of the vehicle to the destination, and the task must be reasonably decomposed. According to the characteristics of the task environment, time span, and space scope, the entire task can be divided into five levels, as follows:

1. Tasks and subtasks

A task is a macro command from the operator received by the driving control system, and its content is: from a certain point A in the real world to another point B. The task usually takes tens of minutes or even days to complete, and its span in space is often very large.

Decompose the task into sub-tasks. Usually the starting point and the end point of the sub-task should be two points on the road with the same structural characteristics as the autonomous car, that is to say, the autonomous car should be able to use the same control strategy when performing a certain sub-task. Typical subtasks are: the autonomous vehicle travels from point A1 to point A2 along a two-lane highway, and then from point A2 to point A3 along a country road. The duration of each subtask generally ranges from a few minutes to a few hours.

2. Behavior

Behavior is a sequence of actions taken by autonomous vehicles to cope with changing traffic conditions. Each behavior should meet some requirements of autonomous vehicles in terms of safety, driving efficiency or traffic rules. The execution period of a behavior is usually a few seconds to a few minutes. Common behaviors include: changing into the left lane and overtaking the vehicle in front, tracking the vehicle in front, stopping at an intersection, etc.

3. Trajectory and planning trajectory

Trajectory is the path and speed sequence of the autonomous vehicle. Planning trajectory is the sequence of paths and speeds the autonomous vehicle expects to pass in a period of time in the future. The length of time for planning a trajectory is generally from a few hundred milliseconds to a few seconds.

4. Action

Action is generated by the driving control system and executed by the low-level instructions of each actuator. For example, the throttle opening is α_i, the front wheel deflection angle is δ_i, etc. The execution time of each action is usually a few milliseconds to tens of milliseconds.

7.3.2 *Four-layer modular autonomous driving control system structure*

1. Four-layer modular autonomous driving system structure

Based on the above-mentioned task hierarchy decomposition, a four-layer modular autonomous driving control system structure as shown in Figure 7.6 is given [10, 93, 95]. Among them, the 4 levels are: task planning, behavior decision-making, behavior planning and operation control. In addition, it also includes two independent function

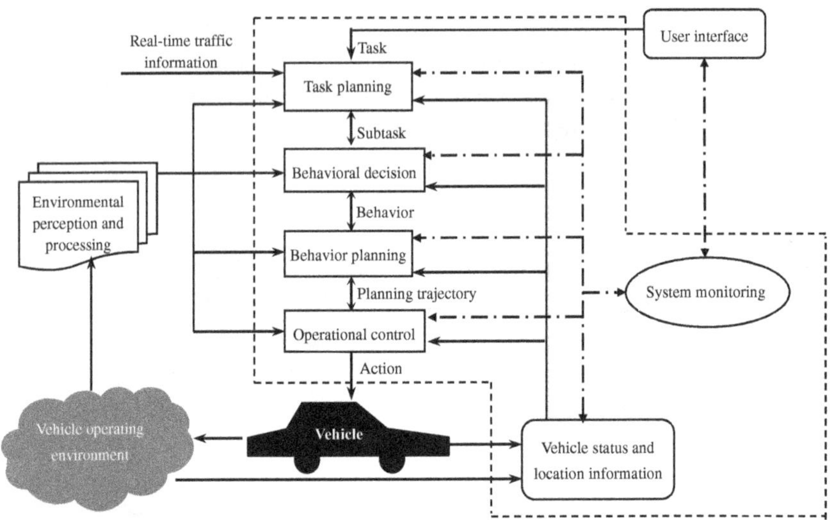

Figure 7.6. Structure of the four-layer modular automobile autonomous driving control system.

modules of vehicle status and positioning information and system monitoring.

The four levels of the control system are respectively responsible for completing tasks of different scales, and the scale of tasks decreases sequentially from top to bottom. Among them: the task planning layer maps from tasks to subtasks; the behavior decision layer maps from subtasks to behaviors; the behavior planning layer maps from behaviors to planned trajectories; the operation control layer maps from planned trajectories to vehicle actions.

Each layer of the system is different in time span, spatial scope, environmental information concerned, logical reasoning method, and responsibility for accomplishing the control goal. As an independent module, the system monitoring module is responsible for collecting system operation information, supervising the operation of the system, and adjusting system operation parameters when necessary. The vehicle status and positioning information module is responsible for the generation of vehicle status and positioning data, and provides decision-making control at all levels of the system.

2. The general internal structure of each layer of the driving control system

The general structure of each layer of the driving control system is shown in Figure 7.7. Each layer includes five basic components: environment modeling, reasoning and decision-making, supervisory control, domain learning, and domain knowledge base. Among them:

(1) Environment modeling component: Extract the information related to the task of this component from the perception information provided by the environment perception system, and generate an environment description related to the task of this component.
(2) Reasoning decision component: According to the current environment description, use relevant domain knowledge to produce the decision result of this component.
(3) Domain knowledge base: The storage and management component of knowledge related to the task of this component.
(4) Domain learning component: Through related machine learning methods, the domain knowledge base is modified to continuously improve the system's environment modeling ability and reasoning and decision-making ability.

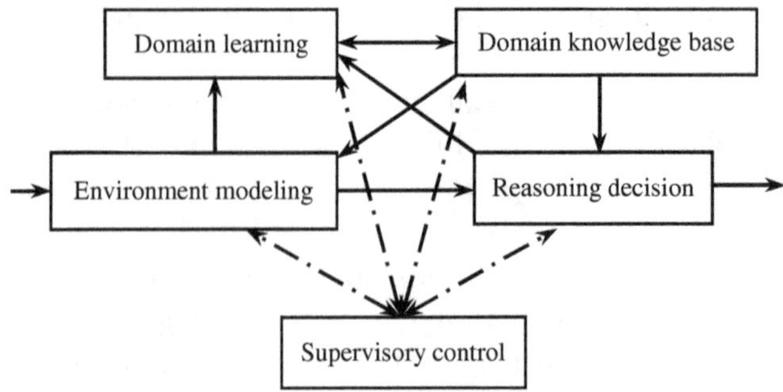

Figure 7.7.   General structure of the intelligent controller in the layer.

(5) Supervision and control components: Evaluate and judge the operation of the remaining components and the operation of the sub-system, and adjust the relevant parameters of each component.

7.3.3 *Structural characteristics of each layer of the driving control system*

Due to the different levels of tasks to be handled by each layer of the driving control system, it is different from the time and space span, the environmental information concerned, the logical reasoning method, and the responsibility for accomplishing the control goal.

(1) Operation control layer

The operation control layer converts the planned trajectory from the behavior planning layer into the actions of each actuator, and controls each actuator to complete the corresponding actions. It is the bottom layer of the entire autonomous driving system. It consists of a series of traditional controllers and logical reasoning algorithms, including vehicle speed controller, direction controller, brake controller, throttle controller, steering controller, and signal light/horn control logic, etc., as shown in Figure 7.8.

The input of the operation control layer is the path point sequence, the vehicle longitudinal speed sequence, the vehicle behavior conversion information, the vehicle status and the relative position information generated by the behavior planning layer. This information is processed through the operation control layer, and

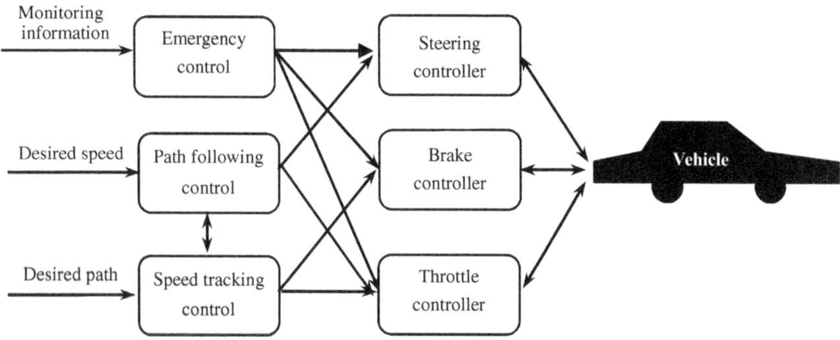

Figure 7.8.   Schematic diagram of main modules of the operation control layer.

finally becomes the action of the vehicle actuator. The modules of the operation execution layer are executed periodically at millisecond intervals to control the movement of the vehicle along the planning results in the previous planning cycle. Tracking control accuracy is an important indicator to measure the performance of the operational control layer.

(2) Behavior planning layer

The behavior planning layer is the interface between the behavior decision layer and the operation control layer, which converts the behavior symbol results generated by the behavior decision layer into the trajectory instructions that the traditional controller of the operation control layer can accept. The input to the behavior planning layer is vehicle status information, behavior instructions, and passable road information provided by the environment perception system. The behavior planning layer includes: behavior supervision execution module, vehicle speed generation module, vehicle expected path generation module and so on.

When the vehicle behavior changes or the information processing results of the passable road surface are updated, the various modules of the behavior planning layer are activated to supervise the execution of the current behavior, and re-plan the behavior based on the environmental perception information and the current state of the vehicle to provide vehicles for the operation control layer expected speed and expected motion trajectory and other instructions. In addition, the behavior planning layer also feedbacks the execution

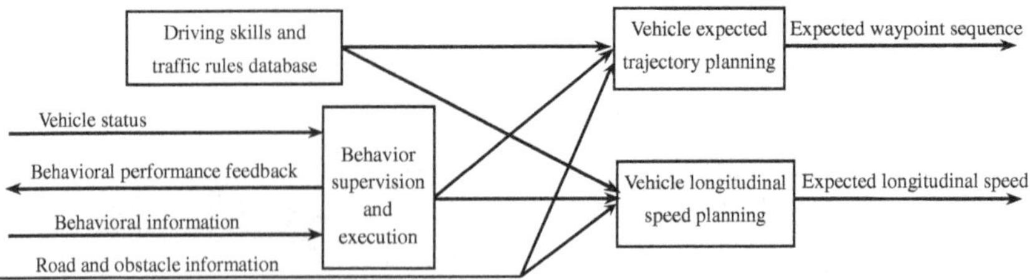

Figure 7.9. Schematic diagram of the main modules of the behavior planning layer.

of the behavior to the behavior decision-making layer. Figure 7.9 is a schematic diagram of a behavior planning layer.

The behavior supervision execution module integrates environmental perception information, vehicle status information and decision-making results of the behavior decision-making layer, determines the timing of behavior transition, and forms feedback on behavior implementation, such as behavior completion ratio, current execution behavior, etc.

The vehicle longitudinal control planning module uses driving skills and knowledge of relevant traffic rules to plan the vehicle's longitudinal control targets before the next planning cycle, including the expected obstacle position, expected speed, and expected acceleration based on the current behavior, current state of the vehicle, and expected driving speed and so on, for the operation control layer to execute in a planning cycle.

The vehicle expected trajectory planning module, based on the current execution behavior, the current state of the vehicle and the road and obstacle information obtained from environmental perception, combined with vehicle dynamics, driving knowledge and traffic rules, plans the sequence of path points that the vehicle should pass in the next time period, as the input of the path tracking module of the operation control layer. The behavior planning layer should be able to make reasonable plans for the various behaviors produced by the behavior decision-making layer.

The vehicle behavior planning module is activated by the environment perception information, and its execution is synchronized with the environment perception module. Each planning result can be executed by the operation execution layer in several or even dozens of cycles. Each planning distance is determined by the perception distance of the environment perception system.

(3) Behavior decision-making layer

Common vehicle behaviors include: starting, stopping, accelerating along the road, advancing along the road at a constant speed, avoiding obstacles, turning left, turning right, reversing, etc. According to the environmental information obtained by the environmental perception system, the current state of the vehicle, and the task goals planned by the task planning layer, the autonomous vehicle should take appropriate actions to ensure the smooth completion of the task; this work is done by the behavior decision-making layer.

The factors that affect the decision-making of autonomous vehicles include road conditions, traffic conditions, traffic signals, mission requirements for safety and efficiency, and mission objectives. How to integrate the above-mentioned factors and make behavioral decisions efficiently is the focus of behavioral decision-making research.

Relevant theories and methods of reasoning and decision-making in artificial intelligence are currently commonly used methods to realize behavioral decision-making. A good behavioral decision-making system should be able to react to changes in the vehicle driving environment in real time, and be able to generate reasonable behaviors and complete tasks according to the user's attention to safety and efficiency.

Figure 7.10 is the general structure of the decision-making layer of autonomous vehicle behavior. Among them, the behavior pattern generation classifies the operating environment according to the

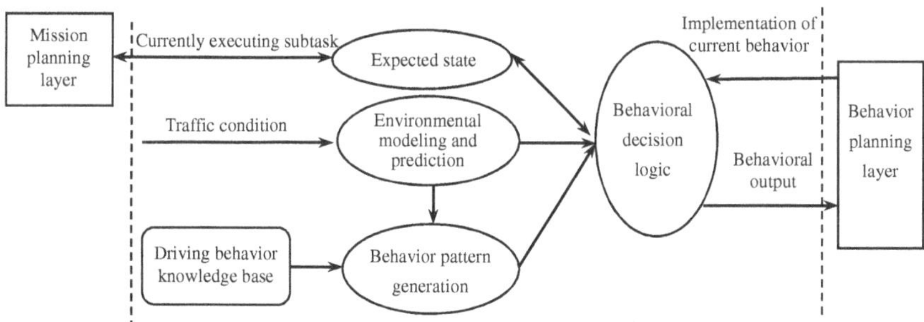

Figure 7.10. Schematic diagram of the main modules of the behavior decision layer.

structural characteristics of the current operating environment of the vehicle, traffic density, etc., and generates the available behavior set and conversion relationship under the current conditions, which is an important basis for behavior decision-making. For example, traffic signals, pedestrians, etc. should be paid attention to on urban highways, but there are no traffic signals on expressways. Therefore, the behavior sets are different in the two cases. The expected state is determined by the current execution of the subtasks, such as the expectation of the speed of the vehicle, the expectation of the safety of the vehicle, and so on. Environmental modeling and prediction according to the perception information of the environmental perception system, some key environmental features that affect behavioral decision-making are modeled, and their development trends are predicted.

Behavioral decision logic is the core of the behavioral decision-making layer, which integrates various types of information and finally issues behavioral instructions to the behavioral planning layer. A good behavioral decision logic is an inevitable requirement to improve the autonomy of the system. The execution time of each behavior of behavior decision is usually a few seconds.

(4) Mission planning layer

The task planning layer is the highest layer of the autonomous driving intelligent control system, so it also has the highest intelligence, as shown in Figure 7.11. The task planning layer receives task requests from users, uses the map database, comprehensively analyzes traffic flow, road conditions and other related factors that affect driving,

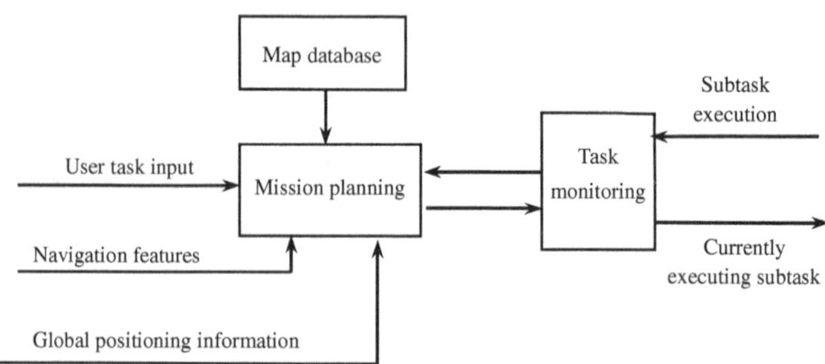

Figure 7.11. Schematic diagram of the main modules of the mission planning layer.

and search for the optimal or suboptimal path from the current point to the target point that meets the task requirements in the known road network. The path usually consists of a series of subtasks, such as: the vehicle travels along the highway A to point X, turns into highway B, travels to point Y, ... to reach the destination, while planning the completion time of each subtask on the path, and sub-tasks requirements for efficiency and safety, etc. The planning results are supervised and executed by the task monitoring module.

The task monitoring module determines the current subtasks to be executed according to the feedback information of the environment perception and the vehicle positioning system, and supervises and controls the execution of the tasks by the lower layer. When the execution of the current subtasks is blocked, the task planning module is required to re-plan.

The task planning problem is generally performed offline. It is a static search problem in a fixed environment. It has been well solved in artificial intelligence research. Commonly used heuristic search algorithms can be used to implement the task planning layer.

(5) System monitoring module

The system monitoring module is a major feature of this control structure and plays an important role in the entire driving control system, as shown in Figure 7.12. Its tasks include interaction with the operator, vehicle error detection, system operation monitoring,

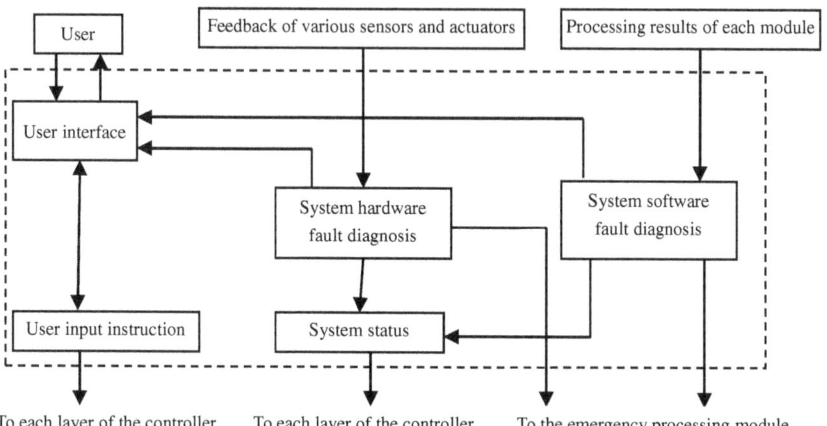

Figure 7.12. Schematic diagram of the system monitoring module structure.

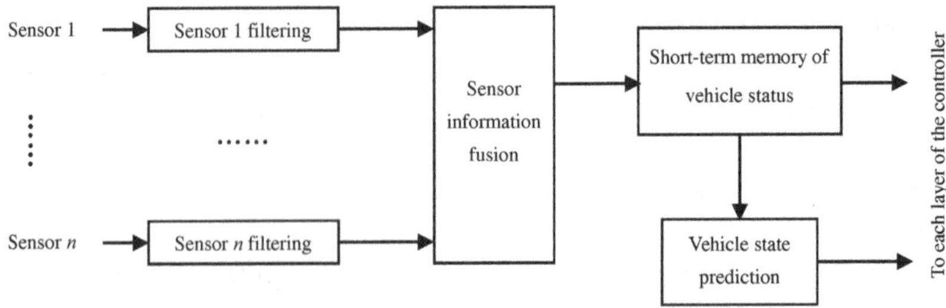

Figure 7.13.   Schematic diagram of the structure of the vehicle state perception module.

etc., and submit the operator's operating instructions and the current system status to each layer of the autonomous driving control system, and each layer will respond accordingly. The emergency handling module of the operation control layer can be directly activated at any time, and emergency handling plans such as emergency stop can be adopted to ensure vehicle safety.

The faults of the system hardware diagnosis and inspection include various hardware faults of various sensors, various actuators, tires, engines, etc. Software fault monitoring monitors the operating results of each module to determine whether there is a fault in the calculation. The system fault status information will affect the operation of each module of the controller.

(6) Vehicle status perception module

The vehicle state perception module performs necessary filtering, fusion, and processing on the signals from the various state sensors of the vehicle, and calculates the current state of the vehicle, including vehicle speed, acceleration, steering wheel angle, position, posture and other related states, and records these states in a short-term memory device, it is prepared for the query of the relevant modules of the control system. In addition, the vehicle state perception module should also be able to predict the future state of the vehicle, as shown in Figure 7.13.

7.4   Dynamic Model of Wheeled Mobile Robot

In the theoretical research on the motion control of wheeled mobile robots, it is generally assumed that the wheel is in point contact with the ground, and there is only pure rolling at the contact point without relative sliding (including lateral and longitudinal sliding). This ideal constraint is essentially a nonholonomic constraint, which makes the wheeled mobile robot under the ideal constraint a typical example of a nonholonomic system [43]. Incomplete systems may be very simple in form, but their control problems are considered "deceptive" and complex.

7.4.1 *Typical mechanism of wheeled mobilebot*

According to Brockett's necessary condition [7]: For nonholonomic systems, there is no smooth time-invariant static state feedback control law that stabilizes the system to a non-singular configuration. This result has now become a classic. In fact, there is no continuous time-invariant static state feedback control law [55]. Pomet extended Brockett's conclusions, pointing out that the smooth (continuous) time-invariant dynamic state feedback law also does not exist [80]. This puts forward an essential constraint for the stabilization control of this type of system: there is no smooth (continuous) time-invariant state feedback law that stabilizes a nonholonomic system. In this way, the possible form of the stabilization control law of nonholonomic systems is essentially restricted. Many mature results in classical linear system theory cannot be directly applied to the stabilization control of nonholonomic systems. New methods and tools must be found to solve this problem. The problem is extremely challenging; at the same time, the nonholonomic wheeled mobile robot is an inherently nonlinear under-driven drift-free dynamic system, so its control is more difficult. It is the needs of practical applications and potential theoretical research value that have greatly promoted people's research on nonholonomic control systems represented by wheeled mobile robots.

At present, the commonly used mobile robot mechanisms include wheel-type mobile mechanism, leg-type mobile mechanism, crawler-type mobile mechanism and wheel-leg-type mobile mechanism. Wheeled mobile mechanisms have a long history and are very mature in mechanical design, so wheeled mobile robots (WMR) are the most common among various practical mobile robots. In practical applications, there are many different wheeled mobile robot motion configurations, including the number and type of wheels, the installation position and driving mode of the wheels, and single or multi-body structures. From the perspective of the number of wheels, three-wheel and four-wheel are more methods, and robots with more wheels are more common in variable-configuration wheeled mobile robots. Among the single robots, the common types include differential drive and synchronal drive two-wheeled robots, such as Super-MARIO developed by the University of La Sapienza in Roma, Italy; car-like robots, such as Nanrover developed by the Jet Propulsion Laboratory (JPL), Nomad and Ratler developed by Carnegie Mellon University (CMU); six-wheeled robots such as Rocky7 [60], Valor and Opportunity for Mars exploration It is a six-wheeled wheel-legged robot. Examples of well-known multi-body wheeled mobile robots include Hilare2-Bis from the LAAS laboratory in France.

7.4.2 *Dynamic model of wheeled mobile robot under nonholonomic constraints*

So far, many documents have taken the kinematic model of wheeled mobile robot as the research object. This approach is not only because the kinematics model is easy to handle, there are deeper theoretical reasons behind it. In this section, the Lagrange equation will be used to establish a dynamic model of an n-dimensional mechanical system (including wheeled mobile robots) constrained by kinematics, and partial feedback linearization will be used to explain the reason. For the derivation of the dynamic model, readers can refer to the classic works [43].

The given system is subject to the following constraints:

$$a_i^T(q)\dot{q} = 0, \quad i = 1,\ldots, k < n, \quad \text{or} \quad A^T(q)\dot{q} = 0.$$

Define the Lagrange function L as the difference between the kinetic energy and potential energy of the system:

$$L(q,\dot{q}) = T(q,\dot{q}) - U(q) = \frac{1}{2}\dot{q}^T B(q)\dot{q} - U(q) \tag{7.1}$$

Where, $B(q)$ is the positive definite inertia matrix of the mechanical system. The Euler-Lagrange equation of the system is

$$\frac{d}{dt}\left(\frac{\partial L}{\dot{q}}\right)^T - \left(\frac{L}{\partial q}\right)^T = A(q)\lambda + S(q)\tau \tag{7.2}$$

Among them, $S(q)$ is the $n \times m$ matrix, which maps the $m = n - k$ external input τ to the force/moment that does work on the displacement q; $\lambda \in R^m$ is the unknown Lagrange multiplier vector. The term $A(q)\lambda$ represents the binding force vector.

The dynamic model of the constrained mechanical system is

$$B(q)\ddot{q} + n(q, \dot{q}) = A(q)\lambda + S(q)\tau \tag{7.3}$$

$$A^T(q)\dot{q} = 0 \tag{7.4}$$

Where,

$$n(q, \dot{q}) = \frac{dB(q)}{dt}\dot{q} - \frac{1}{2}\left(\frac{\partial}{\partial q}(\dot{q})^T B(q)\dot{q}\right)^T + \left(\frac{\partial U(q)}{\partial q}\right)^T$$

Consider the matrix $G(q)$, whose columns are the basis of the zero space of the matrix $A^T(q)$, so there is $A^T(q)G(q) = 0$. Multiplying both sides of equation (7.3) by $G^T(q)$ can eliminate the Lagrange multiplier, thereby obtaining a simplified kinetic model composed of m differential equations:

$$G^T(q)[B(q)\ddot{q} + n(q, \dot{q})] = G^T(q)S(q)\tau \tag{7.5}$$

Now suppose that $\det[G^T(q)S(q)] \neq 0$ this is a realistic assumption in most practical situations, which can be ensured by mechanical design. Then the kinematics and dynamics models are synthesized into a simplified state-space model of $(n + m)$ dimensions, in the form as follows:

$$\dot{q} = G(q)v$$

$$\dot{v} = M^{-1}(q)m(q, v) + M^{-1}(q)G^T(q)S(q)\tau$$

Where $v \in R^n$ is the simplified velocity vector, and:

$$M(q) = G^T(q)B(q)G(q) \; 0$$

$$m(q, v) = G^T(q)B(q)\dot{G}(q)v + G^T(q)n(q, G(q)v)$$

Among them,

$$\dot{G}(q)v = \sum_{i=1}^{m}\left(v_i \frac{\partial g_i}{\partial q}(q)\right)G(q)v$$

Non-linear feedback in the form of "calculated torque"

$$\tau = [G^T(q)S(q)]^{-1}(M(q)a + m(q,v)) \tag{7.6}$$

The simplified state space model can be partially linearized, where $a \in R^m$ is the simplified acceleration vector, and the following system is obtained:

$$\begin{aligned} \dot{q} &= G(q)v \\ \dot{v} &= a, \end{aligned} \tag{7.7}$$

It can be seen that the first n equations are kinematic models, and the last m equations are dynamic extensions. It should be noted that calculating the torque input (7.6) needs to measure the speed vector v of the system. If v can't be measured directly, as long as q and $\dot{q}$ can be measured, the pseudo-inverse of the kinematics model can be calculated as follows:

$$v = (G^T(q)G(q))^{-1}G^T(q)\dot{q}$$

Then define the state vector $x = (q, v) \in R^{n+m}$ and the input vector $u = a \in R^m$, the state space model of the closed-loop system can be written in the following compact form:

$$\dot{x} = f(x) + g(x)u = \begin{bmatrix} \boldsymbol{G}(q)v \\ 0 \end{bmatrix} + \begin{bmatrix} 0 \\ \boldsymbol{I_m} \end{bmatrix} u$$

It is a non-linear control system with drift, also known as the "second-order kinematics model" of a constrained mechanical system.

In short, for nonholonomic systems, it is possible to eliminate dynamic parameters through nonlinear feedback, as long as: (1) The dynamic model is fully known. The entire system state is measurable. If these two conditions are met, the dynamic model of the original system can be transformed into a second-order kinematics model through feedback, so that the control problem can be solved directly at the speed level. In other words, through synthesis v, the kinematics model $q = G(q)v$ has the desired behavior. When v is measurable, as long as v is sufficiently smooth (at least differentiable), because there is $v \approx a$, the control input τ can be calculated at the generalized force level from equation (7.6). It is this reason that motivates the incomplete system in-depth study of smooth stabilizer.

7.5 Stabilization and Tracking of Wheeled Mobile Robots

Since wheeled mobile robots generally work in a dynamic, unknown and complex environment and have a large working area, the research on mobile robots must solve a series of problems, including trajectory planning, motion control, environment modeling, and real-time positioning, etc., among which motion control is the most basic problem in the research of mobile robot systems, because the execution of many tasks of mobile robots is ultimately attributed to the motion of the vehicle body. The motion control problems of wheeled mobile robots mainly include motion planning and feedback motion control. Because the wheeled mobile robot subjected to nonholonomic constraints is a typical nonholonomic system, its trajectory planning and stabilization become quite difficult due to the non-integrable nature of the constraints.

7.5.1 *Stabilization and tracking controller design issues of wheeled mobile robots*

According to different control targets, the feedback control of wheeled mobile robots can be roughly divided into three categories, namely, trajectory tracking, path following, and point stabilization [72].

The trajectory tracking problem is that in the inertial coordinate system, the robot starts from a given initial state, follows and reaches a given reference trajectory. The path-following problem is that in the inertial coordinate system, the robot starts from a given initial state, follows and reaches the specified geometric path. The difference between the two is that the reference trajectory of the former depends on time, while the latter is independent of time variables. For mobile robots, if the robot is only required to track the planned path, and there is no requirement for the time to reach the specified location, then it is a path following problem; and if the robot is required to track a reference trajectory that changes in real time, then it is a trajectory tracking problem. In the path following problem, the established tracking state error is a functional relationship between the system state and the reference path, and it provides the system with a target point on the reference path according to the system state. Because it is not directly related to time, in actual research, the path following problem is not particularly difficult compared to trajectory tracking problem, and it can also be seen as a simplified case of the latter.

The point stabilization problem is that the system starts from a given initial state and arrives at the designated target state. Generally speaking, the balance point of the system is always taken as the target point. For mobile robots, point stabilization can also be called posture stabilization or posture regulation.

Different control objectives lead to different control difficulties. Take the unicycle type mobile robot as an example, its kinematics model has two input variables (u_1, u_2) and three state variables, namely position coordinates (x, y) and direction angle θ. In the path following problem, only two input variables are needed to control two states, so it appears as a so-called "square" problem in form; while in trajectory tracking and point stabilization problems, two input variables need to be used to control three at the same time. State variables are therefore a typical under-drive problem. For incomplete systems, the problem of point stabilization is the most complicated. Although a nonholonomic system is a controllable nonlinear system, it does not satisfy Brockett's necessary conditions, so it cannot be stabilized by smooth state feedback.

7.5.2 *Research on stabilization and tracking controller*

1. Posture stabilization

Balance point stabilization is the most challenging problem of nonholonomic system control, and it has also received the most attention. Because the possible forms of the stabilization control law of nonholonomic systems are essentially constrained, most of the research focuses on how to overcome this difficulty. Since smooth (continuous) time-invariant state feedback cannot be used for system (kinematics and dynamics) stabilization, people will naturally think of the law of discontinuous (time-invariant) feedback stabilization and the law of time-varying feedback stabilization (smooth or Lipschitz continuous), or a hybrid control law of the two. Various methods are used, including the technologies of feedback linearization, irregular static feedback linearization, singular perturbation, sliding mode variable structure control, neural network control, fuzzy control, adaptive control, nonlinear control based on the Lyapunov control function, differential flatness, etc., almost covers the main control technology.

By considering with the different types of mobile robots studied, the differences in the uncertainty of the models, whether to use dynamic models, whether to consider the restrictions on input and turning radius, etc., research in this area has yielded fruitful research results [83, 105].

(1) Discontinuous time invariant stabilization law

One way to circumvent Brockett's necessary conditions is to use discontinuous or piecewise continuous time-invariant feedback control. Its feasibility has been proven by Sontag [88]. This control can result in a non-oscillating trajectory that converges exponentially (or finite time). The existing discontinuous time-invariant control laws can be roughly divided into two categories: piecewise continuous control laws and sliding mode control laws. For a class of nonlinear systems that meet specific conditions, Sussmann proved the existence of the piecewise continuous static state feedback stabilization law [94]; later, Lafferierre further used piecewise smooth control Lyapunov function to give a piecewise expression of continuous stabilization law [58], the stabilization law can achieve global stabilization. Unfortunately, there is no general way to construct such a control Lyapunov function. The construction method of another kind of discontinuous controller is to use discontinuous state transformation, and its purpose is to overcome the obstacles brought by Brockett's necessary conditions. For the transformed discontinuous system, a smooth time-invariant stabilization law can usually be designed, but after the inverse state transformation, the stabilization law obtained in the original coordinate system is still discontinuous, so this processing method does not destroy Brockett's necessity condition.

Luo *et al.* divided the state space into two parts and designed a transition controller with asymptotic characteristics [73]. However, due to the switching behavior of the controller, if the controller switches multiple times in the two state spaces, the system will oscillate. Lee *et al.* combined the concept of "expansion" with σ transformation, and took the homogeneous function of the system state as the augmented state. Under certain assumptions, based on the auxiliary continuous controller, a global exponential convergence controller was obtained [63]. Since the σ transformation is a discontinuous state transformation, the resulting control law is also discontinuous, and there is a single hyper plane. If the initial state of the system is close to the hyper plane, the control input will become very large. Sun *et al.* gave the non-smooth form of the non-regular feedback linearization, and proved that the chain system can be linearized through the non-regular feedback, so that the feedback control design method of the linear system can be used to control the system, and the non-continuous control law is obtained, and makes the closed-loop system exponentially converge [92]. Marchand *et al.*, proposed a new transformation for the single-chain system. The static discontinuous feedback control law is obtained by switching between two smooth static feedbacks. The switching is only related to the system state and not to the time [76]. This control law realizes the real global exponential stability of the closed-loop system, and can be extended to multi-chain systems. Xu *et al.* first transformed the chain system into an extended Brockett non-integral form through state transformation, and then transformed it into a linear time-invariant system through discontinuous coordinate transformation. The control law was designed according to the arrival conditions in sliding mode control, thereby the n-dimensional chain stabilization problem of the formula system is transformed into the pole placement problem of the n-3 dimensional linear time invariant system, which simplifies the control design [5, 109]. Paper studied the stabilization of nonholonomic systems when the nonholonomic constraints were destroyed, and designed the control law using discrete sliding mode control to ensure the robustness and realizability of the system against disturbances [21].

In short, there are many ways to realize discontinuous or piecewise continuous feedback control. The main difference lies in the coordinate transformation used, the processing method of the singularity, the corresponding control law design and the stability technology used. The main problem is how to ensure the boundedness of each signal, how to ensure that the state will not escape the convergence domain, and the globality and rapidity of convergence. When ensuring the globality of convergence, before starting control, it is often necessary to use several switching control laws to adjust the initial configuration to the convergence domain.

(2) Time-varying stabilization law

The main disadvantage of the discontinuous control law is that it is difficult to realize physically, because many ideal discontinuous characteristics are impossible to obtain in reality. Therefore, many documents study continuous or smooth time-varying feedback. The time-varying feedback control law not only depends on the system state, but also explicitly depends on the time variable. Since smooth and time-invariant static state feedback cannot be used to stabilize an incomplete system, can smooth and time-varying state feedback be adequate? In fact, the idea of time-varying feedback control was proposed by Sontag *et al.* as early as the early 1980s [89]. In the mobile robot control, Samson first carried out the practice, and its analysis process is based on the Lyapunov stability and the so-called "signal chasing" technology [86]. Pomet gave a construction process based on the Lyapunov direct method, now called the Pomet method, which has the advantage of not only giving the time period stabilization law, but also giving the closed-loop Lyapunov function [80]. The core of the Pomet method is to impose an additional dissipation term on a stable system to achieve asymptotic stability. Subsequently, Coron and Pomet merged their respective methods and proposed a smooth stabilization law design method suitable for any controllable smooth and drift-free system [20]. Samson proposed a global asymptotic control solution for a class of general nonholonomic systems, aiming at the path tracking problem of point stabilization or fixed reference frame [85]. Literature applies the field-oriented control method, which is widely used in the field of induction motor control, to the stabilization and tracking of incomplete dual-integral systems [31]. The continuous static state feedback controller designed is simple in form and can ensure the system converges exponentially from all initial states except the origin [7].

The biggest disadvantage of the smooth time period state feedback control law is the slower convergence speed, and in general, the time period feedback control is more complicated than the discontinuous strategy, and it will lead to non-smooth oscillation orbits. Therefore, if the convergence speed is to be increased, a non-smooth or non-periodic time-varying law must be used. From the control point of view, smooth, time-varying state feedback does not seem as natural as discontinuous state feedback, but its advantage is to keep the control smooth and easy to analyze and implement. However, for high-dimensional systems, there is no general and constructive design method, which is also an active research direction in the field of mobile robot motion control today.

(3) Hybrid control law

Hybrid control laws are generally divided into upper and lower layers in structure, the lower layer is a number of continuous-time control laws (closed loop or open loop), and the upper layer is a discrete-time or discrete-event strategy. The switching of the lower-layer control law is controlled by the high-layer strategy. For example, Hespanha uses a supervisory control method to design a hybrid control law based on logic switching. When the system model is uncertain, it selects a reasonable control law from multiple candidate control strategies based on logic switching to stabilize the system. Kolmanovsky *et al.* studied the hybrid control of extended power systems. Literature designed a hybrid fault-tolerant adaptive control algorithm to model the robot as a continuous system with a high-level supervisory controller [53]. The design combines qualitative reasoning and model-based quantitative estimation to online detection and fault estimation. Morin *et al.* gave a constructive algorithm that can design an explicit hybrid control law for any smooth controllable drift-free system, and is robust to a large class of external disturbances [77].

(4) Linearization method

Conceptually, it is not appropriate to juxtapose the linearization method with the above three types of control laws. However, when designing nonlinear system control, it is a natural and simple idea to linearize the nonlinear system so as to utilize mature linear control technology. Although the stabilization problem of nonholonomic systems cannot be solved by the conventional Taylor approximate linearization method, several other linearization methods have been successfully applied.

The first is dynamic feedback linearization technology. The dynamic feedback control law inevitably leads to the linearization of the whole state. By introducing more control, each state is controlled, which can ensure the exponential stability. Woernler pointed out that they have a common property (deficiency): when the robot's forward speed is zero, the dynamic feedback control law is singular, that is, once the robot stops, this type of control law will not restart it [103].

The second is the differential flat technique [37]. In order to design a unified control law for the entire motion (including the static position) of the drift-free system, Fliess *et al.* proposed a method that uses the time re-parameterization and the motion planning properties of the differential flat system. The flatness of the output makes the whole state can be linearized. This method has the advantage that state variables and inputs can be represented algebraically as outputs and their derivatives up to a certain order. Combining the reference output and its derivatives, an asymptotically stable tracking control law can be designed through static state feedback, and if I/O linearization is used, dynamic feedback must be used. It has been proved that differential flatness and dynamic feedback linearization are essentially the same [38].

The third method is the non-regular static feedback linearization technique [91]. For example, the previously mentioned literature [92] proposed a non-smooth version of the non-regular feedback linearity, and obtained the non-regular feedback linearization criterion, and then proved that the chain system can be controlled by the non-regular feedback linearization. In this way, the feedback controller can be designed using standard techniques for linear systems.

The fourth method worthy of attention is to achieve homogeneous linearization (or homogeneous approximation) through the homogeneous function method. Because the shortcoming of smooth time-varying feedback is that the convergence rate is too slow, in order to improve its convergence rate, M'Closkey and Murray discussed the application of homogeneous approximation and homogeneous feedback in the control of drift-free systems, and achieved "ρ-exponential stability." Their work has inspired many other studies in this field. Homogeneous system theory gives a satisfactory answer to increasing the system convergence rate, because any analytically controllable affine drift-free system can always be approximated by a homogeneous nilpotent system. In fact, many physical systems subject to nonholonomic constraints satisfy these conditions. Under the action of the feedback law of homogeneous asymptotic stabilization, the homogeneous approximation system is exponentially stabilized in the sense of the homogeneous norm, and the closed-loop system becomes a zero-order homogeneous system. Homogeneous linearization retains the controllability and other properties of the original system, which is significantly different from Taylor's linear approximation.

2. Trajectory tracking and road following

For nonholonomic systems, trajectory tracking and path following are simpler than the stabilization problem, and the control technology used can be borrowed from the stabilization problem [4,61]. Thuilot *et al.* pointed out that as long as the reference trajectory does not contain a static configuration, that is, the system does not stop moving, the tracking problem will not encounter situations that do not meet Brockett's necessary conditions. Therefore, classic nonlinear control methods can be used for nonholonomic system tracking problems, such as Taylor linearization along a reference trajectory, etc. If you need a global result, you can turn to Lyapunov stability technology. Because there are not too many restrictions on the available nonlinear control technology, the literature on trajectory tracking problems is very abundant [66].

Samson *et al.* gave a global time-varying tracking control law under the condition of continuous excitation, which is one of the earliest global results [84]. Jiang *et al.* combined the integrator backstepping and time-varying feedback to respectively propose an exponentially convergent local and global robust tracking control law. Through its local and global controller design process, it can be seen that the design method of backstepping is very flexible. It is to explore the triangular structure or the form of strict feedback. In [56], when the control input is saturated, the global stabilization law and the global trajectory tracking control law are designed based on the kinematics model of the mobile robot. The proposed model-based control design strategy adopts the concept of passivity and regularization method, the saturated and Lipschitz continuous global time-varying feedback control law is obtained. For the trajectory tracking control of general incomplete chain systems, Lefeber *et al.* respectively gave linear controllers based on state feedback and dynamic output feedback, and applied them to wheeled mobile robots [64]. The design process is very clever, and input saturation is taken into account at the same time. The closed-loop system has global. κ-exponential stability performance. Kim *et al.* used the I/O linearization method. Because the number of inputs and outputs are not the same, they used the generalized inverse in the sense of least squares when solving the inverse matrix of the decoupling matrix. In [98], the problem of full-state tracking of nonholonomic wheeled mobile robots was studied, and the output tracking control law was designed. Its control law design is based on the dynamic model derived from the full-state tracking error, which is conducive to a better understanding of the internal dynamic subsystem and the zero dynamic subsystem of the tracking error model. Aiming at the trajectory tracking problem of a four-wheel differential drive mobile robot in a field environment, Caracciolo *et al.* derived a dynamic model that considers wheel skidding, and the controller design uses a dynamic feedback linearization method. Different from the general literature, this literature artificially adds an operable nonholonomic constraint in the controller design, so that the behavior of the instantaneous rotation center of the car body can be predicted to prevent additional sideslip. In [90], an adaptive decoupling control algorithm was given to a differential drive wheeled mobile robot to synchronize the two drive wheels to keep the robot on the desired trajectory. The uniqueness of this method is that it directly controls the movement of the driving wheel, rather than the configuration of the robot in the conventional method, so there is no need to directly consider non-holonomic constraints in the control design. Consider that there are both continuous and discrete signals in the feedback information. The inverse optimal method is also a promising method in the design of tracking controllers, which deserves attention [39]. For other issues regarding trajectory tracking and control, please refer to the literatures [67, 69, 114].

In recent years, the theories and methods of trajectory tracking of mobile robots have achieved fruitful results. Among them are mainly based on sparse representation and model: target tracking based on appearance modeling and sparse representation [17, 116], robust target tracking based on incremental subspace dynamic sparse model [54], using sparse prototypes for online target tracking [99], discriminative subspace learning based on sparse representation view robust visual tracking [107], Meanshift-based visual system real-time target tracking [15], based on spatio-temporal context anti-occlusion visual tracking [71], visual target tracking based on multi-feature hybrid model [59, 117], robust incremental visual tracking based on sample and pixel weighting strategy [22], and adaptive based on multiple features combined real-scene robust tracking [16] and so on. In addition, there are target tracking using learning feature manifold [44], schema online tracking learned by convolutional neural network [47], target tracking based on partial least squares [100] and automatic tracking of traffic signs based on particle filtering [106] and so on. Due to space limitations, no further introduction is given here. Interested readers are invited to refer to relevant literature.

3. Uncertain non-holonomic system control

The nonholonomic dynamic system has a strong application background, strong coupling and highly nonlinear characteristics, and is a very complex multiple-input multiple-output system. Due to the inaccuracy of measurement and modeling, coupled with load changes and external interference, it is difficult to obtain an accurate and complete system model. Therefore, feedback control laws based on accurate models have great limitations in practical applications, and it is more practical to study uncertain non-holonomic systems. The uncertainty of the actual non-holonomic system can be divided into two categories, namely parameter uncertainty and non-parametric uncertainty. The parameter uncertainty mainly includes the unknown or partially unknown parameters such as the mass and inertia of the system; the non-parametric uncertainty mainly refers to the external interference, the dead zone of the actuator, the gap, the measurement noise, and some unmodeled dynamics. From the late 1990s, especially after 2000, many researchers began to pay attention to this problem and devoted themselves to the control research of uncertain incomplete systems. The focus of related work is mainly to solve the practical problems of system model uncertainty, external interference and signal noise pollution, input limitation, and turning radius limitation, and carry out related robust and adaptive control and filter design. The diversity of the literature on the stabilization and tracking of uncertain non-holonomic systems is mainly due to the use of different models for uncertainty or interference, and the use of different processing methods to achieve robustness or adaptability. The main methods for designing uncertain non-holonomic dynamic systems include adaptive control, robust control, robust adaptive control, and intelligent control [23].

(1) Adaptive control

There are many literatures on the use of adaptive control methods to solve the problem of trajectory or path tracking of mobile robots [26, 52]. Lefeber *et al.* took a car-like wheeled mobile robot as an example, and focused on the mathematical description of the adaptive state feedback tracking control problem of a non-holonomic system with uncertain parameters. Colbaugh *et al.* used homogeneous system theory and adaptive control to study the stabilization of uncertain mobile manipulators. The control strategy is divided into two parts: the low-level homogeneous kinematics stabilization strategy provides the desired speed signal, and the high-level adaptive strategy achieves accurate speed tracking. This method can realize the precise stabilization of any desired configuration, and the detailed information of the system dynamics model is not required when the controller is specifically implemented.

One of the foundations of linear adaptive system design is the separation principle, but generally speaking, the separation principle is not applicable to nonlinear systems. The theoretical contribution of [24, 25] is the introduction of modular design, which separates the controller and parameter update law under nonlinear conditions. This processing method brings greater design flexibility and is conducive to the improvement of the performance of the system transition process.

By selecting variables directly related to linear velocity and angular velocity, Huang obtains a simplified dynamic model with good properties. The coupling between the two wheels is considered in the control design, and the performance of the model-based adaptive controller is better than most conventional PID controllers and can achieve more accurate speed tracking [48]. The uncalibrated visual servo system will bring additional parameter uncertainty and make the robot system suffer more complicated interference. Tsai *et al.* considered car-like wheeled mobile robots, based on the homeomorphic transformation of inputs and states, provided sufficient conditions for the conversion of a nonlinear model to an oblique symmetric chain system [85], and designed a multi-level controller to realize the system asymptotic stabilization of the system [96]. In the controller design, adaptive control is used to deal with uncertain dynamic parameters, and sliding mode control is used to overcome the influence of external interference.

Dong and others have carried out fruitful research on uncertain non-holonomic dynamic systems [27–29]. Among them, the literature [29, 30] mainly considers the uncertainty of inertia parameters, assuming that the dynamics is linear to the parameters (most of the current adaptive methods have to make this assumption), and the reference trajectory satisfies the non-holonomic constraints. The adaptive controller is designed through passive nature.

Backstepping method has been a very popular nonlinear control design method in the past decade. Based on the backstepping method, the adaptive controller design of uncertain non-holonomic systems has been paid more attention. In [34], backstepping becomes the link between kinematics and dynamics, and the design of the controller is actually based on neural networks. Fukao *et al.* considered the tracking problem based on the dynamic model [40]. It is pointed out in the article that most of the work of mobile robot control is based on the kinematic model (using speed as speed to be obtained instantaneously. At this time, it is implicitly assumed that all the knowledge of the dynamic model can be obtained without interference; Based on the dynamic model, it is more realistic to use torque as the input. Therefore, they extended the results of [35] to uncertain situations and designed a tracking controller using an adaptive backstepping method. The results in the paper show that as long as an adaptive controller can be designed for a kinematic model with unknown parameters, an adaptive controller can be designed with a dynamic model with unknown parameters. Dixon *et al.* deal with the trajectory tracking of wheeled mobile robots, but the results can also be used for global adjustment of poses with slight modifications. Its characteristic is that the reference trajectory is required to meet the "continuous excitation" condition, and the article explains for the first time how the reference trajectory excitation is used to improve the transient tracking performance. Its theoretical basis is the integrator backstepping technology. Jiang *et al.* studied a chain system with strong nonlinear interference and drift terms, using discontinuous state scaling transformation and integrator backstepping technology, and designed a global exponential convergence dynamic state and output feedback switching control strategy [55]. Tayebi *et al.* used the backstepping method to design a time-invariant adaptive controller for a single-wheel mobile robot whose input was disturbed. Except for the manifold determined by the initial orientation of the robot, the controller is defined everywhere and smooth. The design process is also adaptively modified on the basis of ensuring the stability of the nominal controller index.

It should be pointed out that most of the above-mentioned adaptive controller designs assume that the uncertainty in the system can be linearized with constant coefficients. But for actual systems, this assumption is often difficult to satisfy. Chen *et al.* used adaptive fuzzy cancellation technology; first, without uncertainty and interference, design a nominal control with the required eigenvalue configuration; then, construct a fuzzy logic system to eliminate as effectively as possible non-linear uncertainty to enhance the robustness; finally, a minimax control scheme is used to optimally reduce the worst possible effect of the residual error after fuzzy cancellation to the required level, so as to achieve robustness. When solving the minimax problem, the H_∞ optimization theory based on adaptive fuzzy is adopted. Another feature of this work is that uncertain parameters in kinetics do not have to be parameter linearizable.

For other researches on the adaptive control of mobile robots, please refer to the literature [74, 115].

(2) Robust control

For systems with unmodeled dynamics, unknown external disturbances and other non-parametric uncertain systems, the use of robust control can achieve better results [97, 102]. Lucibello *et al.* gave a robust stabilization strategy for nonlinear systems and applied it to chained systems. This stabilization strategy is called "iterative state drive," which uses appropriate control laws iteratively. Assuming that a certain control law is calculated (not required be a feedback form), the state is driven from any point to a closer equilibrium point within a finite time. Under suitable assumptions, the control law is applied in an iterative manner. In an iterative cycle, the system state shifts from the current point to a point closer to the equilibrium point. In this way, the iterative use of the control law to form a time-varying feedback can stabilize the system uniformly and asymptotically to the equilibrium point, and has an exponential convergence rate. This idea is very simple and has similarities with the rolling horizon (receding horizon) and predictive control. Under the action of this strategy, non-continuous small disturbances have no effect on the system, but continuous disturbances will bring limited errors. The final controller in this document is actually a hybrid control law. In order to achieve precise path following, Chung *et al.* designed the controller into two feedback loops inside and outside. The inner loop tracks the desired speed, and the outer loop compensates for external interference to achieve precise position control. Hamel *et al.* considered the trajectory tracking and parking problems of incomplete robots, and studied the inaccurate configuration feedback information that caused disturbances to the system. The proposed controller has robust stability with respect to the state estimation error, and by determining the surrounding reference the bounded attraction domain of the trajectory and the target ensures the control accuracy. It is different from the traditional method based on extended filter, which uses a bounded error model, which is suitable for deterministic perturbation, and the assumption of dealing with bounded errors is much weaker than dealing with random errors. Because the designed controller is Lyapunov stable in a large area near the origin of the tracking error system, a bounded attraction domain can be defined around the reference trajectory, thereby ensuring and quantifying the tracking accuracy.

Jia considers four-wheel drive vehicles, in which speed, mass, inertia, and road-tire interactions are time-varying, and a dynamic model is used. First, a nonlinear I/O decoupling controller is designed to decouple the system into three second-order subsystems; a new decoupling condition is derived, and it is proved that as long as the acceleration is not too fast or the brakes are too dead, and the front and the rear wheel is not off the ground, then the vehicle with front-wheel brake or rear-wheel drive can always be decoupled. In order to reduce the influence of vehicle parameters on handling performance, a robust control strategy is proposed. The gains of the controller and the observer can be obtained by solving two algebraic Riccati equations, where the form of the observer is appropriately selected, and the robust controller will not destroy the decoupling structure of longitudinal and yaw motion.

Sliding mode control is a mature and robust control method with simple algorithm, good anti-interference performance, and fast system response speed. It has received extensive attention in non-holonomic system control. Although sliding mode control has complete robustness in theory, because its control law itself is not continuous, it is easy to cause chattering in practical applications. The general solution to flutter is to introduce a boundary layer [87]. Of course, this kind of discontinuity is sometimes needed in the field of motor control, so it can be directly applied. In fact, high-order sliding mode controllers can better eliminate chattering, and have better convergence and robustness.

(3) Adaptive robust control

In many cases, the system has both parametric and non-parametric uncertainties. At this time, it is a natural idea to use adaptive control and robust control to compensate the parameter and non-parametric uncertainties of the system respectively [51,82]. In the field of manipulator control, there is already a wealth of literature confirming this idea. However, researches on robust adaptive control for wheeled mobile robots are rare and need to be in-depth.

For uncertain non-holonomic dynamic systems, Dong *et al.* designed a simple adaptive robust stabilization controller with the help of non-singular transformations. It only needs to know the order of the system, the measurement state, and the derivative of the state, instead of detailed dynamics information. For the situation where there are both parameter and non-parameter uncertainties, Dong *et al.* used adaptive backstepping technology and neural network in the controller design. The tracking error converges to the sphere containing the origin, and the radius of the sphere can be adjusted by the control parameters. In [57], Kim *et al.* designed a robust adaptive control law when system parameters such as mass and inertia are unknown or changed. In the design, the kinematics and dynamics model factors of the system are comprehensively considered. In order to compensate for the uncertainty of dynamic parameters and external interference, a robust control term is introduced under the assumption that the uncertainty term of the system satisfies linearization to achieve accurate speed tracking. Ge *et al.* designed a robust adaptive controller for uncertain non-complete chain systems, using adaptive technology to compensate for the system's parameter uncertainty, and sliding mode control to overcome external interference. The controller forces the output of the dynamic system to track the artificially designed bounded auxiliary signal, thereby ensuring that the kinematics subsystem is stabilized to zero.

Compared with a separate adaptive controller and robust controller, the robust adaptive controller naturally has a more complex controller structure, so it puts forward higher requirements on the real-time performance of the system.

(4) Intelligent control method

The intelligent control method represented by neural network, fuzzy control, evolutionary algorithm, etc., due to its good learning function, has good robustness to interference and model parameter uncertainty; coupled with the control law designed by the intelligent control method it may be discontinuous in itself, so it can avoid the necessary conditions of Brockett and provide a new means to solve the control problem of uncertain incomplete systems. Therefore, more and more attention is paid.

(a) Neural network control

Neural network has a powerful learning function, and its fault tolerance and nonlinear approximation capabilities make it widely used in system modeling, identification and control. The neural network control method basically does not depend on the model. The application of neural networks to the control of non-holonomic systems with high nonlinearity can effectively reduce the impact of system uncertainty on the performance of the controller, thereby improving the system's ability to adapt to the environment.

(b) Fuzzy control

Fuzzy control has been widely used in the research of wheeled mobile robot control. Choomuang *et al.* combined Kalman filter and fuzzy control to improve the control accuracy under obstacle avoidance conditions [18]. Fuzzy adaptive control, fuzzy adaptive impedance control and fuzzy adaptive predictive control have been widely used in mobile robot control [81, 112, 113].

(c) Evolutionary algorithm

By simulating the evolutionary process of organisms, evolutionary algorithms enable the system to continuously develop and evolve its own structure and behavior through natural selection of the survival of the fittest, and finally have the ability to adapt to the environment and tasks. Through the genetic process of selection, crossover, and mutation, evolutionary algorithms can obtain optimal or suboptimal solutions to problems.

7.6   Examples for Stabilization and Tracking Control Design

Most of the existing documents deal with the tracking control and stabilization control of mobile robots separately, and design their own controllers. In many practical applications, it is not desirable to switch back and forth between stabilizing and tracking two different controllers. In theory, in order to ensure the stability of the system and good control quality, there is also such a demand. Therefore, the best solution is to design a single controller that is suitable for these two different types of control problems.

7.6.1 *Tracking control law design based on backstepping*

Path-tracking is one of the basic problems in mobile robot control, which is included in the implementation phase of path-planning. For situations with a less demanding accuracy for the tracking control, an open-loop control can be used. Under this control, the robot moves toward the target in a straight line or circular arc path when the heading error is small, and only rectifies its movement when the heading error exceeds a given threshold. There is no feedback control existing in the motion-control layer. Open-loop control is easily realized, but it has a greater error in path-tracking.

Usually, some approximately linearized controllers are used in path-tracking problems, such as the PID control method. In the path-tracking process, the nonlinear characteristics of the system can be ignored when the lateral error is small. Tracking control design methods can also be applied to path-tracking control problems, where the tracking method usually transforms nonlinear problems into linear problems in order to solve them efficiently.

The chained systems obtained by the method of homeomorphic transformation have achieved an ideal control effect in simulation experiments. Since the physical meaning of the control variable is not clear after homeomorphic transformation, practical tuning of the parameter is very difficult. The high-order sliding mode control method introduces higher-order derivative terms in the state variables, which are difficult to obtain in engineering practice.

1. Description of path-tracking problems

First, we transform the kinematic model of the global coordinate into a lateral error state equation, which is in polar coordinates centered on the robot. The mobile robot moves forward along the planned path. As seen in Figure 7.14, the point P is the tangent point of the circle centered on the reference point R with a radius given by the minimum distance to the tracked path; ϕ is the cut angle of the road

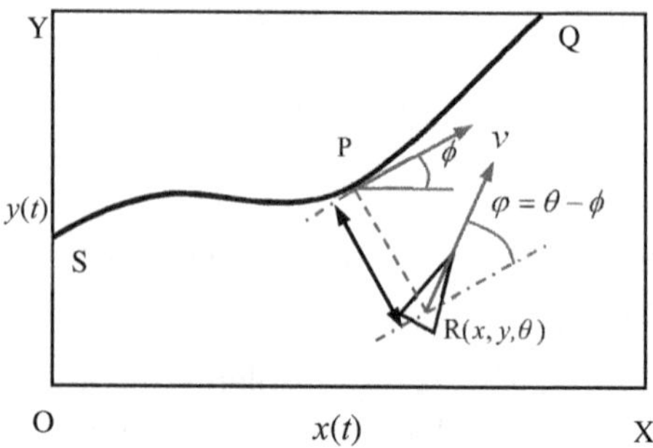

Figure 7.14.　Path-tracing schematic.

curves at the point P; and d is the distance from the point $R(x, y, \theta)$ to the point P, namely the lateral error.

Denote the reference point R as the center. If the direction of the heading angle θ is in polar coordinates, then denote the lateral angle as negative when the robot moves on the left side of the road and as positive when the robot moves on the right side of the road. Define the angle $\varphi = \theta - \phi$, which is the angle between the direction of robot motion and tangential direction of the road. Transform the state variables $\{x, y, \theta\}$ of the global coordinates into the state variables $\xi = \{d.\varphi, \varpi_1, v\}$ of the polar coordinates [36], we can obtain the following equation:

$$\begin{cases} \dot{d} = v \sin \varphi \\ \dot{\phi} = \tilde{\omega}_1 = \dot{\theta} - \dot{\phi} = \tilde{\omega} - \tilde{\omega}_0 \\ \dot{\tilde{\omega}}_1 = \hat{\alpha}_1(t) = \alpha_1(t) - \dot{\tilde{\omega}}_0 \\ \dot{v} = \alpha_0(t) \end{cases} \tag{7.8}$$

Where, $\alpha_0(t)$ is the acceleration control variable for the translational linear velocity and $\alpha_1(t)$ is the acceleration control variable for the rotation angular velocity of the robot body; $\varpi_0 = \dot{\phi}$ is the rate of change of the tangential angle at point P. The lateral error is adjusted by regulating the rotational angular acceleration of the mobile robot under the condition that the translation speed should be above $0\,\text{m/s}$. The control objective is to design a lateral error control law $\alpha_1(t) = \hat{\alpha}_1(t) + \dot{\varpi}_0$ such that a system under control can be stabilized. The state equations of the tracking error are

$$\begin{cases} z_1 = k_1 d \\ z_2 = k_2 v \sin \varphi - \beta_1, \\ z_3 = k_3 \varpi_1 - \beta_2 \end{cases} \tag{7.9}$$

Where, $k_1 > 0$, $k_2 > 0$, $k_3 > 0$ are proportional coefficients, z_1, z_2, z_3 are dimensionless quantities, and the dimensions of k_1 are $(1/m)$, of k_2 are (sec/m), and of k_3 are (sec/ard). Both β_1 and β_2 are dimensionless quantities.

2. Backstepping-based tracking control law

Backstepping method is used to decompose the complex nonlinear system into subsystems so as to construct Lyapunov function for each subsystem. Then, the results are finally extended to the entire system. The key of the method is to construct an appropriate Lyapunov function. The backstepping procedure for the system (3-73) is performed as follows:

Step 1: Construct a Lyapunov function $V_1 = \frac{1}{2} z_1^2$. By the formulas (7.8) and (7.9), the form of V_1 can be written as

$$\dot{V}_1 = z_1 \dot{z}_1 = \frac{k_1}{k_2} z_1 (z_2 + \beta_1) \tag{7.10}$$

Where,

$$\beta_1 = -c_1 \frac{k_2}{k_1} z_1 = c_1 k_2 d \quad \text{with} \quad c_1 = \text{const} \ (c_1 > 0). \tag{7.11}$$

Then

$$\dot{V}_1 = -c_1 z_1^2 + \frac{k_1}{k_2} z_1 z_2 \tag{7.12}$$

Step 2: Define another Lyapunov function as $V_2 = V_1 + \frac{1}{2} z_2^2$. Then

$$\dot{V}_2 = \dot{V}_1 + z_2 \dot{z}_2 = -c_1 z_1^2 + z_2 \left(\frac{k_1}{k_2} z_1 + \dot{z}_2 \right)$$

It follows from (3-73) that $\dot{z}_2 = k_2 \dot{v} \sin \varphi - \dot{\beta}_1 + k_2 v \dot{\varphi} \cos \varphi$.

Considering that $z_3 = k_3 \varpi_1 - \beta_2$ and $\dot{\varphi} = \varpi_1(z_3 + \beta_2)/k_3$, and substituting $\varphi, \dot{z}_2$ into the equation of $\dot{V}_2$, we obtain

$$\dot{V}_2 = -c_1 z_1^2 + z_2 \left(\frac{k_1}{k_2} z_1 + \frac{k_2}{k_3} v z_3 \cos\varphi + \frac{k_2}{k_3} v \beta_2 \cos\varphi + k_2 \dot{v} \sin\varphi - \dot{\beta}_1 \right)$$

Where,

$$\beta_2 = \frac{-c_2 z_2 - \frac{k_1}{k_2} z_1 - k_2 \dot{v} \sin\varphi + \dot{\beta}_1}{\frac{k_2}{k_3} v \cos\varphi} \tag{7.13}$$

Since $c_2 = \text{const}(c_2 > 0)$,

$$\dot{V}_2 = -c_1 z_1^2 - c_2 z_2^2 + \frac{k^2}{k_3} v z_2 z_3 \cos\varphi$$

Step 3: Define another Lyapunov function as $V = V_2 + \frac{1}{2} z_3^2$, then

$$\dot{V} = c_1 z_1^2 - c_2 z_2^2 + z_3 \left(\frac{k_2}{k_3} v z_2 \cos\varphi + \dot{z}_3 \right)$$

$$= -c_1 z_1^2 - c_2 z_2^2 + z_3 \left(\frac{k_2}{k_3} v z_2 \cos\varphi + k_3 \hat{\alpha}_1 - \dot{\beta}_2 \right)$$

$$\tag{7.14}$$

Define

$$\hat{\alpha}_1 = \frac{-c_3 z_3 - \frac{k_2}{k_3} v z_2 \cos\varphi + \dot{\beta}_2}{k_3} \tag{7.15}$$

Where, $C_3 = \text{const}(c_3 > 0)$. Substituting it into (3.78) yields

$$\dot{V}(z_1, z_2, z_3) = -c_1 z_1^2 - c_2 z_2^2 - c_3 z_3^2 \leq 0$$

with $V(z_1, z_2, z_3) = \frac{1}{2}(z_1^2 + z_2^2 + z_3^2) \geq 0$. According to the Lyapunov Theorem, this is an asymptotic stability system. Therefore, the rotational angular acceleration control law can be achieved in terms of (7.15).

In order to analyze the distribution of unstable poles of the control law, substituting intermediate variables β_1, β_2 and $\dot{\beta}_1, \dot{\beta}_2$ into equation (7.15) yields

$$\alpha_1(t) = \hat{\alpha}_1(t) + \dot{\tilde{\omega}}_0(t) = A_t = A_1(t) + A_2(t) + A_3(t)$$
$$+ A_4(t) + A_5(t) + \dot{\tilde{\omega}}_0(t), \tag{7.16}$$

Where,

$$A_1(t) = -\left(\frac{k_2}{k_3}\right)^2 v^2 \sin\varphi \cos\varphi - c_1 \left(\frac{k_2}{k_3}\right)^2 dv \cos\varphi$$

$$A_2(t) = \left[\left(\frac{\dot{v}}{v}\right)^2 - c_3\frac{\dot{v}}{v} - \left(\frac{k_1}{k_2}\right)^2 - c_1 c_2 - c_1 c_3 - c_2 c_3 - \frac{\ddot{v}}{v}\right] \tan\varphi$$

$$A_3(t) = \left[-c_3 - \frac{\dot{v}}{v\cos^2\varphi} - \frac{(c_1 + c_2)}{\cos^2\varphi}\right] \tilde{\omega}_1$$

$$A_4(t) = \left[\left(\frac{k_1}{k_2}\right)^2 + c_1 c_2\right]\left(\frac{\dot{v}}{v^2} - \frac{c_3}{v}\right)\frac{d}{\cos\varphi}$$

$$A_5(t) = \left[\left(\frac{k_1}{k_2}\right)^2 + c_1 c_2\right]\frac{\sin\varphi}{v\cos^2\varphi}d\varpi_1$$

We know that there exist singular points such as $\varphi = (360l \pm 90)^\circ$, $l = 0, \pm 1, \pm 2, \ldots$, and $v = 0$ in the control law. For the singular point $v = 0$, we assume that the translational velocity of the robot is always positive when it is not approaching the end point; if the robot is near the end point, i.e., $d = 0$, then $A_4(t)$, $A_5(t) = 0$.

If the ratio of translational velocity increment to translational velocity in unit cycle is bounded, i.e., when $|\dot{v}/v| < \gamma$ (with γ being a bounded constant), then we can guarantee that $A(t)$ is bounded. For such singular points, $\varphi = (360l \pm 90)^\circ$, $l = 0, \pm 1, \pm 2, \ldots$ the controller can be designed to guarantee that the differential value between the heading angle and cut angle is in the range of $\pm 90^\circ$.

7.6.2 *Trajectory generation method based on differential flatness for a wheeled mobile robot*

1. The concept of differential flatness
(1) Definition and properties of differential flatness
Fliess *et al.* used differential flatness as a differential-algebraic concept in paper [48] for the first time. From the viewpoint of control, for the nonlinear system

$$\dot{x} = f(x, u) \quad x \in \mathbf{R}^n, \ u \in \mathbf{R}^m$$
$$y = h(x) \quad y \in \mathbf{R}^l, \tag{7.17}$$

is differentially flat when we can find a flatness output z satisfying the following conditions:

(C1)   the dimension of z is equal to the dimension of control input u, i.e., dim $z = m$

(C2) z includes the system state variables x, the control input u, and u's finite-order derivative, i.e.,

$$z = F(x, u, \dot{u}, \dots u^{(i)}), \quad z \in \mathbf{R}^m \tag{7.18}$$

(C3)   the elements of z are differentially independent, i.e., there is no function Q such that $Q(z, \dot{z}, \dots, z^{(\alpha)}) = 0$

(C4)   the system state variables x and the input u can be represented by z and its finite-order derivative, i.e.,

$$x = x(z, \dot{z}, \dots, z^{(j)})$$
$$u = u(z, \dot{z}, \dots, z^{(j)}) \tag{7.19}$$

Several properties of the flatness system are given without proof as follows:

(1)   For a given system, the flatness output is not unique;

(2)   A controllable linear system is flat; an uncontrollable system is not flat;

(3)   A nonlinear system is flat if it can be made linear by dynamic (state) feedback.

(4)   There is one-to-one mapping existing among the flatness output z, system state x, and input u such that the motion behavior of the system can be decided by the flatness output.

(5)   is the theoretical basis of calculation of the systematic flatness output. For the following system

$$\dot{x}_{11} = f_1(x, u_1)$$
$$\dot{x}_{21} = x_{22} \tag{7.20}$$
$$\dot{x}_{22} = f_2(x, u_1, \dot{u}_1, u_2)$$

if the rank of the matrix

$$\begin{pmatrix} \dfrac{\partial \boldsymbol{f}_1}{\partial \boldsymbol{u}_1}(\boldsymbol{x},\boldsymbol{u}) & \dfrac{\partial \boldsymbol{f}_1}{\partial \boldsymbol{u}_2}(\boldsymbol{x},\boldsymbol{u}) \\ \dfrac{\partial \boldsymbol{f}_2}{\partial \boldsymbol{u}_1}(\boldsymbol{x},\boldsymbol{u}) & \dfrac{\partial \boldsymbol{f}_2}{\partial \boldsymbol{u}_2}(\boldsymbol{x},\boldsymbol{u}) \end{pmatrix}$$

is equal to the dimensions of the control input (i.e., 2), then $\boldsymbol{z} = \begin{pmatrix} x_{11} \\ x_{21} \end{pmatrix}$ is the flatness output of the system. For most systems, this calculation is not a trivial task. Therefore, in the application, we select a set of outputs satisfying the condition (C1) $\sim$ (C3) according to the physical meaning and understanding to the system dynamics, such as the position of the mechanical systems, and then verify whether condition (C4) is satisfied.

(2) Application of differential flatness in trajectory-tracking control

From the viewpoint of mathematics, differential flatness and feedback linearization are equivalent: the system that can be linearized by feedback (relative degree n) is the differential flatness system; the flatness output is the output of the feedback linearization system [149]. However, when this is applied to specific control issues, it is easy to find that they emphasize different sides of the systems. Feedback linearization is focused on coordinate transformation and finding a nonlinear state feedback to convert the nonlinear system into a linear system that is convenient for system analysis and controller design. Differential flatness is mainly used in the trajectory-tracking control in order to generate a feasible desired trajectory (including the desired state of the trajectory $\boldsymbol{x}_d$ and the desired control input $\boldsymbol{u}_d$). Thus, with this method, we do not need to convert the nonlinear system into a linear one. This application is based on the property 4) of the differential flatness system, from which we know that a desired trajectory $\boldsymbol{z}_d$ in the flatness output of the system (R^m) can be obtained when many constraints such as position, velocity, and the smoothness of the trajectory are considered. Then, the desired state trajectory $\boldsymbol{x}_d$ and the desired control input $\boldsymbol{u}_d$ of the system can be calculated by (7.19). From the aforementioned discussion, we know that the tracking controller design is divided into two parts: the trajectory-generation part (Trajectory Generation, which is the feedforward control) and the feedback-compensation

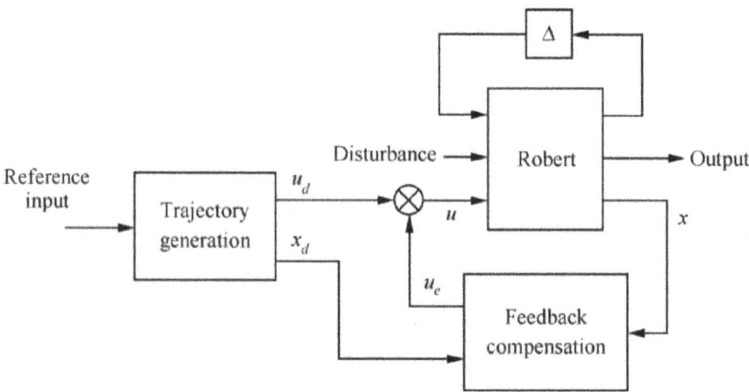

Figure 7.15.   Trajectory-tracking control strategy.

part (Feedback Compensation, which is the feedback control), as shown in Figure 7.15 (with an example of a mobile robot).

The trajectory-generation part can be accomplished by differential flatness. The feedback-compensation part is mainly used to ensure that the closed-loop system is stable, and that there is acceptable tracking performance with respect to uncertainties and disturbances. If control or state constraints exist, then the feedback section is expected to trade off between performance and constraint satisfaction. Then, the control input comprises two parts:

$$u = u_d + u_e.$$

Note 3.1 There are at least two advantages of using differential flatness theory in motion-planning and trajectory-generation problems of robots

(1) The dimension of the flatness output space m is less than the dimension of the state space n such that the trajectory-planning can be conducted in a low-dimensional space.
(2) Since differential flatness is equal to precise feedback linearization, the system dynamics can be completely reproduced in the trajectory-generation process. For a wheeled mobile robot with nonholonomic constraints, these constraints can be satisfied when differential flatness theory is used to generate an expected trajectory.

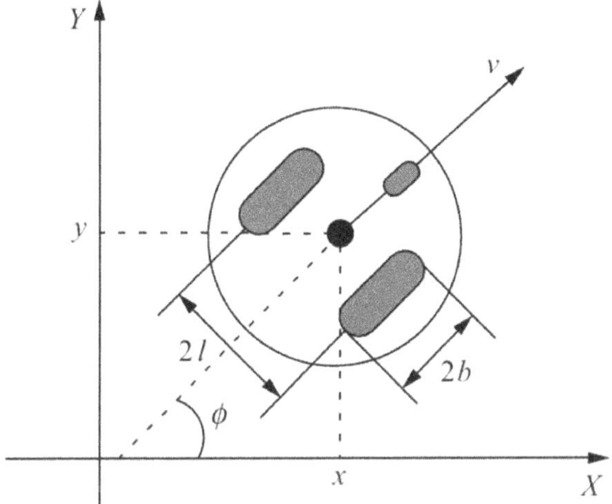

Figure 7.16. Mobile robot system.

Note 3.4 While the control or state constraints are considered, a reasonable z_d should be planned carefully to ensure that u_d and x_d satisfy the constraints.

2. Trajectory generation of a wheeled mobile robot

(1) Wheeled mobile robot model and its flatness

A wheeled mobile robot (hereafter referred to as a robot or a Wheeled Mobile Robot (WMR)) is shown in Figure 7.16. This type of robot is driven by motors with two rear-wheels, and the movement direction of the front-supporting wheel is arbitrary.

According to the kinematic laws of motion, the nonlinear differential equation of the robot can be obtained as follows:

$$\begin{cases} \dot{x} = v\cos\phi \\ \dot{y} = v\sin\phi \\ \dot{\phi} = \omega \\ \dot{v} = \beta_1(\tau_1 + \tau_2) + \beta_2 w_1 \\ \dot{\omega} = \beta_2(\tau_1 - \tau_2) + \beta_4 w_2 \end{cases} \tag{7.21}$$

where (x, y) are the location coordinates of the robot, φ is the pointing angle of the robot (i.e., the angle between the front of the robot and X-axis), and (v, ω) represent, the instantaneous velocity and angular velocity at the center of the vertical axis respectively, (τ_1, τ_2) is the driven torque provided by the motor with the two rear-wheels, and $\beta_1 = \frac{1}{bm}$, $\beta_2 = \frac{1}{bI}$, $\beta_3 = \frac{1}{m}$, $\beta_4 = \frac{1}{l}$, with m and I denoting, respectively, the mass and rotational inertia relative to the vertical axis, $2l$ representing the length of the rear axle shaft, and b representing the radius of the rear wheels; w_1 and w_2 represent the external forces and torques applied in the directions of the linear and angular velocity. respectively. In addition, the uncertainties of the system, e.g., the parametric uncertainties in β_1 and β_2, can be seen as external disturbances.

Since the output of the motor torque is limited, the control constraints are formulated as follows:

$$|\tau_j(t)| \leq \tau_{j,\max}, \quad j = 1, 2. \tag{7.22}$$

In addition, lateral sliding is not allowed during robot motion; hence, the system has the following constraint:

$$\dot{x} \sin \phi - \dot{y} \cos \phi = 0 \tag{7.23}$$

We can verify that the robot system (7.22) is a kind of differential flat system by choosing

$$z = [x, y] \tag{7.24}$$

as a candidate flat output. After a simple derivation, the following equations can be achieved:

$$\begin{cases} \phi = \arctan\left(\dfrac{\dot{y}}{\dot{x}}\right) + k\pi \\[2mm] v = \pm\sqrt{\dot{x}^2 + \dot{y}^2} \\[2mm] \omega = \dfrac{\dot{x}\ddot{y} - \ddot{x}\dot{y}}{\dot{x}^2 + \dot{y}^2} \\[2mm] \tau_1 = \dfrac{u_1 + u_2}{2} \\[2mm] \tau_2 = \dfrac{u_1 - u_2}{2} \end{cases} \tag{7.25}$$

Where

$$u_1 = \frac{\dot{x} \cdot \ddot{x} + \dot{y} \cdot \ddot{y}}{\beta_1 \sqrt{\dot{x}^2 + \dot{y}^2}}$$

$$u_2 = \frac{1}{\beta_2} \left(\frac{\dot{x} \cdot \dddot{y} - \dddot{x} \cdot \dot{y}}{\dot{x}^2 + \dot{y}^2} - \frac{2(\dot{x} \cdot \ddot{y} - \ddot{x} \cdot \dot{y})(\dot{x} \cdot \ddot{x} + \dot{y} \cdot \ddot{y})}{(\dot{x}^2 + \dot{y}^2)^2} \right)$$

$$(7.26)$$

The equations (7.25) and (7.26) show that all the state and input variables can be represented by (3-88) and its finite-order derivatives. Thus, the output of the system $z = [x, y]$ is flat.

(2) Example of Trajectory-Generation Based on a Differential Flat WMR Algorithm

Suppose we expect the WMR to track an "8"-shaped trajectory defined by the following formula

$$x_d(t) = a_1 \sin \frac{t}{a_3}, \quad y_d(t) = a_2 \sin \frac{t}{2a_3}. \tag{7.27}$$

At the same time, the generated trajectory $\boldsymbol{\xi}_d := [x_d, y_d, \varphi_d, v_d, \omega_d]$ satisfies the nonholonomic constraint (7.23) and the generated feedforward control $(\boldsymbol{\tau}_{1d}, \boldsymbol{\tau}_{2d})$ satisfies the constraints (7.22).

According to Note 3.3, by substituting (7.27) into (7.25) and (7.26), we can obtain $\boldsymbol{\xi}_d$ and $(\boldsymbol{\tau}_{1d}, \boldsymbol{\tau}_{2d})$, where $\boldsymbol{\xi}_d$ satisfies the nonholonomic constraint (7.23). As they are functions of (a_1, a_2, a_3, t), whether the feedforward input $(\boldsymbol{\tau}_{1d}, \boldsymbol{\tau}_{2d})$ satisfies (7.22) depends on the selection of parameters (a_1, a_2, a_3) in (7.27). Therefore, an expression for $(\boldsymbol{\tau}_{1d}, \boldsymbol{\tau}_{2d})$ is given as follows:

$$\begin{cases} \tau_{1d} = \dfrac{u_{1d} + u_{2d}}{2} \\ \tau_{2d} = \dfrac{u_{1d} - u_{2d}}{2} \end{cases} \tag{7.28}$$

$$u_{1d} = \frac{\dot{x}_d \cdot \ddot{x}_d + \dot{y}_d \cdot \ddot{y}_d}{\beta_1 \sqrt{\dot{x}_d^2 + \dot{y}_d^2}}$$

$$u_{2d} = \frac{1}{\beta_2} \left(\frac{\dot{x}_d \cdot \dddot{y}_d - \dddot{x}_d \cdot \dot{y}_d}{\dot{x}_d^2 + \dot{y}_d^2} - \frac{2(\dot{x}_d \cdot \ddot{y}_d - \ddot{x}_d \cdot \dot{y}_d)(\dot{x}_d \cdot \ddot{x}_d + \dot{y}_d \cdot \ddot{y}_d)}{(\dot{x}_d^2 + \dot{y}_d^2)^2} \right)$$

$$(7.29)$$

From Figure 7.17, notice that parameter a_1 determines the width of the transverse axis of the "8"-shaped trajectory, a_2 determines the

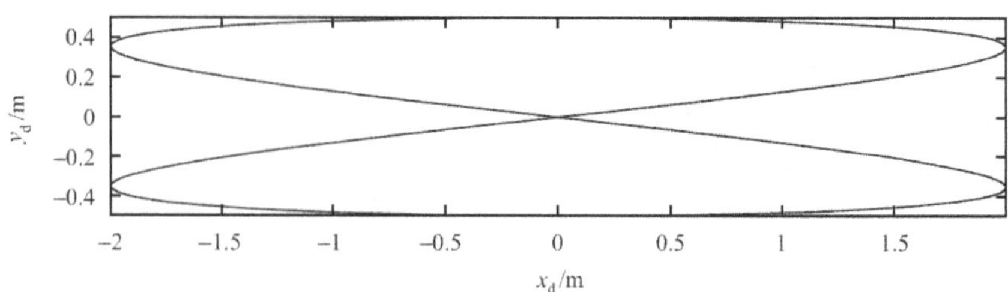

Figure 7.17.   "8"-shaped trajectory.

height of the vertical axis, and a_3 determines how long it will take to complete an "8"-shaped trajectory (7.27).

It is easy to prove that when the parameters (a_1, a_2) are equal, a smaller value of a_3 requires the velocity of the robot to be faster, implying that the control input easily exceeds its constraints (especially in the corner).

The parameters $\boldsymbol{\tau}_{1d}$ and $\boldsymbol{\tau}_{2d}$ in the "8"-shaped trajectory have a symmetrical variation; hence, it is sufficient to study one of them such as $\boldsymbol{\tau}_{1d}$. The allowed minimum a_3 can be obtained by the following procedure:

(1) For given parameters (a_1, a_2), the formula of $\boldsymbol{\tau}_{1d}$ can be obtained from (7.28) and (7.29). For convenience of the derivation, denote $\boldsymbol{\tau}_{1d}$ as

$$\boldsymbol{\tau}_{1d} = \boldsymbol{\tau}(a_3, t). \tag{7.30}$$

(2) Solve the following equation to determine the maximum of $\boldsymbol{\tau}_{1d}$,

$$\frac{\partial \boldsymbol{\tau}_{1d}}{\partial t} = 0, \quad \frac{\partial^2 \boldsymbol{\tau}_{1d}}{\partial^2 t} < 0 \tag{7.31}$$

For simplicity, denote the maximum of $\boldsymbol{\tau}_{1d}$ as $t_{\max}$, where $t_{\max} = T(a_3) > 0$.

(3) Substituting $t_{\max} = T(a_3)$ into (7.30), we can obtain the following equation:

$$\boldsymbol{\tau}_{1d}(t_{\max}) = \boldsymbol{\tau}(a_3, t_{\max}) = \tilde{\boldsymbol{\tau}}(a_3) \tag{7.32}$$

Let $\boldsymbol{\tau}_{1d}(t_{\max}) = \tilde{\boldsymbol{\tau}}(a_3) = \tau_{1,\max}$, then the allowed minimum of $a_{3,\min}$ can be solved.

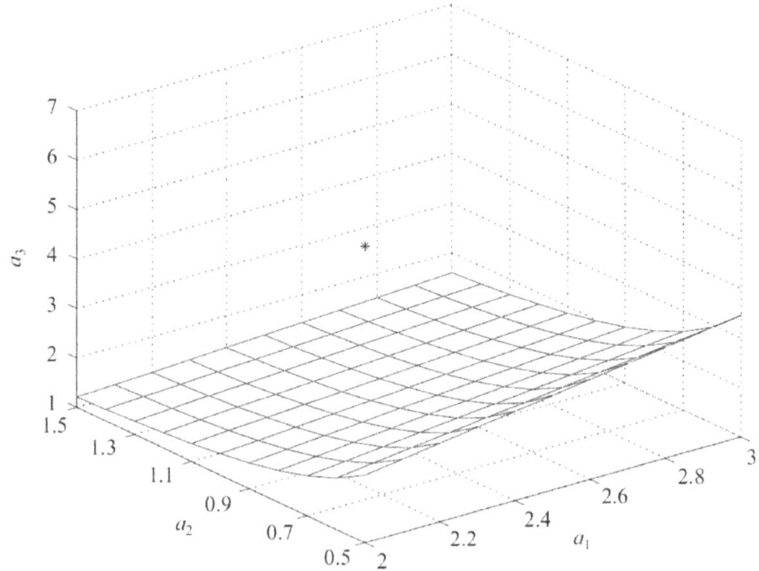

Figure 7.18.   Feasible solution maps of (a_1, a_2, a_3).

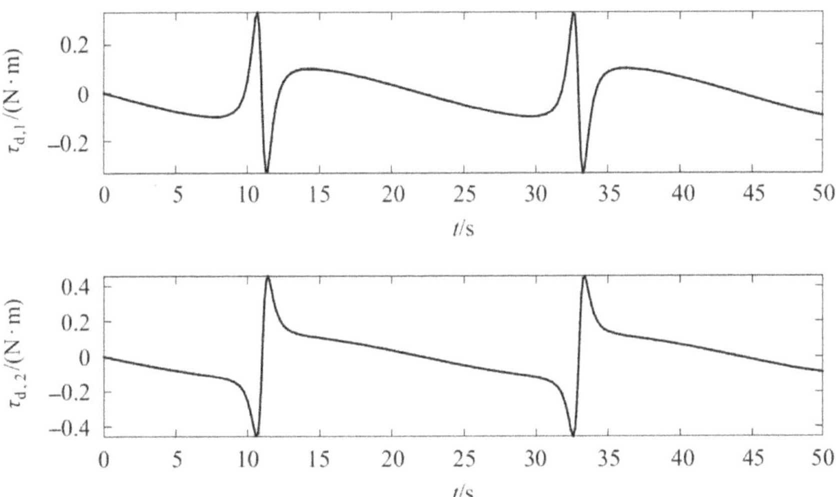

Figure 7.19.   Nominal control input (τ_{1d}, τ_{2d}) as functions of time.

(4) Change (a_1, a_2), and repeat the aforementioned steps to get a group of $(a_1, a_2, a_{3,\min})$. Then any selected $(a_1, a_2, a_3 \geq a_{3,\min})$ force (τ_{1d}, τ_{2d}) can satisfy the control constraints (7.22). For example, suppose $\tau_{1,\max} = \tau_{2,\max} = 4\mathrm{N} \cdot \mathrm{m}$, and select $a_1 = [0.5, 0.6, 0.7 \cdots 1.5]_{1 \times 11}$, $a_2 = [2, 2.1, 2.2 \cdots 3]_{1 \times 11}$. Then, 118 points of $(a_1, a_2, a_{3,\min})$ are generated following the above steps with the combination of $a_{1,i}$ and $a_{2,j}$, and the surface with the 118 points is shown in Figure 7.18.

The control input (τ_{1d}, τ_{2d}) satisfies (7.22), which corresponds to the trajectory (x_d, y_d) determined by the point on the surface. For example, select $(a_1, a_2, a_3) = (2, 0.5, 7)$, which is shown by "*" in Figure 7.18. The related "8"-shaped trajectory determined by the combination is shown in Figure 7.17, and the nominal control input evolutions of τ_{1d} and τ_{2d}, which satisfy (7.22), are shown in Figure 7.19.

7.7　Chapter Summary

In this chapter the architecture and dynamics model of mobile robot are studied. Design of an intelligent control system first needs to consider determining an appropriate architecture to achieve reasonable coordination between system modules, and to have openness and scalability in the software and hardware of the system.

Section 7.1 discusses architecture of mobile robots that is usually divided into three types: hierarchical architecture, reactive architecture, and deliberative/reactive architecture.

Section 7.2 investigates the composition of mobile robot prototype system including the mobile system and sensor system of mobile robot, and the software and hardware structure of the mobile robot system.

Section 7.3 introduces and analyzes the architecture of four-layer hierarchical mobile robot navigation including the navigation task decomposition and navigation system, the structure of the fourth-order modular automobile autonomous driving navigation system, layer structure of driving navigation system, and the structural characteristics of each layer of the driving navigation system.

Section 7.4 analyzes dynamic model of wheeled mobile robot that includes typical mechanism of wheeled mobile robot, and dynamic model of wheeled mobile robot under nonholonomic constraints.

Section 7.5 studies the stabilization and tracking of wheeled mobile robots, that is the stabilization and tracking controller design for wheeled mobile robots, and the research on related issues of stabilization and tracking controllers.

Section 7.6 gives and analyzes stabilization and tracking control design examples involving the design of tracking control law based on backstepping, the wheeled mobile robot trajectory generation based on differential flatness, and the trajectory tracking control of wheeled mobile robot based on T-S fuzzy model.

References

[1] Arkin, R.C. and Balch, T. (1997). Aura: Principles and practice in review. *Journal of Experimental & Theoretical Artificial Intelligence*, 9(2/3):175–188.

[2] Arkin, R.C. (1990). Integrating behavioral, perceptual and world knowledge in reactive navigation. *Robotics and Autonomous Systems*, 6(1–2):105–122.

[3] Arkin, R.C. (1989). Motor schema-based mobile robot navigation. *Int. Journal of Robotics Research*, 8(4):92–112.

[4] Bi, F.K., Lei, M.Y., Wang, Y.P., *et al.* (2019). Remote sensing target tracking in UAV aerial video based on saliency enhanced MDnet, *IEEE Access*, 7:76731–76740.

[5] Bloch, A. and Drakunov, S. (1996). Stabilization and tracking in the nonholonomic integrator via sliding modes. *Systems & Control Letters*, 29(2):91–99.

[6] Bonasso, R.P. (1991). Integrating reaction plans and layered competences through synchronous control. *12th Proc. of the Int. Joint Conf. on Artificial Intelligence*. Sydney, v2: 1225–1231.

[7] Brockett, R.W. (1983). *Asymptotic Stability and Feedback Stabilization*. Boston: Birkhauser.

[8] Brooks, R.A. (1986). Robust layered control system for a mobile robot. *IEEE Journal of Robotics and Automation*, 2(1):14–23.

[9] Cai, Z.X. and Xu, G.Y. (2004). *Artificial Intelligence and Its Applications (Postgraduate's Book)*. Beijing: Tsinghua University Press (in Chinese).

[10] Cai, Z.X. (2004). *Intelligent Control*, 2nd Edition. Beijing: Publishing House of Electronics Industry (in Chinese).

[11] Cai, Z.X., Zou, X.B. and Wang, L., *et al.* (2004). A Research on Mobile Robot Navigation Control in Unknown Environment

Objectives, Design and Experience. 2004 Korea-Sino Symposium on Intelligent Systems. Busan, 10:57–63.

[12] Cai, Z.X., He, H.G. and Chen, H. (2009). *Theories and Methods of Navigation and Control of Mobile Robots in Unknown Environments.* Beijing: Science Press (in Chinese).

[13] Cai, Z.X., Zhou, Xiang., Li, M.Y., *et al.* (2000). Evolutionary control architecture of autonomous mobile robot based on function/behavior integration. *Robot*, 22(3):169–175 (in Chinese).

[14] Cai, Z.X. and Wang, Y. (2014). *Principles, Algorithms and Applications of Intelligent Systems.* Beijing: Mechanical Industry Press (in Chinese).

[15] Chen, K., Fu, S., Song, K., *et al.* (2012). *A meanshift based imbedded computer vision system design for real-time target tracking 7th International Conference on Computer Science & Education*, Melbourne.

[16] Chen, W., Cao, L. and Zhang, J. (2013). An Adaptive Combination of Multiple Features for Robust Tracking in Real Scene. In *Proceedings of the IEEE International Conference on Computer Vision*. Sydney, pp. 129–136.

[17] Chen, F., Wang, Q., Wang, S., *et al.* (2011). Object tracking via appearance modeling and sparse representation. *Image and Vision Computing*, 29(11):786–796.

[18] Choomuang, R. and Afzulpurkar, N. (2005). Hybrid Kalman filter/Fuzzy logic based position control of autonomous mobile robot. *International Journal of Advanced Robotic Systems*, 2(3):197–208.

[19] Connell, J.H. (1992). SSS: A hybrid Architecture Applied to Robot Navigation. ICRA1992. *Nice*, v3:2719–2724.

[20] Coron, J.M. and Pomet, J.B. (1992). A remark on the design of time-varying stabilizing control laws for controllable systems without drift. IFAC Nonlinear Control Systems Design Symposium. (NOLCOS). Bordeaux, France, pp. 397–401.

[21] Corradini, M.L., Leo, T. and Orlando, G. (1999). Robust stabilization of a mobile robot violating the nonholonomic constraint via quasi-sliding modes. Proceedings of the American Control Conference (ACC99). San Diego, CA, USA, pp. 3935–3939.

[22] Cruz-Mota, J., Bierlaire, M. and Thiran, J. (2013). Sample and pixel weighting strategies for Robust incremental visual tracking. *IEEE Transactions on Circuits and Systems for Video Technology*, 23(5):898–911.

[23] Cui, J., Zhao, L., Yu, J.P., *et al.* (2019). Neural network-based adaptive finite-time consensus tracking control for multiple autonomous underwater vehicles, *IEEE Access*, 7:33064–33074.

[24] Dixon, W.E., De, Q.M. and Daoson, D.M. (2002). Adaptive tracking and regulation of a wheeled mobile robot with controller/update law modularity. *Proceedings of the 2002 IEEE International Conference on Robotics and Automation (IROS02)*. Washington DC, United States, pp. 2620–2625.

[25] Dixon, W.E., De, Queiroz, M., Dawson, D.M., *et al.* (2004). Adaptive tracking and regulation of a wheeled mobile robot with controller/ update law modularity. *IEEE Transactions on Control Systems Technology*, 12(1):138–147.

[26] Dogan, K.M., Gruenwald, B.C., Yucelen, T., *et al.* (2019). Distributed adaptive control and stability verification for linear multiagent systems with heterogeneous actuator dynamics and system uncertainties. *International Journal of Control*, 92(11):2620–2638.

[27] Dong, W. and Huo, W. (1999). Adaptive stabilization of uncertain dynamic non-holonomic systems. *International Journal of Control*, 72(18):1689–1700.

[28] Dong, W. (2002). On trajectory and force tracking control of constrained mobile manipulators with parameter uncertainty. *Automatica*, 38(9):1475–1484.

[29] Dong, W., Huo, W., Tso, S.K., *et al.* (2000). Tracking control of uncertain dynamic nonholonomic system and its application to wheeled mobile robots. *IEEE Transactions on Robotics and Automation*, 16(6):870–874.

[30] Dong, W., Xu, W.L., and Huo, W. (1999). Trajectory tracking control of dynamic non-holonomic systems with unknown dynamics. *International Journal of Robust and Nonlinear Control*, 9(13): 905–922.

[31] Escobar, G., Ortega, R. and Reyhanoglu, M. (1998). Regulation and tracking of the nonholonomic double integrator: A field-oriented control approach. *Automatica*, 34(1):125–131.

[32] Estier T., Crausaz Y., Merminod, B., *et al.* (2000). An Innovative Space Rover with Extended Climbing Alilities. *Proc. of Space and Robotics 2000*. Albuquerque, v2:201–206.

[33] Fan, Y. and Tan, M. (1999). Current status and prospects of robot controllers. *Robot*, 21(1):75–80 (in Chinese).

[34] Fierro, R. and Lewis, F.L. (1997). Control of a nonholonomic mobile robot: backstepping kinematics into dynamics. *Journal of Robotic Systems*, 14(3):149–163.

[35] Fierro, R. and Lewis, F.L. (1995). Control of a nonholonomic mobile robot: backstepping kinematics into dynamics. *Proceedings of the 34th IEEE Conference on Decision and Control*. pp. 3805–3810.

[36] Fliess, M., Levine, J., Martin, P., *et al.* (1995). Flatness and defect of non-linear systems: Introductory theory and examples. *International Journal of Control*, 61(6):1327–1361.

[37] Fliess, M., Levine, J., Martin, P., *et al.* (1997). Controlling nonlinear systems by flatness Systems and control in the Twenty-First Century, In *Progress in Systems and Control Theory*, Byrnes, C.I., Datta, B.N., Gilliam, D.S., *et al.* (Eds.), Birkhauser: Springer.

[38] Fliess, M., Levine, J., Martin, P., *et al.* (1995). Flatness and dynamic feedback linearizability: Two approaches. *Proc. of the 3rd European Control Conf. Rome*, Italy, pp. 726–732.

[39] Fukao, T. (2004). Inverse optimal tracking control of a nonholonomic mobile robot. *Proceedings of the 2004 IEEE/RSJ International Conference on Intelligent Robots and Systems (IROS)*. Sendai, Japan, pp. 1475–1480.

[40] Fukao, T., Nakagawa, H. and Adachi, N. (2000). Adaptive tracking control of a nonholonomic mobile robot. *IEEE Transactions on Robotics and Automation*, 16(5):609–615.

[41] Gat, E. (1991). Robust Low-computation Sensor-driven Control for Task-directed Navigation ICRA1991. *Sacramento*, v3:2484–2489.

[42] Gat, E. (1998). Three-Layer Architectures. In *Artificial Intelligence and Mobile Robots*. Kortenkamp, D., Bonasso, R.P., Murphy, R. (Eds.), MIT Press, pp. 195–110.

[43] Goldstein, H. (1980). *Classical Mechanics*. Reading: Addison-Wesley.

[44] Guo, Y., Chen, and Tang, F. (2014). Object tracking using learned feature manifolds. *Computer Vision and Image Understanding*, 118(1):128–139.

[45] Hartley, R. and Pipitone, F. (1991). Experiments with the Subsumption Architecture. ICRA1991. *Sacramento*, 2:1652–1658.

[46] He, H.G., Sun, Z.P. and Xu, X. (2016). *Prospects of autonomous vehicle driving technology under intelligent transportation conditions*. National Science Foundation of China (in Chinese).

[47] Hon, G.S., You, T., Kwak, S., *et al.* (2015). Online tracking by learning discriminative saliency map with convolutional neural network Proceedings of the 32th International Conference on Machine Learning, Lille.

[48] Huang, L. (2005). Speed control of differentially driven wheeled mobile robots — Model-based adaptive approach. *Journal of Robotic Systems*, 22(6):323–332.

[49] Huang, W.L. (2016). Intelligent vehicle environment perception technology and platform construction. *Single Chip Microcomputer and Embedded Application*, (8):9–13 (in Chinese).

[50] Iagnemma, K., Rzepniewski, A., Dubwsky, S., *et al.* (2003). Control of Robotic vehicles with actively articulated suspensions in rough terrain. *Autonomous Robots*, 14(1):5–16.

[51] Ioannou, P.A. and Sun, J. (1995). *Robust adaptive control.* Englewood Cliffs, New Jersey: Prentice-Hall.

[52] Jhong, B.-G., Yang, Z.-X. and Chen, M.-Y. (2018). An Adaptive Controller with Dynamic Gain Adjustment for Wheeled Mobile Robot, Joint 10th International Conference on Soft Computing and Intelligent Systems (SCIS). 19th International Symposium on Advanced Intelligent Systems (ISIS). Toyama, Japan, Dec 05-08, pp. 592–596.

[53] Ji, M., Zhang, Z., Biswas, G., *et al.* (2003). Hybrid fault adaptive control of a wheeled mobile robot. *IEEE/ASME Transactions on Mechatronics*, 8(2):226–233.

[54] Ji, Z., Wang, W. and Xu, N. (2014). Robust object tracking via incremental subspace dynamic sparse model. In *Proceedings of the IEEE International Conference on Multimedia and Expo*. Chengdu, pp. 1–6.

[55] Jiang, Z.P. (2000). Robust exponential regulation of nonholonomic systems with uncertainties. *Automatica*, 36(2):189–209.

[56] Jiang, Z.P., Lefeber, E. and Nijmeijer, H. (2001). Saturated stabilization and tracking of a nonholonomic mobile robot. *Systems & Control Letters*, 42(5):327–332.

[57] Kim, M.S., Shin, J.H., Hong, S.G, *et al.* (2003). Designing a robust adaptive dynamic controller for nonholonomic mobile robots under modeling uncertainty and disturbances. *Mechatronics*, 13(5): 507–519.

[58] Lafferriere, G.A. and Sontag, E.D. (1991). Remarks on control Lyapunov functions for discontinuous stabilizing feedback. Proceedings of the 30th IEEE Conference on Decision and Control, pp. 1115–1120.

[59] Lan, L., Zhang, X. and Luo. Z.G. (2017). Status and development of visual target tracking. *National Defense Science and Technology*, 38(5):12–18 (in Chinese).

[60] Laubach, S.L., Burdick, J. and Matthies, L. (1998). Autonomous Path Planner Implemented on the Rocky 7 Prototype Microrover. ICRA1998. *Leuven*, 1:292–297.

[61] Lee, J.G., Ryu, S.W. and Kim, T.W., *et al.* (2018). Learning-based Path Tracking Control of a Flapping-wing Micro Air Vehicle, 25th IEEE/RSJ International Conference on Intelligent Robots and Systems (IROS), Madrid, Spain, Oct 01–05, pp. 7096–7102.

[62] Lee, S.J. and Lee G.K. (1996). Studies in Guidance, Navigation and Control for an Articulated-body Mars Rover Testbed. Proc. of the

1996 2nd Specialty Conf. on Robotics for Challenging Environments. Albuquerque, 157–163.

[63] Lee, T.C. (2003). Exponential stabilization for nonlinear systems with applications to nonholonomic systems. *Automatica*, 39(6):1045–1051.

[64] Lefeber, E., Robertsson, A. and Nijmeijer, H. (2000). Linear controllers for exponential tracking of systems in chained-form. *International Journal of Robust and Nonlinear Control*, 10(4):243–263.

[65] Li, F.H. and Zhu, D.Q. (2001). *Electrical Engineering*. Beijing: Science Press (in Chinese).

[66] Li, X., Hu, W., Shen, C., *et al.* (2013). A survey of appearance models in visual object tracking. *ACM Transactions on Intelligent Systems and Technology (TIST)*, 4(4):58.

[67] Li, Z.C., Jing, X.J. and Yu, J.Y. (2019). Trajectory Tracking Control of a Tracked Mobile Robot with a Passive Bio-inspired Suspension, IEEE International Conference on Mechatronics (ICM), Tech Univ Ilmenau, Ilmenau, GERMANY, MAR 18–20, pp. 114–119.

[68] Li, Z.J., Luo, Q. and Lu, T.S. (2003). Research on robot agent control structure based on hybrid structure. *Computer Engineering and Application*, 39(21):90–93 (in Chinese).

[69] Liu, C. (2019). Neural-Network-Based Distributed Formation Tracking Control of Marine Vessels With Heterogeneous Hydrodynamics. *IEEE Access*, 7:150141–150149.

[70] Liu, J. (2003). *Research on Navigation Mechanism of Mobile Robot Based on Spatio-temporal Information and Cognitive Model*. Changsha: Central South University (in Chinese).

[71] Liu, W.J., Dong, S.H. and Qu, H.C. (2016). Space-time context anti-occlusion visual tracking. *Chinese Journal of Image and Graphics*, 21(8):1057–1067 (in Chinese).

[72] Luca, A.D., Oriolo, G. and Vendittelli, M. (2001). Control of wheeled mobile robots: An experimental overview. RAMSETE — Articulated and Mobile Robotics for Services and Technologies, Nicosia, S., Siciliano, B., Bicchi, A., *et al.*, Berlin: Springer, pp. 181–216.

[73] Luo, J. and Tsiotras, P. (2000). Control design for chained-form systems with bounded inputs. *Systems & Control Letters*, 39(2): 123–131.

[74] Makavita, C.D., Jayasinghe, S.G., Nguyen, H.D., *et al.* (2019). Experimental study of command governor adaptive control for unmanned underwater vehicles, *IEEE Transactions on Control Systems Technology*, 27(1):332–345.

[75] Mallot, H.A. (1999). Spatial cognition: Behavioral competences, neural mechanisms and evolutionary scaling. *Kognitionswissenschaft*, 8(1):40–48.

[76] Marchand, N. and Alamir, M. (2003). Discontinuous exponential stabilization of chained form systems. *Automatica*, 39(2):343–348.

[77] Morin, P. and Samson, C. (1999). Exponential stabilization of nonlinear driftless systems with robustness to unmodeled dynamics. *Control, Optimization and Calculus of Variations (COCV)*, 4(1):1–35.

[78] Nilsson, J. (1980). *Principles of Artificial Intelligence*. Palo Alto: Tioga Publishing Company.

[79] Phongchai, N. (2003). A Multi-Agent Based Architecture for an Adaptive Human-Robot Interface. Nashville, Tennessee: PhD Thesis, Vanderbilt University.

[80] Pomet, J.B. (1992). Explicit design of time-varying stabilizing laws for a class of controllable systems without drift. *Systems & Control Letters*, 18(2):147–158.

[81] Qu, Z.C., Wei, W., Wang, W., *et al.* (2019). Research on Fuzzy Adaptive Impedance Control of Lower Extremity Exoskeleton, 16th IEEE International Conference on Mechatronics and Automation (IEEE ICMA),Tianjin, Peoples R China, Aug 04-07, 2019, IEEE International Conference on Mechatronics (ICMA), pp. 939–944.

[82] Roy, S., Roy, S.B. and Kar, I.N. (2018). Adaptive-robust control of euler-lagrange systems with linearly parametrizable uncertainty bound. *IEEE Transactions on Control Systems Technology*, 26(5):1842–1850.

[83] Sahu, B.K., Subudhi, B. and Gupta, M.M. (2018). Stability analysis of an underactuated autonomous underwater vehicle using Extended-Routh's stability method. *International Journal of Automation and Computing*, 15(3):299–309.

[84] Samson, C. and Ait-Abderrahim, K. (1991). Feedback control of a nonholonomic wheeled cart in Cartesian space. Proceedings of the IEEE International Conference on Robotics and Automation (ICRA91). Sacramento, CA, USA, pp. 1136–1141.

[85] Samson, C. (1995). Control of chained systems application to path following and time-varying point-stabilization of mobile robots. *IEEE Transactions on Automatic Control*, 40(1):64–77.

[86] Samson, C. (1993). Time-varying feedback stabilization of car-like wheeled mobile robots. *International Journal of Robotics Research*, 12(1):55–60.

[87] Slotine, J.E. and Li, W.P. (1991). *Applied nonlinear control[M]*. Englewood Cliffs, New Jersey:Prentice-Hall.

[88] Sontag, E.D. (1998). *Stability and stabilization: Discontinuities and the effect of disturbances. NATO Advanced Study Institute "Nonlinear Analysis, Differential Equations, and Control."* Montreal, Canada: Kluwer.

[89] Sontag, E.D. and Sussmann, H.J. (1980). Remarks on continuous feedback. Proceedings of the IEEE Conf. on Decision and Control. Albuquerque, NM, pp. 916–921.

[90] Sun, D., Dong, H.N. and Tso, S.K. (2002). Tracking stabilization of differential mobile robots using adaptive synchronized control. Proceedings of the 2002 IEEE International Conference on Robotics and Automation (ICRA02). Washington DC, United States, pp. 2638–2643.

[91] Sun, Z.D. and Xia, X.H. (1997). On nonregular feedback linearization. *Automatica*, 33(7):1339–1344.

[92] Sun, Z.D., Ge, S.S. and Huo W., *et al.* (2001). Stabilization of nonholonomic chained systems via nonregular feedback linearization. *Systems & Control Letters*, 44(4):279–289.

[93] Sun, Z.P. (2004). *Autonomous Driving Car Intelligent Control System*. Changsha: National University of Defense Technology (in Chinese).

[94] Sussmann, H.J. (1979). Subanalytic sets and feedback control. *Journal of Differential Equations*, 31:31–52.

[95] The self-driving car successfully passed 67 times from Changsha to Wuhan, (2011). Sohu News.com. http://news.sohu.com/20110726/n314512981.shtml [2011-07-26] (in Chinese).

[96] Tsai, P.S., Wang, L.S., Chang, F.R., *et al.* (2004). Point stabilization control of a car-like mobile robot in hierarchical skew symmetry chained form. Proceedings of the 2004 IEEE International Conference on Networking, Sensing and Control. Taipei, Taiwan, pp. 1346–1351.

[97] Verginis, C.K., Bechlioulis, C.P., Dimarogonas, D.V., *et al.* (2018). Robust distributed control protocols for large vehicular platoons with prescribed transient and steady-state performance. *IEEE Transactions on Control Systems Technology*, 26(1):299–304.

[98] Wang, D. and Xu, G. (2003). Full-state tracking and internal dynamics of nonholonomic wheeled mobile robots. *IEEE/ASME Transactions on Mechatronics*, 8(2):203–214.

[99] Wang, D., Lu, H. and Yang, M.H. (2013). Online object tracking with sparse prototypes. *Image Processing, IEEE Transactions on*, 22(1):314–325.

[100] Wang, Q., Chen, F., Xu, W., *et al.* (2012). Object tracking via partial least squares analysis. *IEEE Transactions on Image Processing*, 21(10):4454–4465.

[101] Wang, T.R. and Qu, D.K. (2002). The open architecture of industrial robot control system. *Robot*, 24(3):257–261 (in Chinese).

[102] Wang, Y.Y., Chen, B. and Wu, H.T. (2017). Nonlinear Robust Control of Underwater Vehicle-Manipulator System Based on Time Delay Estimation, 14th International Conference on Ubiquitous Robots and Ambient Intelligence (URAI), Jeju, South Korea, Jun 28–JUL 01, pp. 119–123.

[103] Woernler, C. (1998). Flatness-based control of a nonholonomic mobile platform. *Journal of Applied Mathematics and Mechanics*, 78(Suppl.1):43–46.

[104] Wu, X.Y., Dai, X.H. and Meng, Z.D. (2003). Research on the architecture of distributed Robot controller. *Journal of Southeast University (Natural Science Edition)*, 33(s):200–204 (in Chinese).

[105] Wu, Y.H., Han, F. and Zheng, M.H., *et al.* (2018). Attitude control for on-orbit servicing spacecraft using hybrid actuator. *Advances in Space Research*, 61(6):1600–1616.

[106] Wu, R., Zhang, Z., Cheng, W.W., *et al.* (2017). Automatic tracking method of traffic signs based on particle filter. *Metrology and testing technology*, 44(7):64–67 (in Chinese).

[107] Xie, Y., Zhang, W., Xu, Y., *et al.* (2014). Discriminative subspace learning with sparse representation view-based model for robust visual tracking. *Pattern Recognition*, 47(3):1383–1394.

[108] Xu, H., Jia, P.F. and Zhao, Y.N. (2003). Research progress of open robot controller software architecture. *High-tech Communications*, 13(1):100–105 (in Chinese).

[109] Xu, W.L. and Huo, W. (2000). Variable structure exponential stabilization of chained systems based on the extended nonholonomic integrator. *Systems & Control Letters*, 41(4):225–235.

[110] Yang, S., Mao, X.J., Liu, Z., *et al.* (2017). The Accompanying Behavior Model and Implementation Architecture of Autonomous Robot Software, 24th Asia-Pacific Software Engineering Conference (APSEC),Nanjing, Peoples R China, Dec 04-08, pp. 209–218.

[111] Yang, S., Mao, X., Yang, S., *et al.* (2017). Towards a hybrid software architecture and multi-agent approach for autonomous robot software. *International Journal of Advanced Robotic Systems*, 14(4):Jul 4, 2017, DOI:10.1177/1729881417716088.

[112] Yu, C.Y., Xiang, X.B., Wilson, P.A., *et al.* (2020). Guidance-error-based robust fuzzy adaptive control for bottom following of a flight-style AUV With saturated actuator dynamics. *IEEE Transactions on Cybernetics*, 50(5):1887–1899.

[113] Yu, W.-S. and Chen, C.-C. (2018). Cloud Computing Fuzzy Adaptive Predictive Control for Mobile Robots, IEEE International Conference

on Systems, Man, and Cybernetics (SMC), Miyazaki, Japan, Oct 07-10, pp. 4094–4099.

[114] Yue, M., Hou, X.Q., Gao, R.J., *et al.* (2018). Trajectory tracking control for tractor-trailer vehicles: a coordinated control approach. *Nonlinear Dynamics*, 91(2):1061–1074.

[115] Zhao, T., Liu, Y.Y., Li, Z.J., *et al.* (2019). Adaptive control and optimization of mobile, manipulation subject to input saturation and switching constraints. *IEEE Transactions on Automation Science and Engineering*, 16(4):1543–1555.

[116] Zhong, W. (2013). *Target tracking algorithm based on sparse representation*. Dalian: Dalian University of Technology (in Chinese).

[117] Zhu, Y., Mao, X.J. and Yang, Y.B. (2016). Visual target tracking based on multi-feature hybrid model. *Journal of Nanjing University (Natural Science Edition)*, 52(4):762–770 (in Chinese).

Chapter 8

Localization and Mapping of Mobile Robot

The localization of mobile robot, the "where am I" problem, is a fundamental issue of mobile robot navigation [34]. Mapping means how to sense, understand and remember environments by mobile robot while navigating, and then the mapping will help navigation. Simultaneous Localization and Mapping (SLAM) also called Concurrent Mapping and Localization (CML) is the process where an autonomous robot starts at an unknown location in an unknown environment and then incrementally builds a map of this environment while simultaneously using this map to compute the robot location [7,17,36,53]. The first consistent solution to the SLAM problem was published by Smith, Self and Chesseman [46]. The SLAM algorithm has been recently seen a considerable amount of interest from the mobile robotics community as a tool to enable fully autonomous navigation [10,13,29].

This chapter introduces mobile robot mapping and localization by dead reckoning, maximum likelihood estimation mapping by laser radar, and EKF SLAM method. Some other issues in SLAM, such as data association and dealing with dynamic environment, are discussed in detail.

8.1 Introduction to Map Building of Mobile Robot

According to the requirements of the environment model in navigation, the navigation can be divided into three categories: map-based navigation, map-building-based navigation and mapless navigation [22]. The so-called mapless navigation means that the robot recognizes and tracks a certain target to complete the navigation task for a specific purpose. If the environment map already exists, the robot can continuously correct its position by using Kalman filtering and other methods under the guidance of the prior map to achieve precise positioning.

In the unknown environment, there are neither known targets that can be tracked nor *a priori* map. Only the robot's own state information (used for positioning) obtained by internal sensors such as odometer is prone to accumulative errors. Therefore, the robot must use its own sensors to perceive the external environment, automatically or semi-automatically build a map in the continuous exploration of the environment, and then achieve precise positioning and path planning so that to more effectively explore the environment or complete specific tasks. In this case, in order to achieve precise positioning, the robot needs to continuously correct its position using environmental features (maps) with a relatively determined location. In order to determine the location of the environmental features, the robot needs to know its precise location. This type of issue is commonly referred to as CML (Concurrent Mapping and Localization) or SLAM (Simultaneous Localization And Map Building) problems.

Environmental model representation is the first step for the problem-solving of environmental modeling. The essence of environmental modeling belongs to the category of environmental feature extraction and knowledge representation methods, which determines how the system stores, utilizes and acquires knowledge. The purpose of creating a map is for the robot to plan the path, so the map must be easy for the machine to understand and calculate, and when new environmental information is detected, it should be able to be easily added to the map [56]. The classification of environment models commonly used in the field of mobile robot navigation is shown in Figure 8.1, of which several plane models are more common.

1. Measurement model

The coordinate information in the world coordinate system is used to describe the characteristics of the environment, which includes the spatial decomposition method and the geometric representation method. The space decomposition method decomposes the

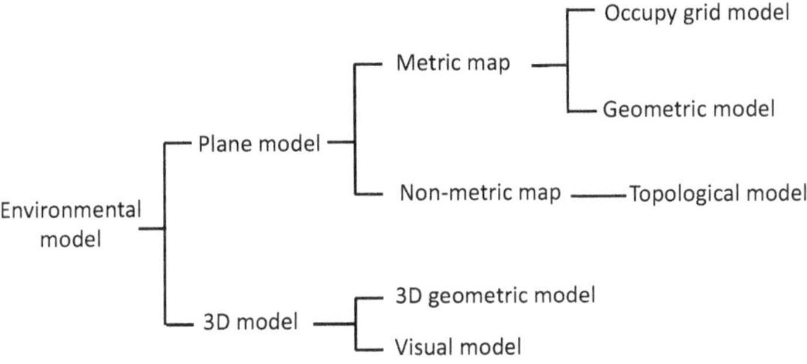

Figure 8.1.   Classification of environmental models.

environment into local units and uses whether they are occupied by obstacles to describe the state [1]. Spatial decomposition often uses grid-based uniform decomposition method and hierarchical decomposition method [18]. Geometric representation uses geometric primitives to represent the environment [6].

Occupancy grid map is a very successful metric map construction method. It was first proposed by HP Moravec and A. Elfes in 1985 [18]. The idea is to divide the environment in which the mobile robot is located into many sizes. For equal grids, the space state is described by the probability that each grid is occupied or not occupied by obstacles. Using distance sensors such as sonar or lidar can easily obtain obstacle information and add it to the map. The information of each grid directly corresponds to a certain area in the environment, so it is easy to create, maintain and understand. The occupied grid map can represent any environment, and the accuracy of the map can be adjusted by adjusting the resolution of the grid. It has good advantages in terms of uncertainty management, sensor characteristics and environmental representation. With the help of this map, the robot can easily complete self-location and path planning. This method uses probability values to represent the uncertainty of the model, can provide more accurate measurement information, and facilitate the fusion of multi-sensor information, so it has been applied in a large number of mobile robot systems, and it has maintained active vitality until now [48]. However, the accuracy of metric information depends heavily on the accuracy of the odometer and the degree of uncertainty processing of the sensor. The amount of data stored and maintained is large, and the heavy computational burden in a large-scale environment is difficult to meet real-time requirements. Especially for vision sensors, unless there is a problem of global vision (such as a ceiling surveillance camera) object occlusion, it will cause great difficulty in establishing a globally consistent grid model. Therefore, the occupancy grid model is rarely used in unknown environments, especially outdoor environments, and is generally used for obstacle avoidance and local path planning.

Geometrical map refers to the mobile robot collecting the perception information of the environment, extracting more abstract geometric primitives or features, such as points, lines, surfaces, etc., and usually assigning global coordinates to represent the environment [6]. Because it can provide the measurement information required for planning and positioning and the storage capacity is relatively small, there have been many studies in recent years, including many CML/SLAM studies. The extraction of geometric features requires additional processing of perceptual information, and a certain amount of perceptual data is needed to get the result. In the indoor environment, the environment can be defined as more abstract geometric features, such as a collection of faces, corners, and edges, or walls, corridors, doors, rooms, etc. Because it is difficult to extract geometric features in outdoor environments, it is limited in outdoor environment applications. Geometric elements based on vision sensors are sometimes called visual features (such as corner point features). In order to obtain measurement information, binocular or trinocular stereo vision technology is commonly used in this type of method [55]. David Lowe proposed a landmark extraction strategy with near and far scale invariance, called SIFT (scale invariant feature transform), which uses trinocular vision to throw away too many SIFT features and calculate the 3D coordinates of the SIFT features that can be used as landmarks [49]. The environment model is a weighted SIFT feature database, and each data is represented as (x, y, z, s, o, l), where (x, y, z) are the coordinates of the feature relative to the robot, (s, o) Is its characteristic description. It is decided whether to discard from the model according to the value of l, and data association is required to predict the appearance of SIFT features through subsequent frames.

Figure 8.2 shows examples of typical raster maps and geometric maps. Figure 8.2(a) is a grid map constructed by distance sensors

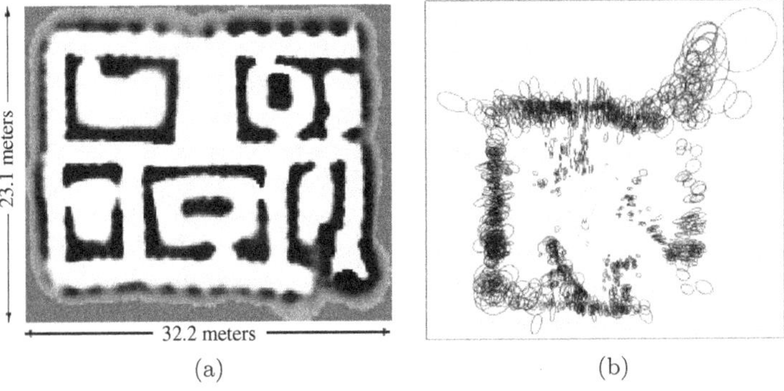

Figure 8.2.   Typical raster map and geometric map.

[13, 43], where the black area represents the grid occupied by obstacles, and the white area represents the barrier-free area, that is, the travelable area. Figure 8.2(b) is a geometric map constructed by visual sensors [49]. According to the geometric coordinates of road signs (geometric primitives) detected in the environment, they are depicted in a plane coordinate system. Each ellipse in the figure represents the uncertainty range of the road sign.

The advantage of the geometric map is that the representation method is more compact, and it is convenient for positioning and target recognition. The difficulty mainly includes how to extract stable geometric elements from sensor data without changing with robot movement, environmental changes, etc., and how to find corresponding matches in the map based on the observed geometric features when positioning and model update that is, the data association problem [48]. This problem is one of the difficult problems in the field of environmental modeling and positioning. Inaccurate data association will completely invalidate some algorithms, leading to failures in modeling and positioning.

2. Non-metric model

The topological map represents the environment in the form of a graph $G(N, L)$, which N represents a set of nodes, and L represents a set of edges connecting nodes (link set). The node corresponds to the characteristic state, location (determined by perception), etc. in the environment, and the connection corresponds to the relationship between them. Nodes and connections are used to describe the relationship between various objects in space or between different environments. The storage space required is small, and the efficiency of path planning by using them is very high (the path may also be sub-optimal), and it is suitable for large-scale environments application. Because it does not require precise distance and other measurement information, it is particularly suitable for monocular vision applications, so it is very popular in the field of visual navigation [52]. Moreover, because the accurate position information of the robot is not required, it is more robust to the position error of the robot, but the spatial dispersion of the topological nodes makes the robot unable to locate in the non-topological node positions.

The process of establishing a topological map is actually the process of identifying nodes. The differences between various topological modeling methods are mainly reflected in how to form nodes, how to identify different nodes, how to ensure the compactness of the model, and how to deal with the uncertainty of sensor observation. The difficulty is mainly reflected in how to identify the previously established nodes when positioning, that is, how to match the observed features with the existing nodes in the map when using map navigation. This problem can also be described as a search problem in the map space. Currently, a depth-first search strategy with depth boundaries is commonly used in a large-scale environment with a large number of nodes. But at the same time, it also brings the identification problem of the ring-shaped area, that is, when the robot is traveling in the ring-shaped area, the node may not be in the search space when it returns to the starting position, which causes the robot to be unable to locate itself accurately [48].

In addition to the above three 2D plane models, Thrun proposed a hybrid method that integrates the respective advantages of the grid model and the topology model in 1998 [43]. The local map is represented by a grid, and on this basis, the topological structure is extracted through methods such as the Voronoi diagram, and the connection of multiple local maps is realized. In path planning, two levels of planning, local and global, are adopted, that is, the planning is divided into regional planning based on occupancy grid and global planning based on topological connections. This hybrid method adopts topological description in the global space to ensure global continuity, while the use of geometric representations in specific local environments is conducive to the use of the advantages of precise positioning of mobile robots. However, this method is generally only suitable for representing indoor environments, and is mainly based on distance sensors such as sonar or lidar, and is rarely used in camera-based modeling and positioning.

3. Three-dimensional model

So far, most environmental modeling studies have been conducted on the above-mentioned three 2D plane models, and the research on 3D environmental models is relatively small and has just started. The 3D model has richer environmental information, but it also brings higher storage requirements, more complex data processing algorithms, and more calculations.

Thrun uses two 2D lidars (one facing forward and one facing upward), designed filtering and simplified algorithms to build a multi-resolution 3D map, and used virtual reality tools to draw an image of the 3D environment [44]. Andrew J. Davison *et al.* proposed SLAM based on extended Kalman filtering for wheeled robots walking on outdoor undulating terrain (i.e. natural environment). The environment model uses a 3D map (environmental features are represented as "points"), and the map includes the position and direction of the robot and 3D coordinates of environmental features [2]. The update of the map adopts the EKF algorithm, and the active vision is designed to reduce the computational burden [47]. The modeling idea of using visual sensors to build 3D maps to contain richer environmental information is becoming a hot spot in robotic navigation research.

8.2   Dead Reckoning Localization

Most mobile robots use a combination of relative and absolute techniques for position estimation [9]. The information acquirement for absolute localization is difficult and the computation is complex, while the method for dead reckoning can quickly estimate the position of mobile robots, reduce the computational complexity and realize the real-time navigation. Dead reckoning has been already extended to the case of a mobile robot moving on uneven terrain [19]. Hence, it is also necessary for mobile robot navigation to design a high-precision dead reckoning system [8].

Dead reckoning provides the dynamic information of the pose for mobile robots by directly measuring the parameters such as position, velocity, and orientation [41]. Generally speaking, mobile robots adopt the wheeled architecture to move in operating environments [20]. Therefore, the most direct approach to tracking their pose is to measure the angular velocity of the wheels and the orientation of the steering architecture [51]. Wheeled mobile robots often use the encoders (optical, inductive, capacitive magnetic) as odometry to measure its displacement and heading [40]. However, the heading depending on encoders would cause rather great errors [33, 38]. In recent years, fiber optic gyros (FOG) have become more affordable. So it is widely used in many mobile robot platforms to supplement odometry, especially in outdoor applications [12, 16]. Furthermore, if the terrain is uneven, the pitch and roll also need to be detected and added into the pose computation of mobile robot to improve the localization performance [30, 35].

With the requirement for the navigation of mobile robots under unknown environments, this section designs a dead reckoning system when developing an experimental platform of mobile robot [9]. Aimed at the real-time navigation of mobile robots on uneven terrain, the difficulty is to compute the displacement in height direction. So this section presents the dead reckoning method based on rigid-body kinematic equations of this robot platform and uses high-precision proprioceptive sensors to estimate the pose of mobile robot in 3D environments. Simulation and experiment using this robot platform demonstrate the reliability of the system and the validity of the method, which enhances the localization precision of navigation control.

8.2.1 *Proprioceptive sensor system*

1. Angular transducer potentiometers

Two high-precision angular potentiometers are attached to the rotating axis of left and right rocker-bogie suspension respectively, and its rotating angles can change from $-15°$ to $15°$ in our design. With the rotation of rocker, the rotating information is transformed to the electrical signal changing from 0V to 5V, then magnified by an amplifier circuit, converted by A/D conversion card (4 channel 10 bit) to digital signal and lastly sent to industrial control computer.

2. Odometry

Odometry is implemented by means of incremental optical encoders that monitor the wheel revolutions or steering angle. Encoders are mounted on the shaft of the drive motor to count the wheel revolutions, and its resolution P is 360 pulses per revolution (PPR). Gear ratio of the reduction gear between the motor and the drive wheel η is 15, and the wheel radii r is 125 mm. After a short sampling period, the left and right wheel encoders show a pulse increment of N_L and N_R respectively. The incremental travel distance for the left and right wheel can be computed according to the product of conversion factor and pulse increment at all sampling period in which the conversion factor that translates encoder pulses into linear wheel displacement is shown in Formula (8.1).

$$\delta = \frac{2\pi r}{\eta P} \tag{8.1}$$

3. Fiber optic gyros

With the development of fiber technique, fiber optic gyros (FOG) has been the important component for measuring the heading of mobile robot and its performance directly affects the heading resolution of mobile robot to a great extent. Similar to the ring laser gyros, FOG employs the principles of Sagnac interferometer in which a light wave is split and passed in opposite directions through a coil to measure angular acceleration. FOG used in our system is KVH E-Core RD1100 gyros which is open loop gyros. RD1100 gyros is a true single-axis rotation rate sensor insensitive to cross-axis motion and errors.

4. Tilt sensor

It must detect the pitch and roll of mobile robot in 3D environment in order both to obtain its unbalanced state and to make dead reckoning on uneven terrain. The CXTIL02EC employs Crossbow's Soft sensor calibration and onboard temperature sensor to compensate internally for temperature induced drift, so it can offer outstanding resolution, dynamic response and accuracy. To measure tilt, also called pitch and roll, CXTIL02EC makes use of two micro-machined accelerometers, one oriented along the X axis and one along the Y axis.

8.2.2 *Design of dead reckoning system*

The architecture of two rocker-bogie suspension and four drive wheels is adopted. The system component is shown in Figure 8.3. In order to determine the pose of mobile robot, angular transducer potentiometers is used for the measurement of the rotating angle of the rocker-bogie suspension relative to the robot body (the left angle β_L, the right angle β_R); incremental optical encoder for the measurement and the computation of the velocity of each wheel (v_{R1}, v_{R2}, v_{L1}, v_{L2}); FOG for the heading θ and its angular velocity ω (rotating around the Z axis of the reference frame of robot platform); and tilt sensor for the pitch α and roll φ of the robot platform relative to the horizontal plane. So the above sensors can measure the ten pose variables of mobile robot $\{v_{R1}, v_{R2}, v_{L1}, v_{L2}, \beta_R, \beta_L, \theta, \alpha, \varphi, \omega\}$.

To describe the motion of mobile robot in operating environment, the world coordinate system and the robot coordinate system are built as follows:

World coordinate system $\{w\}$: on the assumption that the coordinate origin O be the start point of mobile robot, X axis be the original motion direction at the start point of mobile robot, Z axis be vertical to the horizontal plane and Y axis be determined according to the rule of right hand.

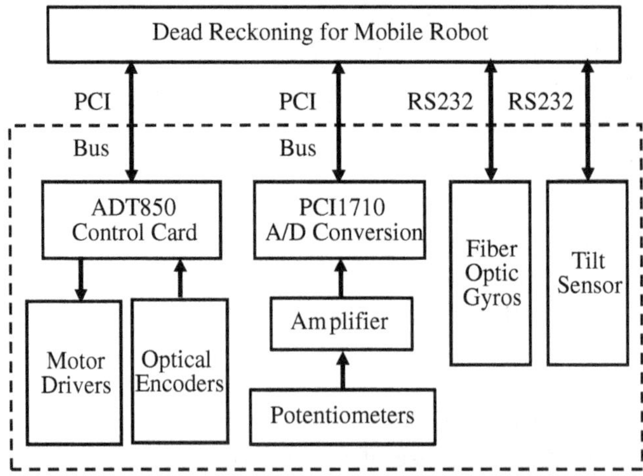

Figure 8.3.   Hardware components of the dead reckoning system.

Robot reference coordinate system $\{r\}$: on the assumption that the coordinate origin O_r be the midpoint of line between left and right shoulder joint (cradle fulcrum), X_r axis be in the plane of robot platform and point at just the front direction of the robot, Z_r axis be vertical to the robot platform and Y_r axis be also determined according to the rule of right hand.

Facing to the motion direction of mobile robot, L_1, L_2 are the left-side drive wheel and R_1, R_2 are the right-side drive wheel respectively that are all the fixed drive wheel no turning. If mobile robot moves in non-ideal environment, the accuracy kinematic model is difficult to be constructed because of the complexity of wheel-ground contact. Assumed that mobile robot works in quasi-static mode and ignores the dynamics influences taking into account of the operating speed of mobile robot is quite slow. The pose of mobile robot in world coordinate system can be described as follows (see Figure 8.4).

x, y, z: Coordinates of reference center O_r of robot coordinate $\{r\}$ in world coordinate system $\{w\}$;

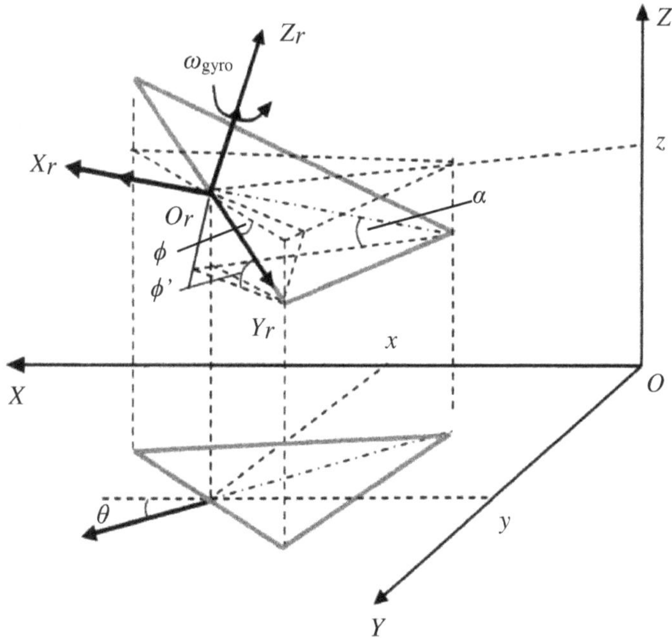

Figure 8.4.   Coordinate transformation in world coordinate system.

θ: the heading that is the angle between X_r axis in robot coordinate $\{r\}$ and X axis in world coordinate system $\{w\}$ and use counterclockwise rotation as positive direction;

α: the pitch that is the angle between X_r axis in robot coordinate $\{r\}$ and $X-Y$ plane in world coordinate system $\{w\}$ and uses the uplift as positive direction;

φ: the roll that is the angle between Y_r axis in robot coordinate $\{r\}$ and $X-Y$ plane in world coordinate system $\{w\}$ and uses right-side rolling as positive direction;

β_L, β_R: the angle between rocker and robot body and use the uplift as positive direction (the subscript L and R denote the left and right rocker respectively);

v: the velocity of reference center O_r of robot coordinate $\{r\}$ in world coordinate system $\{w\}$.

Furthermore, v_L and v_R denote the velocity of the left and right shoulder point; $v_{L,Xr}$ and $v_{L,Zr}$ denote the components of v_L along the X_r and Z_r axis; $v_{R,Xr}$ and $v_{R,Zr}$ denote the components of v_R along the X_r and Z_r axis, respectively. With the rigid-body kinematic constraints, we can obtain the components of v of reference center O_r of robot coordinate$\{r\}$. Generally speaking, the pitch is adopted to represent the tilting condition of robot platform, but β_L, β_R can more accurately discover the real condition in this rocker-bogie suspension architecture. So, the components of v of reference center O_r is obtained by β_L, β_R as follows.

$$\begin{cases} v_{X_r} = \dfrac{v_{R,Xr} + v_{L,Xr}}{2} = \dfrac{v_R \cdot \cos \beta_R + v_L \cdot \cos \beta_L}{2} \\ v_{Y_r} = 0 \\ v_{Z_r} = \dfrac{v_{R,Zr} + v_{L,Zr}}{2} = \dfrac{v_R \cdot \sin \beta_R + v_L \cdot \sin \beta_L}{2} \end{cases} \tag{8.2}$$

By the inverse transformation of robot coordinate $\{r\}$ using world coordinate system $\{w\}$ as the reference frame ${}^{w}_{r}T$, the transform formula is shown as follows.

$${}^{w}_{r}T = \mathrm{Trans}(x, y, z)\mathrm{Rot}(Z\mathrm{new}, \theta)\mathrm{Rot}(Y'\mathrm{new}, -\alpha)\mathrm{Rot}(X_{\mathrm{r}}, \varphi') \tag{8.3}$$

According to the above transform formula, ${}^{w}_{r}T$ can be computed as follows in which s denotes the sine function $\sin(\cdot)$ and c denotes the cosine function $\cos(\cdot)$ in the following transform matrix.

$$
{}^{W}_{r}T = \begin{bmatrix} c\theta c\alpha & -c\theta s\alpha s\varphi' - s\theta c\varphi' & \text{-}c\theta s\alpha c\varphi' + s\theta s\varphi' & x \\ s\theta c\alpha & \text{-}s\theta s\alpha s\varphi' + c\theta c\varphi' & \text{-}s\theta s\alpha c\varphi' - c\theta s\varphi' & y \\ s\alpha & c\alpha s\varphi' & c\alpha c\varphi' & z \\ 0 & 0 & 0 & 1 \end{bmatrix}
\tag{8.4}
$$

where, $\text{Trans}(x, y, z)$ denotes the translation transformation from the origin O of world coordinate system to the current position of mobile robot (x, y, z); $\text{Rot}(Znew, \theta)$ denotes the rotation transformation at θ angle around new Znew axis after translation transformation; $\text{Rot}(Y'new, -\alpha)$ denotes the rotation transformation at $-\alpha$ angle around new Y'new axis after the transformation above. Because we define the pitch to use the uplift as positive direction which is contrary to the rotation direction by right-hand rule, thus the rotation angle is $-\alpha$. $\text{Rot}(Xr, \varphi')$ denotes the rotation transformation at φ' angle around new X_r axis after the transformation lastly where φ' is equivalent roll.

On the assumption that position vector of reference center O_r of robot coordinate $\{r\}$ in world coordinate system $\{w\}$ is $u = [x, y, z, 1]$, then,

$$
\dot{u} = \begin{bmatrix} \dot{x} \\ \dot{y} \\ \dot{z} \\ 1 \end{bmatrix} = \begin{bmatrix} v_X \\ v_Y \\ v_Z \\ 1 \end{bmatrix} = {}^{w}_{r}T \cdot {}^{r}\dot{u} = {}^{w}_{r}T \begin{bmatrix} v_{X\text{r}} \\ v_{Y\text{r}} \\ v_{Z\text{r}} \\ 1 \end{bmatrix}
\tag{8.5}
$$

Hence, the pose of mobile robot can be expressed by the state of reference center of robot platform for 3D environments $(x_i, y_i, z_i, \theta_i, \alpha_i, \varphi_i)$ in which i is the sequence number of sampling period as follows.

$$
\begin{cases} x_{i+1} = x_i + T \cdot v_X \\ y_{i+1} = y_i + T \cdot v_Y \\ z_{i+1} = z_i + T \cdot v_Z \\ \theta_{i+1} = \theta_i + T\dot{\theta} \\ \alpha_{i+1} = \alpha_i + T\dot{\alpha} \\ \varphi'_{i+1} = \arcsin \dfrac{\sin \varphi_i}{\cos \alpha_i} \end{cases}
\tag{8.6}
$$

8.2.3 *Simulation and experiment*

The simulation puts emphasis on analyzing and simulating the motion of Z axis of mobile robot owing to the dead reckoning method in X and Y axis is relatively matured now, and utilizes Visual C++ to design the simulation program of traversing obstacles of mobile robot. In simulation process, it is assumed that the front drive wheels hold the steady thread velocity, the rear wheels get the moving velocity based on rigid-body constraint, and the center position of each wheel obtains according to the curve of the terrain profile.

In addition, supposed that the terrain condition of left and right side is equal so $\beta_L = \beta_R = \beta$ and the roll do not change so $\varphi = 0$. The motion of the rear wheels does not be analyzed because the pose of mobile robot is mainly determined by the front drive wheels.

Figure 8.5 shows the simulation results of mobile robot's traversing the terrain with sine-wave change. It can be seen if the curve of terrain profile is continuous differential, so is the wheel and ground constraint. Dead reckoning for mobile robot is very accurate when the angle of the rocker-bogie suspension relative to the ground (the

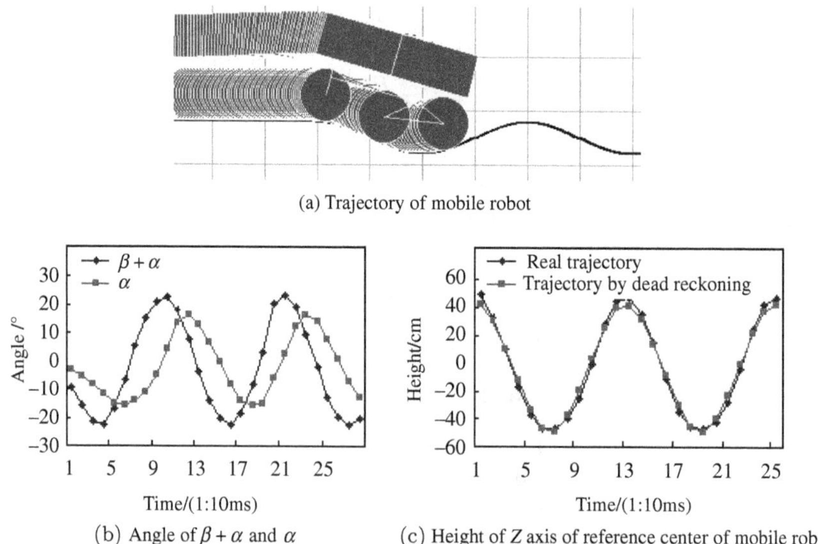

(a) Trajectory of mobile robot

(b) Angle of $\beta + \alpha$ and α (c) Height of Z axis of reference center of mobile robot

Figure 8.5. Simulation of mobile robot's traversing the terrain with sine-wave change.

left angle $\beta_L+\alpha$, the right angle $\beta_R+\alpha$) and the pitch α is continuous differential.

In order to verify the method above, we use mobile robot MORCS-1 (means a MObile Robot made by Central South University) as experimental platform (see Figure 8.6(a)). Proprioceptive sensors described as above are utilized to form the dead reckoning system of MORCS-1. We implemented the experiment with a slope whose angles relative to the ground is $6.5°$ and its length is $660\,\mathrm{cm}$.

In the experiment, robot moves forward at the speed of 20 cm/s, and the whole sampling time of the system is $0.2\,\mathrm{s}$. The angle errors between the rocker-bogie suspension relative to the ground $\beta+\alpha$ and pitch α and the real pitch are shown in Figure 8.6(b) from which it can be seen the errors of dead reckoning according to the angle of rocker-bogie suspension relative to the ground are smaller than the pitch from tilt sensor. As a result, the height of Z axis of reference center of mobile robot is more accurate by this method shown in

(a) Slope experiment

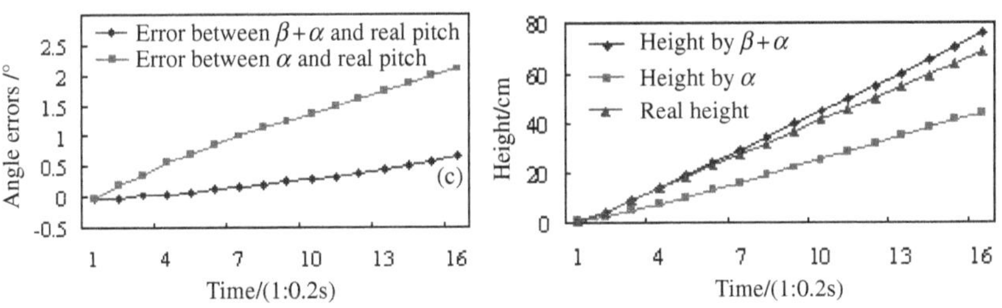

(b) Angle errors between $\beta+\alpha/\alpha$ and real pitch   (c) Height of Z axis of reference center of mobile robot

Figure 8.6.   Experiment of mobile robot's traversing the slope.

Figure 8.6(c). Although the errors between the real and measurement height is inevitable owing to the noise disturbance of proprioceptive sensors, it is quite smaller by many tests and can meet the requirement of dead reckoning to a certain extent.

8.3　Map Building of Mobile Robot

8.3.1　*Map building based on laser radar*

Because laser radar measures distance accurately, it is generally used as the mobile robot navigation sensor. The laser radar LMS291 produced by Sick Corporation and equipped on MORCS-1 can provide 2D scanning range (180°) and has the advantages of high speed, accuracy and anti-jamming. It is based on a Time-Of-Flight (TOF) measurement principle as depicted in Figure 8.7 [45, 59]. A pulsed laser beam is emitted and reflected from the object surface. The reflection is registered by the scanner's receiver. The time between the transmission and the reception of the laser beam is used to measure the distance between the scanner and the object [5, 25]. The pulsed laser beam is deflected by an internal rotating mirror turning at 4500 rpm (75rps) so that a fan-shaped scan is made of the surrounding area.

The distance d of the obstacle equals the half of the product of the light velocity c and the time interval Δt between the beams emitted from the laser radar and reflected by the obstacle.

$$d = \frac{\Lambda t \times c}{2} \tag{8.7}$$

Here, the velocity of light is $3.0 \times 10^8 \text{m/s}$.

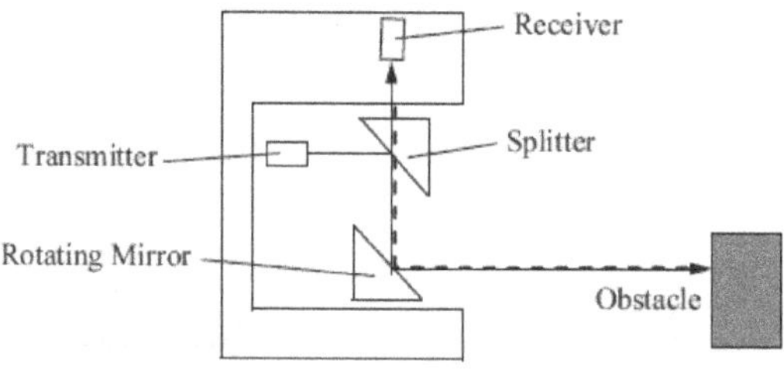

Figure 8.7.　Measurement principle of laser radar.

Table 8.1.   Measurement error of LMS291.

d (cm)	σ_λ (cm)	σ_d (cm)
$d \leq 500$	1.0	3.0
$500 < d \leq 1000$	1.2	3.6
$1000 < d \leq 2000$	1.35	4.05
$2000 < d \leq 4000$	1.7	5.1
$d > 4000$	1.8	5.4

Theoretically speaking, the measurement output of LMS291 is enough to cover the long distance up to 80 m in the cm. mode and up to 8 m in the mm. mode. According to the environment size and the needed accuracy, one of the two modes is chosen. By tests, the values of the standard deviation are obtained in Table 8.1 to the different distance range [62]. σ_λ is the standard deviation of the observations in static environment, whereas σ_d is used in dynamic environment. But the laser radar information is still local and sparse relative to the vision, and it will be affected by the range and the angular interval in measuring.

The angular resolution of LMS291 is selectable at $1°$, $0.5°$ or $0.25°$ and the data transfer rate can be programmed to be 9.6, 19.2, 38.4 or 500 Kbaud. A high-speed communication interface card installed in PCI expansion slot of industrial control computer communicates with LMS291 by RS422 protocol. In our research, LMS291 would get 361 measurement data throughout the $180°$ scanning field with $0.5°$ resolution. Each measurement data has two bytes and the length of data package including start bit and parity bit is 732 bytes. With 500 Kbaud communication rate, the transmission delay time is 13 ms and the scan time of LMS291 is 26.67 ms. Thus we use 40 ms to undergo the scan of LMS291 and deal with the data. At this sampling period, the scanning frequency of LMS291 is about 25 Hz, which means that 9025 (25 multiple 361) measurement data can be obtained per second.

With the requirement for the control on mobile robot navigation, we designed and developed an experimental platform of mobile robot with two rocker-bogie suspension and four drive wheels [27]. The reduction kinematic model of our robot platform is shown in Figure 8.8. In order to estimate robot's pose using dead-reckoning

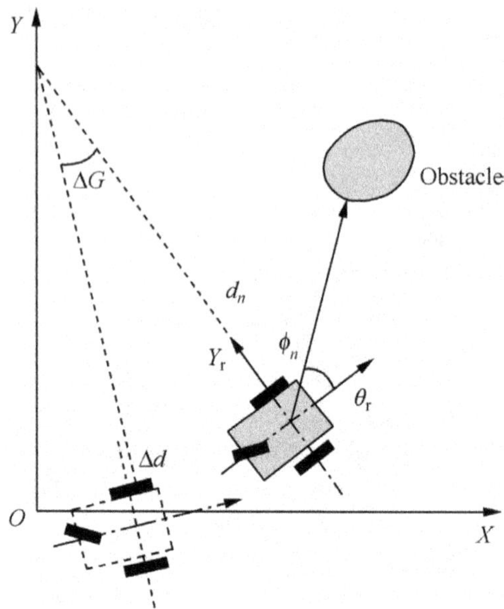

Figure 8.8.   Coordinate transformation from laser radar to global environment.

method, incremental optical encoder for the measurement and the computation of the velocity of each wheel (v_{R1}, v_{R2}, v_{L1}, v_{L2}) and a fiber optic gyros (FOG) for the heading θ and its angular velocity ω (rotating around the Z axis of the reference frame of robot platform). So the robot's pose in a 2D space can be estimated by the above sensors. A laser radar LMS291 is mounted on a high precision rotating table which can rotate from $-150°$ to $150°$ in horizontal direction and from $-50°$ to $15°$ in pitch direction so as to detect 3D environment. Here we only focus our interest in the x-y plane and in the θ direction here.

Let $\boldsymbol{X} = (x_r, y_r, \theta_r)$ denotes the robot's pose in a 2D space, then the equations describing the kinematic model in the discrete time domain got by the above sensors are shown as follows.

$$\begin{cases} x_r(k+1) = x_r(k) + \Delta d_r(k) \cdot \cos(\theta_r(k) + \Delta\theta_r(k)) \\ y_r(k+1) = y_r(k) + \Delta d_r(k) \cdot \sin(\theta_r(k) + \Delta\theta_r(k)) \\ \theta_r(k+1) = \theta_r(k) + \Delta\theta_r(k) \end{cases} \qquad (8.8)$$

where, $(x_r(k+1),\, y_r(k+1))$ and $(x_r(k),\, y_r(k))$ is the current and previous position of robot center, $\theta_r(k+1)$ and $\theta_r(k)$ is the current and previous heading of robot center, $\Delta d_r(k)$ is the estimated values of translation displacement provided by the velocity of each wheel $(v_{R1},\, v_{R2},\, v_{L1},\, v_{L2})$ and $\Delta\theta_r(k)$ is the estimated values of rotation displacement provided by FOG (ω) at time k in an sampling period (ΔT).

The environmental information is the snapshot from LMS291. LMS291 can scan a range to get a sequence of points which represent the contour curve of the local environment, thus a range scan describes a 2D slice of the environment. Points of the range scan can be specified in a polar coordinate system whose origin is the location of LMS291, and the reference axis for the laser beam direction is the home orientation of LMS291. Each scan points is represented by the laser beam direction, and the range measurement along that direction. We refer to (ρ_n, φ_n) $(n = 1, \ldots, 361)$ as the scan pose, where ρ_n is the distance from the origin to the n-th obstacle, φ_n is the angle of the n-th laser beam relative to the home orientation, n is the number of scanning data from LMS291.

The operating environment of mobile robot is described by 2D Cartesian grids in this paper. We take a 2D array as the environment map model and an array value corresponding to its each grid cell is recorded according to the times that the obstacle is observed. It assumed each grid cell can only be in one of two states, empty or occupied, and that has to be estimated from sensor observations. The knowledge that robot at time k about occupancy of a given cell is stored as probability of these two states, given all the prior sensor observations. When an obstacle is observed at such position, the array value is added by 1. Since the size of grids directly make an effect on the resolution of control algorithms, each grid cell corresponding to $5 \times 5\,\mathrm{cm}^2$ in real environment is adopted considering that LMS291 has the advantage of high measurement resolution and quick response time. If doing so, the environmental model of mobile robot can be created in real time and keep the high accuracy.

The conversion from polar to Cartesian coordinate system is written in Equation (8.9).

$$\begin{cases} x_n' = x_r + \rho_n \cdot \cos(\theta_r + \varphi_n) \\ y_n' = y_r + \rho_n \cdot \sin(\theta_r + \varphi_n) \end{cases} \tag{8.9}$$

where, (x'_n, y'_n) represents the coordinate of the n-th obstacle in the global coordinate system; (x_r, y_r) the current center coordinate of mobile robot in the global coordinate system obtained from the Equation (8.7).

Then the project from the coordinate of obstacles in the global coordinate system to the corresponding grid cell in the local map is written by Equation (8.10).

$$\begin{cases} x_{g,n} = \text{int}(x'_n/w) \cdot w + \text{int}(w/2) \\ y_{g,n} = \text{int}(y'_n/w) \cdot w + \text{int}(w/2) \end{cases} \tag{8.10}$$

where, $(x_{g,n}, y_{g,n})$ is the project coordinate of obstacles in the local map, and w is the grid width. Thus the local map is built by occupancy grid.

8.3.2 *Map matching based on maximum likelihood estimation*

As analyzed above, since most objects in operating environment of mobile robot are relatively stationary, a few dynamic obstacles can be treated as the environmental noise, while the inherent uncertainty of environment grids is mainly derived from the accumulated pose errors. The errors between the estimated and the real values of robot's pose is represented by $\delta \boldsymbol{X} = \{\delta x, \delta y, \delta \theta\}$ and statistical characteristics of these parameters can be gathered by dead-reckoning sensors (odometry, FOG) during robot motion. By searching for the optimal parameters δx, δy, $\delta \theta$ under scan matching, we can find the most likely position of mobile robot in the global coordinate system. Therefore the estimation for the parameters δx, δy, $\delta \theta$ is a maximum likelihood estimation problem.

$$(\delta x^*, \delta y^*, \delta \theta^*) = \underset{\delta x, \delta y, \delta \theta}{\arg \max} \, l(\delta x, \delta y, \delta \theta) \tag{8.11}$$

For estimating the parameters δx, δy, $\delta \theta$, the map matching is formulated in terms of maximum likelihood estimation using the distance from the occupied cells in the current local map to their closest occupied cells in the previous map with respect to some relative position between the maps as measurements. Assumed that n occupied grid cells in the local map and its value is often at 361 for a range scan of LMS291 in a structural environment. We denote the distances for these occupied grid cells at some position of mobile robot by $D_1, \ldots, D_n$, we thus formulate the likelihood function for the parameters δx, δy, $\delta \theta$, as the product of the probability distributions of these distances:

$$l(\delta x, \delta y, \delta \theta) = \sum_{i=1}^{n} \ln p(D_i; \delta x, \delta y, \delta \theta) \qquad (8.12)$$

In doing so, we adopted the following criterion for the probability distribution function (PDF) of grid-grid matching $p(D_i; \delta x, \delta y, \delta \theta)$. If a occupied grid cell in the current local map can fit with the corresponding grid cell in the previous map then the PDF values of this cell is evaluated to be 1; if it border on the corresponding grid cell in the top, down, left and right direction then the PDF values is evaluated to be 0.75, or 0.5 in the diagonal direction; if it doesn't map to any grid cells above then the PDF values is evaluated to be 0. That is to say, we limit the map matching in a 3×3 grid area centered by the matching grid cell.

8.3.3 *Self-localization based on feature mapping*

Despite the measurement accuracy of laser radar being rather high, many studies indicate that the actual measuring range of laser radar is influenced by the reflectivity of objects' surface, the incidence angle of the laser beam and so on. Moreover, due to the existence of random noise, dynamic disturbances and self-occlusion (i.e. some area is only visible in one scan but not in the other), it may be impossible to align two scans perfectly even if obtained in the same position. For this reason, fuzzy logic is employed into local map matching to deal with such uncertainty in order to improve the robot localization performance.

Since the scan range of LMS291 is from $-90°$ to $90°$ and the rotating table rotates from $-150°$ to $150°$ in horizontal direction, the perceptual environment around the robot can be completely sensed by 3 scans of LMS291 with the rotation of the rotating table at least which would be divided into corresponding scan sectors in the same position. On account of the analysis above, we can know it is different for the results of maximum likelihood estimation in various sector

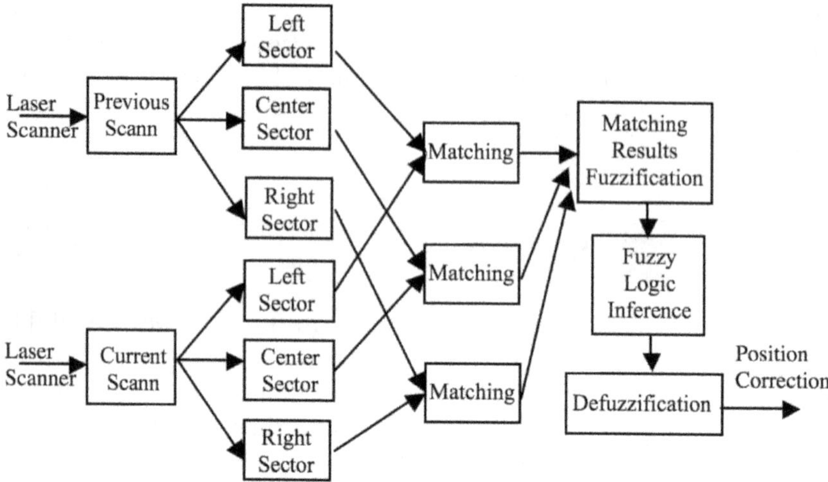

Figure 8.9.   Algorithms description based on fuzzy logic.

whether the robot moves or not. Hence, we assign a fuzzy degree of membership to each estimated parameter between scans and define its membership function is {low, rather low, median, rather high, high}. Next, we make a fuzzy logic inference according to the matching results, and then the crisp number is gained by a defuzzification based on the centroid average method.

The algorithm for the self-localization of mobile robot based on fuzzy techniques is described in Figure 8.9. We first test the initial position of mobile robot given by dead reckoning so that we have an initial position to compare against. The new robot's pose and its associated uncertainty from the previous pose, given dead-reckoning information, are estimated. Then the ambient environment is completely scanned by the laser radar with the rotation of the rotating table. The error between the estimated and the real values of robot's pose is computed by maximum likelihood estimation during matching the map between scans and is fuzzed according to some fuzzy rule. Lastly, we make a fuzzy logic inference of the matching results and a defuzzification by the centroid average method to find a most likely position of mobile robot.

It can realize self-localization to a certain extent by applying map matching based on MLE to estimate the parameters δx, δy, $\delta \theta$ and

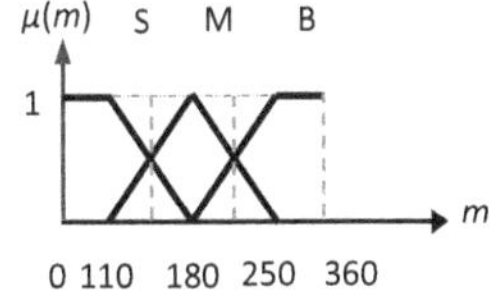

(a) Membership function for the matching number m

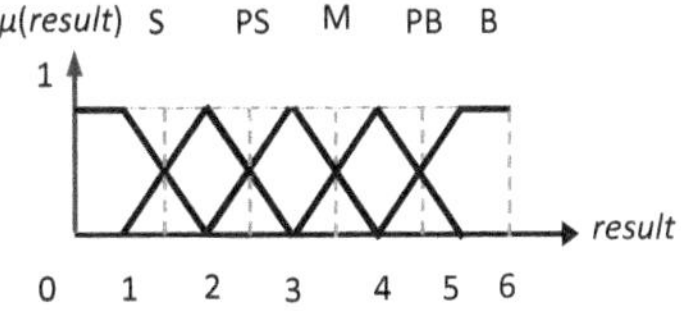

(b) Membership function for the fuzzy matching result

Figure 8.10.   Fuzzy membership function.

by using these to correct the robot's pose and the obstacle's position. However, the ranging data of laser scanner include noisy disturbance from environment such as ambient light, mixed pixel or exist data losing for the smaller reflectivity and the bigger incidence angle. Moreover, due to the existence of random noise, dynamic disturbances, obstacles obstructing or the scanning gap at some fixed scanning resolution. All these may produce the ambiguity information about the ambient environment. Hence, it may be impossible to find the most likely position of mobile robot under unknown environment. For this reason, fuzzy logic is employed into local map matching based on MLE to deal with such uncertainty in order to improve the robot localization performance.

In order to be robust to deal with this uncertainty and to introduce fuzzy logic, the scan area of LMS291 from –90 to 90 degrees can be divided into 3 equal sectors A (left), B(center), C(right). On account of the analysis above, we can know it is different for the results of maximum likelihood estimation in various sector whether the robot moves or not. Hence, we assign a fuzzy degree of membership to the number of likelihood estimation of each sector between scans $m_i(m_i <= 120, i = A, B, C)$ and define its fuzzy set is {S, M, B} (see Figure 8.10(a)). For the result of likelihood estimation of each sector, it can also define the corresponding fuzzy set as {S, PS, M, PB, B} (see Figure 8.6(b)).

In order to feedback roundly the inference information, the centroid average method is used to do a fuzzy decision-making. For the requirement of real-time correspondence, the scanning data of LMS291 is only dealt with map matching at a fixed time, and the preferable matching results is defuzzificated and used to correct the pose of mobile robot in the world coordinate system.

8.3.4 *Experiment*

To verify the method above, MORCS-1 is designed and developed by Intelligent Lab in Central South University (see Figure 8.11). MORCS-1 adopts the 5-wheeled locomotion architecture with two rocker-bogie suspension, four drive wheels and an omnidirectional wheel. It is equipped with odometry and fiber optical gyros that are internal sensor to dead reckoning, and laser radar and cameras that are external sensor to sensing environment. LMS291 is utilized as the exteroceptive sensors to build environment map and ADT-850 control card is used to drive the step motor to control the locomotion of MORCS-1.

We implemented experiments in an office environment. Experimental results are shown in Figure 8.12. Figure 8.12(a) depicts the robot using proprioceptive and exteroceptive sensors to determine its pose and build an initial local map so as to compare against from which we can see black grids are obstacles, a door is on the left and researcher are on the right as dynamic obstacles; Figure 8.12(b) shows the robot moving along the wall to scan the environment in which blue curve is its trajectories and we can know the accumulate errors are increased from it; Figure 8.12(c) and Figure 8.12(d) are the local map before and after correction. Comparing the left and right side of the office, the results using maximum likelihood estimation exhibit discrepancy owing to the dynamic disturbances. After the robot wanders for some time, we can find the pose of the mobile robot in the global coordinate system changed due to the accumulate errors and cause the local map uncertainty from Figure 8.12(a) and

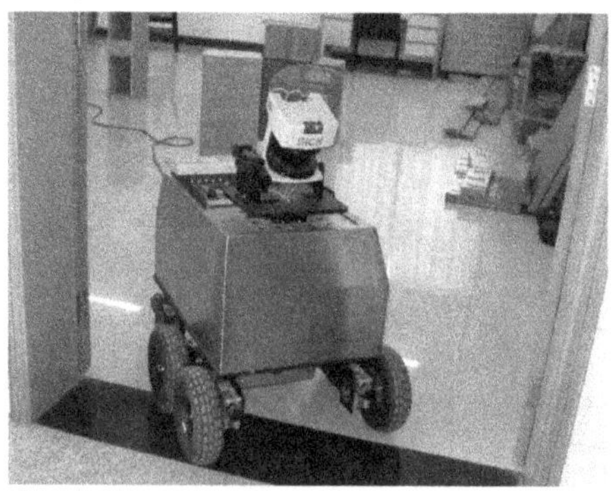

Figure 8.11.   Mobile Robot MORCS-1.

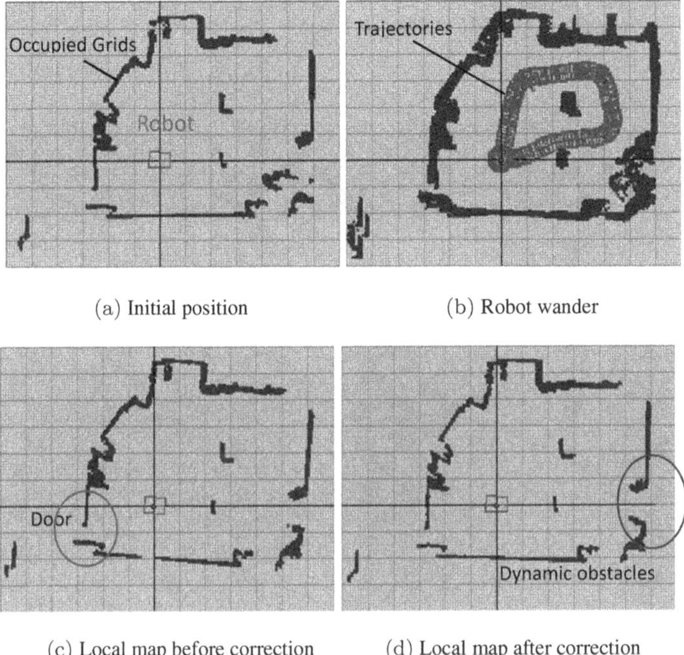

(a) Initial position

(b) Robot wander

(c) Local map before correction

(d) Local map after correction

Figure 8.12. Local map matching.

Figure 8.12(c). With the correction for self-localization by the local map matching using fuzzy logic, the random noise and dynamic disturbances can be reduced related Figure 8.12(a) and Figure 8.12(d).

In order to validate the effectiveness of this method in a larger and more complex environment, we add this algorithm in locomotion planning of mobile robot and the following experiment is carried out. IMR-01 starts at beginning point in office A and moves at the given target that we set in office B as ending point. IMR-01 uses laser radar with the rotation of the rotating table to build local map, acts at the target position, traverses the corridor successfully, and lastly reaches the target in office B. Figure 8.13 shows the navigation of mobile robot IMR-01 from starting point in office A to ending point in office B based on self-localization by the local map matching using fuzzy logic method. Navigation results discovered that IMR-01 can fulfill the navigation task successfully. At the same time, a fine environment map can be built and it is combined with the dead-reckoning information to reduce the accumulated pose errors of mobile robot. The

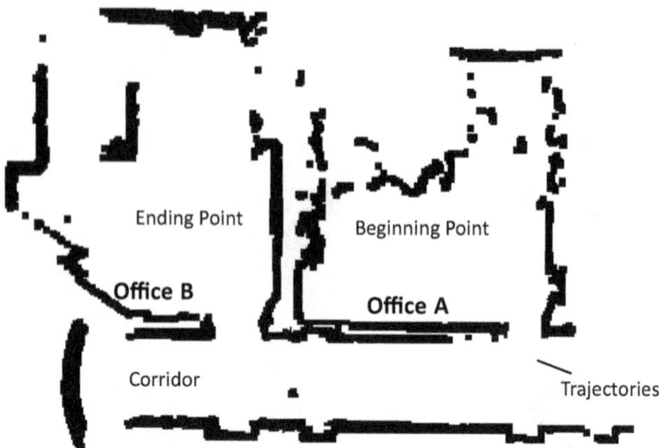

Figure 8.13.   Navigation based on localization correction.

effectiveness of this method in rather complex environment and the adaptability of that to the dynamic environment are also validated.

8.4   Simultaneous Localization and Mapping

Simultaneous Localization and Mapping realizes incrementally build a map of this environment while simultaneously using this map to compute vehicle location.

The solution presented by Smith *et al.* leads to a considerable demand on computation and storage, as identified in several publications [24,46,50]. Several attempts have been made to overcome this problem. One method for improving the algorithm performance is to limit the number of features that are included in the map [23,32]. Another method used the sub-optimal or near optimal covariance to update schemes [24]. Here, an approach based on the local map is advanced. A local frame of reference will be established periodically at the position of the robot, and the position of the robot and landmarks in the local map will be predicted, measured and updated by extended Kalman Filter, and then fused into the global frame of reference. Because of the independence of the local maps, the approach will not cumulate the estimate and calculation errors which are produced by SLAM using Kalman Filter directly. At the same time, it reduces the computational complexity [47].

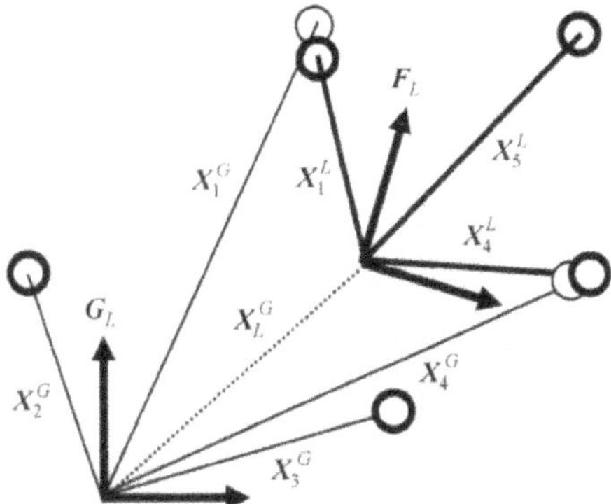

Figure 8.14.   System State.

8.4.1   *System state*

The system state is shown in Figure 8.14. At first, the global frame of reference $\mathbf{G}_L$ is established at time t_0, and the absolute pose of the robot $\mathbf{X}_v^G(t_0)$ is obtained by the odometry, and the absolute positions of the landmarks $\mathbf{X}_m^G(t_0)$ $(m = 1, 2, \dots)$ are observed by the robot sensor. At second, the local frame of reference F_L is established at the position of the robot at time k, and the direction of x-axis is the direction of the robot at time k. Mark the offset and rotation of the local frame in the global frame as $\mathbf{X}_L^G(k)$ and the landmarks as $\mathbf{X}_m^L(k)$ $(m = 1, 2, \dots)$.

The system state can be described as

$$
\mathbf{X}(k) = \begin{bmatrix} \underline{\mathbf{X}_L^G(k)} \\ \mathbf{X}_1^G(k) \\ \mathbf{X}_2^G(k) \\ \vdots \\ \hline \underline{\mathbf{X}_v^L(k)} \\ \hline \mathbf{X}_1^L(k) \\ \mathbf{X}_2^L(k) \\ \vdots \end{bmatrix} = \begin{bmatrix} \mathbf{X}_L^G(k) \\ \mathbf{X}_m^G(k) \\ \mathbf{X}_v^L(k) \\ \mathbf{X}_m^L(k) \end{bmatrix} \tag{8.13}
$$

including the location of the robot and the landmarks in the local frame of reference, $\mathbf{X}_v^L(k)$ and $\mathbf{X}_m^L(k)$ the location of the landmarks which has been observed before in the global frame of reference, $\mathbf{X}_m^G(k)$, and the pose of the local frame of reference, $\mathbf{X}_L^G(k)$ allowed the local states to be transformed into the global frame of reference.

8.4.2   *EKF algorithm with local maps*

Extended Kalman Filter algorithm consists of a recursive, three-stage process comprising predication, observation and update [37, 47]. In this paper, in one period the local frame of reference will be established, then the local map will be getting by EKF algorithm and will be transformed into global map. The essential steps of the filtering process in one period are shown in Figure 8.15.

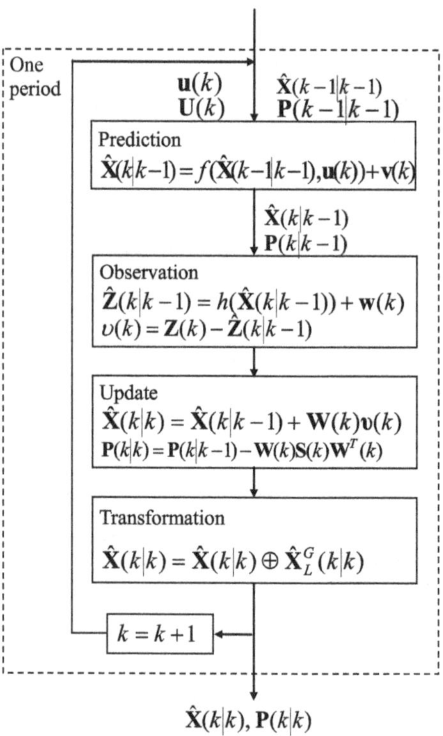

Figure 8.15.   The essential steps of the filtering process in one period.

1. Estimation and prediction

Order the estimate, or the posterior estimate of the system state as

$$\hat{\mathbf{X}}(k\,|k\,) = \begin{bmatrix} \hat{\mathbf{X}}_L^G(k\,|k\,) \\ \hat{\mathbf{X}}_m^G(k\,|k\,) \\ \hat{\mathbf{X}}_v^L(k\,|k\,) \\ \hat{\mathbf{X}}_m^L(k\,|k\,) \end{bmatrix} \tag{8.14}$$

Its covariance matrix is defined as

$$\mathbf{P}(k\,|k\,) = \mathrm{E}[\hat{\mathbf{X}}(k|k)\hat{\mathbf{X}}^T(k\,|k\,)] = \begin{bmatrix} \mathbf{P}^G(k\,|k\,) & 0 \\ 0 & \mathbf{P}^L(k\,|k\,) \end{bmatrix} \tag{8.15}$$

where

$$\mathbf{P}^G(k\,|k\,) = \begin{bmatrix} \mathbf{P}_{LL}^G(k\,|k\,) & \mathbf{P}_{mL}^G(k\,|k\,) \\ \mathbf{P}_{mL}^G(k\,|k\,) & \mathbf{P}_{mm}^G(k\,|k\,) \end{bmatrix} \tag{8.16}$$

$$\mathbf{P}^L(k\,|k\,) = \begin{bmatrix} \mathbf{P}_{vv}^L(k\,|k\,) & \mathbf{P}_{mL}^L(k\,|k\,) \\ \mathbf{P}_{vm}^L(k\,|k\,) & \mathbf{P}_{mm}^L(k\,|k\,) \end{bmatrix} \tag{8.17}$$

The prior estimate of the state at time k given the information up to time $(k-1)$, also referred to as the one-step-ahead prediction, will be written as $\hat{\mathbf{X}}(k\,|k-1\,)$

$$\hat{\mathbf{X}}(k\,|k-1\,) = \begin{bmatrix} \hat{\mathbf{X}}^G(k-1\,|k-1\,) \\ f(\hat{\mathbf{X}}_v^L(k-1\,|k-1\,), \mathbf{u}(k)) + \mathbf{v}(k) \\ \hat{\mathbf{X}}_m^L(k-1\,|k-1\,) \end{bmatrix} \tag{8.18}$$

where, $\hat{\mathbf{X}}(k-1\,|k-1\,)$ is the posterior estimate of the state at time $(k-1)$ conditioned on the information up to time $(k-1)$, f is a model of the rate of change of system state as a function of time, the control input $\mathbf{u}(k) = [\Delta x, \Delta y, \Delta \theta]^T$ is the increment of the robot's pose, $\mathbf{v}(k)$ is a random vector describing both dynamic driving noise and uncertainties in the state model itself. In order to simplify the algorithm, assume that the process noise $\mathbf{v}(k)$ is the white noise with zero mean, and its covariance matrix is $\mathrm{E}[v(k)v^T(k)] = \mathbf{Q}(k)$.

The covariance matrix of the prior estimate is

$$\mathbf{P}(k\,|k-1) = \begin{bmatrix} \mathbf{P}^G(k-1\,|k-1) & 0 & 0 \\ 0 & \mathbf{P}_{22} & \mathbf{P}_{23} \\ 0 & \mathbf{P}_{32} & \mathbf{P}_{33} \end{bmatrix} \qquad (8.19)$$

where, $\mathbf{P}_{22} = \nabla f(k)\hat{\mathbf{P}}^L_{vv}(k-1\,|k-1)\nabla f^T(k) + \mathbf{Q}(k)$, $\mathbf{P}_{23} = \nabla f(k)\hat{\mathbf{P}}^l_{vm}(k-1\,|k-1)$, $\mathbf{P}_{32} = \mathbf{P}^T_{23}$, $\mathbf{P}_{33} = \hat{\mathbf{P}}^L_{mm}(k-1\,|k-1)$, ∇f is the Jacobian of the model.

2. Observations

The actual observations $Z(k)$ is obtained by the robot's sensors, but we can predict the observations $\hat{\mathbf{Z}}(k\,|k-1)$ by the posterior estimate of the system state

$$\hat{\mathbf{Z}}(k\,|k-1) = h(\hat{\mathbf{X}}^L_v(k-1\,|k-1), \hat{\mathbf{X}}^L_m(k-1\,|k-1)) + \mathbf{w}(k) \qquad (8.20)$$

where, h is a model of the observation of system states as a function of time, $\mathbf{w}(k)$ is a random vector describing both measurement corruption noise and uncertainties with zero mean just like $\mathbf{v}(k)$, and its covariance matrix is $\mathbf{R}(k)$.

The difference between the actual observation and the predicted observation is termed "innovation"

$$\upsilon(k) = \mathbf{Z}(k) - \hat{\mathbf{Z}}(k\,|k-1) \qquad (8.21)$$

The innovation covariance $\mathbf{S}$, is computed from the current state covariance estimate $\mathbf{P}^L(k\,|k-1)$, the Jacobian of the observation model ∇h, and the covariance of the observation error $\mathbf{R}(k)$.

$$\mathbf{S}(k) = \nabla h(k)\mathbf{P}^L(k\,|k-1)\nabla h^T(k) + \mathbf{R}(k) \qquad (8.22)$$

3. Update

At time k, the estimate of system state can be computed from the prior estimate of the state, the Kalman filter gain $\mathbf{W}$ and the innovation υ

$$\hat{\mathbf{X}}(k\,|k) = \hat{\mathbf{X}}(k\,|k-1) + \mathbf{W}(k)\upsilon(k) \qquad (8.23)$$

That is to say the prediction can be adjusted to the optimization by the Kalman Filter gain and the innovation, where

$$\mathbf{W}(k) = \hat{\mathbf{P}}(k\,|k)\nabla h^T(k)\mathbf{S}^{-1}(k) \tag{8.24}$$

$$
\begin{aligned}
\hat{\mathbf{P}}(k\,|k) &= \hat{\mathbf{P}}(k\,|k) - \mathbf{W}(k)\mathbf{S}(k)\mathbf{W}^T(k) \\
&= \begin{bmatrix} \hat{\mathbf{P}}^G(k\,|k-1) & 0 \\ 0 & \hat{\mathbf{P}}^L - \mathbf{W}(k)\mathbf{S}(k)\mathbf{W}^T(k) \end{bmatrix}
\end{aligned} \tag{8.25}
$$

Here, because the local frame of reference is independent of the global frame, the estimate in the global frame does not need to update, which will reduce the load of the computation.

4. Reference transformation

The estimate of the pose of the local frame, $\hat{\mathbf{X}}_L^G(k)$, represents the relative transformation between the global frame of reference, $\mathbf{G}_L$, and the current local frame of reference, $\mathbf{F}_L$.

$$
\hat{\mathbf{X}}^G(k\,|k) = \begin{bmatrix} \hat{\mathbf{X}}_L^G(k\,|k) \\ \hat{\mathbf{X}}_m^G(k\,|k) \\ \hat{\mathbf{X}}_v^L(k\,|k) \oplus \hat{\mathbf{X}}_L^G(k\,|k) \\ \hat{\mathbf{X}}_m^L(k\,|k) \oplus \hat{\mathbf{X}}_L^G(k\,|k) \end{bmatrix} \tag{8.26}
$$

where, $\oplus$ denotes a compounding operator used to calculate the resultant relationship from addition between different frames of reference, including offset and rotation [8].

Information about the landmark states may be distributed between the global and local frames of reference. This may result in multiple estimates associated with the same landmark when the local map landmark estimates are transformed to the global frame. These independent estimates of the landmark location can be fused together to recover all of the information by the following constraint

$$\hat{\mathbf{X}}_i^G(k\,|k-1) - (\hat{\mathbf{X}}_L^G(k\,|k-1) \oplus \hat{\mathbf{X}}_i^L(k\,|k-1) = 0 \tag{8.27}$$

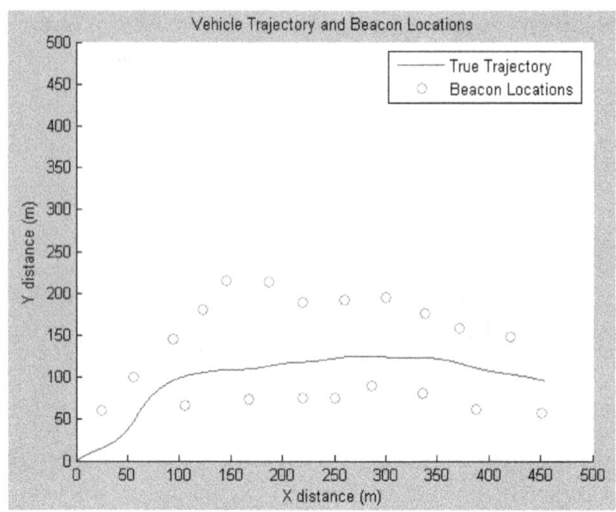

Figure 8.16.   Robot trajectory and landmark locations.

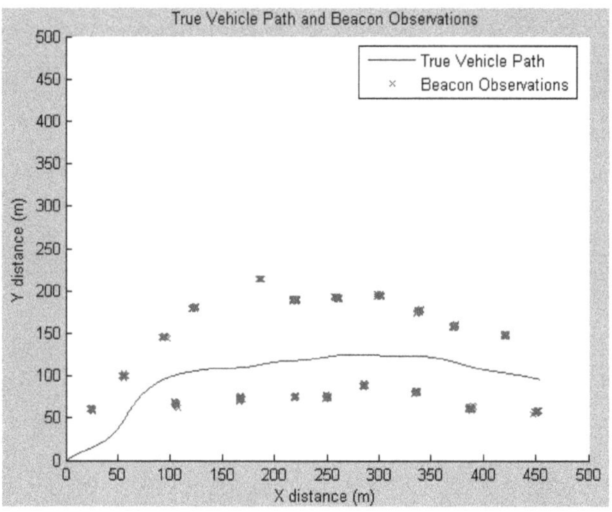

Figure 8.17.   True robot path and landmark observations.

8.4.3 *Simulation*

This section presents the results of the application of the EKF algorithm based on the local maps.

Figure 8.16 shows a map of a simulated environment in which the small circles are landmarks and the curve is the robot trajectory, which are all set by hand randomly. There are 20 landmarks. The distance of the robot path is 453.4 meters, and it takes the robot 231.4s to finish. The robot can only observe the landmarks within 100 meters. The local frame of reference is established every 20 seconds.

Figure 8.17 shows the map which is established by SLAM using the EKF algorithm based on the local maps. The map includes the true robot path and the landmark observations. The landmark near (140,220) is not observed by the robot because it is beyond the range of the sensor.

Figure 8.18 shows the errors of x, y and orientation in the simulation of EKF algorithm based on the local map. From the figure you can find that the true error is below the estimated error and is not cumulated with the time.

8.5 Data Association Approach for Mobile Robot SLAM

Data association in SLAM is the process to establish the correspondence between the sensor measurements (or observations) and the map features to determine whether they originated from the same physical characteristic of the environment [4]. It is the prerequisite and basis for state estimation, and influences the real-time performance and accuracy of SLAM. Therefore, data association plays a very important role in SLAM.

At present, there are many data association algorithm applied in SLAM proposed, and among them the most commonly used are Individual Compatibility Nearest Neighbor (ICNN) [3], Joint Compatibility Branch and Bound (JCBB) [42], and Multiple Hypothesis Tracking (MHT) [14] and *et al.* In addition, Tim Bailey introduced the graph theory into the SLAM data association by looking for two complete graphs the maximum common sub-graph between the observations and the map. This method has strong anti-interference ability, but the problem of searching the largest common sub-graph is difficult. Furthermore, geometric constraints between observations have to be extracted to construct the complete graph. And when the number of the observations increases, the computation will increase significantly. Zhang Sen converts the SLAM data association to 0-1 integer programming problem, and uses multi-dimensional assignment to solve the single frame data association [61]. It is robust, but

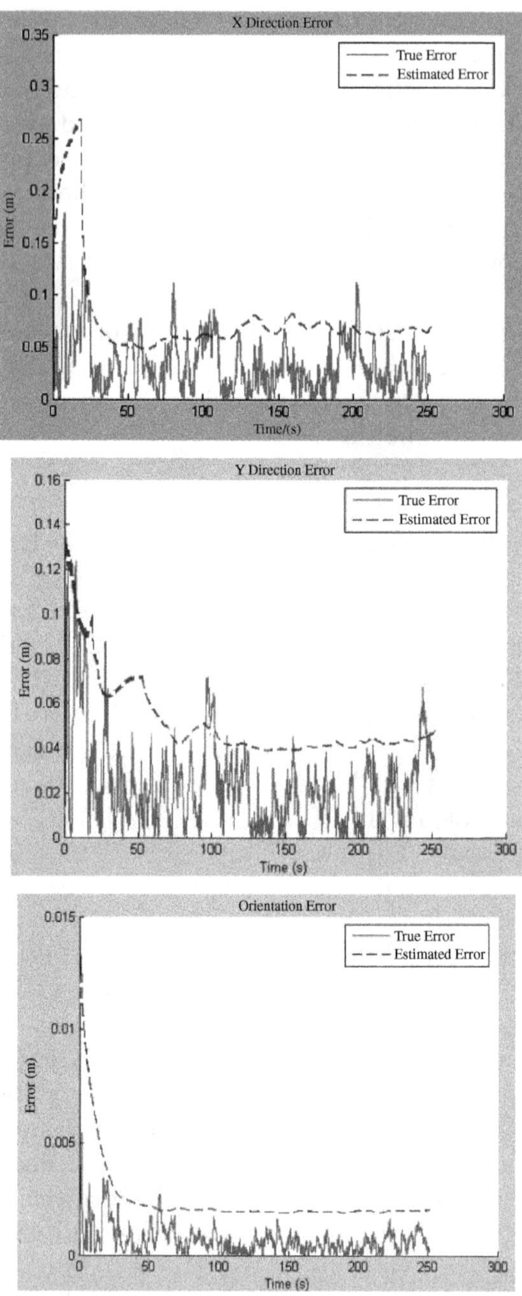

Figure 8.18.   Errors in the local map EKF algorithm.

cannot be applied in real time. Dirk proposed a lazy method which needs to calculate the inverse matrix of large dimension [11]. In other approaches, the evolutionary computation, such as ant colony algorithm, are used in data association [60]. These methods have the same problem that they cannot be applied in real-time or have low accuracy.

The classic ICNN algorithm is simple, easy to realize and real-time, but it has low correct association rate and easily leads to divergent results of SLAM. Another classic JCBB algorithm is of high accuracy but large amount of computation. This paper presents a hybrid approach of data association based on local maps by combining ICNN algorithm and JCBB algorithm, called as LIJ (Local ICNN-JCBB). LIJ method firstly uses ICNN for data association in the local map, and then detect whether the data association results are right or wrong. If wrong, JCBB will be used to correct the result in the local area around mismatched observations. In the simple environment, the method is real-time as ICNN. In the complex environment, the incorrect results are modified by JCBB to ensure the robustness. At the same time, in the whole process of the method only local maps are used to associate or match, so a good time performance can be guaranteed.

8.5.1 *Data association problem in SLAM*

During the process of SLAM, we suppose that there are n features $F = \{F_1, F_2, \ldots, F_n\}$ in the built map and m observations $E = \{E_1, E_2, \ldots, E_m\}$ by sensors. Data association is to set the relationship between the features and the observations, which can be described as K_t

$$K_t = \{k_1, k_2, \ldots, k_m\} \tag{8.28}$$

If the ith observation E_i $(i = 1, 2, \ldots, m)$ is corresponded to the jth map feature $F_j (j = 1, 2, \ldots, n)$, then $k_i = j$ $(k_i \in K_t)$. If the ith observation E_i cannot be corresponded to any features, then $k_i = 0$, which means that the observation E_i may be a new feature or a false observation.

The ith observation at time t denoted $\mathbf{z}_{t,i}$ is given by

$$\mathbf{z}_{t,i} = \boldsymbol{h}_{t,j}(\mathbf{x}_{t|t-1}) + \boldsymbol{\omega}_{t,i} \tag{8.29}$$

where $\boldsymbol{h}_{t,j}()$ is a non-linear observation function which models the relationship between the observation of system states and the states themselves, and $\boldsymbol{\omega}_{t,i}$ is a random vector of measurement noise with zero mean and covariance $\boldsymbol{R}_{t,i}$. The observation function is linearized around the current estimation by

$$\boldsymbol{h}_{t,j}(x_{t|t-1}) \cong \boldsymbol{h}_{t,j}(\hat{\boldsymbol{x}}_{t|t-1}) + \boldsymbol{H}_{t,j}(\boldsymbol{x}_t - \hat{\boldsymbol{x}}_{t|t-1}) \qquad (8.30)$$

where the Jacobian $\boldsymbol{H}_{t,j} = \left.\frac{\partial h_{t,j}}{\partial \boldsymbol{x}_{t|t-1}}\right|_{(\hat{\boldsymbol{x}}_{t|t-1})}$. The distance between the ith observation and the jth map feature can be calculated by the innovation and innovation covariance

$$v_{t,ij} = \mathbf{z}_{t,i} - \boldsymbol{h}_{t,j}(\hat{\mathbf{x}}_{t|t-1}) \qquad (8.31)$$

$$\boldsymbol{S}_{t,ij} = \boldsymbol{H}_{t,j}\boldsymbol{P}_{t|t-1}\boldsymbol{H}_{t,j}^{\mathrm{T}} + \boldsymbol{R}_{t,i} \qquad (8.32)$$

The individual compatibility between E_i and F_j can be determined using an innovation test that measures the Mahalanobis distance as follows:

$$D_{t,ij}^2 = v_{t,ij}^{\mathrm{T}}\boldsymbol{S}_{t,ij}^{-1}v_{t,ij} < \chi_{d,1-\alpha}^2 \qquad (8.33)$$

where $d = \dim(h_{t,j})$ and 1-α is the desired confidence level (usually set to 95%).

The ICNN selection criterion for a given measurement is to choose the one with smallest distance among the features who fall inside the acceptance region according to the formula (8.32).

The ICNN algorithm needs do the association test $m \times n$ times. Thus the computation is linear with the map features, $O(mn)$. The ICNN is easy to realize and low computational complexity, but it has the weak anti-jamming capability and poor robustness. It is only suitable for a case where there are few features or there are large intervals between the features.

The ICNN algorithm can only guarantee the local correct because of not overall consideration. That is to say, the individual compatibility cannot be guaranteed to be jointly compatible to form a consistent hypothesis. Thus, during the SLAM process the state vector will be updated by a set of incompatible measurements, which cause the SLAM to diverge. In order to ensure the SLAM process consistency, all of the measurements should be jointed to find the associated features.

The JCBB algorithm considers all observations at one frame, and the joint predicted observation and its linearization are as follows.

$$\boldsymbol{h}_{K_t}(\hat{\boldsymbol{x}}_{t|t-1}) = \begin{bmatrix} \boldsymbol{h}_{k_1}(\hat{\boldsymbol{x}}_{t|t-1}) \\ \vdots \\ \boldsymbol{h}_{k_m}(\hat{\boldsymbol{x}}_{t|t-1}) \end{bmatrix} \tag{8.34}$$

$$\boldsymbol{h}_{K_t}(\boldsymbol{x}_{t|t-1}) \cong \boldsymbol{h}_{K_t}(\hat{\boldsymbol{x}}_{t|t-1}) + \boldsymbol{H}_{K_t}(\boldsymbol{x}_t - \hat{\boldsymbol{x}}_{t|t-1}) \tag{8.35}$$

where $\boldsymbol{H}_{K_t} = \begin{bmatrix} \boldsymbol{H}_{k_1} & \cdots & \boldsymbol{H}_{k_m} \end{bmatrix}^{\mathrm{T}}$. The joint innovation and innovation covariance are calculated as

$$\upsilon_{K_t} = \boldsymbol{z}_t - \boldsymbol{h}_{K_t}(\hat{\mathbf{x}}_{t|t-1}) \tag{8.36}$$

$$\boldsymbol{S}_{K_t} = \boldsymbol{H}_{K_t}\boldsymbol{P}_{t|t-1}\boldsymbol{H}_{K_t}^{\mathrm{T}} + \boldsymbol{R}_{K_t} \tag{8.37}$$

The JCBB algorithm is to determine the association by the Mahalanobis distance of the joint pairs [6].

$$D_{H_t}^2 = \upsilon_{H_t}^{\mathrm{T}} \boldsymbol{S}_{H_t}^{-1} \upsilon_{H_t} < \chi_{d,1-\alpha}^2 \tag{8.38}$$

When the features in the mobile robot working environment are many and near each other, the JCBB algorithm works better and more reliable than the ICNN algorithm. However, the JCBB algorithm's computation is exponential on the number of the observations. With the size of the environment expanding, its run-time will be significantly increased. So the JCBB algorithm is only fit for a small environment.

8.5.2 *Hybrid data association approach*

ICNN and JCBB approach have their own advantages and disadvantages respectively. We consider making use of their advantages by combining them together. For the sake of time performance of the data association, the hybrid approach is mainly based on the real-time ICNN algorithm. And in order to improve the accuracy, JCBB is used to correct the wrong association pairs by ICNN after checking them out. Here we use a general hypothesis: Different observations originate from different map features, and a map feature has only one matching observation [42, 53, 61].

If k_i is an element of the data association hypothesis K_t, the value of k_i is the serial number of the map feature which matches the ith observation E_i. According to the general hypothesis, all of the K_t elements should have the different values to matching different features except zero values which represent failure to find the matching features or false observations.

So we can obtain a criterion to judge the wrong association pairs: If K_t has the elements with the same non-zero value, these elements are wrong matching pairs.

In the hybrid approach, JCBB will be used after finding the wrong matching pairs. In order to reducing the computational load of JCBB and reducing the time complexity of the hybrid approach, all of the data association are only done in local maps.

The hybrid data association algorithm LIJ mainly has the following steps:

Step 1: ICNN data association based on local map
(1) Establish a circle local map. That is, determine the region in which there are features most likely to match the current observations at time t in the constructed global map. In order to decrease the computational complexity and cover all the possible features, we set the circle center as the location of the mobile robot and radius as r

$$\sqrt{(x_j - x_r)^2 + (y_j - y_r)^2} < r \tag{8.39}$$

which (x_r, y_r) and (x_j, y_j) are the positions of the mobile robot and feature F_j in the global coordinate system respectively.

(2) Calculate the data associate between the features F_j in the circle local map and the observations E_i using ICNN algorithm, and get the association assumption $K_t' = \{k_1, k_2, \ldots, k_m\}$. The value of k_i is the serial number of the feature who matches the observation E_i. For example, $k_i = j$ means that the observation E_i matches the feature F_j.

Step 2: Judgment whether there are association errors or not. If yes, then turn to Step 3, otherwise turn to Step 4.

Here, we use a common assumption in SLAM data association: different observations originate from the different features, and one map feature can produce at most one observation. If there are more than one observation matched with the same feature, which will result in the wrong data association in SLAM. Thus, we build a criterion to evaluate the ICNN result: if there are same non-zero values in K'_t, then the data association result of ICNN algorithm has errors.

Step 3: Modification by JCBB algorithm

(1) Find the observations E_a which should be re-matched.

As JCBB computational complexity increases exponentially with the number of observations, we consider not using all observations to re-match but only those around the map features which have been wrong matched. Firstly, find the non-zero elements $k_c(c \in [1, m])$ of the same value in K'_t. Through k_c we can get the map features $F_b(b = k_c, k_c \in [1, n])$ which have been wrong matched by ICNN algorithm. Secondly, find the observations $E_a(a \in [1, m])$ around the features F_b by satisfying a threshold d

$$\sqrt{(x_{E_a} - x_{F_b})^2 + (y_{E_a} - y_{F_b})^2} < d \tag{8.40}$$

which (x_{E_a}, y_{E_a}) and (x_{F_b}, y_{F_b}) are the positions of the observation E_a and feature F_b in the global coordinate system respectively. The threshold d is an important factor which influences the number of the re-matched observations. If the threshold is set too high, it is not helpful to improve the real time performance. Otherwise too small a setting will reduce the accuracy. It has been proven in the experiment.

(2) Calculate the data association between the features F_j in the circle local map and the observations E_a using JCBB algorithm, and use the result to modify the result K'_t of Step 1.

Step 4: Getting the association result $K_t = K'_t$. Data association is end at time t.

8.5.3 *Experimental results*

We use Jose Neira's simulation data set [40] to do data association experiments with EKF SLAM. In the experiments, the mobile robot moves at linear velocity $0.3125\,\mathrm{m/s}$ and angular velocity $9°/\mathrm{s}$, and the sensor range is 0–$3.5\,\mathrm{m}$ and $-90° \sim +90°$ with sensing period 1s. Because the hybrid approach of data association proposed in the paper is mainly based on local ICNN and JCBB, we refer to it as LIJ (Local ICNN-JCBB). Here, the threshold r is set to $4\,\mathrm{m}$ and d is set to $0.5\,\mathrm{m}$.

Figure 8.19 shows the running times of ICNN, JCBB and LIJ in the same environment. We can see that LIJ's running time is between

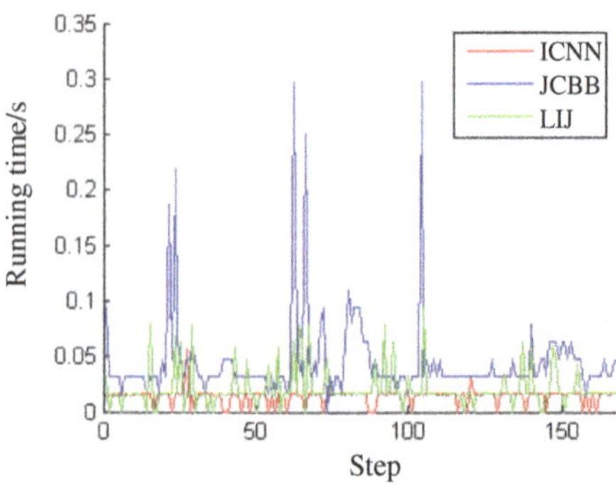

Figure 8.19.　Running time of ICNN, JCBB and LIJ.

ICNN and JCBB. LIJ needs more steps in order to determine the local maps and wrong matching than ICNN, but the test to determine the compatibility between a sensor observation and a map feature is only done in local map. Thus when the features in environment are sparse, LIJ is almost the same as ICNN. However, as the environment become more complex, wrong matching increases. LIJ's running time also increases by using JCBB to correct mistakes. Because these are all done in local map, LIJ still has a good time performance.

Figure 8.20 shows the correct rate of the three association method in the same environment. ICNN is the lowest, while LIJ and JCBB are almost same high.

In order to compare the performance of the three data association methods in different environment, we set a simple simulation environments (features number: 88) and a complex simulation environment (features number: 176), and get the average running time as Table 8.2 and average correct rate as Table 8.3 with different threshold d. We can see that ICNN has the shortest running time with lowest correct rate, and JCBB has the highest correct rate with longest running time, and LIJ's time performance is close to ICNN with correct rate close to JCBB. LIJ has a good time and accuracy performance in different complexity environment. At the same time, we can also find that with the value of d increasing, both of LIJ's average running time and average correct rate increase because of the features involved increasing. The threshold d in LIJ approach has the opposite effect in real-time and accuracy, but in a certain range it can

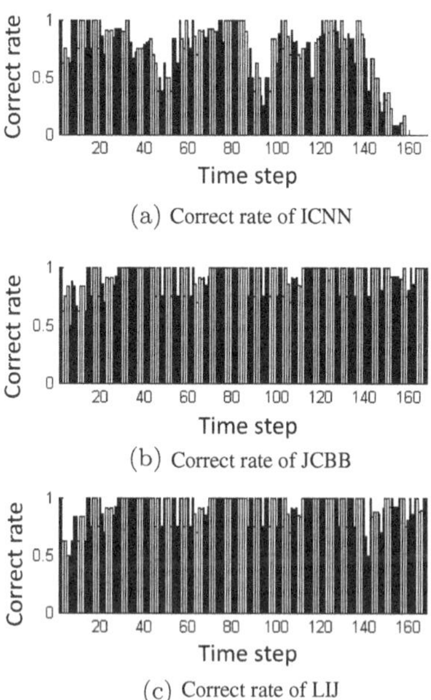

Figure 8.20.   Correct rate of ICNN, JCBB and LIJ.

Table 8.2.   Performance of ICNN, JCBB and LIJ in a simple environment.

Data association methods	Average running time (s)	Average correct ratio (%)
ICNN	0.0113	0.7029
JCBB	0.0235	0.9232
LIJ($d = 0.5\,$m)	0.0138	0.9130
LIJ($d = 1.0\,$m)	0.0140	0.9160
LIJ($d = 1.5\,$m)	0.0159	0.9245

compromise to balance. It can be adjusted according to the different needs in practical application.

8.6   Mobile Robot SLAM in Dynamic Environment

Simultaneous Localization and Mapping (SLAM) is a fundamental and hot problem in mobile robotics. It is one of the most important

Table 8.3. Performance of ICNN, JCBB and LIJ in a complex environment.

Data association methods	Average running time (s)	Average correct ratio (%)
ICNN	0.0151	0.6915
JCBB	0.0446	0.9316
LIJ($d = 0.5\,\mathrm{m}$)	0.0210	0.9194
LIJ($d = 1.0\,\mathrm{m}$)	0.0211	0.9255
LIJ($d = 1.5\,\mathrm{m}$)	0.0234	0.9271

qualifications for mobile robot to realize autonomy. However, most of the methods are based on the assumption that the environments are static. Since most of the real world is dynamic and the presence of moving targets causes errors and compromises the overall quality of the map, it is necessary to research the SLAM approaches in dynamic environments.

In [27], the moving targets in the environment are identified based on the difference of consecutive sensor readings. The objective is filtered out of those sensor readings corresponding to dynamic targets in order to improve the quality of the map. In [26], the EM (Expectation-Maximization) algorithm was used to differentiate dynamic and static parts of the environment off-line. Wang C.C. *et al.* proposed a combination of SLAM and DTMO (Detection and Tracking of Moving Objects) in which scan matching algorithms are used to SLAM and moving targets are detected and tracked [15]. Later, they give a framework for solving the simultaneous mapping, localization, detection and tracking of moving targets in order to improve the quality of the map [54]. Wolf and Sukhatme presented an on-line mapping algorithm capable of differentiating static and dynamic parts of the environment even when the moving targets change position out of field of the view of the robot [58].

Most of the researches can be concluded as two kinds according to the dynamic target processing method. One is to filter out or remove the dynamic information; the other is to track them. Dynamic targets affect the accuracy and complexity of SLAM directly, and then filtering them out will lead the incomplete of the observation and the building map. In addition, there are many researches on some block of SLAM in dynamic environment, but the researches on the overall system framework and its implementation are few. Thus, this chapter introduces the complete system of SLAM in dynamic environment, namely SLAMiDE system. Compared to the traditional SLAM, SLAMiDE adds the dynamic targets processing and applies it to the mapping and localization module.

8.6.1 *Real-time detection of dynamic obstacle by laser radar*

Dynamic obstacle detection is a basic problem of the mobile robot map building in unknown environments. The fake observations produced by dynamic obstacles affect the accuracy of the map [28,31,58]. In order to build the accurate and consistent maps, it is very important to detect the dynamic obstacles and filter them or use them. It is also the problem that must be solved in the process of the mobile robot navigation. Biswas *et al.* detect changes over time in an environment by map differencing technique which is usually used in the machine vision [5]. However the method is not of high reliability. Wolf *et al.* maintain two occupancy grid maps (One models the static obstacles and the other models the dynamic ones) without considering the localization error and uncertain factors [57]. In [15], Simultaneous Localization and Mapping (SLAM) and Detecting and Tracking Moving Objects (DATMO) are integrated to solve both problems simultaneously for both indoor and outdoor applications. In [21], Prassler *et al.* present a concept of Time-Stamp Map which is a projection of range information obtained over a short interval of time onto a two-dimensional grid, where each cell which coincides with a specific range value is assigned a time stamp. In order to enhance the validity and reliability of the real-time detection of dynamic obstacles in unknown environments, this paper presents a detection method indoor by mobile robot. 2D laser radar is navigation sensor to observe environment. Its observations are projected onto the grid map. The method maintains three sequential grid maps and compares the same cell of them to decide whether it is occupied by static obstacle or dynamic obstacle. Eight-neighbor rolling window is used to deal with the uncertain information. Experimental tests have been performed using mobile robot MORCS-1 made by the Intelligent Centre of Central South University.

1. Dynamic obstacle detection

The working environment of mobile robot is described by 2D Cartesian grids. A grid map represents environment by means of a two dimensional evenly-spaced cell $c_{i,j}$. Each grid cell estimates the probability of the corresponding region being an **occupied** ($c_{i,j} = 1$) or **empty** ($c_{i,j} = 0$) space area in the environment.

It is very convenient and direct to describe the distribution of the obstacles especially in the indoor environment by mapping the observations in the grids real time. However, the laser radar alone cannot identify the kind of the obstacles. The history information is needed to help decide whether the cell is occupied by dynamic or static obstacle. In this paper, by differencing three consecutive grid maps the dynamic obstacles are detected.

Order the grid map at time t as M_t. Then M_{t-1} is the grid map at time $t-1$ and M_{t+1} at time $t+1$. If the states of the same cell $c_{i,j}$ in M_{t-1}, M_t and M_{t+1} are all occupied, the cell is occupied by static obstacle, or else by dynamic obstacles potentially. Figure 8.21 shows an example of the dynamic obstacles detection by three OGM.

In Figure 8.21, the grey space in the grid represents an unknown region, the black space represents an occupied region, and the white space represents an empty (or free) region. The cell *C3* is black in M_{t-1}, M_t and M_{t+1}, which means that at three consecutive time

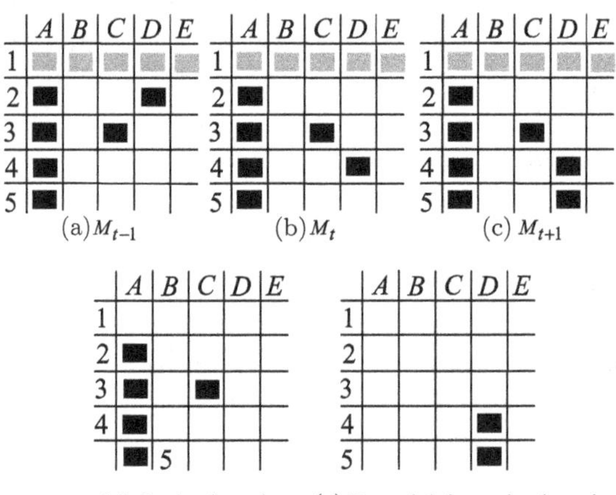

Figure 8.21.   Real-time detection of dynamic obstacles based on maintaining grid maps.

the obstacle is observed still by laser radar in the same region, see Figure 8.21(a), Figure 8.21(b) and Figure 8.21(c). The cell $D2$ is black in M_{t-1} but white in M_t and M_{t+1}, which means the region corresponding to the cell $D2$ is occupied by a potential dynamic obstacle (a dynamic obstacle or a new static obstacle), see Figure 8.21(d) and Figure 8.21(e). In the same way, the cell $D4$ and $D5$ are also potential dynamic obstacles. Figure 8.21(d) and (e) are static and potential dynamic obstacles respectively resulting from the difference of the three consecutive grid maps.

Potential dynamic obstacles are identified by comparing the current grid map and the map of already identified static obstacles. Then after one more scan period the dynamic obstacles or the new static obstacles can be distinguished from the potential dynamic obstacles.

2. Uncertainty treatment

In the process of the mobile robot mapping there are many uncertain factors from laser radar measurement, coordinates transformation, dead-reckoning localization and *et al.* The same static obstacle observed by laser radar would not always at the same cell in the grid map because of the uncertainty. Therefore, it is very important to take the uncertainty into account.

If a cell is black in M_{t-1} and M_t but white in M_{t+1}, the obstacle at this cell maybe static. By analyzing the observations during the mobile robot movement, the angles measured at the consecutive time are of high relativity. The angles of the neighbor scans at one time are relatively high too [1]. The spatiotemporal relativity of the laser radar measurements could help the reliability and validity of the dynamic obstacles detection. Here the eight-neighbor cells are considered. In other words, the eight-neighbor cells should be considered before the cell $c_{i,j}$ is going to be determined static. The eight-neighbor cells are written as

$$
\begin{array}{ccc}
c_{i-1,j-1} & c_{i-1,j} & c_{i-1,j+1} \\
c_{i,j-1} & c_{i,j} & c_{i,j+1} \\
c_{i+1,j-1} & c_{i+1,j} & c_{i+1,j+1}
\end{array}
$$

If the values of the cell $c_{i,j}$ in three consecutive grid maps are consistent as 1 (black) and $c_{i+m,j+n} = 1$ (m, $n = -1$, 0, 1; m and n would not equal 0 at the same time), the cell $c_{i,j}$ is occupied by static obstacle. The simple treatment to the uncertainty enhances

Figure 8.22.　Mobile robot MORCS-1.

the capability of the static obstacle detection and the reliability of the dynamic obstacle detection.

3. Experimental results

In order to validate the ideas presented in this paper, experimental tests have been performed using MORCS-1. The size of MORCS-1 is 800 mm × 700 mm × 900 mm with four drive wheels and an omni-directional wheel as Figure 8.22.

In the experiment, we use cm. mode and set the frequency as 20 Hz, the angular resolution as 0.5°. LMS291 is in serial communication with mobile robot control computer by RS422 protocol with 500 Kbaud communication rate. One cell corresponds to 5 cm × 5 cm region in real environment. The grid maps are set 800 × 600. All of the experiments are performed in the office and people are walking back and forth as dynamic obstacles. The mobile robot MORCS-1 navigates in an unknown dynamic environment and localizes by dead-reckoning assuming that the localization error is neglectable in a short time.

(1) Detection when mobile robot remains still

In this experiment, MORCS is still to detect the dynamic obstacles in an office room with 10 m × 12 m size. The walking speed of the people is about 100 cm/s. When the 26th group of LMS291 observations are processed, the detecting result is as Figure 8.23.

In Figure 8.23, there are two passersby and two still boxes in the room. The blue rectangle represents the robot; black grids are static obstacles and red grids are dynamic obstacles. From the result, we can see the static and dynamic obstacles are detected availably. The

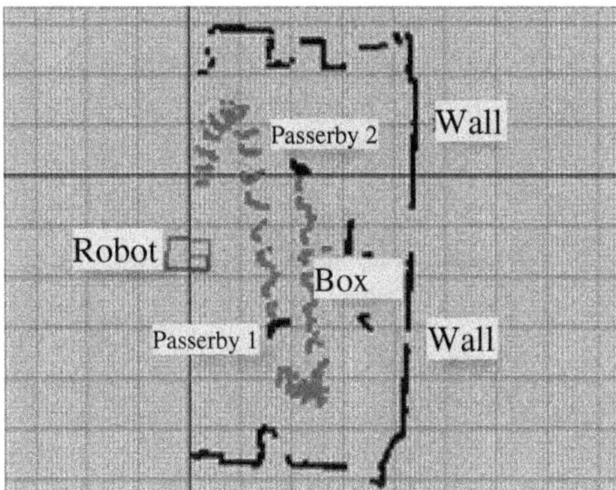

Figure 8.23.   Real-time detection when mobile robot is still.

tracks of Passerby 1 and Passerby 2 are continuous. But the wall is ruptured because the passersby and the boxes blocked the laser beams, which is unavoidable when mobile robot is still. We test 24 times, the average detection time is 30 ms and the dynamic obstacle detection is 97.6%.

(2) Detection when mobile robot is moving
The first set of the experiments is implemented at the corridor with the size of $12\,m \times 2.4\,m$. The speed of the robot is $10\,cm/s$, and the speed of the passersby is about $40\,cm/s$. 60 sets of the returns from LMS291 are processed during the mobile robot navigation.

Mobile robot detects the dynamic obstacles real-time in the corridor as Figure 8.24. The above and below red tracks are Passerby 1 and Passerby 2 respectively. The direction of the passersby and the mobile robot is just opposite. However the tracks of the passersby are not continuous because of the discontinuity of the communication which results from the disturbances in the environment and the limits of the facility. And there are some static points detected as dynamic ones (some red appearing at the black wall). These errors are occurred with the uncertainties. We test 30 times, and the average detection time is 35 ms and 95.4% the right dynamic obstacles are detected.

The second set of the experiments is implemented in the room with the size of $7.5\,m \times 7.2\,m$. There are some pieces of furniture which complicate the robot working environment. The speed of the robot is $15\,cm/s$, and the speed of the passersby is about $50\,cm/s$. Passerby 1 and Passerby 2 are walking across towards the mobile

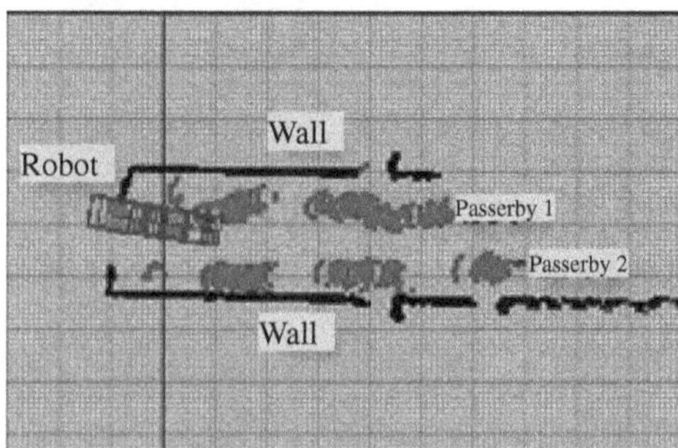

Figure 8.24. Real-time detection in the corridor when mobile robot is moving.

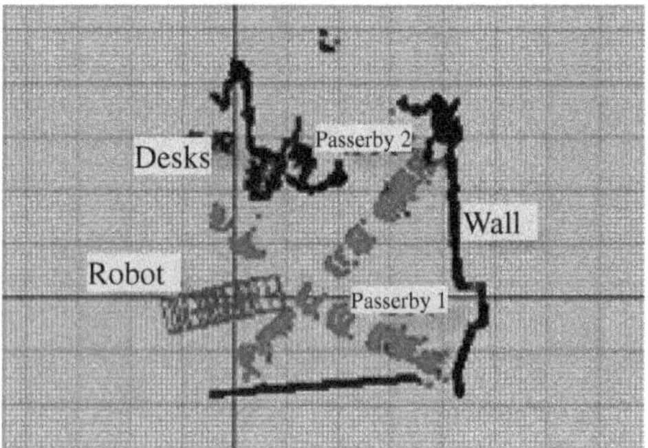

Figure 8.25. Real-time detection in the room when mobile robot is moving.

robot from the above and the below corner respectively. 70 sets of data are returned. As Figure 8.25, the method accurately detects not only the passersby as dynamic obstacles but also the wall and the desks and the chairs as static obstacles. There is some detection leaved out in some scan period due to the communication delay. The experiment is tested 20 times, and the average detection time is 32 ms and 94.8% the right dynamic obstacles are detected.

8.6.2 *Uniform target model*

Because mobile robot SLAM in a dynamic environment involves more types of targets (objects) than in a static environment, building a uniform model of these targets is very important.

After clustering the observation data, the targets are no longer the single point with two coordinate values but a set of points. So the attributes of the points set are used to describe the target. In order to distinguish between the static and dynamic target, the movement velocity and direction of the target are needed; in order to determine whether they are the same targets, the shape or size of the target is needed. Altogether, there are four parameters to set up the target model. Suppose that observation $\mathbf{z}_t$ has m targets at time t, and then the jth target $O_{t,j}$ have four parameters: target coverage area $[x_{t,j}\ y_{t,j}]^T$, target centroid $[\bar{x}_{t,j}\ \bar{y}_{t,j}]^T$, target motion parameter $\mathbf{v}_{t,j}$, and target type $flag_{t,j}$.

$$\mathbf{z}_t = \{O_{t,1}, O_{t,2}, \cdots O_{t,m}\} \tag{8.41}$$

$$O_{t,j} = \{[x_{t,j}\ y_{t,j}]^T, [\bar{x}_{t,j}\ \bar{y}_{t,j}]^T, \mathbf{v}_{t,j}, flag_{t,j}\} \tag{8.42}$$

Here, the coverage area of the target $O_{t,j}$ is the area limited by the furthest points away from each other in the global coordination, and target centroid is calculated by the average measurements of the target, and $flag_{t,j} = 1, -1, 0$ represents the target kind: static, dynamic and unknown respectively.

After calculating the target centroid, the target motion parameter can be estimated by the two continuous sampling as follow:

$$\mathbf{v}_{t,j} = (O_{t,j} - O_{t-1,j})/\Delta t \tag{8.43}$$

Here, Δt is the sampling period. The target motion parameter is used to evaluate, predict and track the target state.

When predicting the dynamic target, the target state can be described as four parameters $(\bar{x}, \bar{y}, vx, vy)$, which $\bar{x}$ and $\bar{y}$ means the target centroid, and vx and vy means the dynamic target velocity on X and Y direction respectively. So we can obtain the kinematics model of the dynamic target as follow:

$$\begin{bmatrix} \bar{x}_{t,j} \\ \bar{y}_{t,j} \\ vx_{t,j} \\ vy_{t,j} \end{bmatrix} = \begin{bmatrix} 1 & 0 & T & 0 \\ 0 & 1 & 0 & T \\ 0 & 0 & 1 & 0 \\ 0 & 0 & 0 & 1 \end{bmatrix} \begin{bmatrix} \bar{x}_{t-1,j} \\ \bar{y}_{t-1,j} \\ vx_{t-1,j} \\ vy_{t-1,j} \end{bmatrix} + \sigma_t \tag{8.44}$$

Error σ_t denotes dynamic noises and model uncertainty. If the sensor is laser radar, the error σ_t mainly includes the laser radar measurement error and the relative motion error between the mobile robot and dynamic targets.

8.6.3   *SLAMiDE system*

1. Principle and architecture of the SLAMiDE system

The mobile robot SLAM in a dynamic environment is the process to estimate the system state through observation information $\mathbf{z}_{0:t}$ and the control input information $\mathbf{u}_{0:t}$. The SLAMiDE system state includes the mobile robot pose $\mathbf{x}_t$, the static map $\mathbf{S}_t$, the dynamic target $\mathbf{D}_t$ and the unknown information $\mathbf{E}_t$ in the observations.

$$p(\mathbf{x}_t, \mathbf{S}_t, \mathbf{D}_t, \mathbf{E}_t \,|\, \mathbf{z}_{0:t}, \mathbf{u}_{0:t}) \tag{8.45}$$

It can be decomposed by Bayesian Rule as

$$p(\mathbf{E}_t \,|\, \mathbf{x}_t, \mathbf{z}_{0:t}, \mathbf{m}_t) \cdot p(\mathbf{D}_t \,|\, \mathbf{x}_t, \mathbf{z}_{0:t}) \cdot p(\mathbf{x}_t, \mathbf{S}_t \,|\, \mathbf{z}_{0:t}, \mathbf{u}_{0:t}) \tag{8.46}$$

Here, $p(\mathbf{x}_t, \mathbf{S}_t \,|\, \mathbf{z}_{0:t}, \mathbf{u}_{0:t})$ is the state estimation in traditional SLAM in static environment, and $p(\mathbf{D}_t \,|\, \mathbf{x}_t, \mathbf{z}_{0:t})$ is the dynamic target tracking estimation and $p(\mathbf{E}_t \,|\, \mathbf{x}_t, \mathbf{z}_{0:t}, \mathbf{m}_t)$ is the estimation process to the unknown information which cannot be determined as static target or dynamic target now.

Observations in current frame usually maybe the features in the map built before or dynamic targets or noises, besides maybe new static targets (new features) and potential dynamic targets. We use the method proposed in [10] to detect the dynamic targets. Then the observations are described as

$$\mathbf{z}_t = \left\{ \mathbf{z}_t^S, \mathbf{z}_t^D, \mathbf{z}_t^E \right\} \tag{8.47}$$

Thus we can obtain the each estimation in expression (8.45) as follows.

$$\widehat{\mathbf{x}}_t^{+} = \arg\max \left\{ p(\widehat{\mathbf{z}}_t^{S} \,\big|\, \widehat{\mathbf{x}}_t^{-}, \mathbf{S}_{t-1}) \cdot p(\widehat{\mathbf{x}}_t^{-} \,|\, \mathbf{x}_{t-1}, \mathbf{u}_t) \right\} \tag{8.48}$$

$$\widehat{\mathbf{S}}_t^{+} = \arg\max \left\{ p(\mathbf{z}_t^S \,\big|\, \widehat{\mathbf{x}}_t^{+}, \widehat{\mathbf{S}}_t^{-}) \cdot p(\widehat{\mathbf{S}}_t^{-} \,\big|\, \widehat{\mathbf{S}}_{t-1}^{+}) \right\} \tag{8.49}$$

$$\widehat{\mathbf{D}}_t^{+} = \arg\max \left\{ p(\mathbf{z}_t^D \,\big|\, \widehat{\mathbf{x}}_t^{+}, \widehat{\mathbf{D}}_t^{-}) \cdot p(\widehat{\mathbf{D}}_t^{-} \,\big|\, \widehat{\mathbf{D}}_{t-1}^{+}) \right\} \tag{8.50}$$

$$\widehat{\mathbf{E}}_t^{+} = \arg\max \left\{ p(\mathbf{z}_t^E \,\big|\, \widehat{\mathbf{x}}_t^{+}, \widehat{\mathbf{E}}_t^{-}) \cdot p(\widehat{\mathbf{E}}_t^{-} \,\big|\, \widehat{\mathbf{E}}_{t-1}^{+}) \right\} \tag{8.51}$$

As seen from the problem description, SLAMiDE system consists of dynamic target detection, mobile robot localization, static map building and dynamic map building. The dynamic and static map is built after detection the dynamic and static targets from the preprocessed observations obtained by external environment sensors (such as camera, laser radar or sonar). At the same time, the mobile robot pose is predicated or calculated by dead reckoning algorithm from the internal sensors (such as fiber optical gyros, odometry etc). The built maps are predicated by the predicated robot pose too, and then do data association with new observations. Finally both of the maps and the robot pose are updated, and the goal of SLAMiDE is achieved.

Figure 8.26 shows the SLAMiDE system framework. The bottom is the sensor layer which mainly has the sensing function with external and internal sensors. The middle layer does some sensor data preprocessing work such as de-noising and clustering and dead reckoning. The top is the localization and mapping layer which is the core of the SLAMiDE system.

Brief description of each functional module in localization and mapping layer are as follows:

(1) Mobile Robot Localization Module
In fact, localization module and static map building constitute the traditional SLAM in static environment. Here we use the particle filter method to localization [13]. The main objective of particle filtering is to "track" a variable of interest as it evolves over time,

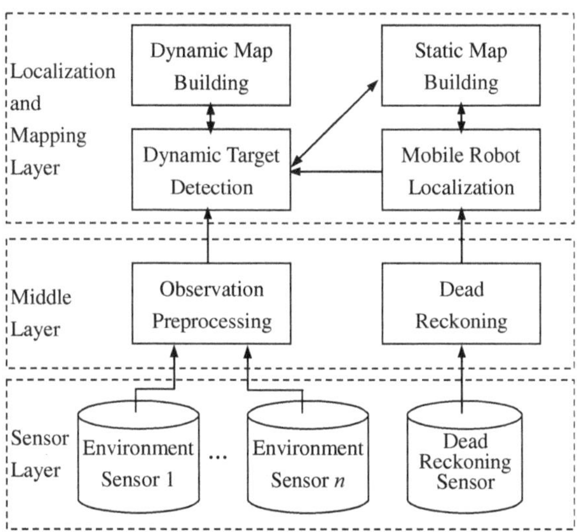

Figure 8.26. SLAMiDE System Framework.

typically with a non-Gaussian and potentially multi-modal posterior distribution function. The key idea of the particle filter for SLAM is to estimate a posterior about potential trajectories of the robot given its observations and its odometry measurements and to use this posterior to compute a posterior over maps and trajectories.

(2) Dynamic Target Detection Module

Dynamic target detection is the key process in SLAMiDE. Its purpose is to differentiate the dynamic, static and unknown target. Here we use the spatially and temporally correlated dynamic target detection method [39].

After filtering out noisy data, sensor readings are mapped into the world coordinate system to build grid map and three consecutive grid maps are maintained. By comparing the state of the same grid cell in the three consecutive grid maps and considering the state of the eight-neighbor cell, static obstacles are identified. Dynamic targets then are identified by comparing the current grid map to the already identified static obstacles. The unknown information set was not determined before because of insufficient information, and then it will be involved in dynamic target detection in current frame.

(3) Dynamic Map Building Module

After the robot current pose predicated, dynamic map building module is predicated the targets' pose in the dynamic map according to the motion parameters of the dynamic targets and the predicated robot pose. The predicated targets are matched with the dynamic targets which are obtained by the dynamic target detection of the observations, and then the dynamic map is updated. In fact, apart from the dynamic targets tracking, the process is the new dynamic targets detected and added in the dynamic map process also.

(4) Static Map Building Module

Static map building module predicates the features in the map too with the predicated current robot pose. The predicated results and the non-dynamic targets from the unknown information determined by dynamic targets detection will do data association, and then the map and the robot pose are updated. Because the current observations are limited by the sensor ranger, the local map is used here to reduce the computational load. The features in the local range of the static map are matched with the observation by JCBB algorithm [34]. After the data association, we will find the features already in the building map or new features or unknown information in the observations.

2. Implementation of the SLAMiDE system

From above, we can see that the four functional modules in the top layer are the four processes to implementation of the SLAMiDE system. The whole process can be described as Figure 8.27. At time t, the system inputs include the previous robot pose $\mathbf{x}_{t-1}$, the current control inputs $\mathbf{u}_t$, the current observation $\mathbf{z}_t$, the previous static map $\mathbf{S}_{t-1}$ and dynamic map $\mathbf{D}_{t-1}$, and the previous unknown information $\mathbf{E}_{t-1}$; the outputs include the current robot pose estimation $\mathbf{x}_t$, the current static map $\mathbf{S}_t$ and dynamic map $\mathbf{D}_t$, and the current unknown information $\mathbf{E}_t$. (Notice: All variables above are the simplified representations of the posterior probability.)

From the vertical perspective, SLAMiDE is an iterative process of prediction and update almost the same as SLAM. From the horizontal perspective, SLAMiDE mainly has four processes including robot localization, dynamic target detection, dynamic map building and static map building. And there is another hidden process about unknown information. These are almost parallel.

SLAMiDE system implementation steps:

(1) According to the previous robot pose and control inputs, the current robot pose is predicated as $\hat{x}_t^-$.

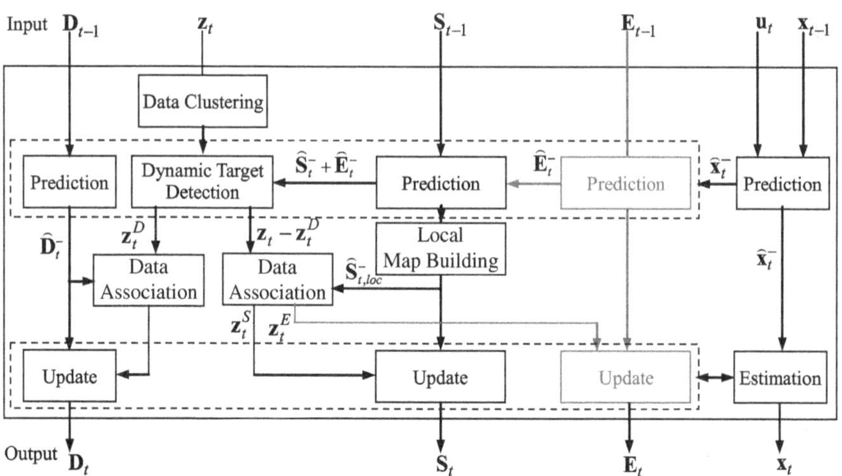

Figure 8.27. SLAMiDE system implementation process.

(2) According to the predicated robot pose $\hat{x}_t^-$, the previous static map $\mathbf{S}_{t-1}$ and unknown information $\mathbf{E}_{t-1}$, the current static map and unknown information are predicated as $\hat{\mathbf{S}}_t^- + \hat{\mathbf{E}}_t^-$.

(3) After observation data clustering, the target model is set up as $\mathbf{z}_t = \{O_{t,1}, O_{t,2}, \cdots O_{t,m}\}$.

(4) Do dynamic targets detection from the observations, and get $\mathbf{z}_t^D$.

(5) Predicate the current dynamic map $\hat{\mathbf{D}}_t^-$ by $\mathbf{D}_{t-1}$. Find out whether there are the same dynamic targets in $\mathbf{z}_t^D$ with the predicated dynamic map. If yes, then tracking them continuously. The new ones are added into the dynamic map and get the current dynamic map estimation $\mathbf{D}_t$.

(6) The remaining observations $\mathbf{z}_t - \mathbf{z}_t^D$ are matched with features in the local static map $\hat{\mathbf{S}}_{t,loc}^-$, and update the static map. The new static targets are added in, and finally the current static map estimation $\mathbf{S}_t$ is obtained.

(7) Still unconfirmed information $\mathbf{z}_t^E$ in observations is added into the unknown information set, and the current unknown information estimation $\mathbf{E}_t$ is obtained.

(8) Finally, according to the updated maps the robot pose estimation is updated as $\mathbf{x}_t$.

Because the SLAMiDE system maintains the mobile robot trajectory, dynamic map, static map and unknown information set, the processing data is huge. In order to reduce the data, SLAMiDE system sets the valid period for the dynamic target and unknown information respectively and only processes the data in the valid period.

8.6.4   *Experimental results*

In order to prove the reliability and correctness of the SLAMiDE system, experimental tests have been performed using MORCS-1 as Figure 8.28 in a simple environment and a complex environment. The speed of MORCS-I is set as 0.05 m/s, and the speed of the pedestrian is about 0.8 m/s–1.6 m/s. The SLAMiDE period is set as 8 times of the system sample period. The valid period of the dynamic target T_d is 5 frames and unknown information T_u is 10 frames.

We can see that the SLAMiDE system handles the dynamic targets well and realizes simultaneous localization and mapping in the simple dynamic environment from Figure 8.28.

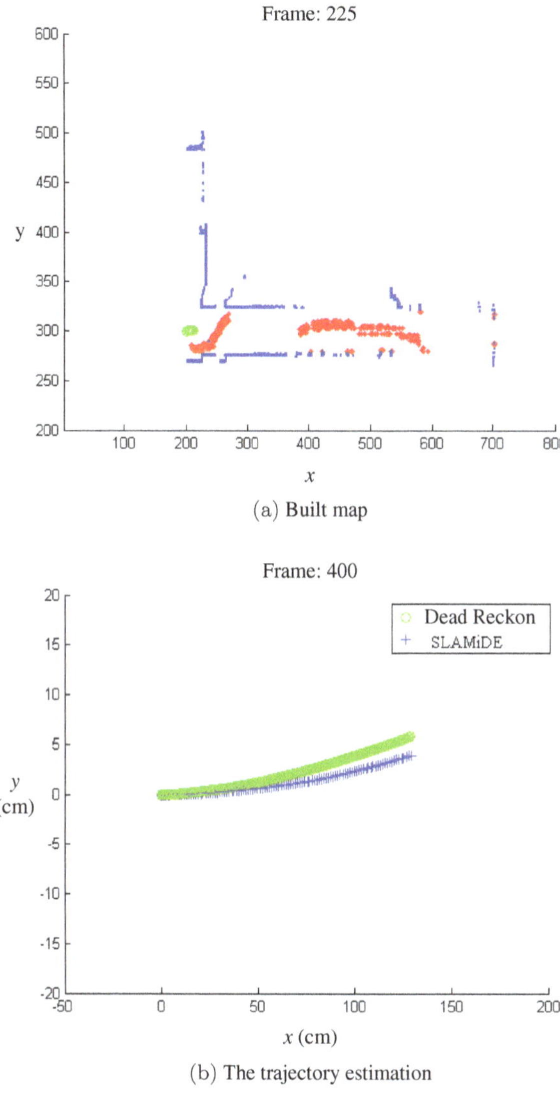

Figure 8.28. Experimental results in corridor environment. (a) The blue is static map, and the red is dynamic map, and the green is the trajectory of MORCS-I. (b) The comparison of the robot trajectory estimation between SLAMiDE and dead reckoning.

In order to verify the valid of SLAMiDE system, the lobby of the same building is chosen to experiment as a more complex environment. There are more complex dynamic targets in the lobby, such

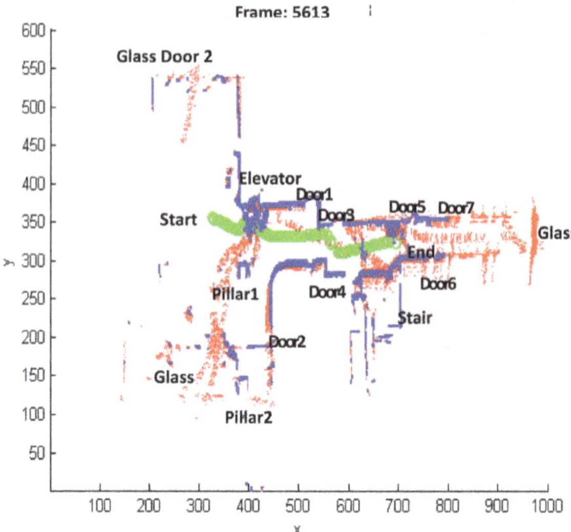

Figure 8.29. Experimental results in lobby. The blue is static map, and the red is dynamic map, and the green is the trajectory of MORCS-I.

as people coming and going, elevator and room doors opening and closing, and people standing still temporarily while waiting for the elevator. The experimental results are shown as Figure 8.29. From the experimental results, we can see that SLAMiDE system can overcome the disturbance from dynamic targets and work well in complex environment too. At the same time we also find out that it is difficult to detect the dynamic targets of temporally static, such as the pedestrians waiting for the elevator. In addition, because of the self-defect, the laser radar is instable to judge some materials such as glass doors.

8.7 Chapter Summary

This chapter studies the localization and mapping of mobile robot, discusses in turn Dead Reckoning Positioning, Simultaneous Localization and Mapping, Data Association Approach of Mobile Robot SLAM, Mobile Robot SLAM, in a Dynamic Environment, and Synchronous localization and mapping of robots based on deep learning. Environmental modeling and localization technology mainly includes two aspects, that is, how to represent the environment and how to build an environmental model and precise positioning.

Section 8.1 briefly introduces the environment model of mobile robot, list the classifications of environmental models commonly used in the field of mobile robot navigation, especially introduce several common plane models, namely metric models, non-metric models and three-dimensional models.

Section 8.2 discusses the dead reckoning positioning including the motion architecture and proprioceptive sensors, design of dead reckoning system, and the simulation and experiment. To describe the motion of mobile robot in operating environment, the world coordinate system and the robot coordinate system are built.

Section 8.3 introduces some typical methods for map building of mobile robots, includes map building based on laser radar, map matching based on maximum likelihood estimation, and self-localization based on feature mapping.

Section 8.4 expounds the simultaneous localization and mapping (SLAM), represents the system space state, analyzes EKF algorithm and the estimation, prediction and observation of the system state, and shows the simulation results.

Section 8.5 probes the data association approach of mobile robot SLAM, gives steps of a hybrid data association algorithm, and then presents experimental results of the running times and the correct rate.

In Section 8.6 a more detailed discussion on mobile robot SLAM in a dynamic environment (SLAMiDE) has conducts. The dynamic obstacle detection is a basic problem of the mobile robot map building in unknown environments as a typical example to demonstrate the lidar real-time detection of dynamic obstacles. Due to the mobile robot SLAM in dynamic environment involves more types of objects than static environment, building a uniform model of these targets is very important; and so discusses the unified target model SLAMiDE and its experimental results.

References

[1] Alexey, S. M., Wang, C. and Andrey, V. S. (2012). Real-time navigation of mobile robots in problems of border patrolling and avoiding collisions with moving and deforming obstacles. *Robotics and Autonomous Systems*, 60(6):769–788.

[2] Andrew, J.D. and Nobuyiki, K. (2001). 3D simultaneous localization and map building using active vision for a robot moving on undulating terrain. *Proc. of IEEE Computer Society Conference on Computer Vision and Pattern Recognition.* Hawaii: IEEE Press.

[3] Badino, H., Huber, D. and Kanade, T. (2012). Real-time topometric localization. *2012 IEEE International Conference on Robotics and Automation*, Saint Paul, Minnesota, USA May 14–18, 2012.

[4] Bailey, T. and Durrant-Whyte, H. (2006). Simultaneous localization and mapping (SLAM): part II, *IEEE Robotics and Automation Magazine*, 13(3):108–117.

[5] Biswas, R., Limketkai, B., Sanner, S. and Thrun, S. (2002). Towards object mapping in non-stationary environments with mobile robots, *Proceedings of the International Conference on Intelligent Robots and Systems*, Lausanne: IEEE Press, pp. 1014–1019.

[6] Borges, G.A. and Aldon, M.J. (2003). Robustified estimation algorithms for mobile robot localization based on geometrical environment maps. *Robotics and Autonomous Systems*, 45(3–4):131–159.

[7] Cai, Z.X., *et al.* (2016). *Key Techniques of Navigation Control for Mobile Robots under Unknown Environment*, Beijing: Tsinghua University Press.

[8] Cai, Z.X., He, H.G. and Chen, H. (2002). Some issues for mobile robot navigation under unknown environments. *Control and Decision*, 17(4):385–391 (in Chinese).

[9] Cai, Z.X., Zou, X.B, Wang, L., *et al.* (2005). Design of distributed control system for mobile robot. *Journal of Central South University (Science and Technology)*, 36(5):727–732 (in Chinese).

[10] Chatterjeea, A. and Matsunob F. (2010). A Geese PSO tuned fuzzy supervisor for EKF based solutions of simultaneous localization and mapping (SLAM) problems in mobile robots. *Expert Systems with Applications*, 37(8):5542–5548.

[11] Chen, B.F., Liu, L.J., Zou1, Z.R. and Xu, X.Y. (2013). A hybrid data association approach for SLAM in dynamic environments. *IEEE International Journal of Advanced Robotic Systems*, 10(118):1–7.

[12] Cho, B.-S., Moon, W.-S., Seo, W.-J. and Baek, K.-R. (2011). A dead reckoning localization system for mobile robots using inertial sensors and wheel revolution encoding. *Journal of Mechanical Science and Technology*, 25(11):2907–2917.

[13] Dasvison, A.J. and Murray, D.W. (2002). Simultaneous localization and map building using active vision. *IEEE Transactions on Pattern Analysis and Machine Intelligence*, 24(7):865–880.

[14] Davey, S.J. (2007). Simultaneous localization and map building using the probabilistic multi-hypothesis tracker, *IEEE Transactions on Robotics*, 23(2):271–280.

[15] David, M.-G. and Michel, D. (2012). Active visual-based detection and tracking of moving objects from clustering and classification methods. *IEEE Advanced Concepts for Intelligent Vision Systems Lecture Notes in Computer Science*, 7517, pp. 361–373.

[16] Duan, Z.H. Cai, Z.X. and Min, H.Q. (2014). Robust dead reckoning system for mobile robots based on particle filter and raw range scan. *Sensors*, 14(9):16532–16562.

[17] Durrant-Whyte, H. and Bailey, T. (2006). Simultaneous localization and mapping: Part I. *IEEE Robotics and Automation Magazine*, 13(2):99–110.

[18] Elfes, A. (1989). Occupancy grids: *A Probabilistic Framework for Mobile Robot Perception and Navigation*. Doctoral Dissertation, Pittsburgh: Carnegie Mellon University.

[19] Fuke, Y. and Krotkov, E. (1996). Dead reckoning for a lunar rover on uneven terrain. *Proc. of the 1996 IEEE Int. Conf. on Robotics and Automation*. New York, NY, USA: IEEE Press, pp. 411–416.

[20] Gao, H.B., Deng, Z.Q., Ding, L. and Wang, M.Y. (2012). Virtual simulation system with path-following control for lunar rovers moving on rough terrain. *Chinese Journal of Mechanical Engineering*, 25(1):38–46 (in Chinese).

[21] Gonzalo, R., Thomas, R.-C.S., Barrientos, A. and MacDonald, B. (2012). A real-time method to detect and track moving objects (DATMO) from unmanned aerial vehicles (UAVs) using a single camera. *Remote Sensing*, 4(4):1090–1111.

[22] Guiherme, N.D. and Avinash, C.K. (2002). Vision for mobile robot navigation: A survey. *IEEE Transactions on Pattern Analysis and Machine Intelligence*, 24(2):1–31.

[23] Guivant, J. and Nebot, E. (2002). Improving computational and memory requirements of simultaneous localization and map building algorithms. *Proc IEEE International Conference on Robotics and Automation*. San Francisco: IEEE Press, pp. 2731–2736.

[24] Guivant, J., Nebot, E. and Durrant-Whyte, H. (2000). Simultaneous localization and map building using natural features in outdoor environments. Pagello E., *Proc. 6th Int. Conference on Intelligent Autonomous Systems*, Venice, Italy: IOS Press, Vol. 1, pp. 581–588.

[25] Günthera, P., Dreiera, F., Pfistera, T., *et al.* Measurement of radial expansion and tumbling motion of a high-speed rotor using an optical sensor system. *Mechanical Systems and Signal Processing*, 25(1): 319–330.

[26] Haehnel, D., Schulz, D. and Burgard W. (2003). Map building with mobile robots in dynamic environments, *Proc. of the IEEE/RSJ International Conference on Intelligent Robots and Systems*, 1557–1563.

[27] Haehnel, D., Schulz, D. and Burgard, W. (2002). Map building with mobile robots in populated environments, *Proc. of the IEEE/RSJ International Conference on Intelligent Robots and Systems*, 496–501.

[28] Hähnel, D., Triebel, R., Burgard, W. and Thrun, S. (2003). Map building with mobile robots in dynamic environments, *Proceedings of the IEEE International Conference on Robotics and Automation*, Taipei: IEEE Press, pp. 1557–1563.

[29] Howard, T.M., Green, C.J., Kelly, A., and Ferguson, D. (2008). State space sampling of feasible motion for high-performance mobile robot navigation in complex environment. *Journal Field Robotics*, 25(6–7):325–345.

[30] Iagnemma, K. and Dubowsky, S. (2004). Traction control of wheeled robotic vehicles with application to planetary rovers. *Int. Journal of Robotics Research*, 23(10):1029–1040.

[31] Jorge, F.-P., José, R.-A. and Juan, M.R.-M. (2015). Visual simultaneous localization and mapping: A survey. *Artificial Intelligence Review*, 43(1):55–81.

[32] Ju, M.-Y. and Lee, J.-R. (2013). Vision-based mobile robot navigation using active learning concept. *2013 International Conference on Advanced Robotics and Intelligent Systems*, May 31–June 2, Tainan, Taiwan.

[33] Kelly, A. (2001). General solution for linearized systematic error propagation in vehicle odometry. *Proc. of the 2001 IEEE/RSJ Int. Conf. on Intelligent Robots and Systems* (lROS'01). New York, NY, USA: IEEE Press, pp. 1938–1945.

[34] Kleinberg, J.M. (1994). The Localization Problem for Mobile Robots. *Proc. of the 35th IEEE Symposium on Foundations of Computer Science* (FOCS). Los Alamitos, CA, USA: IEEE Computer Society Press, pp. 521–531.

[35] Konolige, K., Agrawal, M. and Solà, J. (2011). Large-scale visual odometry for rough terrain. *Robotics Research Springer Tracts in Advanced Robotics*, 66:201–212.

[36] Li, J.G., Meng, Q.-H., Wang, Y. and Zeng, M. (2011). Odor source localization using a mobile robot in outdoor airflow environments with a particle filter algorithm. *Autonomous Robots*, April, 30(3):281–292, Date: 21 Jan 2011.

[37] Li, M.H. and B.R. (2005). A robust method of localization for an indoor mobile robot. *Computer Engineering and Application*, 41(4): 1–8 (in Chinese).

[38] Martinelli, A. (2002). The odometry error of a mobile robot with a synchronous drive system. *IEEE Transactions on Robotics and Automation*, 18(3):399–405.

[39] Mertz, C., Navarro-Serment, L.E., MacLachlan, R., *et al.* (2013). Moving object detection with laser scanners. *Journal of Field Robotics*, 30(1):17–43.

[40] Murphy, R.R. (2004). *Introduction to AI Robotics*. Du J P, Wu L C, Hu J C. Beijing: Publishing House of Electronics Industry (in Chinese).

[41] Myung, H. Lee, H.-K., Choi, K. and Bang, S. (2010). Mobile robot localization with gyroscope and constrained Kalman filter. *International Journal of Control, Automation and Systems*, 8(3):667–676.

[42] Neira, J. and Tardos, J. D. (2001). Data association in stochastic mapping using the joint compatibility test, *IEEE Transactions on Robotics and Automation*, 17(6):890–897.

[43] Se, S. and Lowe, D. (2002). Mobile robot localization and mapping with uncertainty using scale-invariant visual landmarks. *International Journal of Robotics Research*, 21(8):735–738.

[44] Sebastian, T., Wolfran, B. and Dieter, F. (2000). A real-time algorithm for mobile robot mapping with applications to multi-robot and 3D mapping. *Proc. of IEEE International Conference on Robotics and Automation*, San Francisco: IEEE Press.

[45] SICK AG Corp, Technical Description: LMS200/ LMS211/ LMS220/ LMS221/LMS291 Laser Measurement Systems, http.www.sick.com.

[46] Smith, R., Self, M. and Cheeseman, P. (1988). *Estimating Uncertain Spatial Relationships in Robotics*, *Uncertainty in Artificial Intelligence*, New York: Elsevier Science, pp. 435–461.

[47] Teslić, L., Škrjanc, I. and Klančar, G. (2010). Using a LRF sensor in the Kalman-filtering-based localization of a mobile robot. *ISA Transactions*, 49(1):145–153.

[48] Thrun S. Robotics Mapping: A Survey. Pittsburgh: CMU-CS-02-111, School of Computer Science, Carnegie Mellon University.

[49] Thrun, S., Gutmann, J.S., Fox, D., *et al.* (1998). Integrating topological and metric maps for mobile robot navigation: A statistical approach. *Proc. of the 1998 National Conference on Artificial Intelligence*, Madison, USA: AAAI Press.

[50] Tully, S., Kantor, G. and Choset, H. (2014). A unified Bayesian framework for global localization and SLAM in hybrid metric/topological maps. *2013 Journal Citation Reports* Thomson Reuters, 2014.

[51] Tzafestas, S.G. (1999). *Advances in Intelligent Autonomous Systems*. Dordrecht: Kluwer Academic Publishers.

[52] Ulrich, I. and Nourbakhsh, I. (2000). Appearance-based place recognition for topological localization. *Proc. of IEEE International Conference on Robotics and Automation*, San Francisco: IEEE Press.

[53] Vallivaara, I., Haverinen, J., Kemppainen, A. and Roning J. (2011). Magnetic field-based SLAM method for solving the localization problem in mobile robot floor-cleaning task. *The 15th International Conference on Advanced Robotics*, Tallinn University of Technology, Tallinn, Estonia, June 20–23.

[54] Wang, C.C., Thorpe, C. and Thrun, S. (2003). Online simultaneous localization and mapping with detection and tracking of moving objects: theory and results from a ground vehicle in crowded urban areas, *Proc. of the IEEE International Conference on Robotics and Automation*, pp. 842–849.

[55] Wang, L. (2007). *Research on Visual Environment Modeling and Positioning of Mobile Robots in Unknown Environment*. Changsha: Doctoral Dissertation of Central South University (in Chinese).

[56] Wang, W.H., Chen, W.H. and Xi, Y.G. (2001). Research progress of mobile robot map creation based on uncertain information. *Robot*, 23(6):563–568 (in Chinese).

[57] Wolf, D.F. and Sukhatme, G.S. (2004). Online simultaneous localization and mapping in dynamic environments, *Proceedings of the IEEE International Conference on Robotics and Automation*, New Orleans: IEEE Press, pp. 1301–1306.

[58] Wolf, D.F. and Sukhatme, G.S. (2005). Mobile robot simultaneous localization and mapping in dynamic environments, *Autonomous Robots*, 19:53–65.

[59] Ye, C. and Borenstein, J. (2002). Characterization of a 2-D laser scanner for mobile robot obstacle negotiation. *Proceedings of the 2002 IEEE International Conference on Robotics and Automation (ICRA'02)*, Washington DC, USA, pp. 2512–2518.

[60] Zeng, W.J., Zhang, T.D., Xu, Y.R., *et al.* (2009). Data association method of SLAM based on ant colony algorithm, *Journal of Computer Applications*, 29(1):136–148.

[61] Zhang, S., Xie, L.H. and Adams, M. (2004). An efficient data association approach to simultaneous localization and map building, *IEEE International Conference on Robotics and Automation*. New Orleans, USA, pp. 854–859.

[62] Zou X.B. (2004). *Research on Design of Control System and Environment Modeling for a Prototype of Mobile Robot*, Central South University, China.

Chapter 9

Mobilebot Navigation

Intelligent mobilebot (mobile robot) is a kind of robot system that can perceive the environment and its own state through sensors, realize target-oriented autonomous movement in an environment with obstacles, and complete certain tasks. In the past ten years, mobilebot technology has played an important role in many fields such as industry, agriculture, aerospace, navigation, military and space exploration. At the same time, it has shown a wide range of application prospects, so it has become a hot issue of the research and attention for the international robotics academic community [8, 26].

Using the scientific principles and methods of electricity, magnetism, light, mechanics, etc., by measuring the space-time position related parameters of moving objects such as aerial vehicles, ships at sea, submarines in the ocean, ground vehicles and pedestrians, the positioning of moving objects can be realized from the starting point. Following the predetermined route safely, accurately and effectively to guide the moving body to the destination — this is navigation, and this technology is called navigation technology. If the above-mentioned various moving objects in the sea, land and air are unmanned, then they are all kinds of mobilebots.

In the research of mobilebot related technology, navigation technology is its core, and path planning is an important link and subject of navigation research. This chapter mainly studies the navigation of mobile robots, especially the path planning of mobile robots in unknown environments. The so-called path planning means that the mobile robot searches for an optimal or suboptimal path from the initial state to the goal state according to a certain performance index (such as distance, time, energy, etc.). The main problems involved in path planning include: (1) Using the obtained mobilebot environment information to establish a more reasonable environment model, and then using a certain algorithm to find an optimal or nearly optimal collision-free path from the initial state to the goal state; (2) It can deal with uncertain factors in the environment model and errors in path tracking, so as to minimize the impact of external objects on the robot; (3) Using all known information to guide the robot's actions, so as to obtain relatively better behavioral decisions. How to quickly and effectively realize the navigation tasks of mobile robots in complex or even unknown environments is still one of the main directions of robot navigation research in the future [16, 33]. How to integrate the advantages of various navigation methods to achieve better navigation effects is also a problem to be discussed. Multi-agent robot system (MARS) such as soccer robot as a new intelligent technology has gradually become a hot spot [7], it is produced when a single agent machine develops to the condition that requires coordinated operations. The difficulty of this kind of planning lies in the coordination and collision avoidance of multiple robots, which will also be an area that mobilebot technology urgently needs to expand [15, 23].

9.1 Main Methods and Development Trends of Mobilebot Navigation

Mobilebot path navigation methods can be divided into the following three types: navigation methods based on case learning, navigation methods based on environmental models, and behavior-based navigation methods [16].

9.1.1 *Navigation method based on case learning*

Case-based learning navigation method relies on experience to learn and solve problem. A new case is modified from the old one that similar to current situation in the case library. This method applied to mobilebot path planning can be described as follows: First, utilize the information that used or generated by the path planning method to build a case library, any case in the library contains the environment information and path information. These cases can be retrieved by a specific index. Second, retrieve the case from the case library that is the most appropriate one under the planning tasks and environment information. Third, the case was improved, and it is the final result. The navigation of mobilebot requires good adaptability and stability, and the case-based approach can meet this demand. Combining the online match based on case with reinforcement learning can improve the adaptive ability of the robot, and the robot can better adapt to the changes in the environment. The number of cases in the library should be kept certain quantity in the case-based method in order to prevent an increase in online planning time or produce information explosion. Combining the case-based approach with the global planning can improve the efficiency of global planning. Autonomous underwater robots have attracted the attention of various countries due to their advantages in submarine resource detection, but because the underwater environment is very complicated (poor visibility, positioning difficulties, etc.), and the degree of congestion in the underwater environment is relatively low, resulting in general planning methods are difficult to work. The feature that the underwater robots are more likely to work in the same area is just conducive to the application of case-based planning methods, so this method is widely used to solve the navigation problems of underwater robots [14].

9.1.2 *Navigation method based on environment model*

These methods need to build an environment model of robot movement environment firstly. In many cases, the working environment of mobilebots is uncertain (non-structural, dynamic, etc.), such that the mobilebot can not establish a global environment model, but only to establish a local environment model based on the information collected by sensors in real time. The real-time and reliability of local models become the critical factor whether the robot can perform safe, continuous and steady movement. Environmental modeling methods can basically be divided into two categories: network/graph modeling methods, and grid-based modeling methods [53]. The former mainly includes free space method, vertex image method, generalized cone method, etc., which can be used to obtain more accurate solutions in path planning, but the amount of calculation consumed is quite large, which is not suitable for practical applications. The latter is much simpler in implementation, so it is widely used. Its typical representative is the quadtree modeling method and its extended algorithm [6].

The navigation method based on the environment model can be subdivided into global route navigation with completely known environmental information and local route navigation with completely or partially unknown environmental information according to the completeness of the environmental information [3]. Since the environmental model is known, the design standard for global path planning is to maximize the planning effect. There are many mature methods in this field, including the view method, the tangent diagram method, the Voronoi diagram method [24, 30, 67], the topology method, the penalty function method, and the grid method, etc.

As a hot topic of current navigation research, local route navigation has been studied in depth and detail. When the environment information is completely unknown, the robot does not have any prior information. Therefore, the plan is to improve the robot's obstacle avoidance ability as the main goal, and the effect is the second. The incremental D*Lite algorithm uses a heuristic strategy to search for a path from the goal point to the current position of the robot, and replans the path in real time according to the updated information of the local environment during the movement of the robot to the goal, thereby obtaining an optimal path. The planning method based on rolling window has also achieved good results. When the environment is unknown, the navigation methods mainly include artificial potential field method, artificial neural network algorithm, fuzzy logic algorithm, genetic algorithm, simulated annealing algorithm, ant colony optimization algorithm, particle swarm algorithm and heuristic search method, etc. Among them, the research of heuristic methods has made great progress. Its original representative is the Algorithm A*, and its new development is 2 kinds of incremental graph search algorithms (also known as dynamic Algorithm A*): Algorithm D* and Focussed D* Algorithm. In addition, some improved algorithms based on A* have appeared. The method based on the environment model is applied in many fields, especially in space exploration, due to its navigation accuracy and stability. For example, the "Sojourner" rover carried by the "Mars Pathfinder" probe launched by the United States in December 1996 uses the Algorithm D* as a path navigation method, allowing

Sojner to move freely and cautiously walk on the surface of Mars, and can autonomously judge the obstacles on the road ahead, and make action decisions through real-time replanning.

In Chapter 8, the simultaneous localization and mapping (SLAM) method of mobilebots was discussed. As a positioning and perception method, SLAM is a key technology to realize the navigation and control of mobilebot unmanned platforms. In order to realize autonomous positioning and navigation of mobilebots in unknown environments, a design method of mobilebot navigation system is proposed. The Monte Carlo algorithm is used to locate the robot, the grid map scanning and matching method is used to construct an accurate environmental map, and the Dijkstra algorithm is used to plan the path of the current map to realize the autonomous movement of the mobilebot. A set of SLAM and navigation system based on low-cost lidar was designed modularly using robot operating system ROS. The system has high mapping and navigation accuracy, and each functional module has portability and versatility, which can meet indoor robot's navigation requirements. Combining laser-based environmental feature extraction and map construction technology with Beidou satellite navigation, an autonomous navigation system for mobile robots based on feature maps is designed. This method can quickly extract intuitive environmental information, realize autonomous navigation functions, and effectively improve the navigation capabilities of mobilebots. Also proposed a mobile robot SLAM navigation system based on particle swarm filter method and a small-scale SLAM UAV navigation based on real-time graphics in an unknown environment.

9.1.3 *Navigation method based on behavior*

The behavior-based route navigation method was established by Brooks R A in his famous inclusive structure. It is a technology inspired by biological systems to design autonomous robots. It uses a bottom-up principle system similar to animal evolution, try to build a complex system from a simple agent. It is a new development trend to use it to solve the problem of mobilebot path navigation. It decomposes the navigation problem into many relatively independent behavioral units, such as tracking, collision avoidance, and target guidance. These behavior units are complete motion control units composed of sensors and actuators, with corresponding navigation functions. The behavior modes adopted by each behavior unit are different. These units coordinate their work to complete the navigation task.

Behavior-based navigation methods can be roughly divided into three types: reflective behavior, reactive behavior, and deliberate behavior. The reflective behavior is similar to the frog's knee-jumping reflex. It is an instantaneous stress-induced instinctive response that can respond quickly to unexpected situations, such as an emergency stop of a mobilebot during movement. This navigation method is not intelligent and is generally used in combination with other navigation methods. The navigation method based on reactive behavior was first proposed by Brooks. It directly reads sensor data to plan the next action. It can respond to unpredictable obstacles and environmental changes in a stable and timely manner. However, due to the lack of global environmental knowledge, therefore, the generated action sequence may not be globally optimal, and it is not suitable for path planning of mobilebots in complex environments. Deliberate behavior uses the known global environment model to provide an optimal sequence of actions for the agent system to reach a specific goal, which is suitable for planning in a complex static environment. The real-time replanning of a mobile robot in motion is a deliberate behavior. The robot may move backwards to get out of the dangerous area. However, since deliberate planning requires a certain amount of time to execute, it responds slowly to unpredictable changes in the environment. Reactive behaviors and deliberate behaviors can be distinguished by sensor data, global knowledge, reaction speed, inference ability, and computational complexity. In the development of deliberative behavior, a declarative cognitive behavior similar to human brain memory appears. This navigation not only relies on sensors and existing prior information, but also depends on the goal to be reached. For example, for a goal that is far away and temporarily invisible, there may be a behavioral bifurcation point, that is, there are several behaviors to adopt, and the robot must choose the best. This decision-making behavior is declarative cognitive behavior [84]. Using this method for path planning can make mobilebots more intelligent, but due to the complexity of decision-making, this method is difficult to use in practice, and further research is needed [10, 16].

9.1.4 *The development trend of mobile robot navigation*

The requirements for new trends of mobile robot path planning techniques have become more sophisticated with the expansion of mobilebot range of applications. It is not very suitable for single path planning methods to solve some planning problem. The development trend is to combine path planning methods with each other. This can be discussed as follows.

1. Combination of reactive-based behavior planning and deliberate-based behavior planning

Planning methods based on reactive behavior can achieve good planning results under the premise of establishing a static environment model, but it is not suitable for situations where there are some non-model obstacles (such as tables, people, etc.) in the environment. For this reason, some scholars have proposed a of hybrid control structure, that is, combining deliberate behavior with reactive behavior, which can better solve this type of problem. The hybrid control structure combining the two can be used for collision avoidance navigation in unpredictable environments. Each behavior module runs asynchronously. Compared with the general control structure, the module plans a set of points instead of a complete path. The reactive module has the dual functions of collision avoidance and target guidance. A three-layer control structure is proposed, including a control layer, a sequence layer and a deliberate layer. Among them, the control layer uses reactive behavior to quickly sense the environment, the deliberate layer mainly performs processing such as path planning, and the sequence layer connects the above two layers. This structure ensures that the robot can accurately grasp the environment in time. A real-time hybrid control structure is also proposed, which can select reasonable execution actions within a limited time and can adjust the behavior of the robot according to environmental information. It is highly flexible and effectively solves the limited available time and deliberate thinking. The contradiction between behaviors that consume a certain amount of time, satisfies the change requirements of reactive behaviors [28, 61].

2. Combination of global path panning and local path planning

Global planning is generally based on known environmental information, and the scope of adaptation is relatively limited; local planning can be applied to situations where the environment is unknown, but sometimes the response speed is not fast, and the quality of the planning system is relatively high. Therefore, if the two are combined, a better planning effect can be achieved. Another three-level control structure, with global planning as the highest level, local planning of artificial potential field method as the second level, and the lowest level using heading angle method to avoid obstacles, suitable for path planning in a messy environment. There is also a new multi-level structure including global planning, local planning, intelligent supervision, sensor information fusion, path selection and navigation. This method can effectively use the local planning time to obtain a relatively better path.

3. Combination of the traditional planning method and intelligent method

Some new intelligent technologies have been introduced into route navigation, which has promoted the integration and development of navigation methods, such as the combination of artificial potential field, neural network and fuzzy control. For example, a hybrid control structure of neural network, reactive behavior and artificial potential field is proposed. The neural network is first used to classify the environment to avoid falling into a local minimum, and a controller based on reactive behavior is established, and then according to the potential field method selected a specific reactive behavior to guide the mobile robot forward. This structure has good robust stability and dynamic environment adaptability. Another example is the design of a fuzzy controller that combines fuzzy logic and artificial potential field functions to solve the planning problem of unknown dynamic obstacles in the known environment, greatly improving the efficiency of motion planning, and suitable for real-time applications. In addition, the artificial potential field method is used to guide the mobile robot to the target, and then the fuzzy logic method is used to enable the robot to get out of the potential energy trap and generate a flat and safe path.

In recent years, machine learning, especially deep learning technology, has been widely used in mobilebot navigation. Please see the more detailed introduction in Section 8 of this chapter. In addition, artificial cognition and machine thinking, particle swarm optimization and ant colony optimization, evolutionary algorithms and genetic algorithms, immune evolution, knowledge-based and expert systems and other artificial intelligence technologies are also increasingly used in various navigation systems [31, 62].

9.2 Local Navigation Strategy for Mobilebot

9.2.1 *Overview of local navigation methods*

Reactive control controls the movement of the robot based on real-time sensor input information. Typical methods include dynamic window method, visual vertex search method, and potential field method. The artificial potential field method is a classic method in the path planning of mobilebots. It can be applied to both map-based planning and reactive local planning [57].

The advantage of the reactive planning method is that it can respond to immediate changes in the environment, has low computational cost, can run in real time, and can be applied to unknown and dynamic environments. The disadvantage is that the accumulation of global information is ignored, and long-term ineffective wandering and shocks may occur in local traps in complex environments. In order to be able to apply to various complex and unpredictable unknown environments, researchers have done a lot of research on how to design the evaluation function of local behavior, in order to judge whether the robot is in shock and how to use appropriate behavior to get rid of local traps. For example, when there is no ideal path, the state of the robot is monitored by vision to control the robot to use wall-following to find the path. Literature [46, 64] uses neural networks to determine the type of local environment based on sensor information under the premise that the environmental characteristics are a limited set and take corresponding avoidance behaviors for the local trap area. In an unknown environment, when a feasible solution cannot be obtained for local planning, random noise can be applied to make the robot enter the roaming state.

For example, in the schema-based reactive behavior design, a visual potential field is used to characterize the robot's perception schema of the environment. Obstacles and "natural enemies" to be avoided form a repulsive field, while targets form an attractive field. Robot behavior can be represented by the basic schema in the potential field, and the final motion schema is determined by vector synthesis in the potential field.

In Figure 9.1, the robot path planning process based on the schema has experienced basic schemas such as stay-on-path, avoid-static-obstacle, and move-to-goal. In the vector synthesis of the behavioral schema, it is possible to form an area where the resultant force is zero. In this area, the robot will have a speed of 0. At this time, additional noise disturbance is used to get rid of the static area.

For robots, the distribution of obstacles in the unknown environment outside the sensing area can be regarded as a random event. Relying on local perception information to move must have a certain degree of randomness, and its planning is also a random decision-making process. A mobilebot in a random roaming state is like a particle performing Brownian motion in a liquid environment. In most local path planning methods, the constraints of sensor perception information are mainly considered, while the constraints of local traps are not considered much, and the constraints of local

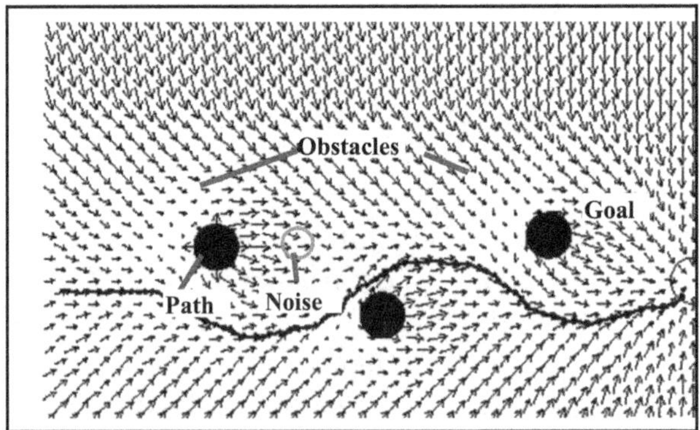

Figure 9.1.   Schema-based reactive path planning.

traps often have a more serious impact on the robot's motion. In an unknown environment, local planning should combine the purposeful target-oriented behavior with the random behavior of target search under roaming disturbance, reflecting the comprehensive constraints of targets, local traps and sensor information on the local behavior of mobilebots.

Literature [68] regards the movement of a mobilebot in an unknown environment as a process of combining the directional movement of liquid particles in a potential field environment with random Brownian motion. In the local planning of the robot, the directionality of the movement to the target direction is combined with the randomness of the feasible path search through Brownian motion. The robot first determines the next movement direction of the robot according to the search principle of priority in the target direction in the local rolling window. When judging that the robot is in a local trap by monitoring the change of the distance potential energy, a disturbance is applied to the target direction angle. After adding an offset to the target direction, it means that the robot will search for a path along the target direction with disturbance noise and use the simulated annealing (SA) method to evaluate the disturbance correction. Through random disturbance, the robot is guided to overcome the local potential energy trap. When the robot runs to a new minimum potential energy area, it means that it has left the original potential energy trap area. At this time, the disturbance state ends and recovers the target trending movement.

9.2.2 *Disturbance rule design based on simulated annealing*

The simulated annealing (SA) algorithm is a random search algorithm based on the principle of similarity between the annealing process of metal substances and the optimization problem [29, 34]. When the substance is heated, the Brownian motion between particles increases, and when it reaches a certain intensity, it is annealed, the thermal motion of the particles weakens, and gradually tends to be orderly, and finally reaches stability. The simulated annealing optimization process is a Markov decision process. Based on the Markov process theory, it can be proved that the simulated annealing algorithm converges to the global optimal value with probability 1.

Simulated annealing is a general algorithm for solving combinatorial optimization problems. As long as the optimization problem can provide an adaptability function or cost function of a candidate solution, simulated annealing can be used to solve it. The simulated annealing method is usually applied to combinatorial optimization problems, such as: TSP problem, large-scale integrated circuit design, etc. In [42], SA is used to optimize the reactive neural network controller of mobilebots. Regarding the random roaming behavior of the robot in an unknown environment as the Brownian motion of particles in the liquid, the SA method can be applied to its random disturbance to guide it to move in the direction of reduced potential energy, so as to realize the unknown environment online dynamic planning, see Figure 9.2. The principle and steps are as follows:

1. Determine the rolling optimization window

The rolling window is centered on the robot's reference coordinates. The size and shape of the window are related to the shape of the robot itself and the sensing area of the sensor. It is often designed as a rectangular, circular or semi-circular area. Search the local optimization solution that satisfies the collision avoidance condition in the scrolling area [59].

2. Calculate potential energy function and direction function

Suppose the current coordinates of the robot are $R(x_\mathrm{r}, y_\mathrm{r})$, the coordinates of the target point are $G(x_\mathrm{g}, y_\mathrm{g})$ $\boldsymbol{R}$ is the vector from $R(x_r, y_\mathrm{r})$

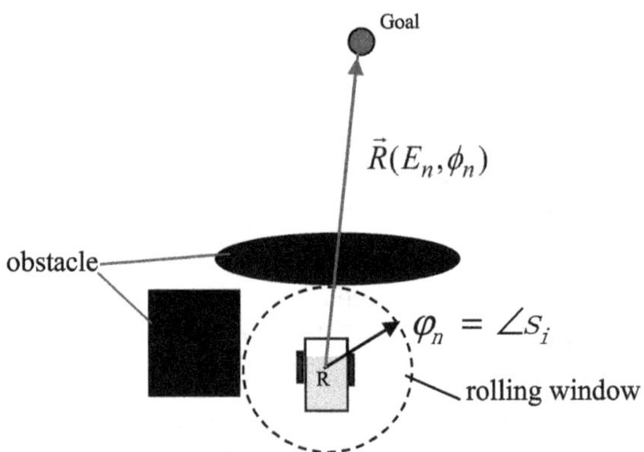

Figure 9.2.   Local planning based on rolling window.

to G (x_g, y_g), and the modulus and direction of the vector R The angle is:

$$E_n = \sqrt{(x_g - x_r)^2 + (y_g - y_r)^2} \tag{9.1}$$

$$\phi_n = \arctan \frac{y_g - y_r}{x_g - x_r} \tag{9.2}$$

In the formula, E_n represents the potential energy of the robot in the potential field of the target distance, and ϕ_n represents the gradient direction of the reduction of the potential energy of the robot. The direction function ϕ_n is continuously modified according to the position of the robot and always points to the target position. The subscript indicates the moment.

3. Establish evaluation function for local optimization

Local planning evaluates the passibility of the robot according to the distribution of obstacles in each direction in the rolling window, and filters out the set $S = \{s_i, i = 1, 2, \ldots\}$ of travel directions that meet the collision-free pass conditions. Evaluate the elements in the feasible S. When advancing in the target direction, an instinctive choice is to proceed as far as possible in the direction with the smallest angle with the target direction. Suppose that at the current moment n, the direction angle from the robot $R(x_r, y_r)$ to the target $G(x_g, y_g)$ is ϕ_n, the heading angle of the robot is θ_n, and the direction of the pending preview target is $\hat{\phi}_n$, and the direction angle when selected $s_i \in S$ as the preview target is set to $\hat{\phi}_{n,s_i}$. Define the objective function as:

$$f(\hat{\phi}_{n,s_i}) = k_1 |\hat{\phi}_{n,s_i} - (\phi_n + \delta)| + k_2 |\hat{\phi}_{n,s_i} - \theta_n| \tag{9.3}$$

In the formula, $|\hat{\phi}_{n,s_i} - (\phi_n + \delta)|$ represents the tendency of the robot to turn to the target direction, δ is the disturbance noise applied in the case of a local potential energy trap, $|\hat{\phi}_{n,s_i} - \theta_n|$ reflects the cost of turning from the current heading to the preview direction, and k_1, k_2 are positive weighting coefficients. According to the objective function of formula (9.3), the current preview direction is determined as:

$$\hat{\phi}_n = \hat{\phi}_{n,s_j}, \; iff., \; f(\hat{\phi}_{n,s_j}) = \min\{f(\hat{\phi}_{n,s_i}, s_i \in S, i = 1, 2, 3, \cdots\} \tag{9.4}$$

Obviously, $(\phi_n + \delta)$ has important guidance for the robot's movement direction. When $\delta = 0$, the robot will preferentially search the target direction ϕ_n to meet the obstacle-free forward direction; when $\delta \neq 0$, the mobilebot will search and forward according to the target direction offset by the additional disturbance.

4. Judgment of local traps

Let $E_{\min} = \min\{E_i, i = 0, \cdots, n - 1\}$, the robot searches in the target direction ϕ_n to meet the obstacle-free forward direction, and will gradually approach the target, resulting E_n in a decrease. Let:

$$\Delta \hat{E} = E_n - E_{\min} \tag{9.5}$$

In a barrier-free environment or an environment with little change in the potential field caused by obstacles, E shows a monotonous decline. At this time, $\Delta \hat{E} < 0$ and $E_{\min}$ is constantly updated by the potential energy value at the current moment. In a more complex environment, local potential energy traps may occur $\Delta \hat{E} > 0$ due to the "siege" of obstacles. According to the change of $\Delta \hat{E}$ to determine whether it is in a local trap, and trigger the robot to enter the state of direction angle disturbance:

$$\delta = \begin{cases} 0 & \Delta \hat{E} < 0 \\ \delta_i & \Delta \hat{E} \geq 0 \end{cases} \tag{9.6}$$

(5) Generation of disturbance angle

Calculate the azimuth angle between the robot and the location of the local potential energy trap:

$$\psi = \arctan \frac{y_{\min} - y_r}{x_{\min} - x_r} \tag{9.7}$$

where $x_{\min}, y_{\min}$ is the coordinates of the location of the local potential energy trap. $\delta = \delta_i$ is generated randomly disturbance, and the target position, local potential energy trap position, sensor perception information and other factors are considered to judge whether to accept the random disturbance.

(a) Target orientation constraints (refer to Figure 9.3)

The target orientation is the orientation angle at which the robot points to the target. The disturbance angle will lead the robot to deviate from the heading angle of $\delta = \delta_i$. Given a disturbance angle δ_i, $\Delta \phi$ is the deviation between the reference running direction and the target orientation after the disturbance is applied. When

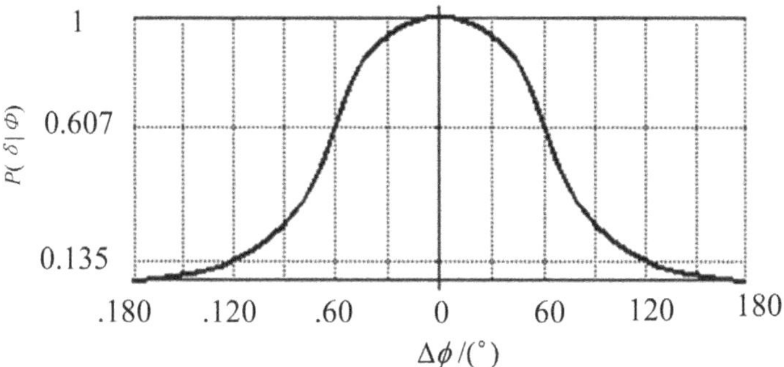

Figure 9.3. Probability $P\{\delta|\phi\}$ distribution of the disturbance being "true" under the target constraint.

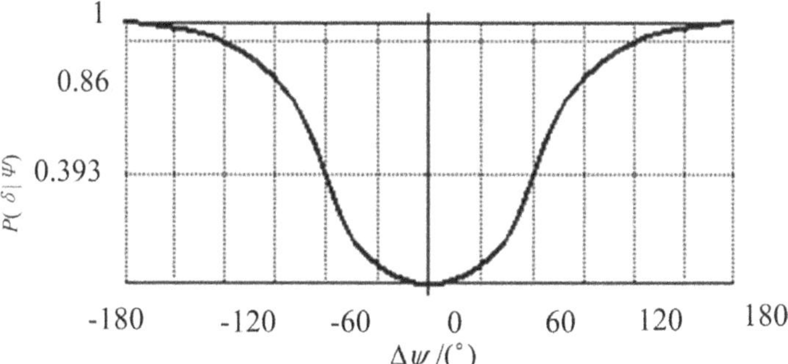

Figure 9.4. $P\{\delta|\phi\}$ Probability distribution curve of the disturbance being "true" under local trap constraint.

$\Delta\phi = \delta_i$, assuming that the conditional probability $P\{\delta \mid \phi\}$ of acceptance $\delta = \delta_i$ is true, and let:

$$P\{\delta|\phi\} = P\{\delta = \delta_i | \Delta\phi = \delta_i\} = 1 - \exp\left(-\frac{\Delta\phi^2}{2 \times 60^2}\right) \quad (9.8)$$

When only considering the target position, it is obviously hoped that the heading of the robot can be close to the target direction. Therefore, starting from the constraint of the target orientation, the closer δ is to 0, the greater the probability of being accepted.

(b) Local trap orientation constraint (refer to Figure 9.4)
In order to prevent the robot from falling into the potential energy trap, the robot's action direction needs to avoid the local trap direction ψ as much as possible. Given a disturbance angle δ_i, $\Delta\psi$ is

the deviation between the reference running direction $\phi + \delta$ after the disturbance is applied and the local trap direction ψ, when $\Delta\psi = \delta_i + \phi - \psi$, set the conditional probability $P\{\delta \,|\, \psi\}$ of acceptance $\delta = \delta_i$ to be true as, and let:

$$P\{\delta \,|\, \psi\} = P\{\delta = \delta_i \,|\, \Delta\psi = \delta_i + \phi - \psi\} = 1 - \exp\left(-\frac{\Delta\psi^2}{2 \times 60^2}\right)$$

$$(9.9)$$

(c) Sensor information constraints (refer to Figure 9.5)

In order to be able to travel safely, mobilebots tend to move away from obstacles. Suppose the obstacle-free distance $s = \rho_{(\phi+\delta)}$ measured by the sensor is in the reference direction $\phi + \delta$ of the mobilebot. Given a disturbance angle δ_i, when $s = \rho_{(\phi+\delta_i)}$, the conditional probability of acceptance $\delta = \delta_i$ is $P\{\delta|\rho\}$, and let:

$$P\{\delta \,|\, \rho\} = P\{\delta = \delta_i \,|\, s = \rho_{(\delta_i+\phi)}\}$$

$$= \begin{cases} 0 & s < L \\ 1 - \exp\left(-\frac{(s-L)^2}{2L^2}\right) & s \geq L \\ 0.5 & s = \text{unknown} \end{cases} \qquad (9.10)$$

Among them, L is the length of a car body. When the direction $\phi+\delta_i$ is in the blind zone of the robot sensor, set $= P\{\delta \,|\, \rho\}0.5$.

Considering the constraints of targets, local traps and sensor information, a disturbance δ_i is randomly generated in the form of

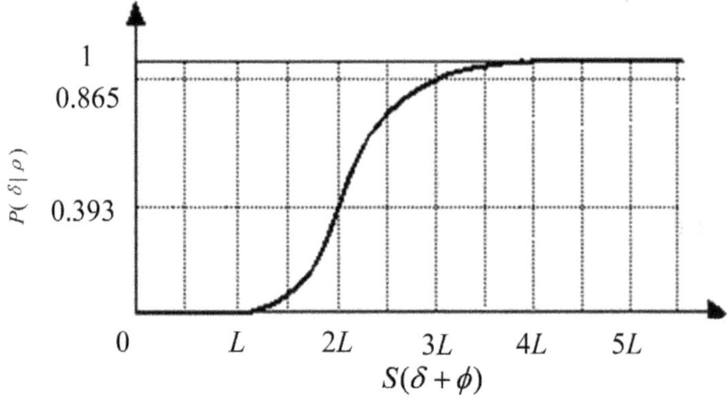

Figure 9.5. Probability $P\{\delta|p\}$ distribution curve of the disturbance being "true" under constraint of sensor information.

roulette, and whether to adopt δ_i as the current disturbance is determined according to the probability $P\{\delta \mid \phi, \psi, \rho\}$. Assuming that the target position ϕ, the local trap position ψ, and the sensor measurement information $\rho_{(\delta_i+\phi)}$ are mutually independent constraints, there are:

$$P\{\delta \mid \phi, \psi, S\} = P\{\delta \mid \phi\}P\{\delta \mid \psi\}P\{\delta \mid \rho\} \tag{9.11}$$

(6) Evaluation of the effects of perturbations $\delta = \delta_i$
Perturbation δ makes the robot off target course and move on to a new direction after the disturbance, there are three possibilities:

♦ Failed to get rid of the obstacles, trapped at a new local potential energy' minimum point.
♦ Get rid of obstacles, but not in the ideal direction, and away from the ultimate goal in the direction of increasing potential energy.
♦ Successfully get rid of obstacles and move in a direction close to the target location.

In order to guide the robot's random disturbance to the direction where the potential energy of the target distance decreases, the *SA algorithm* is used to evaluate the robot's motion as $\delta = \delta_i$. Simulate potential energy E_n as internal energy of particles (robots), and the temperature T is evolved into a time-varying control parameter, then the optimization process is: (1) Generate new solutions δ_i, and assign an initial temperature $T = T_0$ to each new solution; (2) Calculate the generated potential energy difference ΔE_n of robot motion under the control of disturbance δ_i; (3) Determine whether to update δ_i according to the probability value of the Metropolis sampling method in the formula (9.12); (4) This iterative process is repeated continuously, and the T value is attenuated gradually.

$$P\{\delta_i = \text{TRUE} \mid \Delta E_n\} = \min\left\{1, \exp\left(-\frac{\Delta E_n}{bT}\right)\right\} \tag{9.12}$$

$$\Delta E_n = E_n - E_{n-1} \tag{9.13}$$

$$T = c^m T_0 \tag{9.14}$$

where b is the Boltzmann constant, c is the attenuation scale factor of the temperature drop, $c \in (0, 1)$, and n is the step sequence count of the robot movement. When a disturbance is generated according to the rule of step (5), the initial annealing temperature $T = T_0$ is assigned, the annealing step $m = 0$ is counted, and m is increased by using one body length of the robot movement as a unit. During the initial period of the disturbance, because T is larger, it can be allowed to maintain δ_i with a greater probability under a large potential energy disturbance. As the annealing step count m increases, the probability that the increase of potential energy ΔE_n is restricted is:

$$P\{\delta_i = \text{FALSE} \,|\, \Delta E_n\} = 1 - P\{\delta_i = \text{TRUE} \,|\, \Delta E_n\} \qquad (9.15)$$

When the judgment of $\delta_i = \text{FALSE}$ is established, the disturbance quantity δ_i is updated according to the rule of step (5).

In the robot movement, due to the limited local information, the local planning is blind. Therefore, when the solution is solved by a certain rule, it may fall into a local trap under certain environmental constraints and cannot be freed from it. The mobilebot performs a perturbation at each annealing temperature, and when a new perturbation amount δ_i is generated, an annealing process is restarted. In addition, due to the slow convergence speed of the SA algorithm, when the robot reaches a region of distance potential energy $E_n < E_{\min} - \varepsilon, (\varepsilon > 0)$ from the area, it means that the robot has got rid of the local trap. Let $\delta = 0$, that is, end the disturbance control and restore the target tendency behavior.

9.2.3   *Program design of local navigation*

The flowchart of the local planning program based on SA disturbance control is shown in Figure 9.6, and the description of the subroutine is as follows:

1. Initialization parameter (InitialPara)

Preset the initial state $\{x_r, y_r, \theta\}$ and the target position $\{x_g, y_g\}$ of the robot; set the initial position as the position $\{x_{\min}, y_{\min}\} = \{x_r, y_r\}$ of the minimum initial potential energy, calculate the minimum initial potential energy $E_{\min}$; initial disturbance $\delta = 0$, step sequence count $n = 0$; annealing initial temperature $T_0 = 100$, constant term: $b = 0.1$, $c = 0.9$.

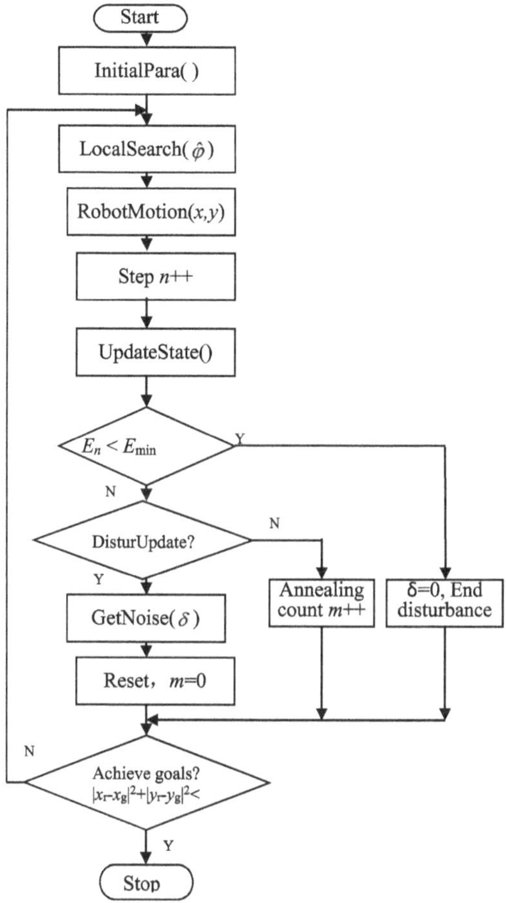

Figure 9.6. Flow chart of local planning program based on SA disturbance control.

2. Searching for the local target direction (LocalSearch)

The rolling window of the local planning of the mobilebot is a semicircle with the length of the robot as the radius, and the local target direction $\hat{\phi}$ is searched in step (3) according to the principle description in 9.3.2.

3. Update state parameter (UpdateState)

After the robot moves one step in the direction $\hat{\phi}$ of the local target, calculates the potential energy E_n, updates the minimum value E_{min}

of the potential energy and the position $\{x_{\min}, y_{\min}\}$ of the minimum value, and updates the sensor information.

4. Calculate the amount of disturbance (GetNoise) and determine the disturbance update (UpdateNoise)

According to the principle described in step (5), at the target position constraints, local potential trap constraints and sensor information constrained, to calculated perturbation δ with probability determined according to the equation (9.11), and in accordance with step 6 of section 6.5.2, to determine whether to update the perturbation.

9.3   Strategies and Implements of Composite Navigation

In this section, lidar is used as the environmental perception sensor, local planning and dynamic deliberate planning methods, and the hierarchical control idea of the inclusive architecture is integrated in the control layer, and the composite navigation control strategy of the mobilebot is designed based on the combination of reflective behavior, reactive behavior and deliberate behavior [55, 66].

9.3.1   *Strategies of composite navigation*

1. Design of deliberative planning strategies for mobilebot

Navigation in an unknown environment means that there is no human-preset environment information before the robot enters the working area, but the working environment of a mobilebot can be gradually recognized through the accumulation of sensory information. This means that the mobilebot runs in the environment, can obtain perception information during the interaction with the environment, and process this information into knowledge that it can understand to guide and optimize its behavior. A robot must have the ability to use it to accumulate knowledge in order to be able to show intelligent behavior. The navigation of a mobilebot in an unknown environment includes a bottom-up cognitive process of the environment, as well as a top-down process in which organized use accumulated knowledge to guide and optimize behavioral decision-making [11]. The bottom-up process is a process of perception fusion, and the top-down process is a process of planning and execution.

The reactive navigation has a small amount of calculation, so it has good real-time response performance and can run in an unknown environment and a dynamically changing environment. The SA-based disturbance control method integrates the target orientation, the local trap orientation, and the sensor information constraint conditions, and generates the kinetic energy that overcomes the local potential energy minimum through the Brownian motion under the probability constraint, and can realize the movement to the target position after a certain disturbance process. However, the reactive navigation mechanism does not establish an environmental map, so it is difficult to make full use of the information accumulated in history. In some special environments, such as a labyrinth environment or in an office, it takes a long period of roaming to find the exit to realize autonomous entry and exit from the door.

To operate in an unknown environment, it is necessary to reasonably handle the contradiction between the large amount of calculation necessary for environmental modeling and global planning and real-time control. Since the establishment of the environment model has a certain lag, and the environment is often dynamically changing, it is often difficult to realize the navigation using the global map in a large-scale environment in actual operation. Regional planning is a deliberate behavior between global planning and behavior control. It only aimed at is only a target in a local area and relies on the accumulation of dynamic environmental knowledge to guide the robot's progress. For an environment where the working area is not too large, the regional planning may already be equivalent to the goal to be achieved by the global planning.

There have been many research results for the path planning of mobilebots in the global or regional environment. The D* algorithm can be applied to many mobilebot examples because it can be applied to complex environments. The path planning algorithm of the US Mars rover uses the D* algorithm, which is applied in the UGV research project of DARPA's unmanned combat vehicle [52] and the tactical mobile robot for combat reconnaissance in the urban environment developed by CMU [44].

However, the D* algorithm is based on the premise that the environment is known or most known. Once an unknown obstacle is encountered, the originally established heuristic distance potential field distribution will be changed, and the potential field distribution needs to be recalculated. In an unknown environment, the robot can often only observe part of the surrounding environment, and often know nothing about the area near the final target. This leads to the initial establishment of the potential field model far from the actual situation. In addition, the D* algorithm is more effective in small-scale regional planning. For a large-scale environment to directly plan the path from the start point to the end point, the calculation cost and the memory space occupied are often large.

In the reverse D* algorithm given in this section, the initial target distance potential field is no longer calculated, but a local area potential field centered on the current position of the robot is established by defining the "Robot distance" method. And by searching the distance target to estimate the intermediate target position with the least cost, it can meet the needs of rolling optimization. Since the direction of the path search is opposite to the D* algorithm, it is called the reverse D* method.

(1) Principles of D* planning

D* planning algorithm is based on Dijkstra and A* algorithm. The main feature of Dijkstra's algorithm is to expand to the outer layer with the target point as the center until it reaches the starting point of the robot. It is a width-first global traversal search method. Dijkstra algorithm can obtain the optimal solution with the shortest path as the goal, but it is inefficient due to the high cost of traversal calculation. The A* algorithm is a heuristic algorithm, in the search process by establishing an evaluation function to determine the direction of the priority search. The A* algorithm can effectively reduce the search space and computational cost in most situations.

The difference between the A* algorithm and the Dijkstra algorithm is whether there is an evaluation value. The evaluation function is expressed as: $f(i) = g(i) + h(i)$, where $g(i)$ is from the target point S_{goal} to the actual cost of the search point i, $h(i)$ is the estimated cost from the search point i to the starting position S_{start}. Generally, the Euclidean distance from the search point i to the starting position S_{start} can be taken. When establishing the target distance potential field, priority is given to searching in the decreasing direction of the evaluation function $f(i) = g(i) + h(i)$ to establish the target distance potential field until the starting position S_{start} is found. Then search according to the gradient direction of the potential energy reduction

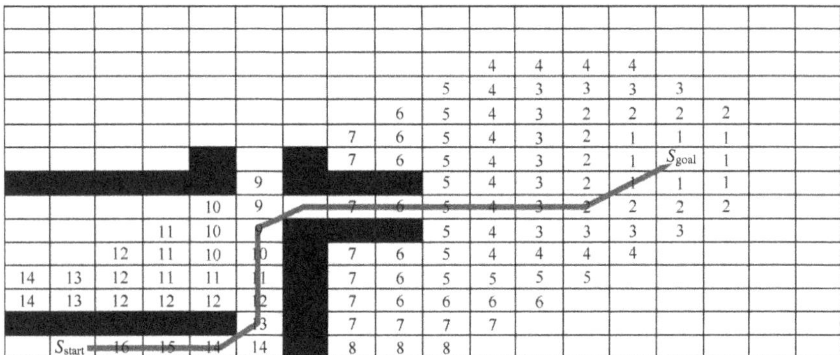

Figure 9.7.   Path planning based on A* search.

of the target distance, and the path of the robot can be planned. In most cases, the search area of the A* algorithm is less than the blind search of the Dijkstra algorithm, so the amount of calculation is greatly reduced, as shown in Figure 9.7.

D* Lite or D* algorithm [47,52] is a dynamic A* algorithm. Firstly, the A* algorithm is applied offline to establish the target distance potential field. The robot advances in the direction where the target distance decreases. When encountering a dynamic obstacle, plan a potential field position from the current position of the robot that can reach the target distance less than the blocked path, and then the robot follows the modified path to bypass the dynamic obstacle. After reaching the originally established target distance potential field area, it resumes searching and advancing in the direction of the gradient where the target distance decreases. As shown in Figure 9.8, the robot moves along the planned path. When it reaches position A, the original planned path is blocked by an unknown obstacle, and the target potential energy value at position A is 9. At this time, the A* method is used to plan the path from position A to the intermediate target L (target potential energy is 7) whose potential energy is less than A, as shown in the dynamic search area marked in Figure 9.8. The D* algorithm makes use of the initial potential field established offline as much as possible, so the area that needs to be dynamically searched at runtime is very small. The main cost of the D* algorithm is to establish the initial potential field. In a large-scale environment, using a global grid map to indicate that it is difficult to calculate and occupy memory.

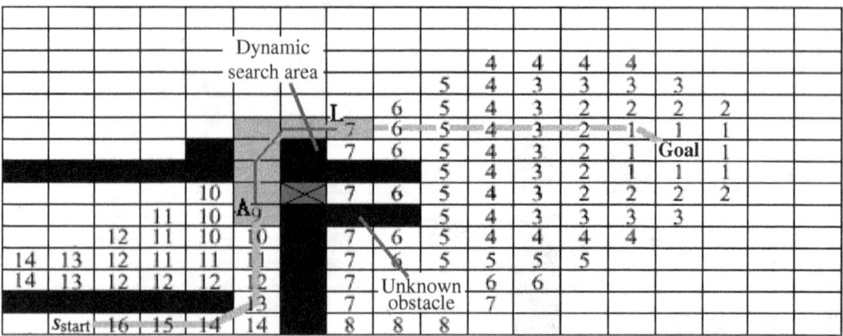

Figure 9.8.   Path planning based on D* search.

(2) The design of Reverse D* Algorithm

The D* algorithm is based on the premise that the environment is known or most known. In an unknown environment, the robot can only observe part of the surrounding environment, and retain the explored environment information through the accumulation of information, but often knows nothing about the area near the final target. As a result, the potential field distribution model of the distance from the target position to the current robot position established in an unknown environment is often far from the actual situation. Moreover, the D* algorithm is generally more effective in small-scale regional planning. For larger-scale environments, directly planning the path from the start point to the end point is often expensive.

Therefore, in the adopted reverse D* algorithm, the initial target distance potential field is no longer calculated, but a local area potential field centered on the current position of the robot is established by defining the "Robot distance" method. By searching for the "departure point" as the intermediate target position, the need of rolling optimization is met. As shown in Figure 9.14, the robot-centered expansion method is used to detect the condition of the "departure point" during the expansion process. The principle and steps of this method are as follows:

(1) The grid method is used to establish an environment map for the sensory information during the robot's traveling process, and the obstacle grid is first pre-expanded. The pre-expansion radius takes into account the robot's travel radius and retains a certain safe collision avoidance distance. After pre-expansion, the robot can be regarded as a "point" robot centered on its reference point.

(2) Let $E_{\min}$ is the minimum target potential energy experienced by the robot during its movement from the starting point $S(x_S, y_S)$ to the target point $G(x_g, y_g)$. The target potential energy function of the free grid node i with coordinates x_i, y_i is:

$$E(i) = \sqrt{(x_i - x_g)^2 + (y_i - y_g)^2} \qquad (9.16)$$

For the rolling optimization process, the current position $R(x_r, y_r)$ of the robot is taken as the center, and the grid node with the target potential energy function less than $E_{\min} - \varepsilon, \varepsilon > 0$ is searched outward, and this position is called the "Leave Point" and represented by $L_j(x_j, y_j)$, and ε is to ensure that the potential energy of $L_j(x_j, y_j)$ is less than the constant term of $E_{\min}$. $L_j(x_j, y_j)$ is an intermediate goal gradually approaching the final goal at the estimated cost of formula (9.17), that is, satisfying:

$$E(L_j) = \sqrt{(x_j - x_g)^2 + (y_j - y_g)^2} < E_{\min} - \varepsilon \qquad (9.17)$$

(3) In the process of searching for intermediate targets $L_j(x_j, y_j)$, the A* method is used to establish an evaluation function to determine the search priority. For the searched node i with coordinates (x_i, y_i), the evaluate function is $f(i) = g(i) + h(i)$. The difference from the D* algorithm is that $g(i)$ is the actual distance cost from the current position of the robot to the search node i, called "robot distance," and $h(i)$ is the estimated cost from the search node i to the target node Goal, the Euclidean distance from the node i to the target node is used to estimate $h(i)$, and the Euclidean distance from node i to the target node $G(x_g, y_g)$ is also regarded as a potential energy function, ie $h(i) = E(i)$.

(4) In the searching area, the robot's distances $g(i)$ are marked on the grid bitmap. When the intermediate destination $L_j(x_j, y_j)$ that meets the conditions of formulas (6-34) is searched, the robot starts from $L_j(x_j, y_j)$ and can go back to the current position along the decreasing gradient direction of robot's distances $g(i)$. Then the reverse process of backtracking process is a path from the robot's current position to the intermediate goal $L_j(x_j, y_j)$.

(5) When the robot starts from the starting point, the scroll dynamic deliberative planning is realized through continuous using of A* algorithm to search for the intermediate goal $L_j(x_j, y_j)$.

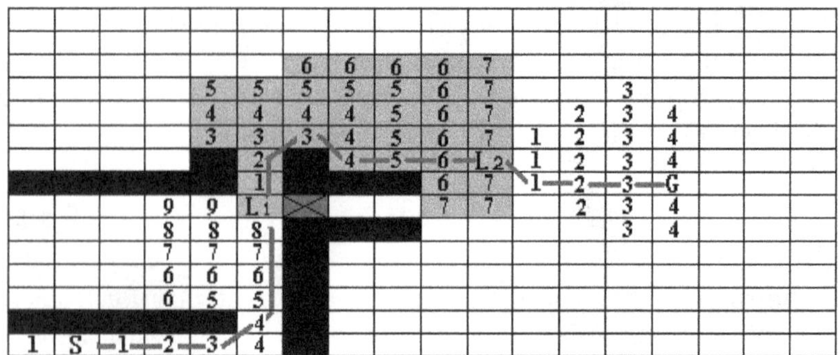

Figure 9.9. Path planning based on reverse D* search.

For example in Figure 9.9, the robot realizes the path planning from the starting point to the target point G with three running of dynamic programming. For the first planning the robot starts from the starting point S, searches for 9 steps, L_1 is found as an intermediate node. When the robot reaches L_1, it carries out the second planning in order to search L_2 as the second intermediate nodes. When the robot arrives at L_2, the third planning is carried out and the robot reaches the final destination point G.

Due to the searching direction of path planning method that is described above is contrary to $\mathbf{D}^*$ algorithm. That is to say, after the intermediate destination point $L_j(x_j, y_j)$ is searched, the robot starts from the point $L_j(x_j, y_j)$ and goes back to the current robot position according to the decreasing gradient direction of the robot distances $g(i)$. The reverse sequence of back journey is a path from the robot's current position to the intermediate target $L_j(x_j, y_j)$, so it is called reverse D* Algorithm. Reverse D* algorithm is no longer to establish the initial target distance potential field, but search "departure point" as the intermediate destination by scrolling.

It can be seen from the contrast of Figure 9.8 with Figure 9.9, comparison of the reverse D* algorithms and D* algorithms, the search area of Reverse D* algorithms is similar to the search area of establishing initial target potential field in the D* algorithms. The two planning approaches are computationally almost the same in statically known environment, but under unknown circumstances, Reverse D* algorithm searches outward for robot-centric, because the surrounding environment information of the robot can be updated in real time by the sense of sensors, therefore, it is more reliable than the target area information that away from the robot and the likelihood of encounter uncertainty obstacles is reduced in the implementation process.

(3) Simulation of deliberative planning
Relative to the D* Algorithms, Reverse D* Algorithm is a rolling optimization algorithm. Since there is no operation process of establishing initial potential field, so it can be applied to scroll-deliberative planning in large scale unknown environment. Meanwhile, this algorithm can be combined with the reactive local planning. In simple environments reactive navigation is used, Reverse D* algorithm searches for the "departure point" from the local potential energy trap in close to the local potential energy traps condition.

In the heuristic searching, let the initial search area is the area of the evaluation function $f(i) = g(i) + h(i) \leq a$, the value a is larger than the distance of the robot to the target. First let preferences $a = 1.5E_n$ in the program, in which E_n is the distance of the robot to the target. The maximum search area is an ellipse whose long shaft is $a = 1.5E_n$, short shaft is $\frac{\sqrt{5}}{2}E_n$. When the intermediate target is not searched in this area, the search area is expanded along the ellipse border, until an intermediate destination is found.

Figure 9.10 is a simulation that is combined by reactive navigation and deliberate planning. The mobile robot starts from the beginning-point, goes forward with responsive navigation method to the target direction. If there are obstacles in the direction of the advancing,

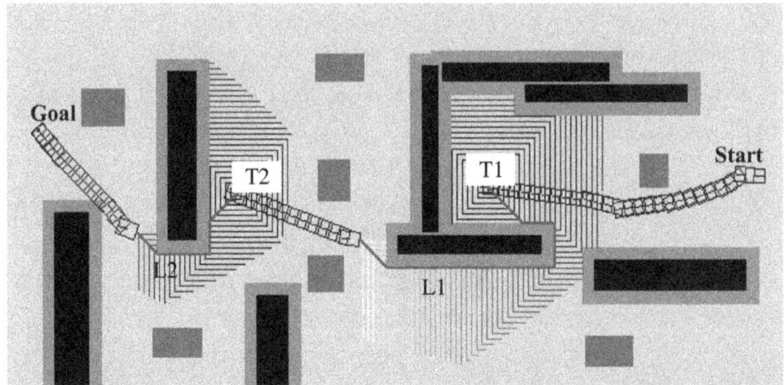

Figure 9.10.   Navigation simulation experiment of reverse D* planning.

which may cause the presence of local trap case, the robot will initiate deliberate planning. In the simulation, reactive navigation is applied from the beginning point Start to the location T_1. A large area barrier is perceived to exist in location T_1 and location L_1 is searched with the deliberative planning. After reach the location L_1, the robot recovers the reactive navigation behavior with the objectives trend. Location L_2 is searched by using the deliberative planning again in location T_2. Finally, the robot reaches the ultimate target goal by means of the reactive navigation. In the simulation the navigation in complex environments with 2 deliberate planning is achieved. The "robot distance" contour of each experienced search area is mapped out in Figure 9.10.

The time of deliberative planning is proportional to the size of the search area. The size of the search area is related to the distribution of obstacles in the environment. After testing, it takes $2 \sim 3$sec time to conduct 300 steps of Reverse D* algorithm running on the computer of PIII 800 M, and this can satisfy the needs of the rolling planning during the operation.

2. Composite navigation based on radar's sensing information

(1) The design of reflective behavior
Lidar (Laser radar) LMS291 not only can transfer distance information to the computer through RS232/422 interfaces, but also can set up three warning areas by the instruction, as shown in Figure 9.11. Each area corresponds to a binary output signal, when an obstacle is detected in the set-up area, the corresponding output signal is asserted. Because the warning message does not require the analysis and processing by a computer, it is very reliable and responsive, and it is particularly effective for the design of the robot's reflexively protective action behaviors.

Region A is the traveling area of the reduced speed, which forces the robot runs into a lower speed. Region B is an area for collision avoidance emergency stop. In order to prevent a collision with the obstacle, emergency stop is taken and it will wait for the robot's decision-making (turn or reverse). Region C is a dangerous escape area in order to protect sensitive equipment, such as laser radar. When there is an obstacle in this region, compulsory retreat behavior is taken to avoid the touching with barriers, especially the dynamic obstacles. Backing behavior is the response behavior

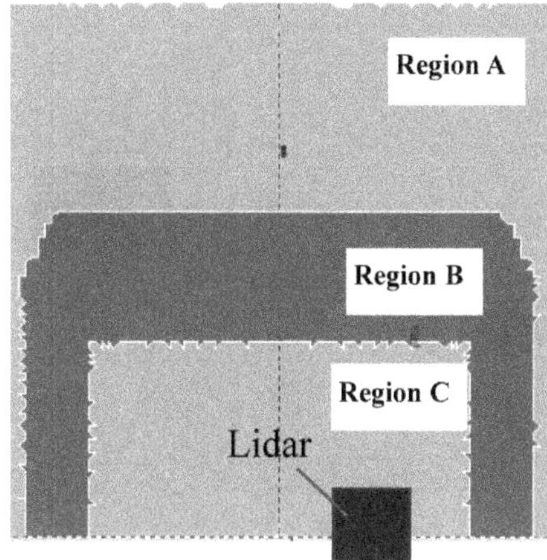

Figure 9.11.   Warning regional settings of LM.

in order to protect the fragile sensors. And it stops until the barrier off the region C. Signals A, B, C are inputted through a digital input channels on the board ADT850 which is responsible for the low-level motion control.

(2) The design of reactive behavior

Reactive behavior is the immediate response for the sensor information in the rolling window. The sensing area of laser radar is within the front range of 180°, so a semi-circular scrolling window is used [60]. Reactive behavior requires the integration of information perceived from the sensors, and the environmental information from the laser radar information processing system IPC1.

In order to reduce the cost of the amount of information processing and communications of reaction control, a scan data (obstacle distance) of laser radar in the range of $-90° \sim +90°$ is divided into $s_0 \sim s_{36}$ a total of 37 scanning areas, as shown in Figure 9.12 [60]. On both sides of the robot, $-90 \sim -87.5°$ is s_0, $87.5 \sim 90°$ is s_{36}, $s_1 \sim s_{35}$ is divided in accordance with 5° angle and obstacle distance in each scanning area is denoted by $|s_i|$. Since laser radar uses the resolution of 0.5° and there are 5 to 11 scan data in each scanning area, so the value of $|s_i|$ is the minimum value in the region. For the in-plant local planning, the resolution of 5° can satisfy the avoidance requirements.

Local planning agent makes reactive behavior planning according to compressed distance information of 37 directions from the laser

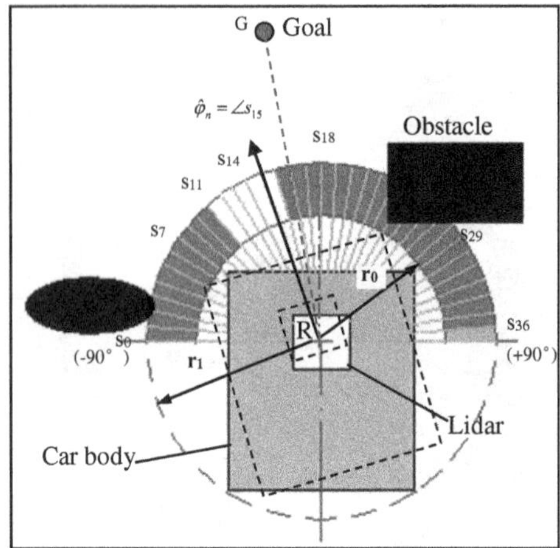

Figure 9.12.   Local rolling planning window based on LMS291.

radar. Local planning only considers how to implement the current path point target of the coordination task queue. Local planning reactive behavior layer contains input, direction assessment, local optimization, disturbance control, steering judgment module. They are shown in Figure 9.13.

Functions of each module for local planning agent are described as follows:

(1) Input: Input information contains parameters and status information of running robot in the local environment and state parameters of the coordinates $\mathbf{R}(x_r, y_r)$, heading angle θ_n of the robot are gotten from the coordinator IPC0. Access barriers distance in the direction $s_0 \sim s_{36}$ and the current trend goals, calculate the target direction angle ϕ_n that are gotten from IPC1. Whether there is a new trend target for the target update module monitors. When the deliberative layer is given a new trend path point, the target direction angle will be updated and the planning module initializes the local variables.

(2) Direction Assessment: the robot traversability of $s_0 \sim s_{36}$ directions is assessed based on obstacle distribution conditions on direction $s_0 \sim s_{36}$ and a direction set S of satisfying the collision-free conditions is selected. In Figure 9.17 the passable direction set $S = \{s11, s12, s13, s14, s15\}$.

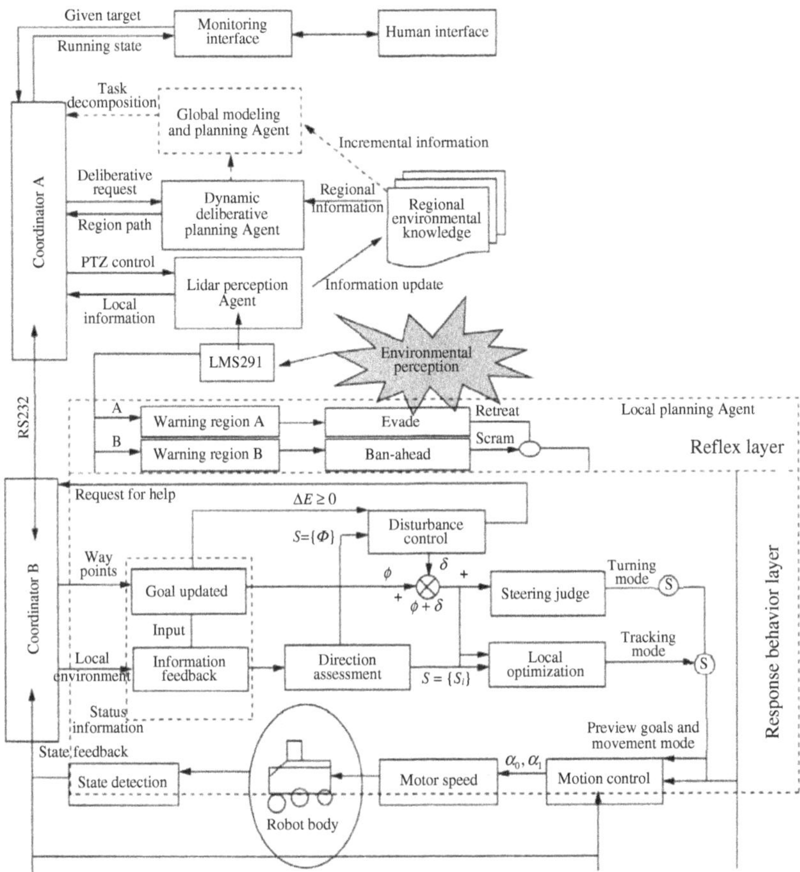

Figure 9.13. Reactive behavior control structure diagram in compound navigation.

(3) Local Optimization: The elements in the feasible set $\boldsymbol{S}$ are evaluated to determine the current "preview point" direction. "preview point" is located in circumference with the radius r_1 away from the robot's observation and rotation center. According to local optimization objective function equation (9.3) with the formula (9.4), the pre-roll the window down the aiming target direction $\hat{\phi}_n$ is determined. k_1 and k_2 are positive weighting factor, taking $k_1 = 5$, $k_2 = 1$. Local scrollable window optimization in Figure 9.12, the direction of travel preview s_{15} is selected, that is to say $\hat{\phi}_n = \angle s_{15}$.

(4) Disturbance Control: According to the results of the assessment of the direction and distance potential changes, the disturbance of target direction angle is determined. The δ_n is determined according to the formula (9.11). And the simulated annealing method is evaluated and updated. (1) the distance between the potential increments $\Delta_{\hat{E}} \geq 0$; (2) due to the observed range of the robot is $-90 \sim +90°$, there is no direction in the robot vision selectable situation, i.e. $S = \{\Phi\}$. Above two kinds of situations will start the disturbance control. When the local planning enters the disturbance control state, also a request for help to deliberative layer will be issued. It is indicated that the information is difficult to determine from the current perception of the proper direction of travel. It is need that the deliberative planning updates the traveling route according to the cumulative historical environmental information.

(5) Steering Judgment: It is to determine whether the robot rotates in order to ensure that the difference between the direction of heading angle θ_n of the robot and the reference direction of travel $\phi_n + \delta_n$ is always less than $90°$, i.e. keeping the facing direction $(\phi_n + \delta_n)$. For the movement of the robot is concerned, under normal circumstances the robot selects only the move-forward (including the left arc forward, right arc forward and straight forward motion) and the spin turn, etc. two behaviors.

Steering behavior: Steering behavior is the pure rotation *in situ*. When the target is not in the sensing range of the radar, the robot's "head" first turns to the target direction. After the preview direction is confirmed, the robot's body turns and "head" homes. Then the vehicle body moves forward.

Move-forward behavior: It contains the arc forward motion and straight forward movement. The preview goal direction is given by means of the local programming. After the motion control agent calculates the line speed control quantity of the vehicle motion $\alpha_0(t)$ and the rotation speed control quantity $\alpha_1(t)$, the vehicle is controlled to move towards the preview goal.

3. Motion control of mobilebot

The given target direction and preview target distance will determine the current track path. To simplify the calculation, a linear tracking mode is used. The preview target from local planning is a polar form of the robot as a reference center. Lateral error $d = 0$, the preview direction error angle of main error term in the tracking process is $\phi = \theta_n - \hat{\phi}_n$. According to the tracking controller determined by the formula (3-156) and (3-162), the control amount $\alpha_1(t)$ of the rotation angular acceleration is calculated. In accordance with the fuzzy controller of the formula (3-163) and Table 3.3, the acceleration $\alpha_0(t)$ of the vehicle line speed in the tracking process is calculated. Let the lower limit of the linear velocity tracking mode $v_{\min}5\,\text{cm/sec}$, upper limit $v_{\max}40\,\text{cm/sec}$.

The local planning makes re-planning cycle around 40 ms. So the preview goal constantly is refreshed, the motion controller gains line speed and the rotation speed of the acceleration control amount based on the control law and the given speed of individual motors is calculated by the formula (3-166). The track of the planned path is achieved.

9.3.2 *Implementation of composite navigation*

The deliberate layer contains several modules such as dynamic deliberate planning and dynamic environment knowledge base. The dynamic deliberate planning is based on the environmental map information in the local area in the dynamic environment knowledge base for path planning, and the amount of calculation is between global planning and reactive planning. The dynamic environment knowledge base uses a grid map to save the obstacle information obtained by the lidar sensing system, and the occupancy grid method is the most commonly used way of representing environment information in mobilebots. Define a grid value of -1 to indicate an unknown area, a value of 0 to indicate a free area, and a value greater than 0 to indicate that obstacles occupy the grid. In dynamic planning, the unknown area is treated as a feasible area. In order to update the knowledge base, in the regional map knowledge base, only the information in the most adjacent time period is retained. When an obstacle grid is detected, the grid value is increased by 2 (the maximum value is set to 9). A forgetting operation is performed on all obstacle grids every 10 seconds, that is, the value of the grid is reduced by 1, so that after multiple forgetting operations, the old map information will be updated.

Dynamic deliberate planning adopts the reverse D* planning method. The planned path sequence is given by careful planning and passed to the control layer by the coordinator. When it is close to the current waypoint, the task manager deletes the current waypoint from the task queue and passes the next waypoint to the control layer. In the planned path, if the way-points are very dense, in order to reach these designated way-points, it will cause the mobilebot to act dull and awkward. If the way-points are sparse, they cannot reflect the obstacles' constraints on the path. Therefore, it is necessary to evaluate the planned path and delete redundant path points:

If the connection direction between the two adjacent nodes before and after a path point meets the conditions that the robot can pass, the path point is a redundant point. As shown in Figure 9.14, path points 1 and 3 are redundant points. After the redundant points are deleted, the number of path points in the task queue is reduced.

The combination of local reactive behavior and dynamic deliberate planning to achieve compound navigation, the key is to rationally schedule reactive behavior and deliberate planning behavior. One method is to run deliberate planning on a regular basis, with a longer interval between two deliberate planning. The mobilebot moves along the planned path in a reactive behavior mode; when all the subtasks in the coordinator have been executed, it starts careful planning and gives the next set of path sequences. The other method is based on reactive navigation. Only when the robot is in a local trap, it is difficult to get out of the predicament, it will turn to deliberate planning. As shown in Figure 9.15, the mobilebot uses lidar information to build an indoor environment map, and uses deliberate planning to search for a path through the office porch.

The occupancy grid method can quickly realize the fusion of the distance between the lidar obstacle and the robot's track estimation

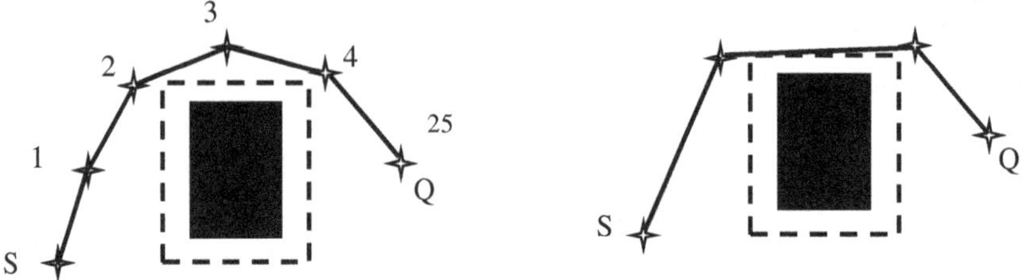

Figure 9.14. Delete operations of redundant path points.

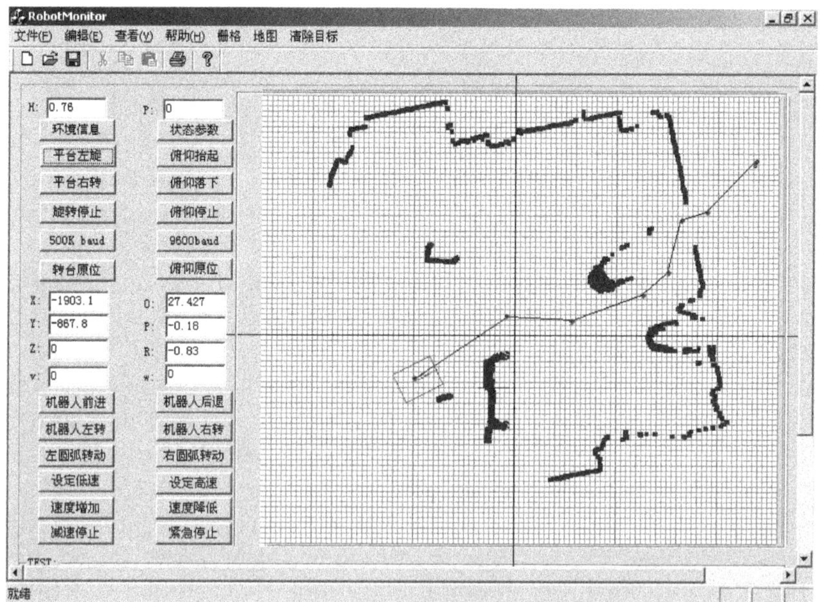

Figure 9.15. Deliberate path planning based on the environment.

information. The reverse D* algorithm is planned in the grid map, and it can complete the path search from 100 steps to 500 steps in a few seconds.

9.4 Mobilebot Path Navigation Based on Ant Colony Algorithm

Many path planning methods, such as path planning algorithms based on evolutionary algorithms and path planning algorithms based on genetic algorithms, have problems such as excessive computational cost and difficult solution construction. It is difficult to design evolutionary operator and genetic operator in complex environments. Ant colony optimization algorithms can be introduced to overcome these shortcomings, but there are also some difficulties in using ant colony algorithms to solve path planning problems in complex environments. This section first introduces the ant colony optimization (ACO) algorithm, and then introduces a mobile robot path planning method based on the ant colony algorithm.

9.4.1 *Introduction to ant colony optimization algorithm*

Biologists have found that ant colonies in nature have some significant self-organizing behavior characteristics during foraging, such as: (1) Ants will release a substance called pheromone during the movement; (2) Released pheromone will gradually decrease over time; ants can detect whether there are similar pheromone trajectories in a specific range; (3) Ants will move along the path with many pheromone trajectories and so on. It is based on these basic characteristics that ants can find a shortest path from the nest to the food source. In addition, the ant colony also has a strong ability to adapt to the environment. As shown in Figure 9.16, when an obstacle suddenly appears on the route that the ant colony passes, the ant colony can quickly find a new optimal path.

This foraging behavior of ant colonies has inspired a large number of scientific workers, resulting in the ant colony optimization algorithm (ACO). This optimization algorithm is a new type of bionic algorithm proposed by Dorigo and Colorni in the early 1990s [20–22]. Although it has only been proposed a decade or so ago, it has gradually attracted the attention of many scholars and has been widely used. The ant colony optimization algorithm was initially successfully used to solve the traveling salesman problem (TSP), and later penetrated other fields [18, 19]. Currently, people are studying how to use ACO to solve many discrete optimization problems.

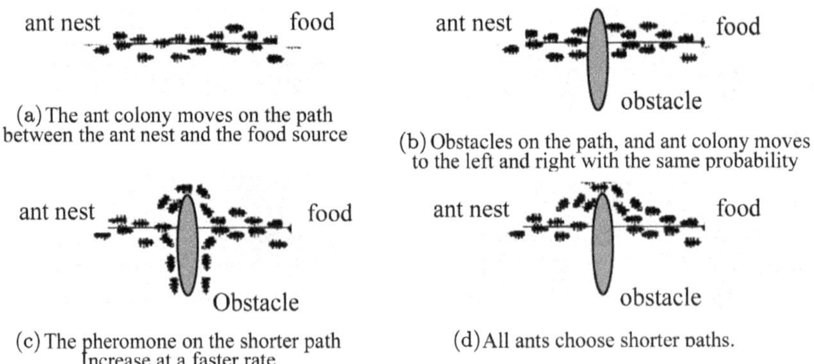

(a) The ant colony moves on the path between the ant nest and the food source

(b) Obstacles on the path, and ant colony moves to the left and right with the same probability

(c) The pheromone on the shorter path Increase at a faster rate

(d) All ants choose shorter paths.

Figure 9.16.   Adaptive behavior of ant colony.

Ant colony algorithm is a simulation of the actual ant colony collaboration process. Each ant independently searches for solutions in the space of candidate solutions and leaves a certain amount of information on the found solutions. The better the performance of the solution, the greater the amount of information left by the ant, and the more likely the solution is to be selected again. In the initial stage of the algorithm, the information amount of all solutions is the same. As the algorithm advances, the information amount of the better solution gradually increases, and the algorithm eventually converges to the optimal solution or the approximate optimal solution. The basic model of ant colony system is explained by taking the TSP problem of n cities in the plane as an example. The TSP problem for n cities is to find the shortest path through n cities once and finally back to the starting point. Let m be the number of ants in the ant colony, d_{ij} $(i, j = 1, 2, \ldots, n)$ represents the distance between city i and city j, and $\tau_{ij}(t)$ represents the information amount of remaining on the connected line ij at time t. During the movement any ant $k(k = 1, 2, \ldots, m)$ determines the transfer direction according to the following formula of the probability transfer rule.

$$p_{ij}^k = \begin{cases} \dfrac{\tau_{ij}^n(t)\eta_{ij}^n(t)}{\sum_{s\in \text{allowed}_k} \tau_{ij}^n(t)\eta_{ij}^n(t)}, & \text{if } j \in \text{ allowed}_k \\ 0, & \text{otherwise} \end{cases}$$

Where p_{ij}^k represents the probability that ant k is transferred from position i to position j at time t; η_{ij} represents the degree of expectation from city i to city j and is generally taken as $\eta_{ij} = 1/d_{ij}$; $\text{allowed}_k = \{0, 1, \ldots, n-1\} - \text{tabu}_k$ (tabu_k is the city set went before by ant k) which represents the city that ants k will choose next. As time goes by, the information left in the past will gradually disappear. The parameter $(1\text{-}\rho)$ is used to indicate the degree of information volatilization. After times n, each ant completes a cycle, and the amount of information on each path is adjusted according to the following formula:

$$\begin{cases} \tau_{ij}(t + n) = \rho(t)\tau_{ij}(t) + \Delta\tau_{ij}, & \rho \in (0, 1) \\ \Delta\tau_{ij} = \sum_{k=1}^m \Delta\tau_{ij}^k \end{cases}$$

Where, $\Delta\tau_{ij}^k$ represents the amount of information indicating that the kth ant left on the path ij in the current cycle, and $\Delta\tau_{ij}$ indicates the increment of the information amount on the path ij in the current cycle.

$$\Delta\tau_{ij}^k = \begin{cases} \frac{Q}{L_k} & \text{Ant } k \text{ passes through grid } i \text{ along } j \text{ direction} \\ 0 & \text{Otherwise} \end{cases}$$

Where Q is a constant and L_k represents the length of the path taken by the kth ant in this cycle; at the initial moment, $\tau_{ij}(0) = C$, $\Delta\tau_{ij} = 0(i, j = 0, 1, \ldots, n - 1)$. The selection of the parameters Q, C, ρ in the basic model of the ant colony system is generally determined by experimental methods, and the stopping condition of the algorithm can take fixed evolution generation number or stop the calculation when the evolutionary trend is not obvious.

9.4.2 *Path navigation based on ant colony algorithm*

The founder of ant colony algorithm M. Dorigo *et al.* summarized the application of ant colony optimization algorithm in the literature [19], including traveling salesman problem (TSP), quadratic allocation problem (QAP), joint scheduling problem (JSP), vehicle routing problem (VRP), graph coloring problem (GCP), etc.

The path planning problem of the robot is very similar to the foraging behavior of the ant. The path planning problem of the robot can be regarded as the process of circumventing some obstacles from the ant nest to find food. As long as there are enough ants in the nest, these ants can avoid the obstacle to find the shortest path from the nest to the food. Figure 9.25 is an example of the ant colony bypassing the obstacle to find a path from the nest to the food.

Path planning based on ant colony algorithm includes the following processes,

1. Environmental modeling
2. Establishment of neighboring areas
3. Establishment of smell zones
4. Constituting of path
5. Adjustment of path
6. Choice of path direction

7. Pheromone update
8. Algorithm Descriptions

The steps of the path planning based ant colony (PPACO) algorithm are as follows:

Step 1: Environment modeling;
Step 2: Establish a neighboring area of the nest and an odor area produced by the food;
Step 3: Place enough ants in the neighboring area;
Step 4: Each ant selects the next walking grid according to the above-mentioned method 6;
Step 5: If an ant has an invalid path, delete the ant, otherwise repeat until the ant reaches the odor zone and finds food in the direction of the odor;
Step 6: Adjust the effective path that the ant walks and save the optimal path in the adjusted path.
Step 7: According to the above method 7 to change the pheromone of the valid path;

Repeat steps $3 \sim 7$ until the number of iterations or the running time exceeds the maximum, ending the entire algorithm.

For the detailed process of path planning based on ant colony algorithm, please refer to the literature [9, 20–22].

9.5 Navigation Strategy Based on Feature Points

At this stage, more and more situations require robots to have the ability to navigate autonomously in unknown environments. For example, search and rescue missions, patrol and surveillance missions, urban combat missions, etc., all require robots to be able to navigate effectively and autonomously. In these situations, if robots can reliably replace humans to perform tasks, it will greatly reduce the risk of casualties [4, 18].

Let's look at a typical example — target reconnaissance mission: a single or a team of robots are dispatched into an unknown building to find and detect the expected target unit. Although the layout of this building is not known, that is, the environment is unknown and unfamiliar, some certain assumptions can be made based on experience: this building has several floors, and each floor has several corridors and rooms. The room will have doors and so on. Therefore, we can use these prior assumptions to design some corresponding navigation strategies for the robot to guide the robot's behavior. This is the research content of mobilebot navigation in unknown indoor environment, and this subject has always been a hot spot for scholars at home and abroad.

9.5.1　*Feature extraction*

For behavior-based robots, the ways of qualitative navigation are simpler and more natural, and we hope the robot can understand such as instructions like "walked down the hallway, turn left at the end, into the first room at right side." However, whether the robot is able to know exactly where the "door" is, where the "corridor" is? If the robot is roaming in the corridors it cannot find the open "door." Then, with no matter what the navigation strategy, the robot cannot walk into the target rooms. And the tasks cannot be completed. So for the needs of mobilebot navigation in unknown indoor environment, the 2D laser radar is used as the primary sensor, a set of effective feature extraction algorithms are designed for typical environmental features such as walls, corners, and exits [36–38].

The algorithm mainly extracts the structural features that can reflect a wide range of environments, such as straight lines and line segments, corners, points, vertical lines, etc., which correspond to features such as walls, corners, corridors, and doors, respectively. Among them, straight lines and line segments are the simplest and most intuitive features, which can easily describe most office environments.

1. Identification of the wall

In a structured environment, the walls have quite obvious geometric features: one linear (or near linear) segment. Therefore, a lot of laser beams will intersect the wall when the lidar is scanning. Once the robot finds the corresponding data points, the wall will be detected. Here Firstly, conduct point-cluster, and then the least square method is used to make these points fit a straight line. A straight line L in the 2D plane coordinate system can be described with only two parameters: ρ and θ. In which ρ is the perpendicular distance from the robot to the straight line L, θ is the angle between the perpendicular and the positive X-axis, as shown in Figure 9.17.

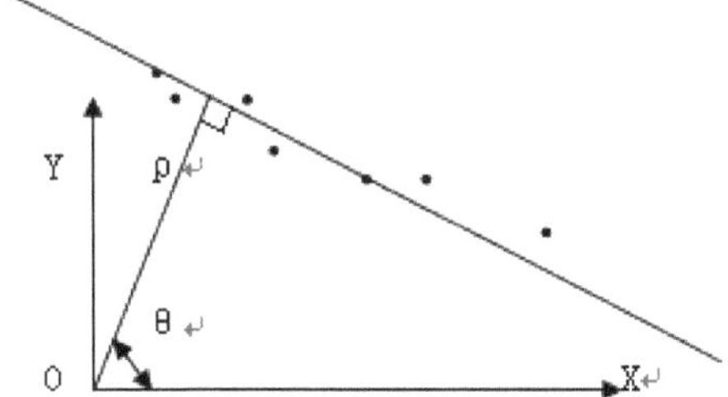

Figure 9.17.   Representation of a line in the robot coordinate system.

By means of the equation (9.18), the straight line is fitted the form like $y = k^*x + b$, or $x = s^*y + t$ by the use of the least square methods. In the formula, let a point set has n points, in which the coordinate of each point is X_i.

$$\begin{cases} k = \dfrac{n*\sum\limits_{i=1}^{n} X_iY_i - \sum\limits_{i=1}^{n} X_i \sum\limits_{i=1}^{n} Y_i}{n*\sum\limits_{i=1}^{n} X_iY_i - \sum\limits_{i=1}^{n} X_i \sum\limits_{i=1}^{n} X_i} \\[2em] b = \dfrac{\sum\limits_{i=1}^{n} Y_i - k*\sum\limits_{i=1}^{n} X_i}{n} \end{cases} \qquad (9.18)$$

In the algorithm, let the number n of points for conducting a straight-line fitting is 7, that is to say, the point set is $[X_i\text{-}3, X_i\text{+}3]$. If there are fewer than 7 points for a straight line to conduct fitting, it will be too sensitive for the environment and easily affected by the noise. If there are more than 7 points, it will result in the reducing of the algorithm effectiveness. The 361 data points that are passed back by liars can be fitted to 355 straight lines. And each line L_i is described by its corresponding (ρ, θ). Then for each straight line use the methods of the angle histogram to vote according to its values of the (ρ, θ). Because the straight lines that were fitted by the point from the same side of the wall and will have the same or similar (ρ, θ) values necessarily. So find those linear with high votes in angle histogram, where must can find the corresponding wall. And thus the lengths of the wall, the distances between the robots, the wall inclinations and other geometric information can be calculated.

2. Identification of the corners
In a structured environment, the corner is formed by the intersecting of two perpendicular line segments. In the previous step, the work

of linear fitting has been completed, so the feature extraction of the corners just as like help things along.

However, in practice, the two line segments of forming the corner that are extracted from the raw data in laser ranging may not have a common angle vertex, and it also cannot guarantee that the two straight lines completely perpendicular. It is supposed that $(X_1, Y_1, X_{c1}, Y_{c1})$ and $(X_2, Y_2, X_{c2}, Y_{c2})$ are two approximately perpendicular line segments of forming the corner, (X_{c1}, Y_{c1}) and (X_{c2}, Y_{c2}) are the segment endpoint coordinates in the corner area. In practical applications, the criterions of the corner angle are as follows:

$$\begin{cases} \left| \left\| \arctan \left[\dfrac{Y1 - Yc1}{X1 - Xc1} \right] - \arctan \left[\dfrac{Y2 - Yc2}{X2 - Xc2} \right] \right\| - 90^0 \right| < e_s \\ \sqrt{(Xc1 - Xc2)^2 + (Yc1 - Yc2)^2} < e_d \end{cases} \tag{9.19}$$

Where threshold values e_s and e_d are empirical values obtained by experiments. Since the value of e_d is small, the actual corner angle endpoint coordinates (X_c, Y_c) can be approximated calculated by the expression (9.25):

$$\begin{cases} Xc = \frac{Xc1 + Xc2}{2} \\ Yc = \frac{Yc1 + Yc2}{2}. \end{cases} \tag{9.20}$$

3. Identification of the doorway

Different from the features of walls and corners, the features of the doorway cannot be represented by specific geometric features, but an abstract description of the open area in the indoor environment. In the indoor environment, there must be a passable path between the relatively closed adjacent local spaces to connect them to each other, which is the characteristic of the doorway. The open door in the indoor corridor environment is one of the most common doorway features, which is also the main symbol of the robot's transition between the corridor and the room environment. Since the lidar reflects the distance information of the obstacle in front of the robot, if the robot is facing a closed obstacle, the distance data it obtains must show a uniform and continuous change. If there is an "opening" in the obstacle (such as a door opened in the wall), then a "jump" phase will inevitably appear in the radar data.

Therefore, the doorway can be found by looking for the above-mentioned "jump" stage among the 361 data points returned by the lidar. This is also in line with the definition of "doorway" in terms of people's cognition: the doorway means opening to an open area. Therefore, lasers at certain angles within 180 degrees in front of the robot can hit the "open area," and a "jump" phase will appear when the distance data is reflected. Find the two points where the "jump" occurs, and the two endpoints of the doorway can be determined.

9.5.2 *Navigation behaviors based on feature points*

In order to satisfy the needs of robot navigation in a complex environment, for the different feature points appropriate navigation behaviors are designed respectively.

1. Navigation behavior along the wall

The two end points which are extracted from the walls are respectively used as the target points of the robot reactive navigation, which constitutes the navigation behaviors along the wall of the robots. The results are shown in Figure 9.18.

2. Corridor navigation behaviors

After the positions of the robot to the centerline of the corridor are adjusted, the robot will be driven forward along this centerline which constitutes the corridor navigation behaviors. The experimental results are shown in Figure 9.19.

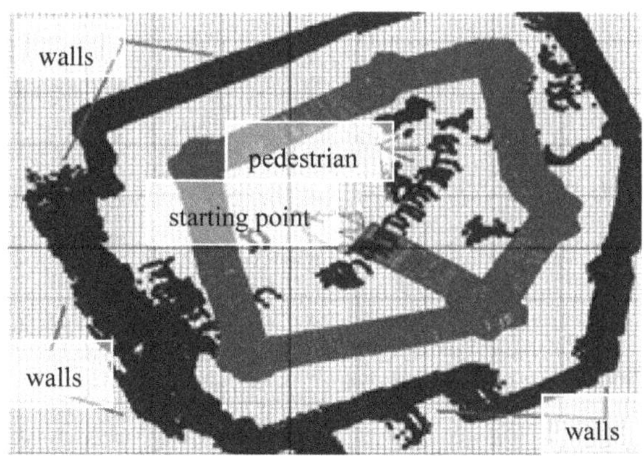

Figure 9.18. Robot navigation behaviors along the wall.

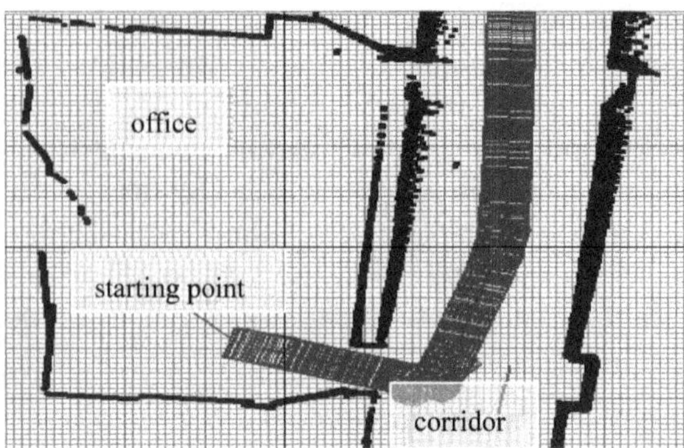

Figure 9.19.   Robot corridor navigation behaviors.

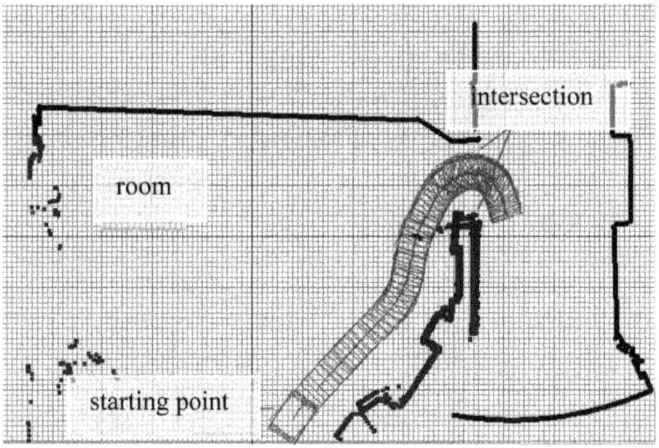

Figure 9.20.   Robot doorway navigation behaviors.

3. Behaviors of doorway navigation

The center point of two endpoints in the doorway is taken as the target point of the robot reactive navigation, which constitutes the doorway navigation behaviors of the robot. The experimental results are shown in Figure 9.20.

9.5.3   *Design and implementation of the navigation strategy*

The design of a finite state automaton (FSA) is used for organical integration of the robots' navigation behaviors. So that the robot can select the corresponding navigation behaviors base on different environmental characteristics and complete autonomous navigation in unknown environments.

1. Design of the state behavior sets

Navigation strategy involves three basic sets of behaviors, where the wall navigation is consisted of three movement patterns: closing to the wall, walking along the wall and avoiding obstacles. The doorway navigation consists of two movement patterns: passing through doorways and avoiding obstacles. The corridor navigation consists of two movement patterns: advancing along the centerline and avoiding obstacles as shown in Figure 9.21.

Three sets of the basic behaviors are: (1) Room navigation: the robot walks along the wall to the left, turns right after finishing the walking of each side of the wall, and then goes forward along the next side of the wall. Throughout the process, if it meets an obstacle, the robot will be reactive to avoid it. (2) Doorway navigation: robot goes forward towards the door, after goes through the door, then turns left. If it meets an obstacle, the robot will avoid it reactively. (3) Corridor navigation: robot goes forward along a corridor centerline, if an obstacle is met, the robot will avoid it reactively.

2. Specific processes of the state transitions

If the robot starts in an unknown environment, it will continuously scan the surroundings with the laser radar to obtain environmental characteristics, and then depending on the different environmental characteristics their navigation states are adjusted until the tasks

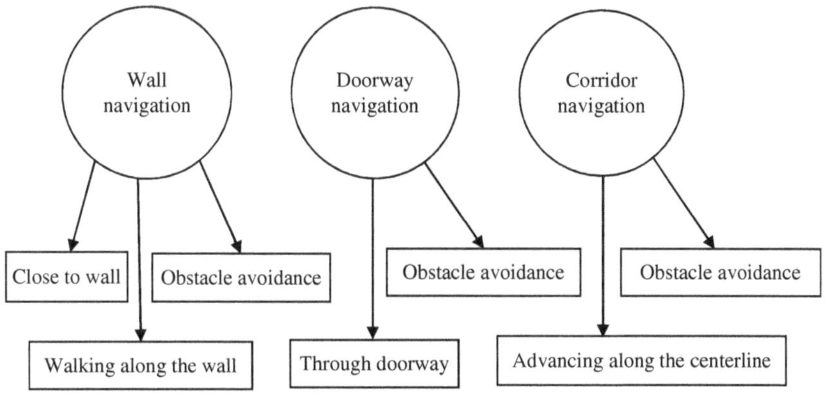

Figure 9.21. State behavior sets of the FSA.

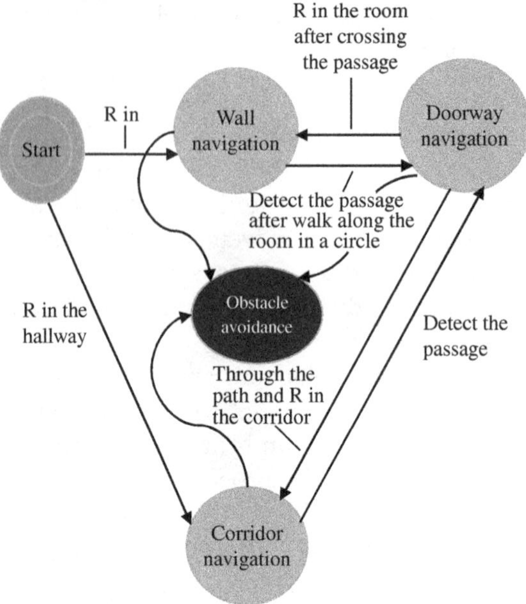

Figure 9.22.   State transition diagram of FSA.

are completed. State behaviors are shown in Figure 9.22, and the processes of specific state transitions are described as follows:

Step 1: After the robot starts, firstly it scans the surroundings. If it finds itself in the corridor, corridor navigation strategy will be started. If it finds itself in the room (or any non-corridor environment), the wall navigation strategy is started.

Step 2: After the robot starts corridor navigation strategy, it advances along the centerline of the corridor. If a passage during the movement is detected, passage navigation strategy is started up.

Step 3: After the robot starts wall navigation strategy, firstly, walking along the wall - surround the room in a circle, after a circle completed if detected the passage, then start passage navigation strategy; if do not detect the passage, then continue to startup executing the behaviors of walking along the wall.

Step 4: Once the robot goes through the path and completes the passage navigation behaviors, firstly it scans the surroundings.

If it is found that it is in the corridor, it starts the corridor navigation strategy. If the robot find that its location is in the room (or any non-corridor environment), it starts the wall navigation strategy.

Step 5: In the above navigation processes, regardless of what kind of navigation state the robot is now in, once encounters the obstacles in the road, it immediately performs obstacle avoidance behaviors. After completing the avoidance, it goes back to the previous navigation state.

Step 6: Termination conditions of the conversion process in the entire states should be determined by actual specific tasks, therefore they are not marked in the Figure 9.22.

3. Experimental results

In order to verify the performances of the algorithm, the mobilebot MORCS-1 is employed as the experimental platform and the mobilebot navigation control laboratory of Railway Institute of Central South University is used as an experimental environment. Algorithm programs are written with VC++6.0, and run on the remote control client computers (CPU 1.7G, memory 256M, Windows2000 operating system) to control the robot to carry out several experiments. And the condition of terminating the navigation is to find the yellow ball. Two results of the experiments are shown in Figure 9.23.

In Figure 9.23, the blue lines are the moving trajectory of the robot, and the black lines are the environmental raster maps which

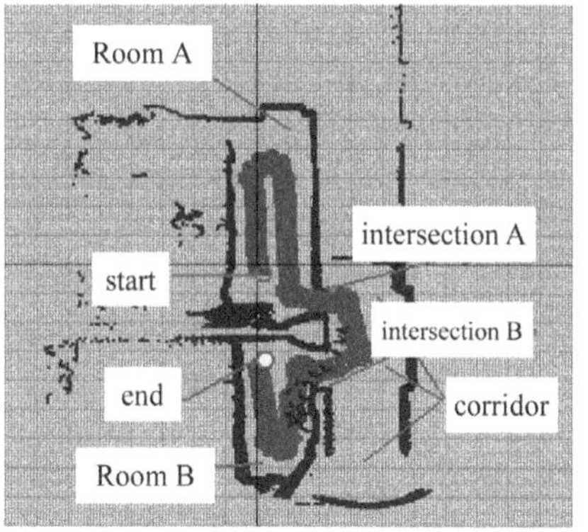

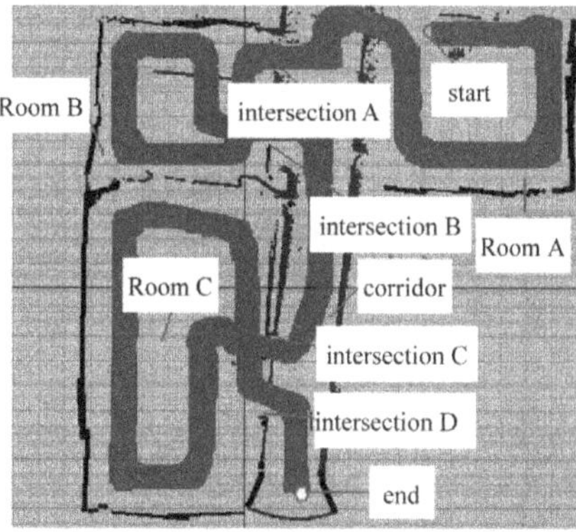

Figure 9.23.   Robot navigation experiment based on feature points.

Table 9.1. Result statistics of navigation experiences.

Number of experiment	Number of success	Number of failure	FRR	Misjudgment rate	Success rate
30	28	2	0%	6.67%	93.33%

are constructed by means of laser radar information. In the experiment, the robot starts from a certain point of the environment, and constantly adjusts its navigation strategies according to the environmental information which detected by liars. It completes the task of finding the yellow ball successfully. Statistics results from the experiment listed in Table 9.1.

It is shown that this robot navigation strategy can satisfy the real-time navigation needs of the robot and it has strong robustness, it also is suitable for various different indoor environments.

9.6 Mobilebot Navigation Based on Machine Learning

This section first introduces the research progress of mobilebot navigation based on machine learning, and then takes an example to discuss the application of mobilebot navigation based on machine learning.

9.6.1 *Advances in intelligent navigation based on machine learning*

In recent years, machine learning, especially deep learning, artificial cognition and machine thinking, particle swarm and ant colony optimization, evolutionary inheritance, immune evolution, knowledge and expert systems and other new artificial intelligence technologies are also increasingly widely used in mobile robot navigation, please see the introduction in Section 8 of this chapter.

In recent years, machine learning has been used more and more in automatic navigation. This subsection briefly introduces the research and application of intelligent navigation based on machine learning; then the research progress of autonomous route navigation for unmanned ships.

1. Overview of intelligent navigation applications based on machine learning

Machine learning has been successfully applied in the fields of pattern recognition, speech recognition, expert systems, and automatic navigation. Deep reinforcement learning (DRL) can effectively solve the path planning problem of continuous state space and action space. It can directly use raw data as input and output results as execution, and realize end-to-end learning mode, greatly improves the efficiency and convergence of the algorithm. In recent years, DRL has been widely used in the fields of planning, control and navigation of robot control, intelligent driving and traffic control.

First, machine learning has been widely used in the navigation of various mobilebots and intelligent moving bodies; for example, patrol robot path planning based on HPSO and reinforcement learning [51]; robot path planning in unknown environments based on deep reinforcement learning [5]; robust capture planning for multi-arm robots based on cloud-based 3D networks and related rewards; deep learning-based fog robot method for object recognition and grabbing planning for deburring surfaces by robot [1]; deep reinforcement learning from virtual to reality for continuous control of mobilebots without map navigation; global path planning for novel planetary rover based on learning [32,62]; autonomous path planning for unmanned ships based on deep reinforcement learning [27]; considering navigation experience rules for intelligent collision avoidance navigation for unmanned ships [40,49]; obstacle avoidance and path planning for agents based on deep reinforcement learning [17,43], etc.

Secondly, intensity deep learning is also more commonly applied to the underlying planning and control of moving bodies (robots); for example, model migration trajectory planning for deep reinforcement learning of intelligent vehicles [58]; a four-rotor aircraft based on deep reinforcement learning low-level control [35] and so on. In addition, machine learning is also applied to non-robot planning; for example, social awareness motor planning with deep reinforcement learning [13], artificial intelligence-assisted planning based on machine learning [29], and integrated planning and planning based on intention networks goal-oriented autonomous navigation of deep learning [25,54], etc.

Reinforcement learning has attracted wide attention in recent years. It can realize the mapping of learning from environment to behavior, and seek the most accurate or optimal action decision through the following "maximizing value function" and "continuous action space."

(1) Maximizing value function

Mnih *et al.* proposed a deep Q network (DQN) algorithm, which opened up the widespread application of DRL [45]. The DQN algorithm utilizes the powerful fitting capabilities of deep neural networks to avoid the huge storage space of the Q table, and uses experience replay memory and target networks to enhance the stability of the training process. At the same time, DQN implements an end-to-end learning method that uses only raw data as input and the output result as the Q value for each action. DQN algorithm has achieved great success in discrete action, but it is difficult to achieve high-dimensional continuous action. If the continuously changing movements are split indefinitely, the number of movements will increase exponentially as the degree of freedom increases, which will cause catastrophic latitude problems and may cause great training difficulties. In addition, discretizing actions remove important information about the structure of the action domain.

Actor-Critic (AC) algorithm has the ability to deal with continuous motion problems and is widely used in continuous motion space [2]. The network structure of AC algorithm includes Actor network and Critic network. The Actor network is responsible for outputting the probabilities of the actions, and the Critic network evaluates the output actions. In this way, the network parameters can be continuously optimized and the optimal action strategy can be obtained; but the random strategy of the AC algorithm makes it difficult for the network to converge. A deep deterministic policy gradient (DDPG) algorithm for solving deep reinforcement learning (DRL) problems in continuous states was provided [39].

(2) Continuous action space

The DDPG algorithm is a model-free algorithm that combines the advantages of the DQN algorithm with empirical replay memory and the target network. At the same time, the AC algorithm based on deterministic policy gradient (DPG) is used to make the network output result have a certain action value, so as to ensure that DDPG is applied to the field of continuous action space. DDPG can be easily applied to complex problems and larger network structures.

A framework for human-like autonomous vehicle following plans based on DDPG was put forwarded [65]. Under this framework, the self-driving car learns from the environment through trial and error to obtain the path planning model of the self-driving car, which has good experimental results. This research shows that DDPG can gain insights into driver behavior and help develop human-like autonomous driving algorithms and traffic flow models.

2. Research progress on autonomous route planning for unmanned ships

Improving the autonomous driving level of ships has become an important guarantee for enhancing the safety and adaptability of ships. Unmanned ships can be more adapted to the complex and volatile environment at sea. This requires unmanned ships to have autonomous path planning and obstacle avoidance capabilities, so as to effectively complete tasks and enhance the ship's comprehensive capabilities.

The research direction of unmanned ships involves autonomous path planning, navigation control, autonomous collision avoidance, and semi-autonomous task execution. As the basis and prerequisite of autonomous navigation, autonomous path planning plays a key role in ship automation and intelligence [41]. In the actual navigation process, ships often meet other ships. This requires a reasonable method to guide ships to avoid other ships and to sail according to targets. Unmanned ship path planning methods can guide ships to take the best course of action to avoid collisions with other ships and obstacles. Traditional path planning methods usually require relatively complete environmental information as prior knowledge, and it is very difficult to obtain information about the surrounding environment in unknown and dangerous marine environments. In addition, the traditional algorithm has a large amount of calculations, which makes it difficult to achieve real-time behavioral decision-making and accurate path planning for ships.

At present, research on autonomous path planning for unmanned ships has been carried out at home and abroad. These methods include traditional algorithms such as APF, speed obstacle method, Algorithm A* and some intelligent algorithms such as ant colony optimization algorithm, genetic algorithm, neural network algorithm and other DRL related algorithms.

In the field of intelligent ships, the application of DRL in unmanned ship control has gradually become a new research area. For example, path learning and maneuvering methods for unmanned cargo ships based on Q learning [12]; autonomous navigation control of unmanned ships based on relative value iterative gradient (RVIG) algorithm [56]; based on the Dueling DQN algorithm to automatically avoid collisions of multiple ships, a behavior-based USV local path planning and obstacle avoidance method is proposed [63], and LiDAR/INS based navigation [50], and so on. DRL overcomes the shortcomings of the usual intelligent algorithms, which require a certain number of samples and have fewer errors and response times.

Many key autonomous path planning methods have been proposed in the field of unmanned ships. However, these methods have focused on small and medium-sized USV studies, while relatively few studies have been conducted on unmanned ships. This paper chooses DDPG for unmanned channel planning because it has strong deep neural network function fitting ability and good generalized learning ability.

This autonomous path planning proposes a DRL-based model to implement intelligent path planning for unmanned ships in unknown environments. Through continuous interaction with the environment and the use of historical empirical data, the model can use DDPG algorithms to learn the best action strategies in a simulated environment. The navigation rules and the conditions encountered by the ship are converted into navigation restricted areas to achieve the safety of the planned path and ensure the validity and accuracy of the model. The data provided by the ship's automatic identification system (AIS) is used to train this path planning model. An improved DRL is then obtained by combining DDPG with artificial potential fields. Finally, the path planning model is integrated into the electronic chart platform for experiments. Comparative experimental results show that the improved model can implement autonomous path planning with fast convergence speed and good stability [12].

9.6.2 *Autonomous navigation based on deep reinforcement learning for unmanned ships*

An application example of deep learning in intelligent planning is introduced below, that is, autonomous path planning for unmanned ships based on deep reinforcement learning. Focus on the discussion of autonomous route planning model for unmanned ships based on deep reinforcement learning [27].

Reinforcement learning has performed well in continuous control problems and is widely used in areas such as path planning. This autonomous path planning proposes a DRL-based model to implement intelligent path planning for unmanned ships in unknown environments. Through continuous interaction with the environment and the use of historical empirical data, the model can use DDPG algorithms to learn the best action strategies in a simulated environment. The navigation rules and the conditions encountered by the ship are converted into navigation-restricted areas to achieve the safety of the planned path and ensure the validity and accuracy of the model. The data provided by the ship's automatic identification system (AIS) is used to train this path planning model. An improved DRL is then obtained by combining DDPG with artificial potential fields. Finally, the path planning model is integrated into the electronic chart platform for experiments. Comparative experimental results show that the improved model can realize autonomous path planning with fast convergence and good stability.

1. DDPG algorithm principle

The following introduces an autonomous path planning model for unmanned ships based on DRL.

DRL is an end-to-end learning method that combines DL and RL, and uses both the DL's perception ability and the RL's decision-making ability [48], which can effectively solve the disadvantages of traditional drones on the air promote the progress of continuous motion control. DDPG is an algorithm in DRL that can be used to solve continuous motion space problems. Among them, the depth refers to the deep network structure, and the policy gradient is a policy gradient algorithm, which can randomly select actions in the continuous action space according to the learned strategy (action distribution). The purpose of determinism is to help the policy gradient avoid random selection and output specific operating values.

(1) AC algorithm
DDPG is based on the AC algorithm, and its structure is shown in Figure 9.24.

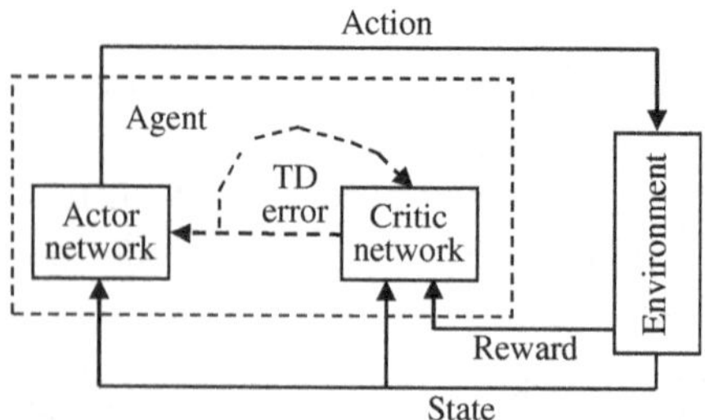

Figure 9.24.   Actor-Critic (AC) algorithm structure.

The network structure of the AC framework includes a policy network and an evaluation network. The policy network is called an actor network and the evaluation network is called a critic network. The actor network is used to select the action corresponding to the DDPG, while the critic network evaluates the advantages and disadvantages of the selected action by calculating a value function. Actor network and critic network are two independent networks that share state information. The network uses state information to generate operations, while the environment feeds back the resulting operations and outputs rewards. The critic network uses status and rewards to estimate the value of current actions and continuously adjusts its own value function. At the same time, the actor network has updated its action strategy in the direction of increasing action value. In this loop, the critic network evaluates the action strategy through a value function, provides a better gradient estimation for the strategy network, and finally obtains the best action strategy. It is very important to evaluate the action strategy in the critic network, it is more conducive to the convergence and stability of the current actor network. The above features ensure that the AC algorithm can obtain the best action strategy with gradient estimation in the case of low variance.

(2) DDPG algorithm structure

Figure 9.25 shows the structure of the DDPG algorithm.

The combination of DDPG and DQN under the premise of AC algorithm further improves the stability and effectiveness of network training, making it more conducive to solving continuous state and action space problems. In addition, DDPG uses DQN's experience replay memory and target network to solve the problem of

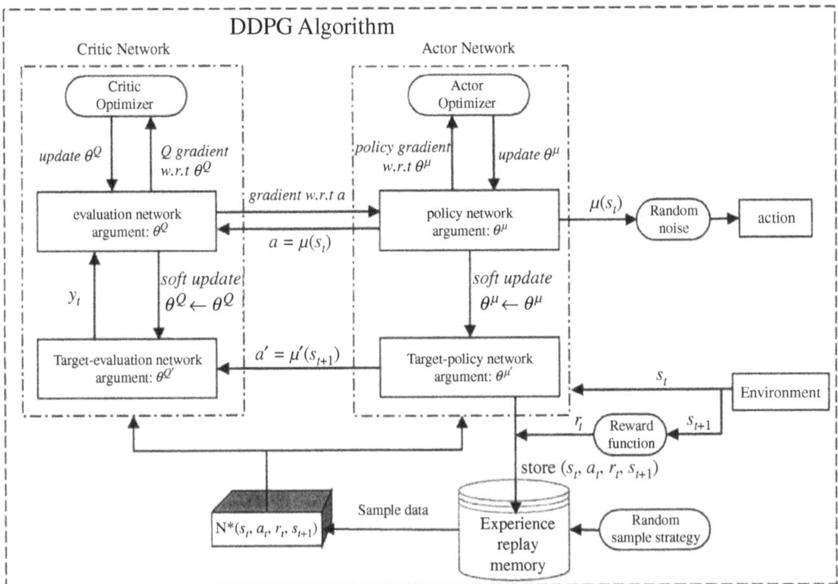

Figure 9.25.   Structure of deep deterministic policy gradient (DDPG) algorithm.

non-convergence when using neural networks to approximate function values. At the same time, DDPG subdivides the network structure into online networks and target networks. The online network is used to output actions in real time through online training, evaluate actions and update network parameters, including online Actor network and online Critic network respectively. The target network includes the target Actor network and Critic network, which are used to update the value network system and Actor network system, but do not perform online training and network parameter update. The target network and online network have the same neural network structure and initialization parameters.

The flow of DDPG algorithm is shown in Figure 9.26.

2. Structural design of autonomous route planning model for unmanned ship based on DDPG algorithm

The DDPG algorithm is a combination of deep learning and reinforcement learning. Based on the algorithm, an autonomous path planning model for unmanned ships is designed. The algorithm structure mainly includes communication algorithms, empirical replay mechanisms and neural networks. The output actions of this model

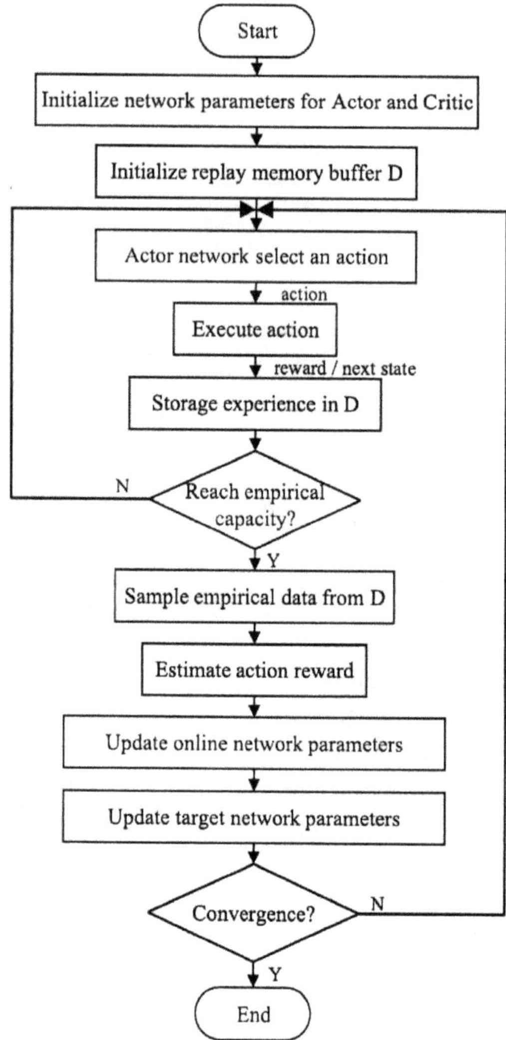

Figure 9.26.   DDPG algorithm flow.

are gradually accurate by using the AC algorithm to output and judge the ship's action strategy. By using the empirical replay mechanism, historical algorithms can output and judge the ship's action strategy. During the experiment, there are many environmental and behavioral states of ships, so it is necessary to use neural networks for fitting and induction. The current state of the unmanned ship

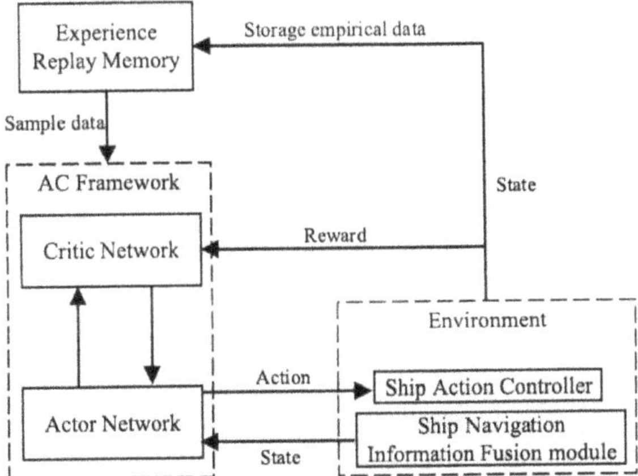

Figure 9.27. Structure of unmanned ship path planning model based on DDPG algorithm.

obtained in the environment is used as the input of the neural network, and the output is the Q value that the ship can perform Q learning actions in the unmanned environment. By continuously training and adjusting the parameters of the neural network, the model can learn the best action strategy in the current state.

Figure 9.27 shows the model structure of path planning for unmanned ship based on DDPG algorithm. The model mainly includes three parts: communication algorithm (AC), environment (ship motion controller and ship navigation information fusion module) and experience replay memory. Among them, the model obtains environmental information and ship status data through the "ship navigation information fusion module" and uses it as the input state of the AC algorithm. By randomly extracting data from the experience buffer pool for repeated training and learning, the optimal ship action strategy is output, which can meet the maximum cumulative return of the ship during the learning process. Finally, unmanned ships can avoid obstacles and reach their destinations with the help of ship motion controllers.

3. Model execution process

An unmanned ship path planning model based on the DDPG algorithm is used to abstract a real complex environment, and then it is transformed into a simple virtual environment through the model. At the same time, the model's action strategy is applied to the electronic chart platform environment to obtain the optimal planned trajectory of unmanned ships, and to implement the end-to-end algorithm learning process in the actual environment. Figure 9.28 shows the execution flow of an unmanned ship autonomous path planning model based on the DDPG algorithm. The unmanned ship model is executed as follows:

(1) Start the unmanned route planning process, and the system reads the ship data through the ship navigation information fusion module.
(2) Call the system trajectory planning model, and use the ship data as the input state, and process and calculate the model to obtain the ship's action strategy in the current state.
(3) The model transforms the action strategy into the actual action that the unmanned ship should take based on the actual movement of the ship.
(4) The ship motion controller analyzes and obtains the action to be performed, and executes the action.

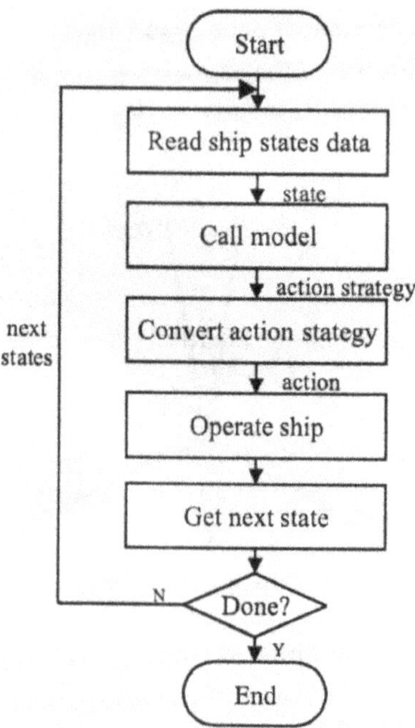

Figure 9.28.   Execution process of the UAV path planning model.

(5) The model acquires the new status information of the unmanned ship at the next moment after the action is performed, and determines whether the state of the ship after the action is completed is the end state.

(6) If it is not the final state, the model continues to use the state information of the ship, and then calculates and judges the action that the ship should take at the next moment and cycles through it. If it is in the end state, it means that the unmanned ship has completed the path planning task, and the model has finished the calculation and call.

4. Conclusion

For traditional path planning algorithms, historical empirical data cannot be recycled and used for online training and learning, which leads to lower accuracy of the algorithm, and the actual planned path is not smooth. This study proposes an autonomous path planning method for unmanned ships based on the DDPG algorithm. First, ship data was acquired based on an electronic chart platform. Then, the model is designed and trained in combination with ship maneuverability and crew experience, and verified under three classic encounter situations on the electronic chart platform. Experiments show that the unmanned ship has taken the best and reasonable action in an unfamiliar environment, successfully completed the task of autonomous route planning, and achieved unmanned operation. Finally, an autonomous route planning method for unmanned ships based on improved DRL is proposed, and the continuous operation output further verifies the effectiveness and reliability of the method.

The research and experimental results show that the autonomous path planning system for unmanned ships based on deep reinforcement learning has higher convergence speed and planning efficiency, the planned route is more in line with navigation rules, and the autonomous route planning and collision avoidance of unmanned ships are realized.

9.7   Chapter Summary

This Chapter 9 discusses the intelligent navigation problem, that is, the mobilebot navigation problem. In Section 9.1, the main methods and development trends of mobilebot navigation are introduced, the methods deal with the navigation method based on case learning, navigation method based on environmental model, and behavior-based navigation methods, then the development trend of mobile robot navigation is overviewed.

Some typical and important mobilebot navigation methods are discussed from Section 9.2 to Section 9.7 respectively. Section 9.2 briefly introduces robot global navigation based on approximate Voronoi diagram. The space representation method of mobilebot environment is presented firstly, then Voronoi diagram and Approximate Voronoi Boundary Network (AVBN) modeling method are mentioned.

Section 9.3 sums up the local navigation strategies of mobilebots, pays attention to the disturbance rule design based on simulated annealing and local navigation program design, and gives the simulation experiment of the local navigation simulation.

Section 9.4 expounds composite navigation strategy and implementation in an unknown environment. The navigation strategic design methods of local planning and dynamic deliberate planning methods are made, and the mobilebot is designed based on the combination of reflective behavior, reactive behavior and deliberate behavior composite navigation strategy.

Section 9.5 introduces the robot navigation based on ant colony algorithm, first briefly explains the ant colony optimization (ACO) algorithm, and then discusses a mobilebot path planning method based on the ant colony algorithm with more detailed.

In order to meet the needs of robots to navigate in complex environments, corresponding navigation behaviors are designed for different feature points. Section 9.6 studies the navigation strategy based on feature points. A finite state automaton (FSA) is designed to organically integrate the navigation behavior of the robot, so that the robot can select the corresponding navigation behavior according to different environmental characteristics, and complete the autonomous navigation in the unknown environment.

Machine learning has been successfully applied in the field of automatic navigation. Deep reinforcement learning can effectively solve the path planning problem of continuous state space and action space. In Section 9.7, an overview of intelligent planning based on machine learning is given, and examples of intelligent planning based on deep learning are listed.

References

[1] Ajay, K.T., Nitesh, M. and John, K. (2019). A Fog Robotics Approach to Deep Robot Learning: Application to Object Recognition and Grasp Planning in Surface Decluttering. *2019 International Conference on Robotics and Automation* (ICRA), Montreal, QC, Canada, pp. 4559–4566.

[2] Bahdanau, D., Brakel, P., Xu. K., Goyal, A., Lowe, R., Pineau, J., Courville, A. and Bengio, Y. (2016). Actor-Critic Algorithm for Sequence Prediction, arXiv 2016, arXiv: 1607.07086.

[3] Benboiuabdallah, Karim. And Supervisor: Zhu Qi. (2013). *Research on Navigation Control and Optimization Technology of Mobile Robot Based on Stereo Vision.* Doctoral Dissertation of Harbin Engineering University (in Chinese).

[4] Blitch, J. (1999). Tactical mobile robots for complex urban environments. *Mobile Robots*, XIV, pp. 116–128.

[5] Bu, X.J. (2018). *Research on Robot Path Planning in Unknown Environment Based on Deep Reinforcement Learning*, Master's Degree Thesis, Harbin Institute of Technology (in Chinese). DOI: CNKI:CDMD:2.1018.896554.

[6] Cai, Z.X. and Zou, X.B. (2004). Research on environmental cognition theory and technology of mobile robot, *Robot*, 26(1):87–91 (in Chinese).

[7] Cai, Z.X., He, H.G. and Chen, H. (2009). *Navigation Control Theories and Methods of Mobile Robots under Unknown Environment.* Beijing: Science Press (in Chinese).

[8] Cai, Z.X., Li, Y., *et al.* (2021). *Autonomous Vehicle Perception, Mapping and Target Tracking* Beijing: Science Press (in Chinese).

[9] Cai, Z.X., Liu, L.J., Chen, B.F. and Wang, Y. (2021). *Artificial Intelligence: From Beginning to Date*, Singapore: World Scientific Publishers.

[10] Cai, Z.X., Zhou, X., Li, M.Y., *et al.* (2000). Evolutionary control architecture of autonomous mobile robot based on function/behavior integration. *Robot*, 22(3):169–175 (in Chinese).

[11] Cai, Z.X., Zou, X.B., Chen, H., *et al.* (2016). *Key Techniques of Navigation Control for Mobile Robots under Unknown Environment* Beijing: Science Press (in Chinese).

[12] Chen, C., Chen, X.Q., Ma, F., *et al.* (2019). A knowledge-free path planning approach for smart ships based on reinforcement learning. *Ocean Engineering*, 189:106299.

[13] Chen, Y.F., Michael, E., Liu, M., *et al.* (2017). Socially Aware Motion Planning with deep reinforcement learning. *2017 IEEE/RSJ International Conference on Intelligent Robots and Systems* (IROS), Vancouver, BC, pp. 1343–1350.

[14] Chen, Z.H. (2000). Study on the case-based learning algorithm for path planning of the lunar probe. *Aeronautical Computing Technology*, 30(2):1–4 (in Chinese).

[15] Chu, X., Wen, G.L. and Lu, Y.Z. (2016). An obstacle avoidance and navigation method for the mobile robot, *Mechanical Science and Technology for Aerospace Engineering*, 35(6):939–945 (in Chinese).

[16] Dai, B., Xiao X.M. and Cai, Z.X. (2005). Current status and future development of mobile robot path planning technology. *Control Engineering*, China, 12(3):198–202 (in Chinese).

[17] Deng, W. (2019). *Research and Application of Intelligent Body Obstacle Avoidance and Path Planning Based on Deep Reinforcement Learning*. Master's Degree Thesis of University of Electronic Science and Technology of China (in Chinese).

[18] Dorigo, M. (1992). *Optimization, Learning and Natural Algorithm* (in Italian), Ph.D. thesis, Dipartimento di Elettronica, Politecnico di Milano, IT.

[19] Dorigo, M., Bonabeaumm, E., *et al.* (2000). Ant algorithms and stigmergy. *Future Generation Computer Systems*, 16:851–871.

[20] Dorigo, M., *et al.*, (1999). Ant algorithms for discrete optimization. *Artificial Life*, 5(3):137–172.

[21] Dorigo, M., Maniezzo, V. and Colorni, A. (1996). The Ant System: Optimization by a colony of cooperating agents. *IEEE Transactions on Systems, Man, and Cybernetics — Part B*, 26(1):1–13.

[22] Dorigo, M., Stutzle T. Translated by Zhang Jun *et al.* (2007). *Ant Colony Optimization*. Beijing, China: Tsinghua University Press (in Chinese).

[23] Fan, Y.S., Sun, X.J. and Wang, G.F. (2019). An autonomous dynamic collision avoidance control method for unmanned surface vehicle in unknown ocean environment. *International Journal of Advanced Robotic Systems*, 16(2):1–11, March 14, 2019. DOI: 10.1177/1729881419831581.

[24] Fu, Z., Wang, S.G. and Wang, J.Y. (2000). Research on the Voronoi diagram generation algorithm of multi-connected domains. *System Engineering and Electronic Technology*, 22(11):88–90 (in Chinese).

[25] Gao, W., David, H.W. and Lee, W.S. (2017). Intention-Net: Integrating Planning and Deep Learning for Goal-Directed Autonomous Navigation. arXiv:1710.05627v2 [cs.AI] October 17, 2017.

[26] Ghallab, M., Nau, D. and Traverso, P. Translated by Jiang, Y.F. *et al.* (2008). *Automatic Planning: Theory and Practice*, Beijing, China: Tsinghua University Press (in Chinese).

[27] Guo, S., Zhang, X.G., Zheng, Y.S. and Du, Y.Q. (2020). An Autonomous Path Planning Model for Unmanned Ships Based on

Deep Reinforcement Learning. *Sensors*, 20(1–35):426; doi:10.3390/s20020426.

[28] Hassan, H., Simo, J., and Crespo, A. (2001). Flexible real-time mobile robotic architecture based on behavioural models. *Engineering Applications of Artificial Intelligence*, 14(5):685–702.

[29] Huang, D.X. (2017). Prospects for artificial intelligence-assisted planning based on machine learning. *Urban Development Studies*, 24(5): 50–55 (in Chinese).

[30] Huang, H., Li, Y.J., Cao, S.L., *et al.* (2000). Comparison of Voronoi sample and ACO galaxy cluster sample. *Journal of Beijing Normal University (Natural Science Edition)*, 36(4):451–457 (in Chinese).

[31] Jiao, L.C. and Wang, L. (2000). A novel genetic algorithm based on immunity. *IEEE Transactions on Systems, Man and Cybernetics — Part A: Systems and Humans*, 30(5):552–561.

[32] Kalogeiton, V.S., Ioannidis, K., Sirakoulis, G.C., *et al.* (2019). Real-time active SLAM and obstacle avoidance for an autonomous robot based on stereo vision. *Cybernetics and Systems*, 50(3):239–260.

[33] Kamil, F., Hong, T.S., Khaksar, W., *et al.* (2017). New robot navigation algorithm for arbitrary unknown dynamic environments based on future prediction and priority behavior. *Expert Systems with Applications*, 86(2017):274–291.

[34] Kirkpatrick, S., Gelatt, C.D. and Vecchi, M.P. (1983). Optimization by simulated annealing. *Science*, 220(4598):671–680.

[35] Lambert, N.O., Drewe, D.S., Yaconelli, J., Calandra, R., Levine, S. and Pister, K.S.J. (2019). Low-level control of a quadrotor with deep model-based reinforcement learning. *IEEE Robotics and Automation Letters*, 4(4):4224–4230.

[36] Li, Q.Z., Wang, Q., Zeng, Y. and Yu, M. (2020). Visual navigation method of mobile robot based on face recognition and optical flow tracking, *Automation & Instrumentation*, 23–27, 65 (in Chinese).

[37] Li, Y.Q. and Chen C.M. (2020). Design of mobile robot automatic navigation system based on ROS and laser radar, *Modern Electronics Technique*, 43(10):176–178, 183 (in Chinese).

[38] Li, L. (2003). *Research on Mobile Robot System Design and Visual Navigation Control.* Beijing: Doctoral Dissertation of Graduate School of Chinese Academy of Sciences (in Chinese).

[39] Lillicrap, T.P., Hunt, J.J., Pritzel, A., Heess, N., Erez, T., Tassa, Y., Silver, D. and Wierstra, D. (2015). Continuous control with deep reinforcement learning, *ResearchGate*, arXiv:1509.02971v1 [cs.LG].

[40] Liu, H., Sheng, X.J. and Zhang, H. (2018). Positioning and navigation system for mobile manipulator based on multi-sensor fusion, *Mechatronics*, 24(3):35–40 (in Chinese).

[41] Liu, Z., Zhang, Y., Yu, X. and Yuan, C. (2016). Unmanned surface vehicles: An overview of developments and challenges. *Annu. Rev. Control*, 41:71–93.

[42] Lucidarme, P. and Liégeois, A. (2003). Learning reactive neurocontrollers using simulated annealing for mobile robots. *IROS2003*. Las Vegas, 674–679.

[43] Luo, R.C. and Chen, C.J. (2017). Recursive neural network based semantic navigation of an autonomous mobile robot through understanding human verbal instructions. *Proc. of IEEE/RSJ International Conference on Intelligent Robots and Systems* (IROS), pp. 1519–1524, Vancouver, Canada, SEP 24–28, 2017.

[44] Matthies, L., Xiong Y., Hogg, R., *et al.* (2000). A portable, autonomous, urban reconnaissance robot. *Proc. of the 6th International Conference on Intelligent Autonomous Systems*. Venice, 163–172.

[45] Mnih, V., Kavukcuoglu, K., Silver, D., *et al.* (2013). Playing atari with deep reinforcement learning. arXiv 2013, arXiv:1312.5602.

[46] Na, Y.K. and Se, Y.O. (2003). Hybrid control for autonomous mobile robot navigation using neural network based behavior modules and environment classification. *Autonomous Robots*, 15(2): 193–206.

[47] Pu, X.C., Tan, S.F. and Zhang, Y. (2014). Research on the navigation of mobile robots based on the improved FAST algorithm, *CAAI Transactions on Intelligent Systems*, (4):419–424 (in Chinese).

[48] Serrano, W. (2019). Deep reinforcement learning algorithms in intelligent infrastructure. *Infrastructures*, 4(3):52; https://doi.org/10.3390/infrastructures-4030052.

[49] Shen, H.Q., Guo, C., Li, T.S., *et al.* (2018) An intelligent collision avoidance navigation method for unmanned ships considering navigation experience rules. *Journal of Harbin Engineering University*, 39(6): 998–1005 (in Chinese).

[50] Song, R., Fang Y.C. and Liu, H. (2020). Integrated navigation approach for the field mobile robot based on LiDAR/INS, *CAAI Transactions on Intelligent Systems*, 15(4):804–810 (in Chinese).

[51] Song, Y. (2019). *Research on Path Planning of Inspection Robot Based on HPSO and Reinforcement Learning*. Master's Thesis of Guangdong University of Technology.

[52] Stentz, A. and Hebert, M.A. (1995). Complete navigation system for goal acquisition in unknown environments. *Autonomous Robots*, 2(2): 127–145.

[53] Urzelai, J. and Floreano, D. (2000). Evolutionary robots with fast adaptive behavior in new environments. *Third International*

Conference on Evolvable Systems: From Biology to Hardware (ICES2000), Berlin: Springer–Verlag.

[54] Vazquez-Santacruz, J.A., Velasco-Villa, M., Portillo-Velez, R.de J., *et al.* (2017). Autonomous navigation for multiple mobile robots under time delay in communication. *Journal of Intelligent and Robotic Systems*, 86(3-4):583–597, June 2017.

[55] Xiang, Y. and Qu K.K. (2020). Research on navigation and path planning of mobile robots. *Auto Expo*, (8):59 (in Chinese).

[56] Yang, J., Liu, L., Zhang, Q., Liu, C. (2019). Research on Autonomous Navigation Control of Unmanned Ship Based on Unity3D. In *Proceedings of the 2019 IEEE International Conference on Control, Automation and Robotics* (ICCAR), Beijing, China, 19–22 April 2019; pp. 2251–2446.

[57] Yi, W.Q., Zhao C.J. and Liu, Y. (2020). Research on autonomous positioning and navigation technology of mobile robots, *Chinese Journal of Construction Machinery*, 18(5):400–405 (in Chinese).

[58] Yu, L.L., Shao, X.Y., Long, Z.W., *et al.* (2019). Model migration trajectory planning method for intelligent vehicle deep reinforcement learning. *Control Theory and Application*, 36(9):1409–1422.

[59] Zhang, C.G. and Xi, Y.G. (2003). A real time path planning method for mobile robot avoiding oscillation and dead circulation. *Acta Automatica Sinica*, 29(2):197–205.

[60] Zhang, C.G. and Xi, Y.G. (2001). Robot path planning based on rolling window when the global environment is unknown. *Science in China (Series E)*, 31(1):51–58 (in Chinese).

[61] Zhang, H., Yi, S., Luo, X., *et al.* (2004). Robot path planning based on ant colony optimization algorithm in complex environment, *Control and Decision*, 19(2):166–170 (in Chinese).

[62] Zhang, J., Xia, Y.Q. and Shen, G.H. (2019). A novel learning-based global path planning algorithm for planetary rovers. *Neurocomputing*, 361:69–76.

[63] Zhang, R.B., Tang, P., Su, Y., Li, X., Yang, G. and Shi, C. (2014). An adaptive obstacle avoidance algorithm for unmanned surface vehicle in complicated marine environments. *IEEE CAA Journal of Automatica Sinca*, 1:385–396.

[64] Zhao, H., Cai, Z.X. and Zou, X.B. (2003). Path planning based on fuzzy ART and Q learning. *Proceedings of the 10th Annual Conference of China Association of Artificial Intelligence*. Guangzhou 2003: 834–838 (in Chinese).

[65] Zhu, M., Wang, X. and Wang, Y. (2018). Human-like autonomous car-following model with deep reinforcement learning. *Transportation Research Part C, Emerging Technologies*, 97:348–368.

[66] Zou, X.B. (2005). *Research on Control System Design and Environment Modeling of Mobile Robot Prototype, Doctoral Dissertation*, School of Information Science and Engineering, Central South University (in Chinese).

[67] Zou, X.B. and Cai, Z.X. (2004). Incremental environment modeling method based on approximate Voronoi diagram. *WCICA2004*. Hangzhou: 4618–4622 (in Chinese).

[68] Zou, X.B., Cai, Z.X. and Liu, J., *et al.* (2001). A local path planning method for mobile robots. *Proceedings of the Ninth Annual Conference of Chinese Society of Artificial Intelligence*. Beijing, pp. 947–950 (in Chinese).

Chapter 10

Intelligent Control of Mobile Robots

This chapter will study mobile robot control issues, including mobile robot tracking control, mobile robot navigation control, mobile robot formation control, and mobile robot control based on neural networks and deep learning. Now, the control of these mobile robots is mostly based on artificial neural networks, and uses a mixture of adaptive control, fuzzy control, and visual control technologies to achieve colorful control of mobile robots in various applications.

This chapter first introduces the basic structure of intelligent control and intelligent control systems and various structures of intelligent control systems; then summarizes various application cases of mobile robot control based on neural networks, involving tracking control, navigation control, formation control and vision control of mobile robots, etc.; Finally, a large number of examples are used to summarize various types of mobile robot control based on deep learning, and to conduct an in-depth analysis of the examples.

10.1 Overview of Intelligent Control and Intelligent Control System

Traditional control technologies (such as open loop control, PID feedback control) and modern control technologies (such as variable structure control, adaptive control, robust control, etc.) have been applied to varying degrees in robot systems, and intelligent control (such as hierarchical control, fuzzy control, neural control) are often the first to be developed on the excellent "test bed" of robots.

This section will first explain the basic concepts of intelligent control and intelligent control systems and the basic structure of intelligent control, and then give examples of several robot intelligent control systems, including robot fuzzy control, neural control, and evolutionary control [3, 9, 10].

10.1.1 *Introduction to intelligent control and intelligent control system*

Automatic control science has made important contributions to the theory and practice of science and technology as a whole and has brought huge benefits to human society. Traditional control theory, including classic feedback control and modern control, has encountered many problems in application. For a long time, robot control has been looking for a new way out. It now appears that one of the ways out is to realize the intellectualization of robot control systems, especially the intelligent control of mobile robots in order to solve the problems faced.

1. Definition and characteristics of intelligent control

There is no universally accepted definition of intelligent control. However, in order to define concepts and technologies, develop new capabilities and methods for intelligent control, and compare the results of different researchers and different countries, some common understanding of intelligent control is required.

Definition 10.1.1. Intelligent machine

A machine that can perform various anthropomorphic tasks in various environments is called an intelligent machine. Or more generally speaking, intelligent machines are machines that can autonomously or interactively replace humans in dangerous, boring, long-distance, or high-precision operations.

Definition 10.1.2. Automatic control

Automatic control is a process that can automatically operate or control a machine or device in accordance with a prescribed procedure. Simply put, control that does not require manual intervention is automatic control. For example, a device that can automatically receive the measured physical variables of the process, automatically calculate, and then automatically adjust the process, it is an automatic control device. Feedback control, optimal control, stochastic control, adaptive control and self-learning control are all automatic control.

Definition 10.1.3. Intelligent control

Intelligent control is the process of driving intelligent machines to achieve their goals autonomously. In other words, intelligent control is a type of automatic control that can independently drive intelligent machines to achieve their goals without human intervention.

Intelligent control has the following two main characteristics:

(1) There are both non-mathematical generalized models represented by knowledge and mixed control processes represented by mathematical models. They are often non-digital process with complexity, incompleteness, ambiguity or uncertainty, and no known algorithms, and use knowledge to reason, and use heuristics to guide the solution process. Therefore, when researching and designing an intelligent control systems, the main focus is not on the expression, calculation and processing of mathematical formulas, but on the description of tasks and world models, symbols and environment recognition, and the design and development of knowledge base and reasoning engine. In other words, the design focus of the intelligent control system is not on the conventional controller, but on the intelligent machine model.

(2) The core of intelligent control is high-level control, that is, organization-level control. The task of high-level control is to organize the actual environment or process, that is, decision-making and planning, to achieve generalized problem solving. In order to achieve these tasks, it is necessary to adopt related technologies such as symbolic information processing, heuristic programming, knowledge representation, intelligent computing, and automatic reasoning and decision-making. The process of solving these problems is similar to the thinking process of the human brain, that is, it has different degrees of "intelligence."

Figure 10.1 shows the general structure of an intelligent controller.

2. Structural theory of intelligent control

Since K. S. Fu proposed intelligent control as the intersected field of artificial intelligence and automatic control in 1971, many researchers

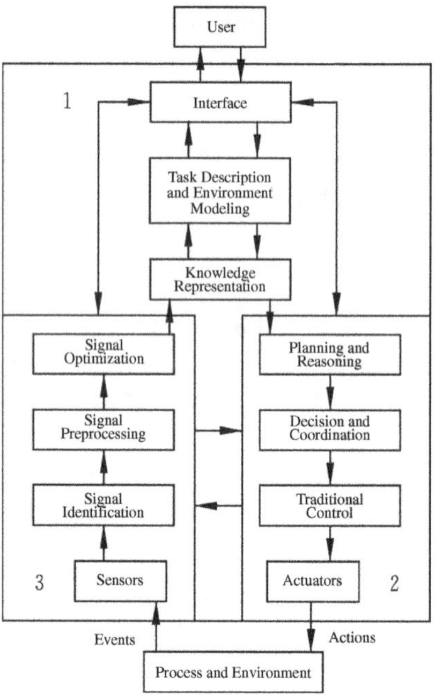

Figure 10.1.　General structure of an intelligent controller. 1 — Intelligent control system; 2 — Multilayer controller; 3 — Multi sensor system.

have tried to establish the new discipline of intelligent control. They put forward some ideas about the structure of the intelligent control system, which will help to further understand the intelligent control.

Intelligent control has obvious characteristics of interdisciplinary (multiple) structure. Here, we mainly discuss the three ideas of intelligent control's binary intersection structure, ternary intersection structure and quaternary intersection structure, which are represented by the following intersections (general sets):

$$IC = AI \cap AC \tag{10.1}$$

$$IC = AI \cap AC \cap OR \tag{10.2}$$

$$IC = AI \cap AC \cap IT \cap OR \tag{10.3}$$

The above-mentioned various structures can also be expressed by a combination of discrete mathematics and predicate formulas

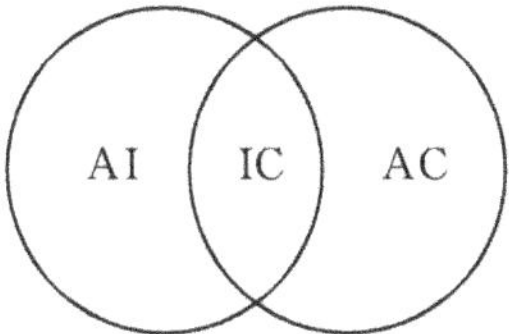

Figure 10.2. Dual structure of intelligent control.

commonly used in artificial intelligence:

$$IC = AI \wedge AC \tag{10.4}$$

$$IC = AI \wedge AC \wedge OR \tag{10.5}$$

$$IC = AI \wedge AC \wedge IT \wedge OR \tag{10.6}$$

In the formula, AI stands for artificial intelligence; AC stands for automatic control; OR stands for operation research; IT stands for information theory or informatics; IC stands for intelligent control; $\cap$ indicates intersection; $\wedge$ indicates the conjunction symbol.

(1) Dual structure

K.S. Fu has conducted research on several fields related to learning control. In order to emphasize the problem-solving and decision-making capabilities of the system, he used "intelligent control systems" to include these areas. He pointed out that "intelligent control system describes the transfer function of automatic control system and artificial intelligence." We can use formula (10.1), formula (10.4) and Figure 10.2 to express this kind of intersection, and call it the dual (two-element) intersection structure of intelligent control.

(2) Ternary structure

In 1977, G.N. Saridis proposed another intelligent control structure, which extended Fu's intelligent control to a ternary structure, that is, regards intelligent control as the intersection of artificial intelligence, automatic control and operations research, as shown in the Figure 10.3. The structure can be described by formula (10.2) and formula (10.5).

Saridis believed that the two elements that constitute a binary intersection structure dominate each other and do not contribute to the effective and successful application of intelligent control.

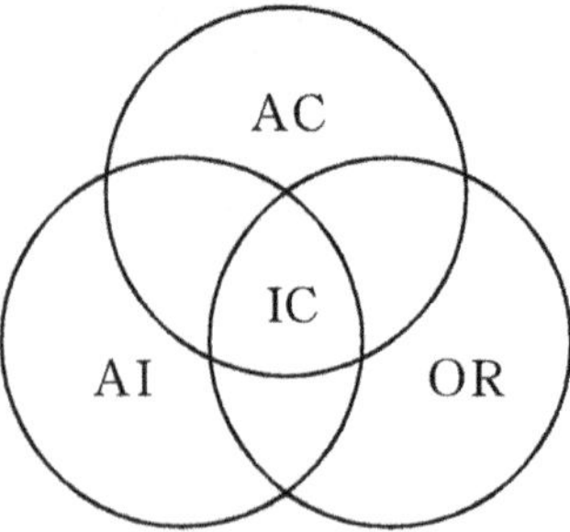

Figure 10.3.   Ternary structure of intelligent control.

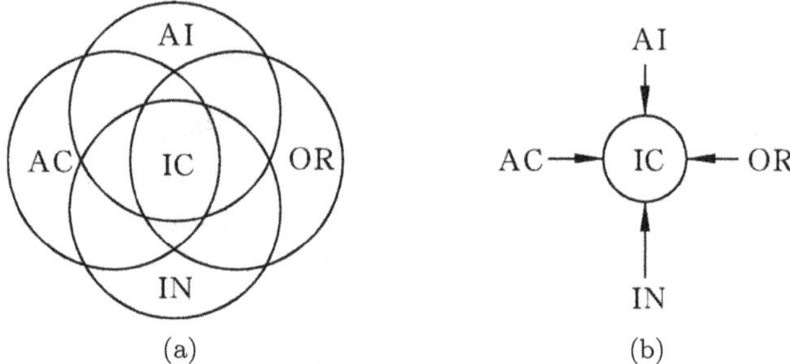

Figure 10.4.   Quaternary structure of intelligent control (a) Four-element intelligent control structure; (b) Simplified diagram of the four-element structure.

The concept of operations research must be introduced into intelligent control, making it a subset of the ternary (three-element) intersection.

(3) Quaternary structure

After in-depth research on the structural theories, knowledge, informatics, and intelligence definitions of the aforementioned various types of intelligent control, as well as the internal relationships of related disciplines, Z.X. Cai proposed the four-element intersection structure of intelligent control in 1986, and looked at intelligent control as the intersection of the four disciplines of automatic control, artificial intelligence, information theory, and operations research, as shown in Figure 10.4(a), the relationship is described in equations (10.3) and (10.6). Figure 10.4(b) shows a simplified diagram of this four-element intersection structure.

The information theory (or informatics) as a key subset of the intelligent control structure is based on the following reasons [3]:

1. Information theory (informatics) is a means to explain knowledge and intelligence;
2. Cybernetics, systematics and informatics are closely interacting;
3. Informatics has become a tool for controlling intelligent machines;
4. Information entropy becomes a measure of intelligent control;
5. Informatics participates in the whole process of intelligent control and plays a central role in the execution level.

10.1.2 *The basic structure of the intelligent control system*

Various intelligent control systems will be introduced below. The systems to be studied include hierarchical control systems, expert control systems, fuzzy control systems, neural control systems, learning control systems, and evolutionary control systems. In fact, several methods and mechanisms are often combined and used in an actual intelligent control system or device to establish a hybrid or integrated intelligent control system.

1. Hierarchical control system

As a unified cognition and control system method, the hierarchical intelligent control proposed by Saridis and Mystel is based on the principle of increasing precision with decreasing intelligence (IPDI). This principle is commonly used in hierarchical management systems.

When discussing the ternary structure of intelligent control above, the hierarchical intelligent control system is composed of three basic control levels, and its cascaded interactive structure is shown in Figure 10.5. The figure shows the online feedback signal f_E^C is from the execution level to the coordination level; the offline feedback signal f_C^O is from the coordination level to the organization level;

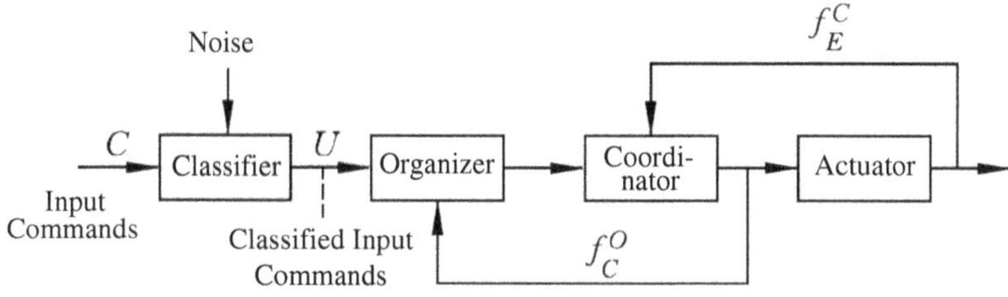

Figure 10.5.   Cascade structure of hierarchical intelligent machine.

$C = \{c_1, c_2, \ldots, c_m\}$ is the input command; $U = \{u_1, u_2, \ldots, u_m\}$ is the output signal of the classifier, that is, the input signal of the organizer.

This hierarchical intelligent control system is a whole, which transforms qualitative user instructions into a sequence of physical operations. The output of the system is realized by a set of specific instructions applied to the drive. Among them, the organizational level represents the dominant thought of the control system, and artificial intelligence plays a control role. The coordination level is the interface between the upper (organization) level and the lower (execution) level. Execution level is the bottom layer of hierarchical control, requiring higher precision and lower intelligence. It controls according to cybernetics and performs appropriate control functions on related processes.

2. Expert control system

As the name implies, the expert control system is a control system that applies expert system technology, and is also a typical and widely used knowledge-based control system.

Hayes-Roth *et al.* proposed an expert control system in 1983. They pointed out that the entire behavior of the expert control system can be adaptively controlled. To this end, the control system must be able to repeatedly explain the current situation, predict the future, diagnose the cause of the problem, formulate a remedial (correction) plan, and monitor the execution of the plan to ensure success. The first report on the application of expert control system was in 1984. It is a distributed real-time process control system for oil refining. K. J. Åström *et al.* published their paper entitled *"Expert Control"* in 1986. Since then, more expert control systems have been developed and applied. Expert systems and intelligent control are closely related. They have at least one thing in common, that is, both are based on imitating human intelligence, and both involve certain uncertainties.

The structure of the expert control system may be different due to different application occasions and control requirements. However, almost all expert control systems (controllers) include knowledge bases, inference engines, control rule sets and/or control algorithms.

Figure 10.6 shows the basic structure of the expert control system. From the point of view of performance indicators, the expert control system should provide the same or very similar performance indicators for the control target as the expert operators.

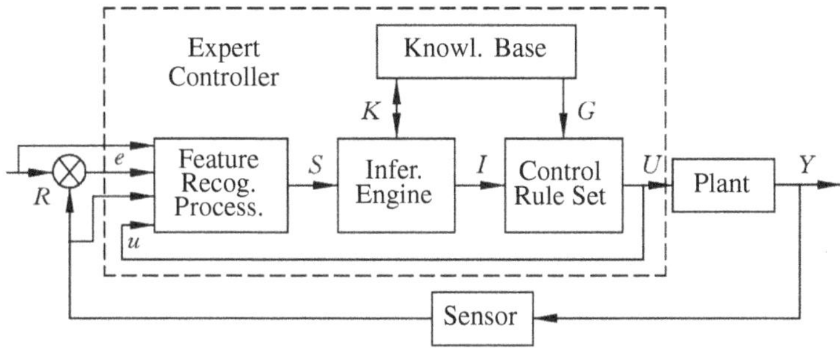

Figure 10.6.   Typical structure of expert controller.

This expert control system is an industrial expert controller (EC), which is composed of knowledge base, inference engine, control rule set and feature recognition information processing unit. The knowledge base is used to store the domain knowledge of industrial process control. The reasoning engine is used to memorize the adopted rules and control strategies, so that the entire system works in a coordinated manner; the reasoning engine can reason, search and derive conclusions based on knowledge.

3. Fuzzy control system

In the past 30 years, fuzzy controllers and fuzzy control systems have been a very active application research field of intelligent control. Fuzzy control is a kind of control method that applies fuzzy set theory. The effectiveness of fuzzy control can be considered from two aspects. On the one hand, fuzzy control provides a new mechanism for realizing knowledge-based (rule-based) or even language-based control laws. On the other hand, fuzzy control provides an alternative method to improve nonlinear controllers. These nonlinear controllers are generally used to control devices that contain uncertainties and are difficult to handle with traditional nonlinear control theories.

The basic structure of the fuzzy control system is shown in Figure 10.7. Among them, the fuzzy controller is composed of four basic units: fuzzy interface, knowledge base, inference engine and fuzzy decision interface.

4. Learning control system

Learning control system is one of the earliest research fields of intelligent control. In the past 20 years, the study of learning control for dynamic systems (such as robot operation control and aircraft

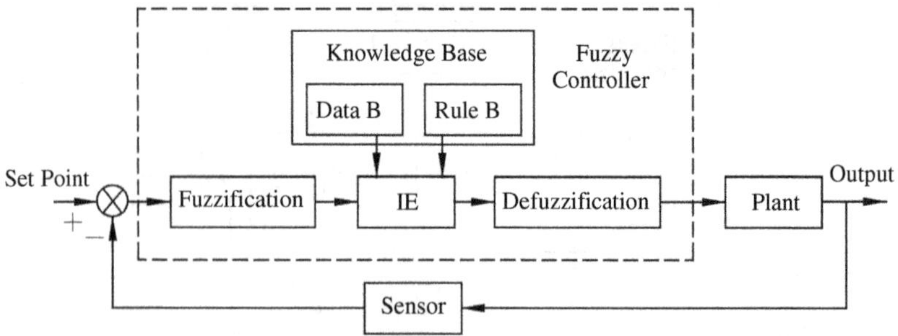

Figure 10.7.   Basic structure of fuzzy control system.

guidance, etc.) has become an increasingly important research topic. Many learning control schemes and methods have been researched and proposed, and better control effects have been obtained. These control schemes include:

(1) Learning control based on pattern recognition;
(2) Iterative learning control;
(3) Repetitive learning control;
(4) Connectionist learning control, including reinforcement learning control;
(5) Rule-based learning control, including fuzzy learning control;
(6) Anthropomorphic self-learning control;
(7) State learning control.

Learning control has 4 main functions: search, recognition, memory and reasoning. In the early stage of the development of the learning control system, there were more researches on search and recognition, while the research on memory and reasoning was relatively weak. Learning control systems are divided into two categories, namely online learning control systems and offline learning control systems, as shown in Figures 10.8(a) and 10.8(b) respectively. In the figures, R represents the reference input; Y is the output response; u is the control function; s is the transfer switch. When the switch is turned on, the system is in an offline learning state.

5. Neural control system

ANN-based control, referred to as neurocontrol or NN control, is a new research direction of intelligent control.

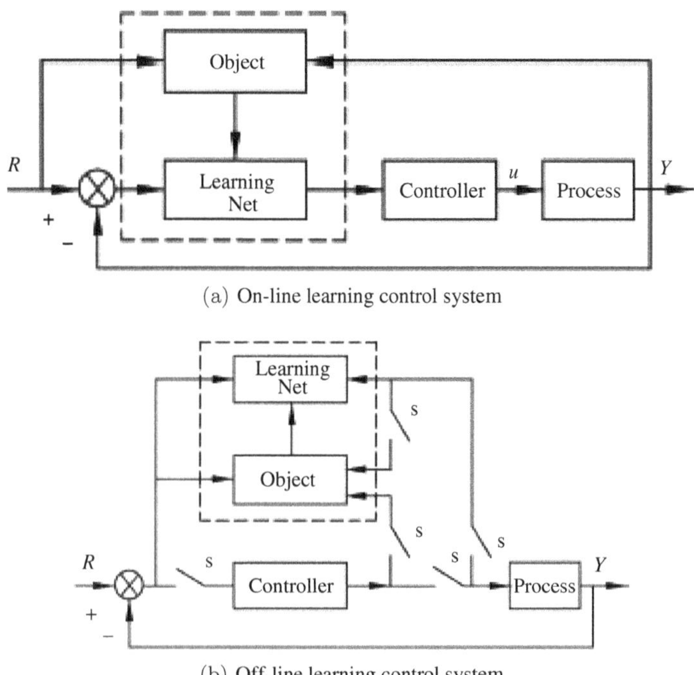

(a) On-line learning control system

(b) Off-line learning control system

Figure 10.8.   Schematic diagram of learning control system.

Since Widrow and Hoff took the lead in using neural networks for automatic control research in 1960, the research on this subject has struggled to make some progress.

Due to the different classification methods, the structure of the neural controller is naturally different. There are many structural schemes of neural control that have been proposed, including NN learning control, NN direct inverse control, NN adaptive control, NN internal model control, NN predictive control, NN optimal decision control, NN enhanced control, CMAC control, hierarchical NN control and multi-layer NN control, etc.

When the dynamics of the controlled system is unknown or only partially known, it is necessary to try to explore the regularity of the system in order to effectively control the system. Rule-based expert systems or fuzzy control can achieve this kind of control. Supervised neural control (SNC) is another way to achieve it.

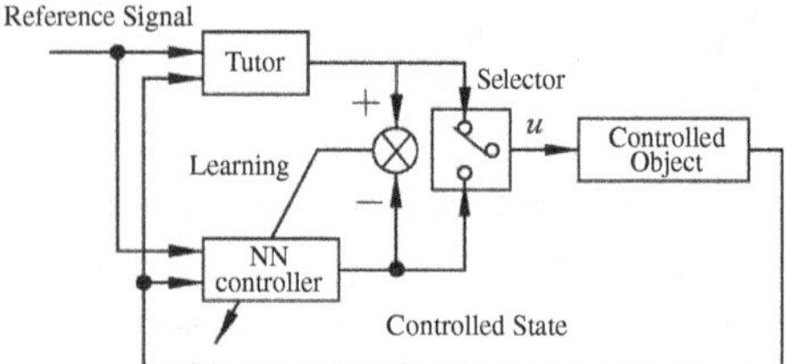

Figure 10.9.   Structure of the supervised learning NN controller.

Figure 10.9 shows the structure of the supervisory neural controller. In the figure, there is a tutor and a trainable controller. The realization of SNC includes the following steps.

6. Evolutionary control system

Evolutionary control is a new control scheme, which is based on evolutionary computation (especially genetic algorithm) and feedback mechanism.

Evolutionary control comes from the evolutionary mechanism of organisms. At the end of the 1990s, 20 years after the evolutionary computing ideas such as genetic algorithm were proposed, there were signs of research on evolutionary control in the biomedical and automatic control circles. In 1998, Ewald, Sussmann, and Vicente *et al.* applied the principles of evolutionary computation to the control of viral diseases. From 1997 to 1998, Z. X. Cai and X. Zhou proposed the evolutionary control idea of electromechanical systems and applied it to the navigation control of mobile robots, and obtained preliminary research results [11].

A variety of evolutionary control system structures have been proposed, but the evolutionary control method combined with fuzzy control has been tested so far, and it is used for the swing and stability control of the single pendulum.

The first one can be called a direct evolutionary control structure, which is a genetic algorithm (GA) directly acting on the controller to form an evolutionary controller based on GA. The evolution controller controls the controlled object, and then forms an evolution control system through feedback. Figure 10.10(a) shows the structural principle diagram of this evolutionary control system.

The second type can be called indirect evolutionary control. It consists of an evolutionary mechanism (evolutionary learning) acting on the system model, and then integrating the system state output

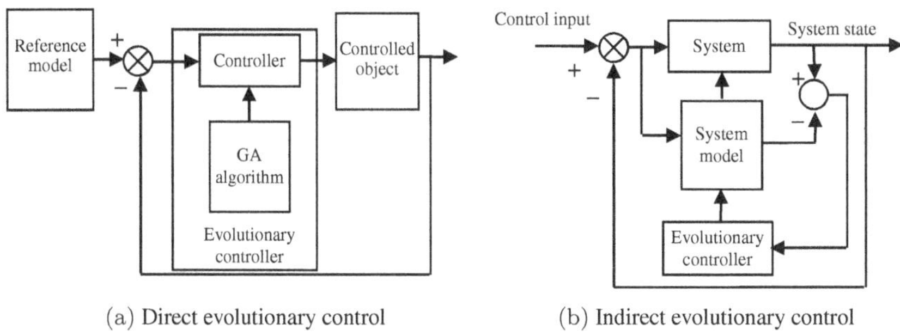

Figure 10.10.   Basic structure of an evolutionary control system.

and system model output on the evolutionary learning controller. Then, the evolutionary control system is constituted by the general closed-loop feedback control principle as shown in Figure 10.10(b). Compared with the first structure, this structure is more complicated, and its control performance is better than the former.

10.2   Mobile Robot Control Based on Neural Network

The control of mobile robots involves kinematics modeling, dynamic modeling, conventional control, affine model-based control, invariant manifold-based control, model reference adaptive control, sliding mode control, fuzzy and neural control, vision-based control, path and motion planning, positioning and mapping, as well as control architecture and software architecture and other issues [5]. Among them, some issues, such as kinematics modeling, dynamic modeling, conventional control, adaptive control, path and motion planning, positioning and mapping, etc., have been introduced in the relevant chapters of the book, while fuzzy and neural control, tracking control and vision-based control will be discussed in this chapter. This section mainly discusses neural control of the mobile robots.

Fuzzy logic (FL) and neural network (NN) have been widely used in the identification, planning and control of mobile robots. Fuzzy logic provides a unified approximation (language) method that uses uncertainty rules to draw conclusions from uncertain data. Neural networks provide the possibility of autonomous (unsupervised learning) or non-autonomous (supervised learning) learning and training,

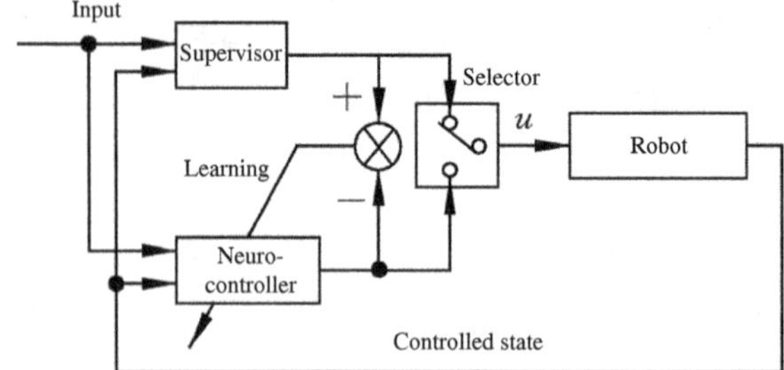

Figure 10.11.   Structure of a robot controlled by a NN with supervised learning.

or evaluate its performance through reinforcement learning. In many practical cases of mobile robots, hybrid neuro-fuzzy systems (NFS) are used to provide better performance [4, 39].

The general structure of a neural control robot with supervised learning is shown in Figure 10.11 [40].

Mobile robot control involves mobile robot tracking control, mobile robot navigation control, and mobile robot formation control etc. Most of them are based on neural network structure.

10.2.1   *Tracking control of mobile robots*

Mobile robot control is mainly tracking control, including path tracking control, trajectory tracking control, target tracking control and speed tracking control etc. The following describes these tracking control cases for readers and researchers to refer to.

1. Path tracking control of mobile robot

Aiming at the tracking problem of wheeled mobile robot (WMR) discrete-time nonlinear state and input delay system, Li *et al.* proposed an adaptive control algorithm based on reinforcement learning. The typical model of WMR is transformed into an affine nonlinear discrete-time system, and the delay matrix function and Lyapunov-Krasovskii function are introduced to overcome the delay problems caused by state and input respectively. In addition, through the approximation of the radial basis function neural network, an adaptive controller, a critic neural network and an action neural network adaptive law are defined to ensure that the unified limit of all signals in the WMR system is bounded and tracking, the error converges to a small compact setting [21].

Lu *et al.* discussed the speed tracking of wheeled mobile robots (WMR) with repeatable trajectories and different initial errors, deduced three mathematical models of WMR kinematics model, dynamics model and DC motor drive model, and proposed adaptive iterative learning control (FNN-AILC) strategy based on fuzzy neural network, including fuzzy neural network components, approximation error and feedback. The proposed scheme can handle MIMO systems, which is different from previous research work [28].

Zeng *et al.* proposed a robust dynamic surface ship control design based on a composite neural network for the path tracking of unmanned surface ships in the presence of nonlinear parameter uncertainties and unknown time-varying disturbances. Different from the existing neural network-based dynamic surface control methods that generally only use the tracking error to update the neural network weights, this scheme uses both the tracking error and the prediction error to construct an adaptive law. Therefore, the system dynamics can be recognized more quickly and the tracking accuracy can be improved. In particular, a prominent advantage of the proposed neural network structure is simplicity. No matter how many neural network nodes are used, only one adaptive parameter needs to be adjusted online, which effectively reduces the computational burden and helps the realization of the controller [51].

X.B. Zou and others also conducted in-depth research on the design method of the path tracking controller for non-holonomic mobile robots, and looked forward to the application of the tracking controller [58].

2. Trajectory tracking control of mobile robots

Boukens *et al.* solved the trajectory tracking control problem of a mobile robot system with nonholonomic constraints in the presence of time-varying parameter uncertainties and external disturbances. This requires an accurate controller to meet as many control goals as possible. Therefore, this case uses artificial intelligence control technology to design a robust controller to meet the control goal. The design of intelligent controller is based on optimal control theory, adaptive neural network system and robust control technology. Use the optimal control method to solve the trajectory tracking problem. Since the non-holonomic wheeled mobile robot is strongly nonlinear, the neural network system is used in the optimal control law to approximate the nonlinear function. Then, the robust controller is applied to adaptively estimate the time-varying parameter uncertainty, external disturbance and the unknown upper limit of the approximation error of the neural network system. The optimal control theory and Lyapunov stability analysis prove the stability of the closed-loop robot system [2].

The trajectory tracking ability of mobile robots is affected by uncertain interference. Wu *et al.* proposed an adaptive control system composed of a new self-organizing neural network controller for mobile robot control. The newly designed neural network includes the typical brain emotion learning controller network and the key mechanism of self-organizing radial basis function network. In this system, input values are passed to a sensory channel and an emotional channel, and these two channels interact with each other to generate the final output of the proposed network. The proposed network has the ability to generate and eliminate fuzzy rules online to achieve the best neural structure. The parameters of the proposed network can be adjusted online through the brain emotion learning rules and the gradient descent method; in addition, the stability analysis theory is used to ensure the convergence of the proposed controller [42, 43].

D. Ma *et al.* proposed a neural network adaptive proportional differential (PD) tracking control strategy based on radial basis function (RBF) for wheeled mobile robots under wheel slip conditions. First, a dynamic model of the wheeled mobile robot under slip conditions is established. Secondly, the backstepping method is used to design the kinematics controller, and the PD controller is designed based on the dynamic model. The RBF neural network with parameter adaptation is used to analyze the parameter and non-parametric uncertainties in the dynamic model under slip. The stability of the closed-loop system is proved by using Lyapunov stability theory. Experimental results show that the proposed control method can better compensate for the uncertainty of the mobile robot wheel slipping, and improve the robustness of wheeled mobile robot trajectory tracking [30].

Guo *et al.* reported on iterative learning trajectory tracking control of wheeled mobile robots [19]

3. Target tracking control of mobile robots
Mobile robots that can track human targets detect the distance between the target and the robot with the assistance of various

sensors to avoid collisions or maintain a constant distance, requiring a powerful control system. In order for a robot to follow a person or target with variable speed, it must have the decision-making ability to decide when to accelerate or decelerate according to the situation. The research work carried out by Ankalaki *et al.* involves the use of a nonlinear controller to provide stable speed control of a mobile robot. Artificial neural networks can be used to reduce the burden on the controller and reduce the computational complexity involved in controlling the speed of the motor. All sensors and actuators are provided with neural network database input, which is used in the learning phase to provide classification into multiple actions. Then use the categories and weights obtained during the learning phase of the neural network to change the speed of the mobile robot. Then the speed will be stabilized by using the linear quadratic regulator control algorithm, thus providing a complete solution for setting a smooth and stable mobile robot, which can track and follow the target under any conditions [1].

Elhaki *et al.* solved the problem of target tracking control for under-driven autonomous underwater vehicles (AUVs) with prescribed performance. To this end, the distance and azimuth angle of the AUV relative to the underwater target are converted into a second-order open-loop error dynamic model by using the specified performance boundary technology. Then, a new tracking controller is proposed, which makes the tracking error converge to an arbitrarily small limit boundary, and their transient performance is guaranteed by the pre-specified maximum overshoot and convergence speed. In order to overcome the unmodeled dynamics and external disturbances imposed on the underwater vehicle by wind waves and ocean currents, a multilayer neural network and an adaptive robust controller are used. The Lyapunov stability synthesis shows that all signals of the control system are bounded, and the tracking error converges to a small area including the origin, with good prescribed performance [17].

Aiming at incomplete wheeled mobile robots with unknown wheel slip, model uncertainty and unknown bounded interference, Nguyen *et al.* proposed an adaptive tracking controller based on a 3-layer neural network and an online weight adjustment algorithm. The online weight adjustment algorithm is modified from back-propagation, and an e modification item is needed to ensure that the NN weights are bounded. The initial offline training of the neural network is not necessary for the weights. The proposed controller can achieve the required target tracking performance regardless of their initial value when the position tracking error converges to an arbitrarily small neighborhood of the origin. According to the Lyapunov theory and LaSalle extension, the stability of the entire closed-loop system is guaranteed to obtain the required tracking performance [32].

In addition, Duan *et al.* also researched and developed robust position tracking control for mobile robots with adaptive evolutionary particle filters [16]; and H. Su *et al.* summarized the research progress of mobile robot moving target tracking technology in unknown environments [36].

4. Speed tracking control of mobile robot

The performance of the wheel drive control scheme of wheeled mobile robots is an important aspect of the application of vehicle-mounted robots, which can not only give full play to the traction capabilities of the robot, but also save energy. This is especially true in environments where robots must traverse unknown and unpredictable deformable terrain. The planetary exploration environment considered in the research by You *et al.* in order to compensate for the interference caused by terrain deformation, and at the same time use the control advantages provided by the pseudo-driving wheel (PDW) concept, propose a control method based on artificial neural network, develop and realize the PDW speed tracking the network algorithm required for active follow control. In order to deal with the considered complex and highly uncertain wheel-terrain interactions, an online sequential forgetting update method for neural networks and an improved online sequential extreme learning machine (OS-ELM) are used in combination with the proportional integral derivative (PID) controller constructs an efficient and high-performance hybrid OS-ELM-PID control system. The actual experimental results show the feasibility and effectiveness of the control system [49].

5. An example of robot path control based on RBF tracking control neural network

We have discussed the kinematics model of wheeled mobile robots, especially two-wheeled mobile robots, in Chapter 7. The following introduces a design case of a mobile robot path tracking controller.

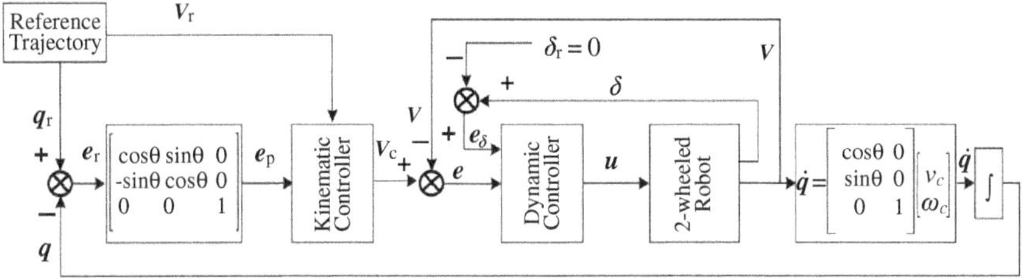

Figure 10.12. The system structure of a two-wheeled mobile robot.

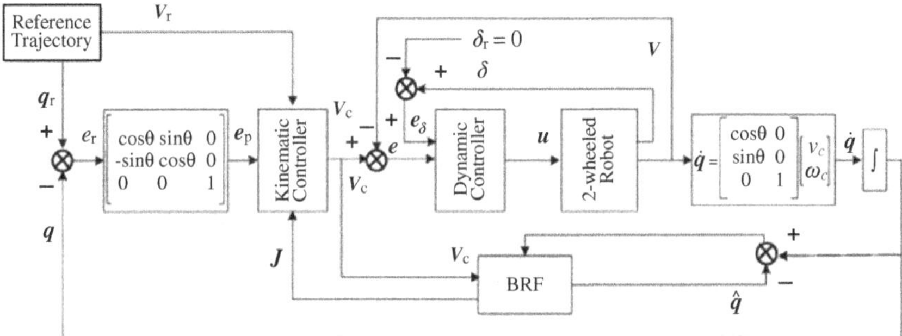

Figure 10.13. Adaptive system structure based on RBF neural network.

The system structure of the two-wheeled mobile robot is shown in Figure 10.12. Among them, $\boldsymbol{q}_r = [x_r, y_r, \theta_r]^T$ is the input reference pose of the system; $\boldsymbol{e}_p = [e_x, e_y, e_\theta]^T$ is the input error of the kinematics controller; $\boldsymbol{V}_c = [v_c, \omega_c]^T$ is the kinetic input speed and angular velocity of the controller; $\boldsymbol{V}_r = [v_r, \omega_r]^T$ is the desired tracking speed and angular velocity; $\boldsymbol{e} = [e_v, e_\omega]^T$ is the tracking error of the dynamic controller; $\boldsymbol{e}_r = [\boldsymbol{q}_r, -\boldsymbol{q}]$; e_δ is the control error of the robot balance angle δ_r. It is hoped to keep vertical during the robot movement, so $\delta = 0$ $\boldsymbol{u} = [u_v, u_\omega]^T$ is the output voltage of the dynamic controller; $\boldsymbol{V} = [v, \omega]^T$ is the actual speed and angular velocity of the robot; $\boldsymbol{q} = [x, y, \theta]^T$ is the actual pose of the robot.

Readers who are interested in the specific design of kinematic controllers, dynamic controllers and adaptive path tracking controllers, please refer to the literature [48]. The structure of the adaptive control system based on RBF neural network is shown in Figure 10.13.

In the design, the gradient descent method is used to modify the gains of the inversion controller based on the kinematics model, and the RBF neural network is used to identify and calculate the J value of the Jacobian matrix. Experimental results show that the adaptive control of mobile robots based on RBF neural network has better tracking characteristics than traditional methods [48].

10.2.2 *Navigation control and formation control of mobile robots*

1. Navigation control of mobile robot

Chen *et al.* proposed an adaptive fuzzy neural network (AFNN) based on the multi-strategy artificial bee colony (MSABC) algorithm to realize the navigation control of mobile robots. In the navigation control process, the AFNN input is the distance between the ultrasonic sensor and the angle between the mobile robot and the target, and the AFNN output is the left and right wheel speed of the mobile robot. The fitness function in reinforcement learning is defined to evaluate the navigation control performance of AFNN. The proposed MSABC algorithm improves the shortcomings of poor development and utilization of traditional artificial bee colonies (ABC), and uses differential evolution mutation strategies to balance exploration and development. The experimental results show that compared with the traditional ABC method, the proposed MSABC method improves the performance of average fitness, navigation time and travel distance by 79.75%, 33.03% and 10.74%, respectively. In order to prove the feasibility of the proposed controller, experiments were carried out on the actual PIONEER 3-DX mobile robot, and the proposed navigation control method was successfully completed [13].

H. Zhang and others introduced the learning and decision-making model based on emotion and cognition into the behavior-based mobile robot control system, and designed a new autonomous navigation control system. The environmental cognitive state in the learning and decision-making algorithms is used to learn online emotion and environmental cognition to form a reasonable behavioral coordination mechanism. The emotional and dynamic system methods were applied to basic behavior design, and using ART2 neural network to realize the classification of continuous environmental perception states. Using the classification results as environmental cognition can significantly improve the efficiency of learning and decision-making processes, and improve behavior-based mobility, and increase the autonomous navigation capability of the robot in an unknown environment [52].

Luo combined fuzzy logic technology and biologically inspired neural network model, and proposed a new fuzzy logic tracking control method based on hybrid biologically inspired neural network to realize real-time navigation control of incomplete mobile robots. The tracking control algorithm comes from the error dynamics analysis of the mobile robot and the stability analysis of the closed-loop control system. The Lyapunov stability theory guarantees the stability of the robot control system and the convergence of the tracking error to zero. Different from some existing tracking control methods for mobile robots with speed jumps, the proposed neural dynamics-based method can generate smooth and continuous robot control signals with zero initial speed. In addition, the proposed fuzzy logic and biologically inspired neural network methods solve the problem of large tracking errors [29].

Cai *et al.* comprehensively studied the navigation control problems of mobile robots in unknown environments, combined with research projects to conduct experimental research, and their research results have important reference value [6, 7].

2. Formation control of mobile robots

Aiming at the convergence problem of formation control of a group of incomplete mobile robots with multiple uncertainties, Y. D. Li *et al.* proposed a multivariable fixed-time leader-follower formation control algorithm for mobile robots based on radial basis function neural networks. RBFNN compensates for the system. It is subject to multiple uncertainties and eliminates the chattering phenomenon of robust control. The control algorithm design is based on the fixed time theory and the Lyapunov method, so that the proposed control method guarantees convergence in the global fixed time of all signals in the formation control system, under the initial conditions of any system, in a fixed time through parameter design, the formation of robots can reach the desired formation, which shows the effectiveness of the proposed algorithm [22].

Xiao *et al.* proposed a non-linear model predictive control (NMPC) strategy based on a general projection neural network to control a multi-robot formation system. The multi-robot formation system is composed of the trajectory tracking of the leader robot and the leader following the formation control. Their motion error system can be reformulated as a convex nonlinear minimization problem through the NMPC method, and can be transformed into a constrained quadratic programming (QP) optimization problem. To effectively solve this QP problem online, a general projection neural network (GPNN) is used to obtain the optimal solution, which includes the optimal input of the formation system. Compared with other existing leader-follower methods, the proposed nonlinear MPC method can consider the input and state constraints of the system. Finally, the effectiveness of the developed strategy is verified by trajectory tracking and multi-robot formation experiments [45].

Aiming at the formation control problem of incomplete mobile robots, L. Zhu *et al.* proposed a new control strategy combining a kinematic controller and an adaptive neural sliding mode controller based on the leader-follower control structure. The radial basis function neural network is used to estimate the nonlinear uncertain part of the follower and the leader's dynamics online, and the adaptive robust controller is used to compensate the neural network modeling error. The proposed method not only solves the parametric and non-parametric uncertainties of mobile robot formation control, but also ensures that the robot formation can track the specified trajectory in the desired formation. The design process based on the Lyapunov method ensures the stability of the control system [57].

10.2.3   *Visual control of mobile robots*

The vision device is a powerful robot sensor that can measure the environment without physical contact [40]. Visual robot control or visual servo is a feedback control method that uses one or more vision sensors to control the movement of the robot. The robot's control input is generated by processing image data (usually extracting contours, features, corners, and other visual primitives). In the manipulator, the purpose of visual control is to control the posture of the robot end effector relative to the target object or a set of target features. In mobile robots, the task of the vision controller is to control the attitude of the mobile carrier (vehicle, ship or aircraft) relative to some ground, underwater or space targets. Only when the visual sensor signal delay is small enough and/or the dynamic model of the robot has sufficient accuracy, can the stability of the movement tracking be guaranteed. Over the years, many techniques have been developed to compensate for this delay in the vision system in robot control. A large amount of literature has conducted in-depth research on the problems related to vision-based control of non-complete systems.

The vision-based robot controller (VRC) depends on whether the vision system provides set points as input to the robot joint controller or directly calculates the joint level input, and whether the error signal is determined in task space coordinates or directly based on the image to determine the feature.

Therefore, VRC is divided into the following three categories [40].

(1) Dynamic observation and movement system: Here, the robot joint controller is replaced by a visual servo controller, which directly calculates the input of the joints, and uses only visual signals to control the robot. In fact, most VRCs implemented are observation and movement types, because the internal feedback with a high sampling rate provides an accurate axis dynamic model for the vision controller. In addition, observation and movement control separate the kinematic singularities of the system from the vision controller.

(2) Position-based visual robot control (PBVRC): Here, features are extracted from the image and used together with the geometric model of the target and the available camera model to determine the pose of the target relative to the camera. Therefore, the error in the estimated pose space is used to close the feedback loop.

(3) Image-based visual robot control (IBVRC): Here, image features are used to directly calculate the control signal, which reduces the calculation time, does not require image interpretation, and eliminates errors in sensor modeling and camera calibration. However, due to the complex nonlinear dynamics of the robot, its realization is more difficult.

Self-driving cars include driverless cars, self-driving cars, and robotic cars, as well as other platforms that can perceive and interact with the environment and navigate without human assistance. On the other hand, semi-autonomous vehicles achieve partial automatic driving through manual intervention, such as in driver-assisted vehicles. The self-driving car first uses the installed sensors to interact with the surrounding environment. Generally, vision sensors are used to acquire images, and computer vision technology, signal processing, machine learning, and other technologies are used to acquire, process, and extract information. The control subsystem interprets sensory information to identify the appropriate navigation path to the destination and an action plan to perform the task. It also gets feedback from the environment to improve its behavior. In order to improve sensing accuracy, self-driving cars are equipped with many sensors (light detection and ranging (LiDAR), infrared, sonar, inertial measurement unit, etc.), as well as communication subsystems. Autonomous vehicles face many challenges, such as unknown environments, blind spots (invisible fields of view), non-line-of-sight scenes, poor sensor performance due to weather conditions, sensor errors, false alarms, limited energy, limited computing resources, and algorithmic complexity, human-machine communication, size and weight restrictions. To solve these problems, several algorithmic methods have been implemented, covering the design of sensors, processing, control, and navigation. Tawiah *et al.* commented and provided the latest information on the requirements, algorithms, and major challenges of using machine vision-based technology for autonomous driving car navigation and control. It also introduces applications that use land-based vehicles as an IoT platform for pedestrian detection and tracking [38].

Wheeled mobile robots have been widely used in industry due to their large working space and flexible movement. The visual servoing of wheeled mobile robots has always been one of the most popular research topics in the robotics field. Due to the different requirements for various applications and specific properties of the system, designing a visual servo control strategy faces many challenges. Li *et al.* elaborated on the development overview of mobile robot visual servoing in the vision and control modules, discussed the research trend of mobile robot visual servoing, analyzed the vision module and its various uncertainties, and provided feedback signals for the servo controller. Associate image space with motion space for further control [20].

After more than ten years of hard work, teams of Cai and Li have conducted comprehensive research on sensing, mapping and object-tracking technologies of autonomous vehicles and smart vehicles, including lidar mapping and vehicle state estimation, and vision-based target tracking and multi-external sensor joint calibration technology, etc., has high reference value [8].

10.3 Mobile Robot Control Based on Deep Learning

With the in-depth development of machine learning research that has been widely applied, more and more machine learning algorithms, especially deep learning and deep reinforcement learning algorithms, are in the field of robot control.

10.3.1 *Overview of mobile robot control based on deep learning*

The following summarizes the research and application examples of mobile robot path and position control, mobile robot trajectory control, mobile robot target tracking control, footed robot walking control and gait planning, and mobile robot motion control based on deep learning [12].

1. Robot path and position control based on deep learning
Z. L. Li *et al.* invented a blind-zone-free sweeping robot based on deep learning algorithms, using target detection algorithms to detect blind spots and 3D imaging algorithms, to perform 3D reconstruction of the cleaning area, to obtain the types and orientations of the blind spots in the cleaning area, and to eliminate the blind spots [23]. S. J. Song *et al.* proposed a fixed depth control method for underwater autonomous robots based on reinforcement learning. The state variables, control variables, and transfer models of the underwater autonomous robot fixed depth control were obtained respectively; the final decision-making network of depth control realizes the fixed depth control of the underwater autonomous robot when the dynamic model of the underwater autonomous robot is completely unknown [35]. H. Liu *et al.* proposed a deep learning control planning method for robot motion path in an intelligent environment. By separately establishing a global static path planning model and a local dynamic obstacle avoidance planning model, using the nonlinear fitting characteristics of deep learning, the global optimal path can be quickly found. It avoids the problem of falling into local optimality

in common path planning [26,27]. In addition, S. Z. Yang *et al.* studied the robot arm control problem based on the deep reinforcement learning strategy. Combining deep learning and deterministic strategy gradient reinforcement learning, design the deterministic strategy gradient (DDPG) deep learning step, so that the robot arm has high environmental adaptability after training and learning, and can quickly and accurately find the moving target point in the environment [47]. L. T. Qian invented a robot control system based on deep learning to deploy and control the entire system to achieve all-round control of the robot's speed, angle and strength, while ensuring the smooth and safe operation of the system [33].

2. Robot trajectory control based on deep learning

Q. X. Ma *et al.* proposed a method of using deep reinforcement learning to achieve optimal trajectory control of underwater robots. First, establish an underwater robot control model based on two deep neural networks (Actor network and Critic network); secondly, construct a suitable reward signal to make the deep reinforcement learning algorithm suitable for the dynamic model of the underwater robot; finally, propose a dynamic model based on the successful evaluation condition of network training that rewards the standard deviation of the signal enables the underwater robot to ensure stability while ensuring accuracy [31]. H. J. Zhang and others merged reinforcement learning and deep learning methods, and proposed an end-to-end control method for robots based on deep Q network learning, which improves the accuracy of collision-free movement of the robot without obstacle maps or sparse lidar data. The model trained by this method effectively establishes the mapping relationship between the lidar data and the robot's motion speed, so that the robot selects the action with the largest Q value to execute in each control cycle, and can smoothly move to avoid obstacles [56]. C. Y. Tang *et al.* invented a robot obstacle avoidance control method and device based on deep learning, which can accurately predict the position information of moving obstacles, quickly generate control instructions for controlling the robot to avoid moving obstacles, and quickly control the robot to complete obstacle avoidance, improve the accuracy of obstacle avoidance [37]. Q. X. Mag *et al.* invented an underwater robot trajectory control method and control system based on deep reinforcement learning, which can realize the precise control of the underwater robot's trajectory [31].

3. Robot target tracking control based on deep learning

J. N. Xu *et al.* reinforcement learning based on the continuous "trial and error" mechanism. Through pre-training, path planning can be realized without map. The current deep reinforcement learning algorithms are researched and analyzed, and low-dimensional radar data and a small amount of location information are used. To achieve effective dynamic target tracking strategies and obstacle avoidance functions in different smart home environments [46]. K. Y. You and others invented an aircraft route tracking method based on deep reinforcement learning. Construct a Markov decision-making process model for aircraft trajectory tracking and control, and obtain the expressions of state variables, control variables, transition models, and one-step loss functions for aircraft flight tracking control; establish a strategy network and evaluation network; make the aircraft on the flight path through reinforcement learning. The final strategy network for route tracking control is obtained in the tracking control training [50]. G. J. Chen *et al.* proposed a target tracking method for underwater robots based on deep learning and monocular vision. For each incoming video frame and an environment without prior knowledge, a previously trained convolutional neural network was introduced to calculate the transfer map, providing with depth-related estimation, the target area can be found and a tracking direction can be established [14]. Y. Z. Zhang *et al.* proposed a visual following method for mobile robots based on deep reinforcement learning. Adopting the architecture of "simulated image supervised pre-training + model migration + RL," the robot can perform following tasks in the real environment, combined with the reinforcement learning mechanism, so that the mobile robot can follow and improve control performance of the direction in the process of environmental interaction [55]. Y. Zhang *et al.* published an intelligent robot visual tracking method based on deep learning, combined with the TLD framework and GOTURN algorithm, so that the overall tracking situation can have strong adaptability under conditions of drastic changes in illumination [54].

4. Robot motion control based on deep learning

Y. K. Wang *et al.* invented a method for small football robots to actively control the ball sucking based on deep reinforcement learning, which enables the robot to adjust itself through interaction with the environment and continuously improve the ball sucking effect. The invention can improve the stability and success rate of the robot in sucking the ball [41]. H. J. Wu and others provided an adaptive motion control method for hexapod robots based on deep reinforcement learning in complex terrain, allowing the robot to adaptively adjust the motion strategy according to the complex changes in the environment, and improve the "survival rate" and adaptability in the complex environment [42]. S. L. Zhang believes that the technology represented by convolutional neural networks can be trained according to different control requirements to improve the control effect of the system. It has been widely used in robot control, target recognition and other fields. With the complexity of the robot application environment, the design is based on the convolutional neural network robot control algorithm to achieve precise object grasping in an unstructured environment, and a complete robot automatic grasping planning system is established [53]. In addition, H. W. Ge *et al.* aiming at the problem that traditional mechanical control methods are difficult to effectively control the behavior of the yellow peach core digging robot, proposed a method based on deep reinforcement learning to control the behavior of the yellow peach core digging robot with visual function. The invention exerts the deep learning perception ability and the reinforcement learning decision-making ability, so that the robot can use deep learning to recognize the state of the peach core, and guide the single-chip microcomputer to control the motor to dig out the peach core through the reinforcement learning method to complete the core digging task [18].

5. Footed robot walking control and gait planning based on deep learning

G. M. Song *et al.* proposed a fall self-reset control method for quadruped robots based on deep reinforcement learning, which uses deep reinforcement learning algorithms to enable the robot to automatically reset on flat ground under any fall posture without pre-programming or human intervention, promote the intelligence, flexibility and environmental adaptability of the robot [34]. H. Y. Liu and others invented a gait control method for a humanoid robot based on a deep Q network, including constructing a gait model, learning and training the deep Q network based on training samples, obtaining the state parameters of the humanoid robot in the action environment, and the gait control of the humanoid robot is carried out by using the constructed gait model, and the deep Q network is updated by generating the reward function. The invention can improve the walking speed of the humanoid robot, and realize the fast and stable walking of the humanoid robot [27].

6. Robot navigation control based on deep learning

J. Chen *et al.* proposed an end-to-end control method based on deep reinforcement learning for mobile robots' unmapped navigation in unknown environments. The robot needs to rely only on the RGB image of the visual sensor and the relative position with the target as input to complete the navigation task and avoid obstacles along the way without a map. In an arbitrarily constructed simulation environment, the robot based on learning strategy can quickly adapt to the unfamiliar scene and finally reach the target position without any human markers [15]. J. T. Lin *et al.* proposed an end-to-end distributed multi-robot formation navigation method based on deep reinforcement learning. This method is based on deep reinforcement learning and obtains a control strategy through trial and error, which can safely and efficiently navigate the geometric center point of the multi-robot formation to the target point, and ensure the connectivity of the multi-robot formation during the navigation process. Through a centralized learning and distributed execution mechanism, this method can obtain a control strategy that can be executed in a distributed manner, making the robot have higher autonomy [25].

In addition, a voice interaction and control method for intelligent industrial robots based on deep learning is also proposed [24].

10.3.2 *Example of mobile robot control based on deep learning*

Y. X. Wu and B. Zeng in the paper "Mobile robot trajectory tracking and dynamic obstacle avoidance based on deep reinforcement learning," aiming at the problem of error-prone and unstable problems of mobile robot trajectory tracking and dynamic obstacle avoidance in a locally observable nonlinear dynamic environment, proposed a visual perception and decision-making method based on deep reinforcement learning [44].

In order to solve the multi-task decision-making problem of mobile robots in a nonlinear dynamic environment, a method of environment visual perception and multi-task decision-making based on deep reinforcement learning is proposed. The algorithm adopts an end-to-end learning method, combining the feature extraction capabilities of deep convolutional neural networks with the decision-making capabilities of reinforcement learning. The local dynamic environment around the mobile robot is visually perceived as the network input, and the network output is the robot's actions. The control of the visual perception of the environment input to the direct output of the action forms a direct closed-loop control of the system's environment perception and decision-making.

This method uses a deep convolutional neural network to extract features of visual perception image information, and has the characteristics of maintaining feature invariance when the image is displaced, zoomed, or deformed. The calculation of the i-th feature map of the n-th layer of convolutional neural network is defined as

$$x_i^n = f\left(0, \sum_{j \in M_i} |x_j^{n-1} k_{ji}^n + b_i^n\right) \tag{10.7}$$

In formula (10.7), M_i is the set of feature maps, k_{ji}^n is the i-th convolution kernel of the n-th layer, b_i^n is the i-th bias of the n-th layer, $f(\ )$ is the activation function, and the correction linear unit (RLU) as the activation function.

The mobile robot trajectory tracking and dynamic obstacle avoidance method based on deep reinforcement learning proposed by this system, the robot system does not need to understand how the environmental dynamics model works, but only focuses on the value function, first evaluates the Q value of each state action, and then according to The Q value solves the optimal strategy, and the strategy function is obtained indirectly from the value function. The network model training data comes from the interaction between reinforcement learning and the environment dynamics model, making a closed loop between robot situation awareness and decision control. Reinforcement learning is based on the strategy in state s_t, the state s_{t+1} and the timely return r_t after taking action a_t are only related to the current state and action, not to the historical state.

The observed current state information completely determines the characteristics required for decision-making, is a partially observable Markov decision process.

This method is called the time difference Q-learning method based on the difference strategy, and the action strategy for generating the action data is different from the strategy that needs to be evaluated. The action strategy uses a small probability random strategy and a guided strategy of using random actions when the threshold is lower than the threshold. The strategy to be evaluated and improved is a greedy strategy with the maximum function corresponding to the action each time. The system structure of the mobile robot trajectory tracking and dynamic obstacle avoidance algorithm based on deep reinforcement learning is shown in Figure 10.14.

Mobile robot trajectory tracking and dynamic obstacle avoidance algorithms mainly use deep convolutional neural networks to extract features from the image data of robot situation awareness. Even when the image information is displaced, zoomed, or deformed, it can still maintain the characteristics immutability of the relative position of the robot to obstacles and trajectories. The network input first pre-processes the RGB image to grayscale in order to reduce the dimensionality of the input data.

The deep convolutional neural network automatically extracts features directly from the input image data, and fits the Q value end-to-end, and can usually learn better generalization capabilities than manually designed features. The approximator of the value

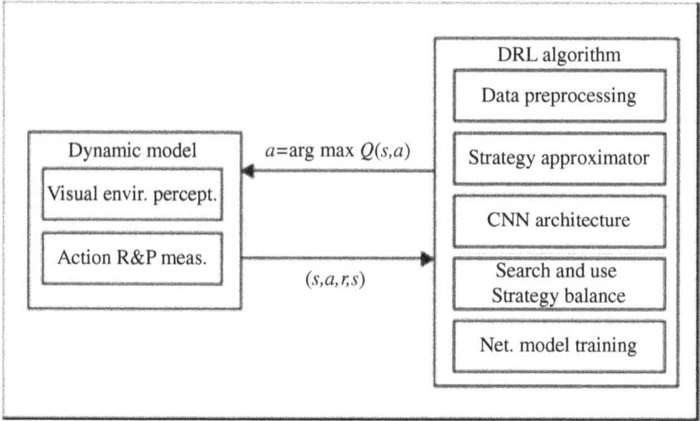

Figure 10.14.   Algorithm framework of deep reinforcement learning.

function or strategy of this algorithm chooses to use a deep convolutional neural network, and the optimization objective function of the model is

$$L(\theta) = E\left[\left(r + \gamma \max_{a'} Q(s'a'|\theta) - Q(s,a|\theta)\right)^2\right] \qquad (10.8)$$

The network model adopts 3 layers of convolutional layer (Conv1 to Conv3) and 2 layers of fully connected layer (FC4, FC5). The network input is the state, and the network output is the Q value corresponding to each action. The action corresponding to the maximum Q value is selected to interact with the environment, and the network calculation cost is proportional to the action space. The network model is shown in Figure 10.15, and the network model parameter settings are shown in Table 10.1.

When the deep convolutional neural network is trained, it is assumed that the training data is independently distributed, and there is a correlation between the data collected from the environment. Using these data for sequential training, the algorithm model will have instability problems. Through experience playback, the

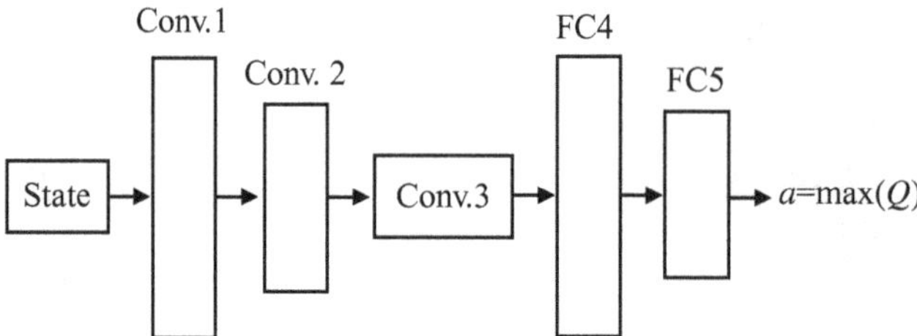

Figure 10.15.   Convolutional NN model with deep reinforcement learning.

Table 10.1.   Deep reinforcement learning convolutional neural network parameters.

Layer model network	Input pixel	Layer parameter setting	Output pixel
Conv 1	$80 \times 80 \times 4$	8×8, 4, Relu	$19 \times 19 \times 32$
Conv 2	$19 \times 19 \times 32$	4×4, 2, Relu	$9 \times 9 \times 32$
Conv 3	$9 \times 9 \times 64$	3×3, 1, Relu	$7 \times 7 \times 64$
FC4	$7 \times 7 \times 64$	Relu	512×1
FC5	512	Linear	5

trained model can be converged and stabilized. In the process of interaction between reinforcement learning and dynamic environment, a part of the state action sequence data is stored at the same time, and then the data is extracted from the storage database by uniform random sampling, and the convolutional neural network optimizes the objective function for the gradient parameter adjustment training through the model. The average value of the trained model behavior distribution exceeds the previous state, smoothing the learning, and avoiding the oscillation or divergence in the parameters. The network model training process is shown in Figure 10.16.

In the multitasking of robot trajectory tracking, local path planning and real-time dynamic obstacle avoidance, it is very difficult to design a reward and penalty function that is easy to train and can make correct decisions. If you want the robot to learn how to track the trajectory, the most natural reward and penalty function is to make the robot system get a reward of 1 when it reaches the required final trajectory configuration, and the reward for other results is -1. Reinforcement learning cannot train the robot to complete multiple tasks such as moving along the trajectory at the desired speed and performing local path planning and dynamic obstacle avoidance. It cannot obtain too many valuable positive returns for this target

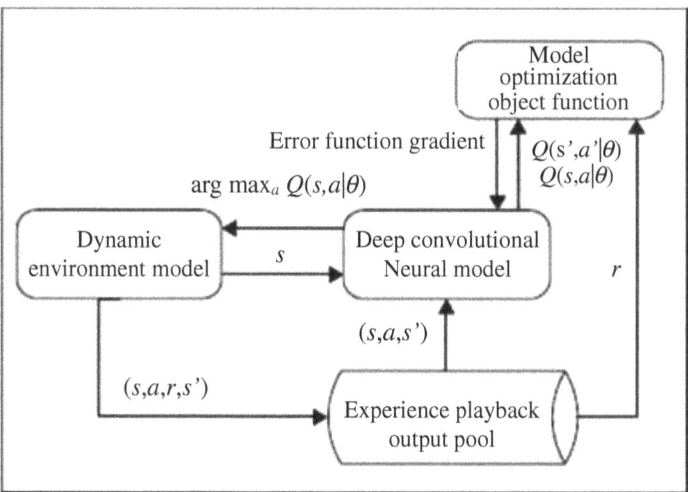

Figure 10.16.   Training flowchart of deep reinforcement learning algorithm.

task. To this end, the environment dynamics model robot is defined to walk along the trajectory and the reward and punishment function is defined, so that the mobile robot can track the trajectory and dynamically avoid obstacles while continuously approaching the target point, and increase the incentive to approach the target on the basis of the reward and punishment function.

In order to verify the effectiveness of the above methods, simulation experiments of mobile robot trajectory tracking and dynamic obstacle avoidance were carried out in a two-dimensional environment. Experimental results show that this method can meet the requirements of multi-task intelligent perception and decision-making, and it can better solve the problem of traditional algorithms that are easy to fall into local optimality, oscillate in similar obstacle groups and cannot identify paths, swing in narrow passages, and problems such as unreachable targets near obstacles, and greatly improve the real-time and adaptability of robot trajectory tracking and dynamic obstacle avoidance.

10.4   Chapter Summary

This chapter studies the control system of mobile robots. Intelligent control is a completely new control method, which has been widely used in mobile robots. Section 6.1 introduces the basic concepts of intelligent control and intelligent control system, discusses the definition of intelligent control and the structural theories of intelligent control. On this basis, the main intelligent control systems are introduced, including hierarchical control, expert control, fuzzy control, neural control, learning control and evolutionary control systems.

Section 6.2 discusses the control of mobile robots based on neural networks, involving mobile robot tracking control, mobile robot navigation control, mobile robot formation control, and mobile robot visual control. Among them, the tracking control of the mobile robot based on neural network includes path tracking control, trajectory tracking control, target tracking control and speed tracking control. A large number of cases illustrate the effective application of neural networks in the control of mobile robots.

Section 6.3 discusses the control of mobile robots based on deep learning, and first reviews the research and application of robot control based on deep learning in recent years. Deep learning has achieved rich research and application results in mobile robot path control and position control, mobile robot trajectory control, mobile robot target tracking control, footed robot walking control and gait planning, and mobile robot motion control. Then, taking a "mobile robot trajectory tracking and dynamic obstacle avoidance based on deep reinforcement learning" as an example, the robot's trajectory tracking and dynamic obstacle avoidance principles, algorithms, and deep reinforcement learning convolutional network model structure are analyzed.

References

[1] Ankalaki, S., Gupta, S.C., Prasath, B.P., *et al.* (2018). Design and implementation of neural network based nonlinear control system (LQR) for target tracking mobile robots. *Proc. 7th International Conference on Computing, Communications and Informatics (ICACCI)*, pp. 1222–1228, Bangalore, Sep. 19–22, 2018.

[2] Boukens, M. and Boukabou, A. (2017). Design of an intelligent optimal neural network-based tracking controller for nonholonomic mobile robot systems. *Neurocomputing*, 226:46–57, Feb 22.

[3] Cai, Z.X. (2019). *Intelligent Control: Principles and Applications*, 3rd Edition. Beijing: Tsinghua University Press (in Chinese).

[4] Cai, Z.X. (1997). *Intelligent Control: Principles, Techniques and Applications*. Singapore/New Jersey: World Scientific Publishers.

[5] Cai, Z.X., *et al.* (2016). *Key Techniques of Navigation Control for Mobile Robots under Unknown Environment*, Chapter 3. Beijing: Science Press.

[6] Cai, Z.X., He, H.G. and Chen, H. (2009). *Theories and Methods of Navigation and Control of Mobile Robots in Unknown Environments*, Chapter 8. Beijing: Science Press (in Chinese).

[7] Cai, Z.X., He, H.G. and Timeofeev, A.V. (2003). Navigation Control of Mobile Robots in Unknown Environment: A Survey. *Proc. 10th Saint Petersburg Int. Conf on Integrated Navigation Systems*, 2003:156–163.

[8] Cai, Z.X., Li, Y., *et al.* (2021). *Sensing, Mapping and Object-tracking Technologies of Autonomous Vehicle*, Chapter 6. Beijing: Science Press (in Chinese).

[9] Cai, Z.X. and Xie, B. (2021). *Fundamentals of Robotics*, 3rd Edition, Chapter 5: Beijing: Mechanical Industry Press (in Chinese).

[10] Cai, Z.X. and Xie, B. (2021). *Robotics*, 4th Edition, Chapter 6. Beijing: Tsinghua University Press (in Chinese).

[11] Cai, Z.X., Zhou, X., Li, M.Y., *et al.* (2000). Evolutionary control architecture of autonomous mobile robot based on function/behavior integration. *Robot*, 22(3):169–175 (in Chinese).

[12] Chen, C., Chen, X.Q., Ma, F., *et al.* (2019). A knowledge-free path planning approach for smart ships based on reinforcement learning. *Ocean Engineering*, 189, 106299.

[13] Chen, C.-H., Jeng, S.-Y. and Lin, C.-J. (2020). Using an adaptive fuzzy neural network based on a multi-strategy-based artificial bee colony for mobile robot control. *Mathematics*, 8(8):1223, Aug 2020.

[14] Chen, G.J. and Chen, W. (2019). A target tracking method for underwater robots based on deep learning and monocular vision (invention patent). Public/Announcement Date: 2019-09-17 (in Chinese).

[15] Chen, J., Cheng, S. and Shi., L. (2019). Mobile robot navigation control based on deep reinforcement learning. *Electronic Design Engineering*, 27(15):61–65 (in Chinese).

[16] Duan, Z.H., Cai, Z.X. and Yu, J.X. (2007). Robust position tracking for mobile robots with adaptive evolutionary particle filter. *Third international conference on natural computation*, IEEE Press: Haikou, China, v4:563–567.

[17] Elhaki, O. and Shojaei, K. (2018). Neural network-based target tracking control of underactuated autonomous underwater vehicles with a prescribed performance. *Ocean Engineering*, 167:239–256, Nov 1.

[18] Ge, H.W., Lin, J.J., Sun, L. and Zhao, M.D. (2018). A reinforcement learning based on the depth of digging Peach nuclear robot behavior control method (patent). Public/announcement: 2018-04-20 (in Chinese).

[19] Guo, Y. and Hu, W. (2004). Iterative learning control of wheeled robot trajectory tracking. *Proceedings of the 8th International Conference on Control, Automation, Robotics and Vision* (ICARCV 2004). Kunming, pp. 1684–1689.

[20] Li, C.P., Li, B.Q., Wang, R.H., *et al.* (2021). A survey on visual servoing for wheeled mobile robots. *International Journal of Intelligent Robotics and Applications*, 5(2):SI 203–218.

[21] Li, S., Ding, L., Gao, H.B. *et al.* (2020). Reinforcement learning neural network-based adaptive control for state and input time-delayed wheeled mobile robots. *IEEE Transactions on Systems, Man and Cybernetics — Systems*, 50(11):4171–4182.

[22] Li, Y.D., Zhu, L., Guo, Y., *et al.* (2019). Multivariable fixed-time formation control of mobile robots based on radial basis function neural network. *Information and Control*, 48(6):649–657 (in Chinese).

[23] Li, Z.L. and Chen, J.B. (2018). A kind of no-blind area sweeping robot based on deep learning algorithm and its cleaning control

method (invention patent). Public/Announcement Date: 2018-11-23 (in Chinese).

[24] Li, Y.Y. and Xiao, N.F. (2017). A voice interaction and control method for intelligent industrial robots based on deep learning (invention patent). Public/Announcement Date: 2017-06-27 (in Chinese).

[25] Lin, J.T., Cheng, H., Yang, X.Y. and Zheng, P.W. (2019). An end-to-end distributed multi-robot formation navigation method based on deep reinforcement learning (invention patent). Public/Announcement Date: 2019-08-20 (in Chinese).

[26] Liu, H., Li, Y.F., Huang, J.H., *et al.* (2017). A deep learning control planning method for robot motion paths in an intelligent environment (invention patent). Public/Announcement Date: 2017-11-21 (in Chinese).

[27] Liu, H.Y., Yuan, W., Tao, Y. and Liu, X.Y. (2020). Optimal control method of humanoid robot gait based on deep Q network (invention patent). Public/Announcement Date: 2020-02-07 (in Chinese).

[28] Lu, X.C., Fei, J.T. and Huang, J. (2017). Fuzzy neural network based adaptive iterative learning control scheme for velocity tracking of wheeled mobile robots. *6th IEEE Data Driven Control and Learning Systems Conference* (DDCLS), Chongqing, PEOPLES R CHINA, May 26–27, 2017.

[29] Luo, C.M. (2017). Neural-network-based fuzzy logic tracking control of mobile robots. *13th IEEE Conference on Automation Science and Engineering* (IEEE CASE), Xian, PEOPLES R CHINA, Aug 20–23, 2017.

[30] Ma, D., Dong, L.Y., Wang, L.L., *et al.* (2020). Adaptive PD tracking control of mobile robot RBF neural network. *Control Engineering*, 27(12):2092–2098 (in Chinese).

[31] Ma, Q.X., Yu, R.S., Shi, Z.Y., *et al.* (2018). Optimal trajectory control of underwater robots based on deep reinforcement learning. *Journal of South China Normal University (Natural Science Edition)*, 50(1):118–123, in Chinese.

[32] Nguyen, T. and Le, L. (2018). Neural network-based adaptive tracking control for a nonholonomic wheeled mobile robot with unknown wheel slips, model uncertainties, and unknown bounded disturbances. *Turkish Journal of Electrical Engineering and Computer Sciences*, 26(1):378–392.

[33] Qian, L.D. (2019). A robot control system based on deep learning (invention patent). Public/Announcement Date: 2019-05-17 (in Chinese).

[34] Song, G.M., He, M., Wei, Z. and Song, A.G. (2020). A self-reset control method for a quadruped robot based on deep reinforcement

learning (invention patent). Public/Announcement Date: 2020-03-06 (in Chinese).

[35] Song, S.J., Wu, H. and You, K.Y. (2018). A fixed depth control method for underwater autonomous robots based on reinforcement learning (invention patent). Public/Announcement Date: 2018-03-02 (in Chinese).

[36] Su, H. and Cai, Z.X. (2004). Research progress of mobile robot tracking technology in unknown environment. *Journal of Huazhong University of Science and Technology*, 2004, 32(s):24–27 (in Chinese).

[37] Tang, C.Y., Chen, Y., Duan, X., *et al.* (2020). A method and device for robot obstacle avoidance control based on deep learning (invention patent). Public/Announcement Date: 2020-04-17 (in Chinese).

[38] Tawiah, T.A.-Q. (2020). A review of algorithms and techniques for image-based recognition and inference in mobile robotic systems. *International Journal of Advanced Robotic Systems*, 17(6), 1729881420972278, DOI:10.1177/1729881420972278.

[39] Tzafestas, S.G. (Ed.). (1997). *Soft Computing and Control Technology*. Singapore/London: World Scientific Publishers.

[40] Tzafestas, S.G. (2018). Mobile robot control and navigation: A global overview, *Journal of Intelligent & Robotic Systems*, 91:35–58.

[41] Wang, Y.K., Chen, Z.X., Huang, Z.Y., *et al.* (2019). Active control method for sucking the ball of a small soccer robot based on deep reinforcement learning (invention patent). Public/Announcement Date: 2019-10-25 (in Chinese).

[42] Wu, H.J. and Lin, X.Q. (2018). A hexapod robot adaptive motion control method based on deep reinforcement learning in complex terrain (invention patent). Public/Announcement Date: 2018-09-14 (in Chinese).

[43] Wu, Q.X., Lin, C.-M. and Fang, W.B. (2018). Self-organizing brain emotional learning controller network for intelligent control system of mobile robots. *IEEE ACCESS*, 2018, 6:59096–59108.

[44] Wu, Y.X. and Zeng, B. (2019). Mobile robot trajectory tracking and dynamic obstacle avoidance based on deep reinforcement learning. *Journal of Guangdong University of Technology*, 36(1):42–50 (in Chinese).

[45] Xiao, H.Z., Chen, C.L.P., Li, T.S., *et al.* (2017). General projection neural network based nonlinear model predictive control for multi-robot formation and tracking. 20th *World Congress of the International Federation of Automatic Control* (IFAC), Toulouse, France, Jul 09-14, 2017.

[46] Xu, J.N. and Zeng, J. (2019). Research on robot dynamic target point following based on deep reinforcement algorithm. *Computer Science*, 46(z2):94–97 (in Chinese).

[47] Yang, S.Z., Han, J.Y., Liang, P., *et al.* (2019). Robot arm control based on deep reinforcement learning. *Fujian Computer*, 35(1):28–29 (in Chinese).

[48] Yang, X.-M., Li, W.-J., Zhu, J., *et al.* (2015). Path tracking control of mobile robot via RBF neural network. *Journal of Hefei University of Technology*, 38(11):1477–1483.

[49] You, B., Qi, H.N., Ding, L., *et al.* (2021). Fast neural network control of a pseudo-driven wheel on deformable terrain. *Mechanical Systems and Signal Processing*, 152, 107478, May 1, 2021.

[50] You, K.Y., Dong, F. and Song, S.J. (2020). An aircraft route tracking method based on deep reinforcement learning (invention patent). Public/Announcement Date: 2020-02-18 (in Chinese).

[51] Zeng, J.F., Wan, L., Li, Y.M., *et al.* (2018). Robust composite neural dynamic surface control for the path following of unmanned marine surface vessels with unknown disturbances. *International Journal of Advanced Robotic Systems*, 15(4):1729881418786646, Jul. 16, 2018.

[52] Zhang, H.D. and Liu, S.R. (2008). Autonomous navigation control of mobile robot based on emotion and environmental cognition. *Control Theory & Applications*, 25(6):995–1000 (in Chinese).

[53] Zhang, S.L. (2019). Robot system control based on convolutional neural network algorithm. *Journal of Changchun University (Natural Science Edition)*, 29(2):14–17 (in Chinese).

[54] Zhang, Y., Yu, J., Li, C. and Liu, Q.H. (2018). Intelligent robot visual tracking method based on deep learning (invention patent). Public/Announcement Date: 2018-11-06 (in Chinese).

[55] Zhang, Y.Z., Wang, S., Pang, L.Z., *et al.* (2019). A mobile robot visual following method based on deep reinforcement learning (invention patent). Public/Announcement Date: 2019-08-02 (in Chinese).

[56] Zhang, H.J., Su, Z.B. and Su, B. (2018). Robot end-to-end control method based on deep Q network learning. *Chinese Journal of Scientific Instrument*, 39(10):6–43 (in Chinese).

[57] Zhu, L., Li, Y.D., Sun, M., *et al.* (2014). Neural network sliding mode control of mobile robot formation. *Electric Machines and Control*, 18(3):113–118 (in Chinese).

[58] Zou, X.B. and Cai, Z.X. (2004). Design and application of nonholonomic mobile robot road tracking controller, *Control and Decision*, 19(3):319–322 (in Chinese).

Part III
Applications and Prospect of Robotics

Chapter 11

Application and Market of Robotics Technology

Today, the most widely-used robots are service robots and industrial robots. According to statistics at the end of 2008, there were more than one million industrial robots in operation around the world, with Japan, the United States, and Germany were among the highest users at the time. Since 2016, China's total number of industrial robots installed has ranked first in the world. Its sales of industrial robots have increased from 57,100 units in 2014 to 137,920 units in 2017. In 2019, a total of 783,000 units have been installed, ranking first in Asia. [4, 19]. Due to the large and increasing number of industrial robots being used to replace humans in various manual labor and part of mental labor, as early as the 1980s, the International Labor Organization (ILO) regarded industrial robots with "blue-collar workers" (industrial workers) and "white-collar workers" (technicians) are juxtaposed, and industrial robots are called "steel-collar workers." Now, more and more countries are promoting "labor conversion," that is, replacing ordinary workers with industrial robots. This new industrial army will grow stronger and play a greater role. They have become mankind's right-hand man and friend, and have an increasing influence on the economy and various fields of human life in various countries.

11.1   Application Fields of Robotics

Robots have been widely used in industrial and agricultural production, sea and air exploration, medical rehabilitation and military fields. In addition, robots have gradually been promoted and applied in entertainment, home and many service industries, and their development is very rapid.

Robotics has an extremely wide range of research and application fields. These fields reflect a wide range of disciplines, involving many topics, such as robot architecture, mechanism, control, intelligence, sensing, robot assembly, robots in harsh environments, and robot languages.

11.1.1 *Industrial robot*

Whether the robot is used in conjunction with other machines or not, compared with traditional machines, it has two main advantages:

(1) Almost complete automation of the production process. It brings higher quality finished products and better quality control, and improves the ability to adapt to changing user needs, thereby improving the competitiveness of products in the market.
(2) High adaptability of production equipment. It allows the production line to quickly switch from one product to another, for example, from producing one type of car to producing another type of car. When a failure makes a component on the production equipment unable to move, the equipment also has the ability to adapt to the failure.

The above-mentioned adaptive production equipment is called Flexible Manufacturing System (FMS). A flexible production unit is composed of a few robots and some supporting machines. For example, a robot designed to be matched with a lathe, together with an automatic lathe, forms a flexible unit. Many flexible units operate together to form a flexible workshop. FMS has become one of the core technologies of intelligent manufacturing.

The industrial robots are now mainly used in the automotive, electromechanical (including telecommunications), general and construction machinery, construction industries, metal processing, casting and other heavy and light industry sectors. Among the many manufacturing fields, the auto parts manufacturing industry sees the widest use of industrial robots.

The industrial application of robots is divided into four areas, namely, material processing, parts manufacturing, product inspection and assembly. Among them, material processing is often the simplest. Parts manufacturing includes forging, spot welding, mashing and casting. Inspection includes explicit inspection (inspecting the completeness of the product surface image and geometry, parts and dimensions during or after processing) and implicit inspection (inspecting the quality or surface integrity of parts during processing). Assembly is the most complex application area because it may include processes such as material processing, online inspection, parts supply, matching, extrusion and fastening.

In agriculture, robots have been used for fruit and vegetable grafting, harvesting, inspection and sorting, sheep shearing, and milking. The application of autonomous (unmanned) mobile robots to farmland cultivation, including seeding, field management and harvesting, is a potential application field of industrial robots. The industry where mobile robots are most used should be agriculture.

With the development of science and technology, the application fields of industrial robots are also expanding. At present, industrial robots are not only used in traditional manufacturing, such as machinery manufacturing, mining, metallurgy, petroleum, chemistry, shipbuilding and other fields, but also have begun to expand to high-tech fields such as nuclear energy, aviation, aerospace, medicine, and biochemistry.

11.1.2 *Explore robot*

In addition to being widely used in industry and agriculture, robots are also used for exploration, that is, to perform tasks in harsh or unsuitable environments for human work. For example, work in underwater (ocean), space, and radioactive, toxic, or high-temperature environments. In this environment, autonomous robots, semi-autonomous robots, or remote-controlled robots can be used.

(1) Autonomous robots. Autonomous robots can perform programming tasks in harsh environments without human intervention.

(2) Remote control robot. A remote control robot places the robot (called a driven device) in a dangerous, harmful or harsh environment, and the operator controls the active device from a distance, so that the driven device follows the operation of the active device to realize remote control.

The following discusses the overview and applications of the two main exploration robots-underwater robots and space robots.

1. Underwater robot

With the advancement of marine development, general diving technology has been unable to meet the needs of high-depth comprehensive survey and research and to complete a variety of operations. Therefore, many countries have paid great attention to underwater robots.

Underwater robots can be classified according to different characteristics. According to the way they move in the water, they can be divided into:

(1) Floating underwater robots;
(2) Walking underwater robots;
(3) Mobile underwater robots.

With the need for marine survey and development in recent years, the application of underwater robots has become increasingly widespread, and the speed of development has exceeded people's expectations. Underwater robots have been widely used worldwide at present, and their application areas include underwater engineering, salvage and life-saving, marine engineering and marine scientific investigations.

At 8:12 a.m. on November 10, 2020, China's "Struggle" manned submersible successfully sat on the bottom of the Marianas Trench in the North-West Pacific, with a depth of 10,909 meters, becoming a major milestone achievement for China in the field of deep diving [16].

2. Space robot

In recent years, with the research and development of various intelligent robots, so-called space robots that can operate in space have become a new research field and have become an important part of space development [8].

At present, the main tasks of space robots can be divided into two aspects:

(1) Complete pioneer exploration under non-human habitation conditions such as the moon, Mars and other planets.
(2) Substituting astronauts in space for satellite services (mainly capturing, repairing and supplying energy), space station services (mainly installing and assembling the basic components of the space station, various payload operations, EVA support, etc.) and space environment application experiment.

At 12:41 on July 23, 2020, China's "Tianwen-1" probe was ignited and launched at the Wenchang Space Launch Site and sent into a predetermined orbit. After landing on Mars, the rover carried by the Mars rover landed on the surface of Mars and carried out exploration operations. The surface morphology, soil characteristics, material composition, water ice, atmosphere, ionosphere, magnetic field, etc. of Mars will be scientifically explored to realize China's technological leap in the field of deep space exploration.

11.1.3 *Service robot*

With the development of network technology, sensor technology, bionic technology, intelligent control and other technologies, as well as the cross integration of electromechanical engineering and biomedical engineering, the development of service robot technology has shown three major trends: First, service robots are equipped with from simple mechatronics to electromechanical integration and intelligence; the second is the development of service robots from a single task to group collaboration, distance learning and network services; the third is the development of service robots from the development of a single complex system to its core technology and core modules embedded in advanced manufacturing related systems. Although the range of service robots is broad, including cleaning robots, medical service robots, nursing and rehabilitation robots, household robots, firefighting robots, monitoring and exploration robots, etc., a complete service robot system usually consists of three basic parts: mobile agencies, perception system and control system. Therefore, the key technologies of various service robots include autonomous mobile technology (including map creation, path planning, and autonomous navigation), perception technology, and human-computer interaction technology.

The robots that are closest to humans that can be seen in real life may be considered household robots. The domestic robot can clean the floor without touching the furniture. Domestic robots have begun to enter homes and offices to replace people in cleaning, washing, guarding, cooking, taking care of children, receiving calls, printing documents, etc. Hotel sales and restaurant service robots, cooking robots and robot nannies are no longer a dream. With the improvement of the quality of domestic robots and the substantial reduction of the cost, domestic robots will be increasingly widely used [3].

Robots developed to treat patients, care for patients, and assist in the rehabilitation of the disabled can greatly improve the condition of the disabled, as well as the living conditions of the paralyzed (including the lower limbs and quadriplegics) and the amputees.

Service robots also include messenger robots, tour guide robots, refueling robots, construction robots, agricultural and forestry robots, etc. Among them, wall-climbing robots can be used for cleaning and construction, and entertainment robots include entertainment, singing and dancing robots and sports robots.

The global service robot market is expected to reach US$9.46 billion in 2019, and it will grow rapidly to exceed $13 billion in 2021. In 2019, the global domestic service robots, medical service robots, and public service robots are estimated to be US$4.2 billion, US$2.58 billion, and US$2.68 billion, respectively, of which the domestic service robot market accounted for up to 44% [6, 17].

11.1.4　*Military robots*

Like any other advanced technology, robotics can also be used for military purposes. Such robots used for military purposes are military robots. Military robots can be used on the ground, underwater (ocean) and in space. Among them, the development of ground-based military robots is the most mature and its applications are also more common.

1. Ground military robot

Ground military robots are divided into two categories: one is intelligent robots, including autonomous and semi-autonomous vehicles;

the other is remote-controlled robots, which are remote-controlled unmanned vehicles for various purposes. Intelligent robots rely on the machine intelligence of the vehicle itself to drive or fight autonomously without human intervention. The remote control robot is remotely controlled by humans to complete various tasks. Remote-controlled vehicles have been equipped with troops in some countries, and autonomous ground combat vehicles have also begun to enter the battlefield.

2. Marine military robot

The Navy is not far behind and has achieved success in the development and application of marine (underwater) military robots.

Another major use of underwater robots in the Navy is mine clearance, such as the MINS underwater robot system, which can be used to find, classify, and remove underwater debris and trapped mines.

France has always been a world leader in military minesweeping robots. ECA has sold hundreds of PAP-104 ROVs for mine clearance to 15 national navies since the mid-1970s. The latest V-type is equipped with new electronic instruments and telemetry transmission devices, which can clear mines that cannot be cleared by manual or other mine clearing tools.

3. Space military robot

Strictly speaking, all the space robots discussed above can be used for military purposes. In addition, UAVs can be regarded as space robots. In other words, UAVs and other space robots may become space military robots.

Micro-planes are used to fill blind spots that military satellites and reconnaissance aircraft cannot reach, and to provide frontline commanders with specific enemy conditions within a small area. This type of aircraft is small and light and can be carried by soldiers' backpacks. It can be equipped with a solid-state camera, infrared sensor or radar, and can has a flight range of several kilometers.

It is very difficult to develop a suitable micro-UAV, and a series of technical problems must be solved. This kind of airplane is not a toy. It must meet military requirements while achieving low cost and low price. This miniature military reconnaissance aircraft has already flown to the battlefield and embarked on a practical road.

11.2　Status Quo and Forecast of Robot Market

This section studies the development status of robot markets and their forecasts.

1. Development status and forecast of the world market from the perspective of robot market size [9–12, 15]

According to statistics from the International Federation of Robotics (IFR), the global robot market reached more than 30 billion U.S. dollars in 2019, of which, industrial robots were 13.8 billion U.S. dollars and service robots were 16.9 billion U.S. dollars. Figure 11.1 shows the scale of the global robot market in 2019 [20].

(1) Industrial robots

The total number of industrial robots in operation in the world now exceeds 2.7 million. Since 2014, the market size of industrial robots has continued to grow at an average annual rate of 8.3%. The IFR report shows that the total sales of major countries such as China, Japan, the United States, South Korea and Germany in 2018 exceeded 3/4 of global sales. The demand for industrial automation transformation in these countries has activated the industrial robot market and has also increased the global industrial robot usage density significantly. Currently, in the global manufacturing industry, the use density of industrial robots has reached 85 units per 10,000 people. In 2018, global industrial robot sales reached 15.48 billion U.S. dollars, of which Asia sales reached 10.48 billion U.S. dollars, Europe sales reached 2.86 billion U.S. dollars, and North America sales reached 1.98 billion U.S. dollars. In 2019, with the further

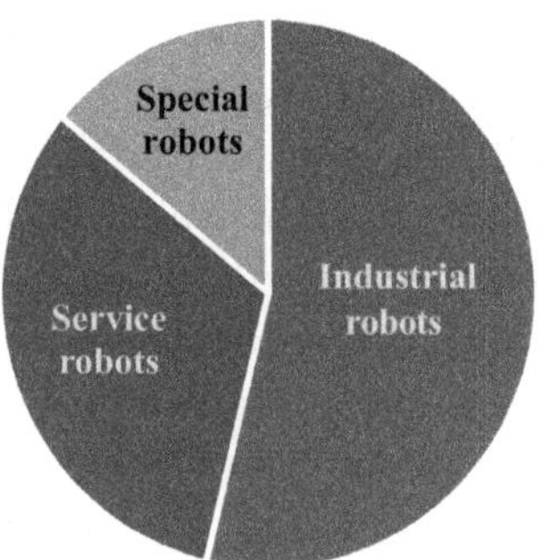

Figure 11.1.　Scale of the global robot market in 2019 [20].

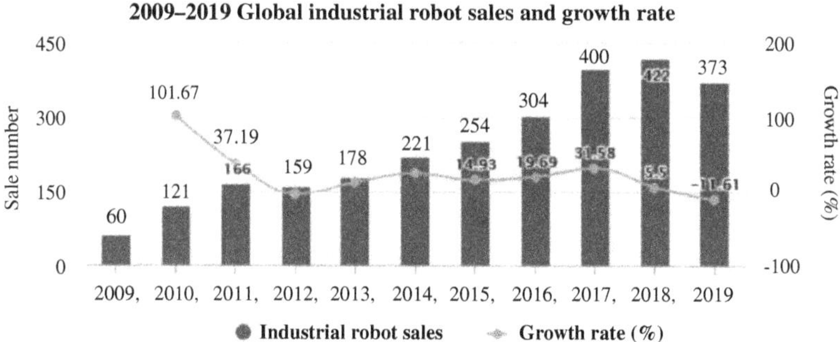

Figure 11.2. Statistics and forecast of global industrial robot sales and growth rate (2009–2019).

Source: IFR.

popularization of industrial robots, sales are expected to be close to 16 billion U.S. dollars, of which Asia was still be the largest sales [6]. Figure 11.2 shows the statistics and forecasts of global industrial robot sales and growth rates from 2014 to 2020.

(2) Service robots

With the rapid development of information technology and the rapid popularization of the Internet, marked by the introduction of the deep learning model in 2006, artificial intelligence ushered in the third rapid development of artificial intelligence. At the same time, relying on artificial intelligence technology, the application scenarios and service modes of intelligent public service robots are continuously expanding, driving the rapid growth of the service robot market. Figure 11.3 shows the statistics and forecasts of global service robot sales and growth rates from 2014 to 2020. Since 2014, the global service robot market has grown at an average annual rate of 21.9%. In 2019, the global service robot market is expected to reach 9.46 billion U.S. dollars, and it will grow rapidly to exceed 13 billion U.S. dollars in 2021. In 2019, the global domestic service robots, medical service robots, and public service robots are estimated to be US$4.2 billion, US$2.58 billion and US$2.68 billion, respectively, of which the domestic service robot market accounts for up to 44%.

(3) Special robots

The overall performance of global special robots has continued to improve in recent years, and has continuously spawned emerging markets. Since 2014, the global special robot industry has grown

Figure 11.3. Statistics and forecast of global service robot sales and growth rate (2014–2020) [2, 15].

at an average annual rate of 12.3%. In 2019, the global special robot market will reach 4.03 billion U.S. dollars; by 2021, the global special robot market is expected to exceed 5 billion U.S. dollars. Among them, the United States, Japan and the European Union lead the world in innovation and market promotion of special robots.

2. From the perspective of annual robot installations [1, 2]

(1) Industrial robot

The number of global industrial robot installations exceeded 400,000 units in 2018, reaching 422,271 units, an increase of approximately 6% over 2017, and the cumulative installation volume was 24,39,543 units, an increase of approximately 15% over 2017. Among them, the automotive industry is still the main purchaser of industrial robots, accounting for 30% of the total global installations of industrial robots, the electrical/electronics industry accounts for 25%, the metal and machinery industries account for 10%, the plastics and chemical industries account for 5%, and the food and the beverage industry accounts for 3%.

With the development of automation technology and the continuous innovation of industrial robot technology, the global demand for industrial robots has been significantly accelerated since 2010. Figure 11.4 provides the annual global industrial robot installations

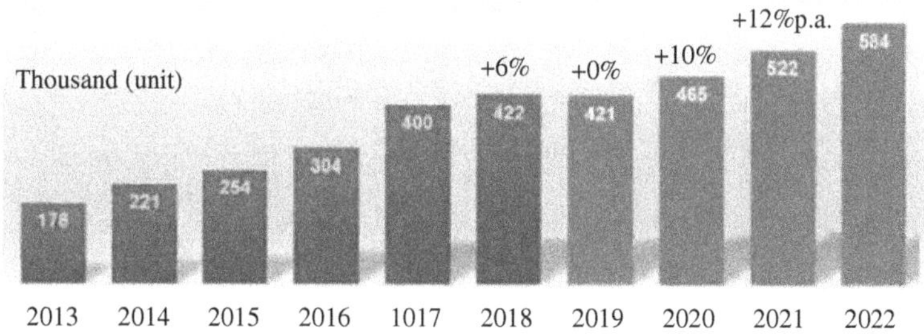

Figure 11.4. 2013–2018 (statistics) and 2019–2022 (forecast) global annual installations of industrial robots.

from 2013–2018 (statistics) and 2019–2022 (forecast). From 2013 to 2018, the annual compound growth rate of global robot sales was approximately 19%. From 2005 to 2008, the global average annual robot sales was about 115,000 units. However, in 2009, due to the financial crisis, the sales of robots fell sharply. In 2010, robot sales were 120,000 units. Until 2015, the global installation of industrial robots more than doubled to nearly 254,000 units. In 2016, the number of industrial robots installed exceeded 300,000 units. The installed robot number soared to nearly 400,000 units in 2017, and more than 420,000 units in 2018.

In 2018, China, Japan, the United States, South Korea, and Germany, the five major global markets of industrial robot, accounted for 74% of robot sales. Asia is still the region with the highest robot sales volume at present, with 283,080 installations in 2018 (see Figure 11.5), an increase of only about 1% over 2017 and a record high, accounting for about 2/3 of the total global robot installations. 2013–2018, the average annual growth rate of global robot installations was approximately 23%. According to IFR statistics, from 2013 to 2018, China, Japan, South Korea, the United States, and Germany accounted for more than 70% of the total global installations.

Japan has changed from the first to the second largest robot market, and its robot sales have increased by approximately 21% to 55,240 units, a record high. From 2013 to 2018, the average annual growth rate of robot sales in Japan was 17%.

The number of robot installations in Korea decreased by 5% to 37,807 units, making it the fourth largest robot market. The main reason is that the electrical/electronics industry has reduced

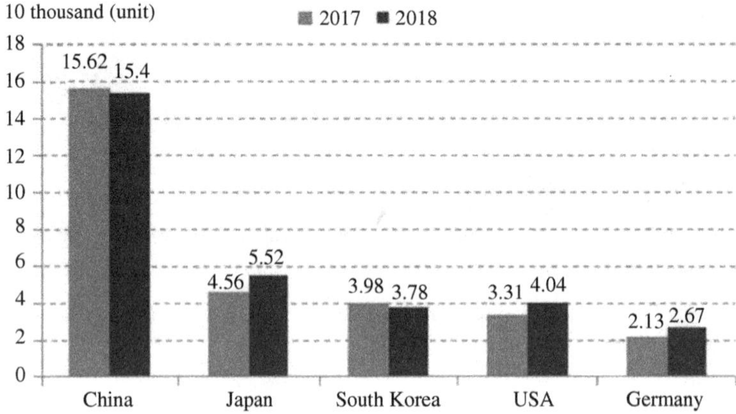

Figure 11.5. Industrial robot sales volume in the world's top five countries in 2017–2018.

investment in robots, resulting in a decrease in robot sales in 2018. From 2013 to 2018, the average annual growth rate of robot sales in Korea was about 12%.

Europe has become the second largest region in robot sales, its industrial robot sales increased by about 14% to 75560 units, and was a record high for six consecutive years. From 2013 to 2018, the average annual growth rate of robot sales in Europe was about 12%.

Germany has become the world's fifth largest robot market, an increase of about 26% over 2017, reaching 26,723 units, a record high. The demand of the automotive industry is the main driving force for its growth.

Compared with 2017, the number of installed industrial robots in the Americas increased by more than 20% in 2018, reaching a new peak of 55,212 units. From 2013 to 2018, the average annual growth rate of robot sales in the Americas was about 13%. Among them, the United States has become the third largest robot market, and its industrial robot installations have increased by about 22% to 40,373 units, setting a record high for eight consecutive years.

2. Statistical data analysis of China's industrial robot market [13, 14, 18]

According to statistics from the International Federation of Robotics, in 2017, China's industrial robots installed 137,920 units, an increase of about 59% over 2016, and it continues to be the world's largest robot market. Among them, in terms of usage, handling robots account for about 45%, welding robots for about 26%, and assembly robots for about 20%. In terms of application industries, the electrical/electronics industry accounts for approximately 35%, and the automotive industry accounts for approximately 31%. From 2012 to 2017, the compound annual growth rate (CAGR) of China's industrial robot installations was approximately 43%, and the average annual growth rate of sales was approximately 33%. In 2017, China's industrial robot sales were approximately US$4.9 billion [7].

According to statistics, in 2017, China is still the world's largest robot market and the fastest growing country in the global robot market. Since 2016, the cumulative installed volume of industrial robots in China has ranked first in the world, and the sales volume of industrial robots has increased from a peak of 57,100 units in 2014 to 137,920 units in 2017. In 2019, 140,500 units were newly installed and 783,000 units were installed in total, ranking first in Asia in total, with an annual growth rate of 12% [5].

By the end of 2017, the cumulative installed volume of industrial robots in China reached 473,429 units, an increase of about 39% over 2016. From 2012 to 2017, the cumulative installation of industrial robots in China increased by an average of 37% annually [4].

11.3   Chapter Summary

This chapter introduces the application of robots, analyzes and predicts the development of the robot technology market.

Section 11.1 comprehensively introduces the application of robots. More than 2.7 million industrial robots worldwide are operating in the automotive industry, electromechanical industry and other industrial sectors, contributing to human material production. Among them, welding robots and assembly robots are the two main application areas.

Compared with industrial robots, the number of robots other than service robots is still relatively limited, but their importance cannot be ignored, and their development prospects are also very promising. In addition to special robots under severe working conditions, exploration robots are mainly space exploration robots and marine (underwater) exploration robots. With the development of space science and marine engineering, more and more various space robots and marine robots are required to participate in the exploration of space and the development of oceans.

Service robots have developed rapidly in recent years, and their number has greatly exceeded that of industrial robots, and is increasing year by year. With the improvement of cost performance and the increase of service workers' labor costs, service robots will soon enter thousands of households, appearing in homes, hospitals, hotels, restaurants, and playgrounds, providing satisfactory first-class services to humans.

Military robots are the product of the use of robotics technology in wars. They are one of the focal points of competition between national power, economic power, technical power, and military power. Military robots are mostly ground-based military robots, and their technology is relatively mature. Marine military robots and space military robots are difficult to be completely separated from civilian marine robots and space robots; their technology can be used for peace-building missions and military activities.

Section 11.2 introduces and analyzes the current situation of domestic and foreign robot markets.

In 2019, the global robot market reached more than 30 billion U.S. dollars, of which, industrial robots were 13.8 billion U.S. dollars and service robots were 16.9 billion U.S. dollars. The total number of industrial robots in operation in the world now exceeds 2.7 million. Since 2014, the market size of industrial robots has continued to grow at an average annual rate of 8.3%.

Since 2014, the average annual growth rate of the global service robot market has reached 21.9%. In 2019, the global service robot market is expected to reach 9.46 billion U.S. dollars, and it will grow rapidly to exceed 13 billion U.S. dollars in 2021.

The overall performance of global special robots has continued to improve in recent years, and has continuously spawned emerging markets. Since 2014, the global special robot industry has grown at an average annual rate of 12.3%. In 2019, the global special robot market reached to 4.03 billion U.S. dollars; by 2021, the global special robot market is expected to exceed 5 billion U.S. dollars.

It can be seen from the above data that the international robot market has maintained rapid and stable development in recent years, and its development prospects are promising.

References

[1] Analysis of global robot market statistics. (2012). 2012-06-29. http://www.robot-china.com/news/201206/29/1790.html

[2] Analysis of the latest statistics of the global robot market. (2020). 2020-02-24 10:05, https://www.sohu.com/a/375391513_320333

[3] Bill Gates. Bill Gates predicted: Everyone will have robots in the future. 2007.02.01 09:17:47 http://people.techweb.com.cn/2007-02-01/149230.shtml

[4] Cai, Z.X. and Xie, B. (2021). *Robotics*, Chapter 1, 4th Edition. Beijing: Tsinghua University Press.

[5] China Business Intelligence Network. (2019). The current situation of the global robotics industry segmentation market in 2019 and the forecast of future development trends. 2019-09-02 11:58, https://baijiahao.baidu.com/s?id=1643534568013791442&wfr=spider&for=pc

[6] China Commercial Industry Research Institute. (2020). Analysis of the status quo of global industrial robots in 2020: General industry has gradually become the main force in the new market. 2020-06-22 16:02, https://www.askci.com/news/chanye/20200622/1602311162338.shtml

[7] China leads the global robot market. (2014). 2014-07-04. http://www.ciqol.com/news/economy/809405.html

[8] Ding, X.L., Shi, X.Y., Robetta, R. *et al.* (2008). The development and prospects of lunar exploration robotics. *Robotics and Applications*, 2008(3):5–13.

[9] IFR. (2013). Executive Summary: World Robotics 2013 Industrial Robots. 2013-09-18. http://www.ifr.org/index.php?id=59&df=Executive_Summary_WR_2013.pdf

[10] IFR. (2013). The robotics industry is looking into a bright future 2013-2016: High demand for industrial robots is continuing. 2013-09-18. http://www.ifr.org/news/ifr-press-release/the-robotics-industry-is-looking-into-a-bright-future-551/

[11] IFR. (2014). International Federation of Robotics 2012 global industrial robot statistics. 2014-02-18. http://wenku.baidu.com/link?url=ZVNynuFZU2w7M_4f_4Nfbta0Vg6vFaum5DI2JsAkMCbYfa9Yk463Hjh-B9-pm2zKSbsZ9B7x1guP1Rwnl5iI_AW_KN5vqUFy6OZw-6uLjxK

[12] Li, L., Ye, T., Tan, M., *et al.* (2020). Research status and future of mobile robot technology. *Robots*, 24(5):4752480.

[13] Liang, W.L. (2014). The fast-growing Chinese robotics market. *Robot Technology and Applications*, (3):2–7.

[14] Liang, W.L. (2019). Analysis of China's industrial robot market statistics. *Robot Technology and Applications*, (3):47–48.

[15] Lu, B. (2020). Analysis of global robot market statistics. *Robot Technology and Applications*, (3):41–42.

[16] Post No. 1. (2020). China's manned submersible surpassed 10,000 meters, setting a new world record. 2020-11-10 17:51, https://baijiahao.baidu.com/s?id=16829664482460626225&wfr=spider&for=pc

[17] Secret Network. Robot development strategies in countries around the world (2019).2019-10-17, https://www.wenmi.com/article/pzhj7s03vm24.html

[18] Wei, M. (2013). Statistics on China's industrial robot market. *Robot Technology and Applications*, (3):8–12.

[19] Yan, G.C. (2014). China has become the world's largest industrial robot market. 2014-06-17. http://gb.cri.cn/42071/2014/06/17/6891s4580547.htm

[20] Yi Smart Internet of Things. (2019). Interpretation of the 2019 Global Industrial Robot Market Report. 2019-09-19 15:28:49, http://www.openpcba.com/web/contents/get?id=3887&tid=15

Chapter 12

Robotics Outlook

Since the 1990s, the application of traditional industrial robots with general functions has become saturated, and many advanced production and special applications require robots with various degree of intelligence, which has promoted the rapid development of intelligent robots. From an international or domestic perspective, an important way to recover and continue to develop the robotic industry is to develop various intelligent robots in order to improve the performance of robots and expand their functions and application areas. This is a great opportunity for the vast number of scientific and technological workers engaged in the research and application of intelligent robots to display their talents.

12.1 Development Trend of Robotics

Looking back on the development history of robotics at home and abroad in the past 20 years, the following characteristics and development trends can be summarized [5].

1. Sensing intelligent robots develop rapidly

As the basis of sensor-based robots, robot sensing technology has made new developments, and various new types of sensors continue to appear. Multi-sensor integration and fusion technology has been applied in intelligent robots. In the research of multi-sensor integration and fusion technology, the application of artificial neural network is particularly eye-catching and has become a research hotspot.

2. Development of new smart technologies

Intelligent robots have many attractive new research topics, and new breakthroughs are brewing in the concept and application of new intelligent technologies. The telepresence technology can measure and estimate the person's anthropomorphic motion and biological state of the predicted target, display on-site information, and use it to design and control the motion of anthropomorphic mechanisms. Virtual reality (VR) technology is a newly researched intelligent technology. It is a technology that decomposes and recombines the reality of events in time and space. Shape memory alloy (SMA) is known as "smart material" and can be used to perform driving actions, complete sensing and driving functions. Reversible shape memory alloy (RSMA) is also used in micro machines.

Multi-agent Robot System (MARS) is another intelligent technology that has begun to be explored in recent years. It is produced when a single intelligent machine has developed to a condition that requires coordinated operations. Multiple robot bodies have a common goal and complete interrelated actions or tasks.

Among many new intelligent technologies, the development and application of recognition, detection, control and planning methods based on artificial neural networks occupies an important position. Robot planning based on expert systems has obtained new developments. In addition to task planning, assembly planning, handling planning and path planning, it is also used for automatic grasping planning.

With the in-depth development of machine learning research, more and more machine learning algorithms, especially deep learning and deep reinforcement learning algorithms, are in the field of robot control, such as robot path and position control, robot trajectory control, robot target tracking control, foot style robot walking control and gait planning, as well as robot motion control, have been widely used. Robot planning based on machine learning, especially robot planning based on deep learning and deep reinforcement learning, involves areas such as intelligent driving and transportation planning and navigation.

3. Adopt modular design technology

The structure of intelligent robots and advanced industrial robots should be simple and compact, and the design of their high-performance components and even all mechanisms has been developed in the direction of modularization; its drive adopts AC servo motors, which are small and high-output; its control devices are small-sized and intelligent development, using high-speed CPU and 32-bit chips, multi-processors and multi-function operating system to improve the robot's real-time and rapid response capabilities. The modularization of the robot software simplifies programming, develops offline programming technology, and improves the adaptability of the robot control system.

4. The upward trend of networking and intelligence of robotic engineering systems

The application of robots in the production engineering system enables the development of automation into comprehensive and flexible automation, and realizes the intellectualization and robotization of the production process. In recent years, the robot production engineering system of many industries and enterprises has been continuously developed. When developing new products, the automobile industry, construction machinery, construction, electronics and electrical industries, and home appliance industries introduce advanced robotics technology, adopt flexible automation and intelligent equipment, and transform the original production methods, so that the development of robots and their production systems is on the rise trend.

Currently, leading companies in the robotics field are increasing their R&D efforts, focusing on industrial internet applications and smart factory solutions, focusing on product development such as unmanned vehicles, humanoid robots, post-disaster rescue robots, and deep-sea mining robots, constantly innovating product forms and optimizing product performance, seizing the opportunity for the development of intelligent robot applications. In terms of industrial robots, the industrial internet has become the focus of the layout, and smart factory solutions are accelerating the implementation; in terms of service robots, unmanned vehicles have received high attention from the technological leaders, and the research and development of humanoid robots ushered in breakthroughs; in terms of special robots, the development of post-disaster rescue robots as a hot spot, mining robots began to expand into the deep-sea space.

5. There have been breakthroughs in the research of miniaturized, lightweight, and flexible robots

Some people call micro machines and micro robots one of the cutting-edge technologies of the 21st century. A miniature mobile robot with the size of a finger has been developed, which can be used to enter small pipelines for inspection operations. They can directly enter human organs for diagnosis and treatment of various diseases without harming human health.

"Nanorobot" is an emerging technology in robotics. The development of nanorobots belongs to the category of "Molecular nanotechnology (MNT)." It designs prototypes based on biological principles at the molecular level, and can be designed and manufactured. The "functional molecular devices" operate in nanospace. Nanorobots are the application of biological principles at the nanometer scale, the development of programmable molecular robots, and the development of "in-body" biological computers or cellular robots, thus giving rise to nanorobots, thus giving rise to nanorobots. Nanorobots have been widely used in the medical field.

In terms of industrial robots, the development of light weight and flexibility is accelerating, and human-machine collaboration continues to deepen; there are also small robots between the large and medium-sized robots and the micro-robot series. Miniaturization is also a trend in the development of robots. The small robot moves flexibly and conveniently, with fast speed and high precision, and is suitable for entering large and medium-sized work pieces for direct operation. Ultra-micro robots that are smaller than micro-robots, using nanotechnology, will be used for medical and military reconnaissance purposes. In terms of service robots, cognitive intelligence has made certain progress, and the process of industrialization continues to accelerate. In terms of special robots, combining perception technology and new bionic materials, intelligence and adaptability continue to increase.

6. Research and develop heavy-duty robots

In order to meet the needs of intelligent and unmanned large-scale and heavy-duty equipment, the development of heavy-duty robots should be a new direction for robotics research and development.

7. Application areas expand to non-manufacturing and service industries

In order to open up new markets for robots, in addition to improving the performance and functions of robots and developing intelligent robots, expansion to non-manufacturing industries is also an important direction. The development of robots adapted to work in unstructured environments will be a long-term direction for robot development. These non-manufacturing industries include aerospace, marine, military, construction, medical care, services, agriculture and forestry, mining, electricity, gas, water supply, sewer engineering, building maintenance, social welfare, home automation, office automation, and disaster relief. Service robots will take the lead in the development and application of medical, education, entertainment and other fields to benefit human, and their development will gradually show the characteristics of intelligence, networking, humanization, and diversification [2].

8. Research on walking robots attracts attention

In recent years, the research on mobile robots has been paid more attention, so that the robot can move to the predetermined target that the fixed robot cannot reach, and complete the set operation task.

Walking robot is a type of mobile robots, including walking robots (two-legged, four-legged, six-legged, and eight-legged) and crawling robots. Autonomous mobile robots and mobile platforms are one of the most studied. Mobile robots have broad application prospects in industry and national defense, such as cleaning robots, service robots, patrol robots, anti-chemical reconnaissance robots, underwater autonomous operation robots, flying robots, etc. A large number of achievements have been made in the research of mobile robots in my country.

9. Develop agile manufacturing production system

Industrial robots must change the past "component development method" and give priority to the "system development method." With the continuous expansion of the application range of industrial robots, robots have evolved from the original flexible loading and unloading device to a programmable highly flexible processing unit.

With the resolution of high rigidity and micro-drive problems, the era of robots as high-precision, high-flexibility and agility processing equipment will arrive sooner or later. No matter what role the robot plays in the production line, it always exists as a member of the system. The development of industrial robots should be considered from the point of view of composing an agile production system.

From the system point of view, we must first consider how to easily connect and communicate with other devices. The communication between the robot and the local database is a field bus from the perspective of development, while the distributed database uses the ethernet. From a system point of view, the design and development of robots must consider the ability to interconnect and coordinate work with other devices.

10. Military robots will equip troops

Here only discuss the development trend of army robots. Due to the small size of the micro-robot, the survivability is particularly strong, and it has a wide range of application prospects. In the future, the networking of semi-autonomous robots is an important application. The combination of moving sensors can provide an overall picture of the battlefield space. For example, dozens of small and inexpensive systems can be used to collect the bullets of the cluster bullets on the ground and pile them up. The research on network robots has become a hot spot of interest. A concept called "Robot Attached Force" has been proposed. The core of this type of force is a manned system, surrounded by various unmanned systems equipped with weapons and sensors.

Humanization, heavy-duty, miniaturization, networking, flexibility, and intelligence have become the main development trends of the robotics industry.

12.2   Robot Development Plans in Various Countries

In the past 10 years, many advanced industrial countries have competed to formulate "robot roadmaps," planning to carry out research on intelligent robots at a higher level, in more fields and on a larger scale, in order to better develop the economy and benefit the people of all countries [1, 12–16, 18].

The United States began to implement the "Advanced Manufacturing Partnership Program" in 2011, which clearly requires the development of industrial robots to revitalize the US manufacturing industry, and relying on the advantages of information network technology to invest US\$2.8 billion to develop a new generation of intelligent robots based on mobile Internet technology. In 2012, the "National Strategic Plan for Advanced Manufacturing" was released, which raised the promotion of advanced manufacturing to a national strategic level, clearly proposed the implementation of US advanced manufacturing strategic goals, and stipulated short-term and long-term indicators for measuring each goal. It demonstrates the determination and vision of the U.S. government to revitalize the manufacturing industry. A roadmap for U.S. robotics (From Internet to Robotics) released by the United States on March 20, 2013. The research direction of the roadmap covers six aspects of robots as economic engines, manufacturing, medical and health, service industries, space applications, and defense applications. It emphasizes the important role of robotics in the US manufacturing and health care fields, and also describes the potential of robotics in creating new markets, new jobs and improving people's lives. In 2012, the National Science Foundation, the National Institutes of Health, the National Aeronautics and Space Administration and the Department of Agriculture have jointly established the National Robotics Research Program and invested US\$50 million in soliciting robotics research projects. The new investment of US\$100 million this time will promote the widespread application of US robotics in various fields and help strengthen the US's leading position in robotics.

In 2012, the National Science and Technology Commission of the United States issued the National Strategic Plan for Advanced Manufacturing, which promoted the development of advanced manufacturing to a national strategic level. The plan objectively describes the development trend of the global advanced manufacturing industry and pointed out the challenges faced by the US manufacturing industry, and clearly puts forward the strategic goals or tasks for the implementation of the US advanced manufacturing industry. In the same year, US President Barack Obama proposed the creation of the "National Manufacturing Innovation Network (NNMI)" with a view to revitalizing the competitiveness of the US manufacturing industry.

The "National Strategic Plan for Advanced Manufacturing Industries" of the United States clarifies three major principles: the first is to improve the advanced manufacturing innovation policy; the second is to strengthen the construction of "industrial commons"; and the third is to optimize government investment. The five goals proposed in the report are: first, to accelerate investment in small and medium-sized enterprises; second, to improve labor skills; third, to establish a sound partnership; fourth, to adjust and optimize government investment; fifth, to increase R&D investment.

The official website of the **European Union** was released on June 3, 2014. 180 companies and R&D institutions affiliated to the European Commission and the European Robotics Association (euRobotics) jointly launched the world's largest civilian robot R&D program "SPARC" ("Spark" program). According to the plan, by 2020, the plan will invest 2.8 billion euros (of which, the European Commission will invest 700 million euros, and euRobotics will invest 2.1 billion euros) to promote robot research and development. The research and development content includes the application of robots in various fields such as manufacturing, agriculture, health, transportation, safety and home. The European Commission predicts that the plan will create 240,000 jobs in Europe, increase the annual output value of the European robot industry to 60 billion euros, and increase its share of the global market to 42%. The development of the robotics industry can stimulate employment and improve the quality of human life and production safety. The robots under development are all advanced robots that integrate modern science and technology. With the support of a new generation of information technology represented by big data, cloud computing, and mobile Internet, advanced robots have stronger independent learning capabilities and independent problem-solving capabilities. Among them, in 2012, **Germany** promoted the "Industry 4.0" plan centered on "smart factories," and its overall goal is to achieve "green" intelligent production. Industry 4.0 covers many aspects of manufacturing, service industry and industrial design, and aims to develop new business models and tap the huge potential of industrial production and logistics models.

As early as 2001, the **Japan** Robot Industry Association published the "Long-term Development Strategy for Robotics," formulated a long-term development strategy for robotics, emphasized the importance of robotics as a high-tech industry, and proposed to vigorously develop robots used in manufacturing and biological industries. The plan regards the robotics industry as one of the seven key industries supported in the Japanese "New Industry Development

Strategy." In the field of humanoid robots alone, it plans to invest a total of US\$350 million in 10 years.

In January 2015, Japan's National Robot Revolution Promotion Group issued the "New Robotics Strategy," which intends to achieve three strategic goals through the implementation of a five-year action plan and six important measures, so that Japan will achieve a robot revolution to response the increasingly prominent aging, working population reduction, frequent natural disasters and other issues, and enhance the international competitiveness of Japanese manufacturing industry, and obtain the global competitive advantage in the era of big data.

The South Korean government enacted the "Intelligent Robot Promotion Act" in March 2008, and announced the "Basic Plan for Intelligent Robots" in April 2009. The plan believes that through this series of active training policies and technology research and development efforts, the competitiveness of Korea's domestic robot industry will be gradually improved.

In 2009, the "Service Robot Industry Development Strategy" was released, and the development goal of becoming one of the world's three largest robot powers was put forward. In December 2010, South Korea released the plan to achieve the goal of becoming one of the world's three largest robot powers — the "Service Robot Industry Development Strategy," hoping to actively cultivate the service robot industry and open up new markets to narrow the gap with developed countries, strengthen the global competitiveness of the robotics industry. In October 2012, the "Future Robotics Strategy Outlook 2022" was released, focusing the policy on expanding the Korean robot industry and supporting domestic robot companies to enter overseas markets. In 2015, a series of policies and measures to support the development of the robotics industry were successively introduced. The mid-to-long-term strategy is "Robots Future Strategy 2022," hoping to realize the vision of "Robots in every corner of society."

The Ministry of Science and Technology of China issued the "12th Five-Year Special Plan for the Development of Intelligent Manufacturing Technology" and the "12th Five-Year Special Plan for the Development of Service Robot Technology" in April 2012. During the "12th Five-Year Plan" period, China will conquer a batch of intelligent high-end equipment, develop and cultivate a batch of high-tech core enterprises with an output value of more than 10 billion Chinese

Yuan; at the same time, it will focus on cultivating and developing emerging industries of service robots, and focus on the development of public safety robots, medical rehabilitation robots, bionic robot platforms and modular core components. The "12 Five-Year Plan for Intelligent Manufacturing Technology Development" proposes to focus on breakthroughs in basic technologies and components in the basic theories and common key technologies in the intelligentization of the design process, the intelligentization of the manufacturing process, and the intelligentization of manufacturing equipment; a number of breakthroughs should be made in intelligent manufacturing basic technologies and components, research and develop a batch of common basic technologies closely related to national security and industrial safety, focus on breaking through a batch of core basic components of smart manufacturing, and lay the foundation for the "13th Five-Year" manufacturing process intelligent equipment and manufacturing process intelligentization Technical basis, mainly in manufacturing informatization, basic components, sensors, automated instrumentation, safety control systems, and embedded industrial control chips [10, 11, 17].

It can be seen from the vigorous development and application of intelligent robots in various countries described above that since the beginning of the 21st century, especially in the past decade, the world's major robot powers are ambitious and scrambling to develop intelligent robot technology, which will surely promote international robotics research and application entering a new era with application, pushing robotics technology to a new level.

12.3 Social Problems Caused by the Application of Robots

Many major inventions and important technologies in history have brought certain negative effects after they have brought well-being to mankind. For example, chemical and biochemical technologies have created various chemical and biochemical products for mankind to meet the needs of people's lives and improve the quality of life. At the same time, they are also used to manufacture chemical and biochemical weapons, causing major threats and harm to mankind. For another example, nuclear energy technology can provide the world with clean energy and cure diseases, but the manufacture and service of atomic bombs and hydrogen bombs have presented mankind with the threat of extinction. These examples illustrate that any "high-tech" has two sides, and it may become a double-edged sword [14].

Robotic technology is also a double-edged sword. While bringing huge benefits to mankind, there are also some problems, especially safety issues. Taken together, the safety issues of robotics involve psychological, social, ethical, legal, and military issues, etc. [3,6–9].

1. Robots cause changes in social structure

In the past few decades, the structure of human society has undergone quiet changes. People used to deal directly with machines, but now they have to use smart machines to deal with traditional machines. This means that the original traditional social structure of "human-machine" has gradually been replaced by the new social structure of "human-robot-machine." People have felt and will see more about artificial intelligence "doctors," "secretaries," "reporters," "editors" and robot nurses," "waiters," "traffic police," "security guards," and "operators," "cleaners" and "nanny," etc., whose job will all be performed by intelligent systems or intelligent robots. In this way, humans must learn to live in harmony with artificial intelligence and intelligent robots to adapt to this new social structure. As early as 2007, Bill Gates predicted that "every family will have robots in the future," and his prediction has begun to come true [2].

Therefore, people will have to learn to get along with robots and adapt to this kind of coexistence. Since dealing with robots is different from dealing with people after all, people must change their traditional concepts and ways of thinking.

2. Robots pose a psychological threat to humans

The intelligence of robots will surpass that of humans, so that they are anti-object-oriented, and humans are required to obey its dispatch. This kind of worry has become very common with the spread of science fiction and movies, television, and the Internet. There are two reasons for this worry: one is that humans do not know enough about robots in the future, which leads to "distrust"; the other is the psychological reaction of people to the contradictions in modern society, for example, the use of robots in Western society later, it was caused by the fear of unemployment brought to workers. When discussing the intelligence of robots, some people worry that the intelligence of robots will surpass that of humans, and that one day robots will be anti-guest-oriented and dominate humans. If this fear of artificial intelligence is not channeled, it may develop into a mental panic disorder. In addition, the widespread use of artificial intelligence and robots gives people more opportunities and time to work with or accompany with intelligent machines, which will increase the loneliness, isolation and anxiety of related personnel.

First, for a long time, people believed that the development of robots and the evolution of humans are completely different in nature, at least in the foreseeable future. Robots must be designed and manufactured by humans. They are neither living creatures nor biological institutions. They are not caused by living substances, but are merely electronic mechanical devices. Even with intelligent robots, their intelligence is different from human intelligence, not a life phenomenon, but a mechanical imitation of non-life.

Certain functions of future high-intelligence robots are likely to surpass humans, but on the whole, robot intelligence cannot surpass human intelligence. At least, it seems so for now.

3. Employment issues

Robots can replace humans in various physical and mental labors, and are called "steel collar" workers. For example, using industrial robots to replace workers for welding, painting, handling, assembly and processing operations, using service robots for medical care, babysitting, entertainment, secretarial, cleaning, and fire-fighting, and using exploration robots to replace astronauts and divers for space and deep-sea exploration and rescue. Therefore, some people may give up their jobs to robots, causing them to be laid off and re-employed, and even cause unemployment. A study report from the University of Oxford in the United Kingdom in 2013 pointed out that more than 700 occupations will be replaced by smart machines, of which the first will be sales, administration and service.

Someone proposed a task and timetable that artificial intelligence will surpass humans [14]:

Translation language 2024
Writing Essays 2026
Driving truck 2027

Retail jobs 2031
Write a bestseller 2049
Autonomous surgery 2053

To solve this problem, on the one hand, it is necessary to expand new industries (such as the tertiary industry) and service projects, and to enter production and service with breadth and depth; on the other hand, it is necessary to strengthen continuing vocational education and training for workers and technicians. Make them adapt to the new social structure and continue to contribute to society in the new industry.

Figure 12.1 shows the average robot price index and labor compensation index curve in the United States from 1990 to 2007, with the index in 1990 being 100. It can be seen from the figure that the average price of robots has dropped by a factor of two in 17 years, while labor compensation has increased by a factor of two; that is to say, the ratio of labor compensation index to robot price index has increased by 4 times, or the robot price index and labor compensation ratio of the index is about 0.25, a decrease of 4 times. Therefore, the number of robots installed has been on the rise. In the past 10 years, my country has also seen a similar situation as shown

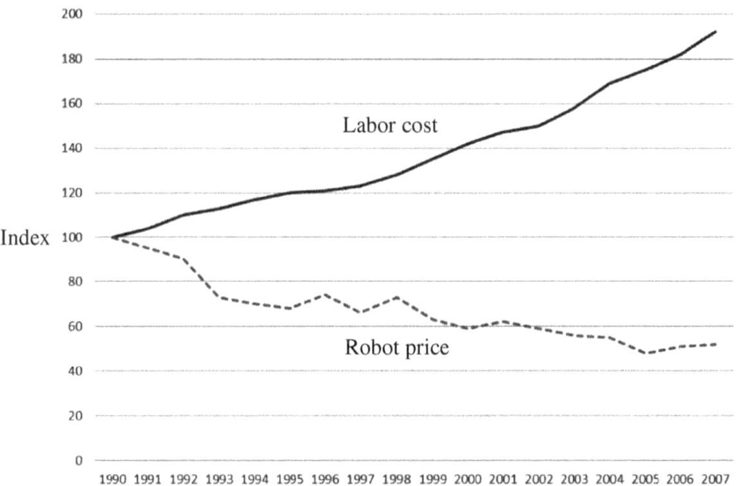

Figure 12.1. Average robot price index and labor compensation index in the United States.

Source: IFR World Robotics 2008.

in Figure 12.1. The demand for labor is in short supply, especially the rapid increase in labor compensation. It is expected that the average labor wage in my country will increase by 5 to 6 times in the next 10 years.

According to the investment and return cycle of industrial robots, that is, the theory of repayment period, if industrial robots are now equipped, investors can recover their investment costs within one to three years; if they invest in industrial robots in 2022, the investment cost will be recovered within 6 months. With the significant decline in the price of industrial robots, the rapid increase in labor compensation, and the emergence of "labor shortage," countries in the world (including China) will inevitably apply more robots (including industrial robots and service robots, etc.) to replace manual labor. This has become an inevitable trend in the 21st century. This may also be an urgent social issue that deserves the attention and in-depth study of scholars and leaders including sociologists, economists, government policy-making officials, and even family planning experts.

4. Ethical issues

The ethical issues of intelligent robots have attracted the attention of the whole society, and the advancement of intelligent robot technology may bring major risks to human society. For example, in the field of service robots, the risks and ethical issues that people worry about mainly involve the care of children and the elderly, and the development of autonomous robotic weapons. The companion robot can provide children with pleasant feelings and stimulate their curiosity. However, the child must be taken care of by an adult, and the companion robot is not eligible to be the child's caregiver. If children spend too much time with the companion robot, they will lose their social skills and cause different levels of social isolation.

The application of military robots also raises some ethical issues. For example, in combat, armed robots on the ground fired guns or unmanned drones fired missile shells, causing casualties to soldiers and even innocent people. Weapons are generally used to carry out fatal strikes under human control; however, military robots can automatically lock the target and destroy them.

Ethics provides a necessary framework of common activities for human society and realizes coexistence with an acceptable common rule. Many experts advocate the establishment of ethical guidelines on the development of intelligent robot technology. Researchers of intelligent machines and artificial intelligence systems need to formulate ethics into corresponding algorithms under the guidance of established ethical standards to guide the behavior of intelligent machines and seek solutions to problems to ensure the absolute friendliness and safety of intelligent machines to humans. Humans still cannot give intelligent machines the sense of responsibility, sense of shame, guilt, and the ability to judge between right and wrong. Unless the ethics of robots can be regulated by programming, advanced intelligent robots may have anti-human and anti-social tendencies. When human beings grant certain rights to smart machines, they should strictly restrict the rights of smart machines.

5. Legal issues

The development and application of intelligent robot technology has brought about many unprecedented legal problems, and traditional laws are facing severe challenges.

What new legal problems have they caused? Please see the example below.

Who is legally responsible for an accident involving a smart-driving car? Traffic laws may be fundamentally rewritten. For another example, does it violate international conventions when robots are used on the battlefield to shoot and kill people? With the improvement of the thinking ability of intelligent machines, they may put forward opinions on society and life, and even political opinions. These problems may bring danger to human society and cause anxiety.

The "robot judge" can automatically generate the optimal judgment result through the analysis of the existing data. Teachers, lawyers and artists also face the threat of unemployment as judges. Today, many works can be created by smart machines, and even press releases can be written by reporter robots. The intelligent software system can also compose music and paint. The existing laws related to the protection of intellectual property rights may be subverted. In the era of artificial intelligence, the law will also reshape the requirements of the profession, and the legal concept will be rebuilt. Soon, "no harm to humans" will likely be written into the labor protection law at the same time as "no abuse of robots."

In addition, in the medical field, the liability of medical accidents caused by the use of medical robots and the security problems of the robot police performing police functions in the field of law enforcement; how should these problems be considered and handled? Therefore, many related legal issues need to be resolved. Many liability issues and safety issues of artificial intelligence products in the legal field need to arouse great attention from countries around the world and smart machine developers.

Close attention should be paid to the above-mentioned legal and safety issues related to the application of intelligent robots. The issue of smart robot safety legislation has been put on the agenda. Relevant laws should be able to regulate the development of smart machines, establish a robot identification and tracking system for smart machines, and ensure the effective control and safe use of smart machines by humans. Robot developers must bear relevant legal responsibilities for their smart products.

The purpose of formulating laws related to smart machines is to make full use of the capabilities of smart machines through legislation, guide smart machines into the right track, prevent their possible negative effects, and ensure that artificial intelligence and smart machines make positive contributions to human society and realize the long-term stability of intelligent machines and human society.

6. Military issues

With the continuous development of artificial intelligence and intelligent machines, research institutions and military organizations in some countries have tried to use intelligent robotics (unmanned systems) for military purposes, and the development and use of intelligent weapons have caused extremely significant security threats for human society and world peace.

For example, in the wars in Iraq and Afghanistan, more than 5,000 remote-controlled robots and some heavy armed robots for reconnaissance, demining, and direct combat were used. Ground-armed robots and intelligent armed drones not only killed many enemy soldiers, but also caused many casualties of innocent civilians.

12.4   The Challenge of Cloning Technology to Intelligent Robots

The debate on "whether robot intelligence can surpass human intelligence" is not over. With the progress of biological genetic engineering and the successful breeding of asexually reproduced animals (such as cloned sheep, cloned dog and cloned cattle, etc.), people are worried about the emergence of cloned humans.

If one day artificial humans, namely clones, appear, regardless of whether they conform to reason and law (there is no such law yet), then many of our concepts of robots will be shaken or even fundamentally changed. These concepts involve important issues such as the definition of robots and robotics, the evolution of robots, the structure of robots, the intelligence of robots, and the relationship between robots and humans. It is necessary for us to re-understand and discuss these issues, in order to learn from each other's strengths and brainstorm through discussions and even debates and controversies to reach a consensus, so that robotics can continue to develop in a healthy direction and be conducive to the correct use of cloning technology.

With the evolution of robots and the development of robot intelligence, we may make necessary modifications to the definition of robots, and even need to redefine robots. The category of robots should include not only "human-like machines made by humans," but also "creatures made by humans" and even "artificial humans." It seems that there is no uniform definition of a robot, and it will be more difficult to give it an exact and generally accepted definition in the future! [4–7].

1. The evolution of robots

From science fiction, craftsmanship to industrial robots, from programmed robots, sensory robots, interactive robots, semi-autonomous robots to autonomous robots, from operating robots, biological robots, bionic robots to anthropomorphic robots and dirty robots, robots have gone through a long "evolution" process.

For a long time, people have held a relatively optimistic view of the evolution of robots. They believe that the evolution of robots is fundamentally different from the evolution of humans. Robots need to be designed and manufactured by humans, and they cannot reproduce on their own; they are neither living beings nor even biomachines. They are not the result of living substances such as cells, but merely mechanical electronic devices. Even for intelligent robots, their intelligence is different from human intelligence. They are non-living mechanical imitations rather than live phenomena.

However, this view is facing new challenges. With the rapid progress of science and technology and bioengineering research, it has been possible to manufacture many artificial organs, such as limbs, hearts, kidneys, vision, hearing, blood, and even liver, pancreas and brains. Some artificial organs have been implanted in the human body and become part of the human body. Some people call these people planted with artificial organs "machine dirty people." Such dirty people have become "semi-robots." For example, a person with an artificial heart implanted or a disabled person who controls the movement of an artificial arm through a bioelectric pulse signal sent by the human body is classified as a dirty person.

Robotic technology is developing rapidly along with the progress of modern science and technology. The capabilities of robots are getting stronger and stronger, approaching human capabilities step by step. With each evolution of robots, the difference between robots and humans gradually decreases. So, to what extent will this difference be reduced?

In the past, the view that "robots can think like humans and have emotions" was regarded as a fantasy. However, the birth of cloned organisms forces us to rethink these issues. Are these cloned cattle, cloned sheep, and cloned mice not just more advanced robotic cattle, robotic sheep, and robotic mice? If the definition of robot includes man-made machines and creatures, then the differences in abilities (including physical and intellectual abilities) between robots and humans may no longer exist; to say the least, the differences are extremely small. Therefore, the debate on robot intelligence will also be carried out at the same time as the definition of robots and the evolution of robots, and corresponding conclusions will be drawn at the same time.

It seems unnecessary to worry that robots will be more intelligent than humans, so that they are anti-guest-oriented and that humans should obey its dispatch. There is no technology to make "Superman," and it is not allowed to use this technology. However, will robot intelligence be similar to human intelligence, or even surpass human intelligence in some respects? Practice will give an authoritative answer to this.

2. The structure of robotics

Robotics is a subject that studies the principles, technology and applications of robots, and it is also a highly intersecting frontier subject. Among many related disciplines (science), we believe that the most closely related are mechanics, anthropology and biology. Figure 12.2 shows the internal relationship between them, that is, the discipline structure diagram; we call it the ternary intersection structure diagram of robotics [4, 7].

Most industrial robots and walking robots are robots or robots that imitate the functions of human upper and lower limbs, and are the research field of robotics. The application of biological engineering (including genetic engineering) technology to study human life and reproduction problems is the research category of artificial life engineering; if creature is regarded as a special mechanical device containing DNA (deoxyribonucleic acid) chains, then artificial life engineering is the study of living robots. The intersection of biology and mechanics gave rise to biomechanical engineering, the study of bionic machinery and robots.

The intersection of anthropology, biology and mechanics, the study of the use of bioengineering methods and technologies to create anthropomorphic robots, that is, clonal bio-robots or clonal humans, can also be called cloned man. This should be a forbidden area for research, and it is also a research field that people are paying great attention to and fiercely controversial.

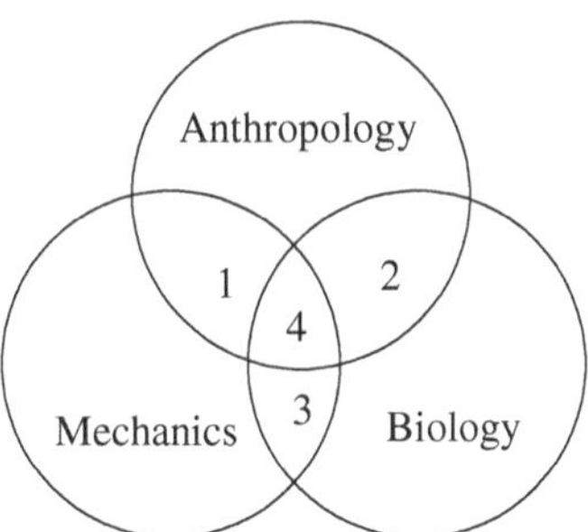

Figure 12.2. Ternary intersection structure diagram of robotics.
Note: 1 — Robot Engineering; 2 — Artificial Life Engineering; 3 — Biomechanical Engineering; 4 — Clone Biorobot.

3. Relationship between robots and humans

From the early times when the robot was restless in the mother and fetus of the human economy and society, humans were a little uneasy when they expected its birth. People hope that robots can replace human beings in various kinds of labor and serve humanity, but they are worried that the development of robots will cause new social problems and even threaten the survival of human beings.

Nearly 60 years have passed since the first robot came out. As of the end of 1999, there were nearly one million industrial robots of all kinds operating in all walks of life in the world. By 2020, the total installed industrial robots exceeded two million, and there will be tens of millions of service robots in operation. The arrival of robots on the social and economic stages has caused quiet changes in the social structure. The human-machine social structure has gradually changed to a human-robot-machine tripolar social structure. People will have to learn to get along with robots, and adapt to this kind of coexistence, and even have to change their traditional concepts and ways of thinking. However, this kind of coexistence is very friendly, and robots have become human assistants and friends. Although in some countries, the development of robots has caused some workers to lose their jobs or change jobs, but this is not the mainstream. Robot intelligence is far from reaching the level of competing with or threatening humans.

At the end of the 20th century, the successful breeding of asexually reproduced mammals caused an uproar all over the world. People are worried that this asexual reproduction technology (i.e. cloning technology) will be abused by irresponsible madmen for human reproduction. If one day, this worry becomes a reality, not only will many new issues concerning social ethics and laws appear, but also the existing relationship between humans and robots will be changed. The three rules for robots formulated by Asimov will be more difficult to implement and supervise. A human-robot relationship that is still unimaginable may be about to emerge.

4. Control cloning technology

The word "clone" is an English word, which means asexual or asexual cell line. As an important part of genetic engineering, cloning technology has been successfully applied. In medical and biological research, asexual reproduction has been widely used, including cuttings, layering, grafting, and single-cell reproduction used in agriculture. The success of using cloning technology to breed mammals makes people naturally think of whether there will be someone has used cloning technology to copy humans?

From a theoretical point of view, it is possible to use this technology to replicate humans; technically speaking, there is no essential difference between the production of cloned humans and the production of cloned sheep and cows, and it is already feasible. Cloning technology is the latest research progress in artificial life sciences, and if applied properly, it will benefit mankind a lot. For example, cloning technology can be used to improve the manufacture of new animal and plant species for the benefit of mankind.

The emergence of any new technology may have negative effects; if it is used improperly or irresponsibly, there will be unknown dangers. The greatest danger of new technology is that humans lose control of it, or that it falls into the hands of madmen who try to use new technologies against humans. People are now worried that cloning technology will be used to copy humans, threatening human security and the stability and development of human society.

We cannot prohibit the research and application of cloning technology because of its hidden dangers. According to a foreign opinion poll, more than half of people support the application of cloning technology to breed animals for medical research. However, if the application of cloning technology to nurture people, it is estimated that they will be opposed by the vast majority of people. Why?

First of all, if the reproduction of humans (human workers) is allowed, it will bring about ethical crises, moral decline, and shake-up of the concepts of marriage and family in human society. Just imagine, those clones who have no biological parents, what kind of family is there to speak of? What impact and consequences will this have on society?

Secondly, the abuse of cloning technology will affect the natural ecological environment and destroy the ecological balance. Even for animal husbandry, the widespread application of clonal reproduction technology may disrupt the ecological balance and cause the large-scale spread of some diseases. If it is used for the reproduction of mammals and humans, there are similar problems.

Again, be wary of the appearance of "black sheep." As is well-known, genetic engineering can reproduce genes, and can also crossover and mutate genes in different cells. Once crossover or mutation operations are implemented, new biological species will be produced and new populations will be formed. This has certain positive significance for the improvement of livestock, fruit trees and crop varieties. But if it is improperly applied or with the wrong techniques, then monsters may be created. This monster, whether it is a plant or an animal, may have an adverse effect on humans. If this monster is a human worker, it would be even more terrifying.

It is worth pointing out that the danger of making human workers is also different from making weapons. Weapons are to be tested and used, and it is difficult to keep them secret; and human workers, as long as they are not significantly different from modern people, they may not be known to anyone if they are copied. As long as the maker does not say anything, then no one knows who is a clone.

5. Conclusion

Human beings have gone through more than three thousand years of history from the fantasy of being able to create machines like humans to the reality of the "robot kingdom" or "robot family" of millions of robots. From the birth of the first industrial robot to the emergence of cloned mammals, it only took a short period of some 30 years. This is enough to illustrate the rapid development of modern science and technology. Faced with the reality that real "artificial man" or human cloning may be created, we have to reconsider and study some fundamental issues of robotics. What is a robot anyway? Do machine cats (e.g. Doraemon) and cloned sheep belong to the category of robots? Is there an essential difference between the evolution of robots and the evolution of humans? Can robot intelligence be compared with human intelligence? Will the relationship between robots and humans undergo a fundamental change? What is the subject of robotics? What about its subject structure? All these issues are worth discussing and studying, and drawing conclusions or reaching consensus.

12.5 Chapter Summary

The last chapter of this book, Section 12.1, looks forward to the development trend of robotics since the beginning of 21st century, which is clearly moving towards the direction of intellectualization, including the evolution of robots to intelligent robots and the realization of robotization production systems. Specifically, sensor-based intelligent robots develop rapidly, and new intelligent technologies have been developed and applied on machines, using modules to further promote robot engineering, pay attention to the development of micro and small robots, pay attention to the development of walking robots, develop non-manufacturing robots and service robots that work in non-structural environments, develop agile manufacturing systems, and military robots will be used to equip troops etc.

Section 12.2 summarizes the robot development plans of major industrial countries in the world in recent years. Since entering the 21st century, especially in the past decade, the major robotic powers of the world have been ambitiously scrambling to develop intelligent robotics technology, which will surely promote the international robotics research and application into a new era and push robotics technology to a new level.

The emergence and large-scale application of robots promoted the development of technology and production, enriched the civilized life of mankind, and also caused a series of social problems. These safety issues are discussed in Section 12.3: Robots make the human-machine social structure quietly change to the human-robot-machine social structure; people must change their ideas and ways of thinking, learn to deal with robots, and live in harmony. Robots pose a psychological threat to some human members. They are worried that the intelligence of robots will surpass that of humans. Therefore, one day, they will be anti-guest and dominate human beings.

Faced with the reality that real "artificial man" or human cloning may be created, robotics faces severe challenges, and people have to reconsider and study some fundamental issues of robotics. Section 12.4 puts forward the challenges of cloning technology and human cloning to robotics and humans. It is hoped that through serious and in-depth discussions, conclusions can be reached and consensus can be reached to ensure that robotics continues to develop in a healthy direction in the 21st century, so that robots can further benefit mankind and become mankind's eternal assistants and friends.

References

[1] Anonymous. (2020). The development pattern and trend analysis of robots in various countries in the world. *China Robot Network*, 20200319 11:21, http://www.elecfans.com/jiqiren/1185798.html

[2] Bill Gates. (2007). Bill Gates predicted: Everyone will have robots in the future. 2007.02.01 09:17:47 http://people.techweb.com.cn/2007-0 2-01/149230.shtml

[3] Cai, Z.X., Liu, L.J., Chen, B.F. and Wang, Y. (2021). *Artificial Intelligence: From Beginning to Date*, Chapter 12. Singapore: World Scientific Publishers.

[4] Cai, Z.X. (1997). Cloning technology challenges intelligent robot technology. *High-tech Communications*, 7(11):60–62.

[5] Cai, Z.X. and Xie, B. (2022). *Robotics*, Chapter 10, 4th Edition. Beijing: Tsinghua University Press.

[6] Cai, Z.X. and Xie, B. (2021). *Robotics*, Chapter 1, 4th Edition. Beijing: Tsinghua University Press.

[7] Cai, Z.X. (2000). *Robotics*, Chapter 10. Beijing: Tsinghua University Press.

[8] Cai, Z.X. (2017). The social issue of artificial intelligence. *Unity*, (6):20–27.

[9] Cai, Z.X. and Guo, F. (2013). Several Issues in the Development of Industrial Robots in China. *Robot Technology and Applications*, (3):9–12.

[10] Cao, X.K. and Cun, X. (2008). The development history of my country's robots. *Robotics and Applications*, (5):44–46.

[11] China Commercial Industry Research Institute. (2020). Analysis of the status quo of global industrial robots in 2020: General industry has gradually become the main force in the new market. 2020-06-22 16:02, https://www.askci.com/news/chanye/20200622/160231116233 8.shtml

[12] IFR. (2013). Executive Summary World Robotics 2013 Industrial Robots. 2013-09-18. http://www.ifr.org/index.php?id=59&df=Execu tive_Summary_WR_2013.pdf

[13] IFR. The continuing success story of industrial robots. 2012-11-11. http://www.msnbc.msn.com/id/23438322/ns/technology_and_sci ence-innovation/t/japan-looks-robot-future/

[14] Lasse Rouhiainen. *Artificial Intelligence: 101 things you must know today about our future*. LASSE ROUHIAINEN, 2019.

[15] Secret Network. Robot development strategies in countries around the world. (2019). 2019-10-17, https://www.wenmi.com/article/pzhj7s03 vm24.html

[16] Sohu. (2017). Robot Development Strategies of Countries in the World (2017). 2017-03-10 13:27, https://www.sohu.com/a/12843191 8_411922

[17] Wang, D.S. (2013). The prospects for the development of the world's industrial robot industry are promising, and China has the greatest growth potential. 2013-10-31. http://www.hyqb.sh.cn/publish/portal 0/tab1023/info10466.htm

[18] Zhou, L. (2019). Analysis of the current development status of the robotics industry in various countries. 09:15, July 13, 2019 (Electronics Network). http://www.elecfans.com/jiqiren/992424.html

Index

E

robust stabilization, 333
robust tracking control, 328–329
robustness, 324, 330, 333, 470, 504
Rocky7, 318
Rodrigue rotation equation, 90
roll, 66, 296
roll angle, 204
rolling optimization, 444, 447
rolling optimization window, 434
rolling planning, 450
rolling window, 403, 426, 433–435, 441, 451
rolling window-based planning, 260
ROPES, 253–254, 256, 258, 278
Rosam's Universal Robot, 2
rotary joints, 154
rotating joints, 63, 97
rotating load, 175
rotation angle, 65, 295
rotation coordinate transformation, 37
rotation matrix, 35–37, 41, 52, 57, 112, 204
rotation sequence, 65–66
rotation transformation, 23, 35, 41, 44–46, 105–106, 145, 373
rotation transformation matrix, 57
rotation vector, 73
rotational angular acceleration, 338
rotational angular velocities, 164
rotational homogeneous coordinate transformation, 57
rotational movements, 13
rotations, 13, 65, 70, 158, 295
rotor, 175
route tracking control, 515
RPY, 66
rule, 239
rule deduction, 278
rule-based expert system, 499–251, 256
rule-based learning control, 498
rule-based robot planning expert system, 250
rule-based system, 250, 253
rules, 251, 254

S

2D space, 378
safety, 276, 307, 473
safety issues, 557
Samson, C., 328
Saridis, G. N., 493, 495
satellite services, 535
scalability, 293, 350
scale invariant feature transform (SIFT), 364
scale transformation, 105
scan matching, 380
scanning data, 379
scanning gap, 383
scanning laser rangefinder, 296
scene knowledge representation, 302
scheduling of robots, 22
schema-based behavioral architecture, 292
schema-based reactive behavior, 291
scrolling window, 451
search, 250, 498
search graph, 242
search tree, 242
searching, 254
second-order kinematics model, 320
second-order Lagrange equation, 131
second-order system, 187
selection, 94, 269, 336
Self, M., 361
self-driving car, 473, 512
self-driving cars, 511
self-learning control, 491
self-localization, 382, 385, 417
self-location, 363
self-organizing, 504
self-organizing behavior, 458
self-organizing neural network controller, 504
self-tuning adaptive control, 216
self-tuning adaptive controller (STAC), 218, 231
self-tuning control, 182
semi-autonomous mission execution, 276

CPSIA information can be obtained
at www.ICGtesting.com
Printed in the USA
JSHW042251190922
30483JS00001B/1